Applications, Biochemistry and Management of Insecticides and Pesticides

Applications, Biochemistry and Management of Insecticides and Pesticides

Editor: Bobby Jones

Callisto Reference
www.callistoreference.com

Callisto Reference,
118-35 Queens Blvd., Suite 400,
Forest Hills, NY 11375, USA

Visit us on the World Wide Web at:
www.callistoreference.com

ISBN: 978-1-63239-904-5 (Hardback)

Printed in China.

Cataloging-in-Publication Data

Applications, biochemistry and management of insecticides and pesticides / edited by Bobby Jones.
p. cm.
Includes bibliographical references and index.
ISBN 978-1-63239-904-5
1. Insecticides. 2. Pesticides. 3. Pests--Control. 4. Biochemistry. I. Jones, Bobby.
SB951.5 .A67 2018
632.951 7--dc23

Table of Contents

Preface

This book was inspired by the evolution of our times; to answer the curiosity of inquisitive minds. Many developments have occurred across the globe in the recent past which has transformed the progress in the field.

Insecticides and pesticides are used for the control and regulation of unwanted organisms that hinder plant growth and destroy plants. This book on the applications, biochemistry and management of insecticides and pesticides discusses the production and methods of insecticides and pesticides usage. Integrated pest management is aimed at eradication and regulation of pests and harmful insects. The extensive content of this book provides the readers with a thorough understanding of the subject. It elucidates new techniques and their applications in a multidisciplinary approach. It is appropriate for students seeking detailed information in this area as well as for experts.

This book was developed from a mere concept to drafts to chapters and finally compiled together as a complete text to benefit the readers across all nations. To ensure the quality of the content we instilled two significant steps in our procedure. The first was to appoint an editorial team that would verify the data and statistics provided in the book and also select the most appropriate and valuable contributions from the plentiful contributions we received from authors worldwide. The next step was to appoint an expert of the topic as the Editor-in-Chief, who would head the project and finally make the necessary amendments and modifications to make the text reader-friendly. I was then commissioned to examine all the material to present the topics in the most comprehensible and productive format.

I would like to take this opportunity to thank all the contributing authors who were supportive enough to contribute their time and knowledge to this project. I also wish to convey my regards to my family who have been extremely supportive during the entire project.

Editor

Identification and Characterisation of *Aedes aegypti* Aldehyde Dehydrogenases Involved in Pyrethroid Metabolism

Nongkran Lumjuan[1]*, Jureeporn Wicheer[1], Posri Leelapat[1], Wej Choochote[2], Pradya Somboon[2]

1 Research Institute for Health Sciences, Chiang Mai University, Chiang Mai, Thailand, **2** Department of Parasitology, Faculty of Medicine, Chiang Mai University, Chiang Mai, Thailand

Abstract

Background: Pyrethroid insecticides, especially permethrin and deltamethrin, have been used extensively worldwide for mosquito control. However, insecticide resistance can spread through a population very rapidly under strong selection pressure from insecticide use. The upregulation of aldehyde dehydrogenase (*ALDH*) has been reported upon pyrethroid treatment. In *Aedes aegypti*, the increase in ALDH activity against the hydrolytic product of pyrethroid has been observed in DDT/permethrin-resistant strains. The objective of this study was to identify the role of individual ALDHs involved in pyrethroid metabolism.

Methodology/Principal Findings: Three ALDHs were identified; two of these, ALDH9948 and ALDH14080, were upregulated in terms of both mRNA and protein levels in a DDT/pyrethroid-resistant strain of *Ae. aegypti*. Recombinant ALDH9948 and ALDH14080 exhibited oxidase activities to catalyse the oxidation of a permethrin intermediate, phenoxybenzyl aldehyde (PBald), to phenoxybenzoic acid (PBacid).

Conclusions/Significance: ALDHs have been identified in association with permethrin resistance in *Ae. aegypti*. Characterisation of recombinant ALDHs confirmed the role of this protein in pyrethroid metabolism. Understanding the biochemical and molecular mechanisms of pyrethroid resistance provides information for improving vector control strategies.

Editor: Gordon Langsley, Institut national de la santé et de la recherche médicale - Institut Cochin, France

Funding: This study was supported by grants from the National Research Council of Thailand (NRTC) to NL and Thailand Research Fund (TRF) to WC and NL (TRF Senior Research Scholar: RTA5480006). The funders had no role in study design, data collection and analysis, decision to publish, or preparation of the manuscript.

Competing Interests: The authors have declared that no competing interests exist.

* Email: nklumjuan@yahoo.com

Introduction

Pyrethroids, synthetic insecticides analogous to natural pyrethrin, have been widely used throughout the world for the control of insects. Pyrethroids are divided into two groups based on their chemical structures. Type I pyrethroids, such as permethrin, lack an α-cyano group, whereas type II pyrethroids, such as deltamethrin and cypermethrin, contain an α-cyano group. However, the extensive use of these insecticides has led to insecticide resistance in insect populations [1,2,3]. Resistance to pyrethroids can be divided into two main mechanisms: an alteration in the target site of the insecticide or increased expression of metabolic detoxification enzymes. Pyrethroids act by targeting sodium channels, leading to neurotoxic effects [4]. Several point mutations in the voltage-gated sodium channel gene are associated with DDT and pyrethroid resistance [5,6,7,8,9]. In metabolic resistance, enhanced activity of enzymes in metabolic pathways in insects leads to insecticides being detoxified or sequestered before they reach the target site. Overexpression of detoxification enzymes such as cytochromes P450 (CYPs), glutathione S-transferases (GSTs) and carboxylesterases (CEs) have been well documented in pyrethroid resistance in insects [10,11,12].

Pyrethroids are mainly metabolised through the hydrolysis of the ester linkage followed by the oxidation of their component alcohol and acid moieties [13]. Pyrethroids have been extensively studied in humans and rats, indicating that both types are mainly hydrolysed by CEs to produce 3-phenoxybenzyl alcohol (PBalc) [14,15], whereas they are mainly oxidised by P450s, alcohol dehydrogenases (ADHs) and aldehyde dehydrogenases (ALDHs) [16,17]. ALDHs have been investigated as enzymes that are important in the oxidation of permethrin in mammals for their oxidation of intermediate products of pyrethroid to carboxylic acid [18]. In the mosquito *Anopheles gambiae*, the up-regulation of ALDH after exposure to permethrin has been reported [19]. Enzyme-based metabolite assays also indicated that the catalytic activity of P450s, ADHs and ALDHs were increased in microsomal fractions of a DDT/permethrin-resistant strain (PMD-R) of *Aedes aegypti* from Thailand [20]. In our preliminary study using a proteomic approach, crude homogenates of 4th instar larvae of *Aedes* mosquitoes were partially purified using glutathione agarose columns. Bound fractions were collected, concentrated and separated by 2-dimensional gel electrophoresis. The result indicated that a detoxification enzyme, ALDH (AAEL014080 in VectorBase), was upregulated in the PMD-R strain relative to the laboratory susceptible strain (unpublished data). However, the ability of individual ALDHs isoforms to metabolise permethrin in mosquito has not yet been investigated.

The present study aimed to identify the ALDH genes responsible for permethrin resistance in *Ae. aegypti*. The individual ALDHs that are involved in permethrin resistance were characterised, and their expression patterns were analysed. Recombinant proteins were produced, and the in vitro metabolism of permethrin and its hydrolysis products were determined.

Materials and Methods

Materials

Cis/Trans-permethrin was purchased from Chem Service (West Chester, PA). Permethrin metabolites, 3-phenoxybenzyl alcohol (PBalc, 98% purity), 3-phenoxybenzylaldehyde (PBald, 98% purity) and 3-phenoxybenzoic acid (PBacid, 98% purity) including β-Nicotinamide adenine dinucleotide (NAD^+) were purchased from Sigma (St. Louis, MO).

Mosquito strains

The PMD and PMD-R strains originated from Chiang Mai Province, Thailand [21]. The PMD strain was resistant to DDT, whereas the PMD-R strain was resistant to both DDT and permethrin. The New Orleans strain was an insecticide-susceptible laboratory strain of *Ae. aegypti*.

Database search and sequence alignment

A preliminary study using 2-dimensional gel electrophoresis demonstrated that expression of ALDH (AAEL014080) was increased in the PMD-R strain relative to the NO and PMD strains at the larval stage (unpublished data). The protein sequence of a known ALDH (AAEL014080) was used as a query for a BLAST search of the *Aedes aegypti* sequences in VectorBase. Deduced amino acid sequences of ALDHs were aligned using ClustalW [22].

Identification of *ALDH* genes

The oligonucleotide primers were designed based on the sequences of ALDH in VectorBase (Table S1). The full-length cDNAs of ALDH genes from *Ae. aegypti* were amplified using Taq DNA polymerase (Qiagen) as described by the manufacturer's protocol. PCR parameters consisted of 35 cycles of 30 s at 95°C, 30 s at 55°C, and 1.5 min at 72°C. PCR products were cloned into the pGEM-T easy Vector (Promega) and then transformed into JM109 competent *Escherichia coli* cells. The plasmid DNA was submitted to 1st BASE Laboratories (Malaysia) for sequencing to verify the integrity of genes.

Quantitative PCR analysis

Total RNA was extracted from 3 biological replicate sets (10 mosquitoes per replicate) of 4th instar larvae, pupae, and one-day-old adult males or females from each of the three strains using the TRIzol plus RNA Purification System (Invitrogen). Complementary DNA was synthesised using SuperScript III reverse transcriptase (Gibco) as described in the manufacturer's protocol. Quantitative PCR was performed as previously described, using QuantiFast SYBR Kit's protocol (Qiagen) [23]. The primers used are shown in Table S2. The PCR parameters consisted of 2 steps of 95°C for 5 min and 35 cycles of 95°C for 10 s, 60°C for 35 s, followed by a dissociation step.

Construction of plasmids and expression of ALDHs

Total RNA was extracted from whole mosquitoes of the PMD-R strain using Trizol reagent (Sigma). Complementary DNA was synthesised using SuperScript III reverse transcriptase (Gibco) as described in the manufacturer's protocol. PCR products generated with ProofStart DNA polymerase (Qiagen) using gene specific primers (Table S1) were cloned into the pET 100-D/TOPO vector using the Champion pET directional TOPO Expression kit according to the manufacturer's instruction (Invitrogen). The construct was verified by DNA sequencing. The plasmids containing the ALDH genes were transformed into *E. coli* BL21 Star (DE3). The recombinant proteins were produced after induction with isopropyl β−D-thiogalactoside at 37°C or room temperature for 4 h.

Protein purification

The pET 100-D/TOPO vector encodes an N-terminal polyhistidine (6xHis) fused to the recombinant protein. Protein purification was performed using HisTrap Ni affinity column (GE Healthcare) as described previously [23]. The protein purity was verified by 12.5% polyacrylamide gel electrophoresis and Coomassie staining. The protein concentration was determined by the Bradford method using the Bio-Rad protein-assay dye reagent and bovine serum albumin as a standard [24].

Western Blot analysis

Western blot analysis was performed as previously described [21]. The membrane was probed with 1:50,000 and 1:100,000 dilutions of polyclonal antibodies against ALDH9948 and ALDH14080, respectively. The bound antiserum was detected by incubation with a 1:50,000 dilution of Peroxidase-labelled Anti-Rabbit Antiserum followed by visualisation using ECL Advanced Blotting Detection Kit (Amersham Bioscience).

Enzyme activity

ALDH activity against PBald was measured as described previously [16]. Briefly, the substrate mixture contained 1 mM EDTA, 0.1 mM pyrazole and 2.5 mM $NAD(P)^+$ in 33 mM Phosphate buffer, pH 8.2. The enzyme was incubated with the substrate mixture at 37°C, and the reaction rate was determined by the formation of NAD(P)H at 340 nm in 4 min. The esterase assay was conducted as described previously, by measuring the hydrolysis of p-nitrophenyl acetate (pNPA) to the products p-nitrophenol (pNP) and acetate [25]. Kinetic studies were performed by varying the concentration of PBald in the presence of NAD^+. The results were analysed by non-linear regression analysis using GraphPad Prism 4 software.

PBald oxidation by recombinant ALDHs

ALDH activity was measured by the oxidation of PBald to PBacid, as detected by HPLC. The assay was modified from the method described previously [17]. Briefly, 20 μg of recombinant ALDHs were incubated with 0.4 mM PBald in the presence of 3 mM NAD^+ in 0.1 M Tris-Cl buffer, pH 7.4 at 37°C for 10 min. Pyrene was then added as an internal control. The reaction mixture was extracted with 1.5 ml of chloroform. This procedure was repeated in triplicate. The chloroform extracts were then pooled, air-dried and analysed with HPLC.

HPLC was performed with a Shimadzu LC 20-A Series (Shimadzu) using a Nova-Pak C18 column (3.9×150 mm; Waters). The extract was resuspended in 200 μl of acetonitrile. The mixture (10 μl) was injected into the column at a flow rate of 1 ml/min. The gradient elution was performed at 35°C, and the detection wavelength was 230 nm. Peaks were integrated into peak area with the LC Solution (Shimadzu). ALDH activity was calculated as the formation of PBacid/min/mg protein. The concentration of PBacid was determined by comparison with a known concentration of PBacid.

Results

Identification of *Ae. aegypti* ALDHs

The DNA sequence of ALDH (AAEL014080) in *Ae. aegypti* was retrieved from VectorBase (http://www.vectorbase.org), and it is located in supercontig 1.1002. Close paralogues of ALDH (AAEL014080), ALDH (AAEL009948) and ALDH (AAEL009029) were included in the experiment to expand for genes of interest that were found on supercontigs 1.440 and 1.363, respectively (Figure S1). The deduced amino acid sequences of these three ALDHs are shown in Figure 1.

Quantitative PCR analysis

To determine whether the ALDH genes were overexpressed at the transcriptional level, real-time PCR was performed in three *Ae. aegypti* strains, the NO susceptible strain and the PMD and PMD-R strains, at three developmental stages. *ALDH14080* was significantly upregulated in the larvae and females of the PMD-R strain relative to the NO strain (Figure 2). *ALDH9948* mRNA levels were significantly upregulated in all life stages except the adult male ($p<0.001$ in larva and adult female, and $p<0.05$ in pupa) when compared to the PMD strain. *ALDH14080* expression was significantly higher in the larval stage of PMD-R only relative to PMD. In contrast, there is no evidence of upregulation of *ALDH9029* mRNA in the PMD-R strain when compared to the PMD strain in any life stage (Table S3). These results show that upregulation of *ALDH9948* and *ALDH14080* may confer resistance to permethrin.

```
ALDH9948     -------------------------MANANPEIKYTQLFINNEFVDAVSGKVFPTVNPST  35
ALDH14080    -------------------------MANPNQEIKYTKLFINNQFVDSQSGKTFPTLNPAT  35
ALDH9029     MLRVLLKGLPKPGVGTQIARYSSIPAPKTSPEILYTGIFINNEWHKSIGGKVFPTLNPAN  60

ALDH9948     GKKIVDIAEGDKADVDLAVQAAKAAFQRSSKWRQMDASARGKLIYKLADLMERDMHQIAS  95
ALDH14080    GQKIVDVAEGDKADVDIAVQAAKTAFARSSAWRQMDASARGKLLHKLADLMERDINVLAN  95
ALDH9029     EQVIAEIQQGQKADIDAAVGAARDAFKLGSPWRRMDASKRGQLLYRLADLMERDRVYLAS  120

ALDH9948     LESLDNGKP-YMSAVYDVYGSMNCLRYYAGWTDKVCGETVPSDGPHLTYTRKEPFGVVGQ  154
ALDH14080    LESLDNGKA-FGDSVFDMNCAIDTFRYYAGWADKIHGSTVPSDGPVMTYIRKEPVGVVGQ  154
ALDH9029     LETLDNGKPYFMSYNVDVPMAINNLRYYAGWADKNHGKVIPMDGEFFVYTRHEPVGVCGQ  180

ALDH9948     IIPWNYPLLMLAWKWGPALAAGCTIVMKPAEQTPLTALYMCSLVKEAGFPPGVVNMVPGY  214
ALDH14080    IIPWNYPVLMLTWKWAPALATGCTLVLKPAEQTPLSALYMAALSKEAGFPDGVINVVNGY  214
ALDH9029     IIPWNFPILMAAWKFGPALATGNTIVLKPAEQTSLTALYMAQLVKEAGFPPGVVNVVPGF  240

ALDH9948     GPTAGNAITMHPDIRKVAFTGSVEVGKIVMAG-AASNLKKVSLELGGKSPLVICDDVDVN  273
ALDH14080    GPTVGAAIVNHAEIRKVAFTGSVETGRLITEGSSKSNLKRVSLELGGKSPLVVFDDFDVD  274
ALDH9029     G-DAGAALVEHNDVDKVAFTGSTEVGKKIQQGAGLSNLKRTTLELGGKSPNIILSDADMK  299

ALDH9948     EAAQIAYTGVFENMGQCCIAATRTFVQEGIYDAFVQKATELAKGRKVGNPFSQGIQHGPQ  333
ALDH14080    EAVEIAHNAIFANHGQNCCAGSRTFVQEGVYDKFVAKAAEMAKARKVGDAFAEGTQQGPQ  334
ALDH9029     HAVETSHFGLFFNMGQCCCAGSRTFIEDKIYDEFVERSAERAKKRTVGNPFDLTTEHGPQ  359

ALDH9948     IDDTQFKKILGFIETGKKEGAKLETGGVQ-VGEEGYFIEPTVFSNVTDEMTIAKEEIFGP  392
ALDH14080    VDEEQLNKILGFFESASKEGAKLQTGGKR-HGNVGYFVEPTVYSDVTDEMRIAREEIFGP  393
ALDH9029     VDKAQYDKILSLIDTGKKQGAKLVAGGKKYEGLPGYFIEPTVFADVKDDMTIAREEIFGP  419

ALDH9948     VQSIIKFKTLDEAIERANATSFGLAAGIVTKNLNNALTFSNAVEAGSVWVNTYLAVSNQA  452
ALDH14080    VQSILKFKTLDEVIERANRTEYGLAAGVLTNNLNNALVFSNAVEAGSVWVNCYDYVMPTT  453
ALDH9029     VQQLIRFKSLDEVIERANQSEYGLAAAVFSNDIDKVNYLVQGLRAGTVWVNTYNVLSAQA  479

ALDH9948     PFGGYKQSGVGREMGKEGNEEYLETKTVSIKLPSKV-  488
ALDH14080    PFGGFKQSGHGRELGYDGIELYTETKTVTIKLPSKV-  489
ALDH9029     PFGGYKMSGHGRENGEYGLQAYTEVKSVITRIPVKNS  516
```

Figure 1. Deduced amino acid sequences of *Ae. aegypti* ALDH9948 and ALDH14080. Sequences shown are from the PMD-R strain. The amino acid sequences were aligned using ClustalW. Letters in bold indicate 100% conservation between the 3 sequences. Dashes are used to denote gaps introduced for maximum alignment.

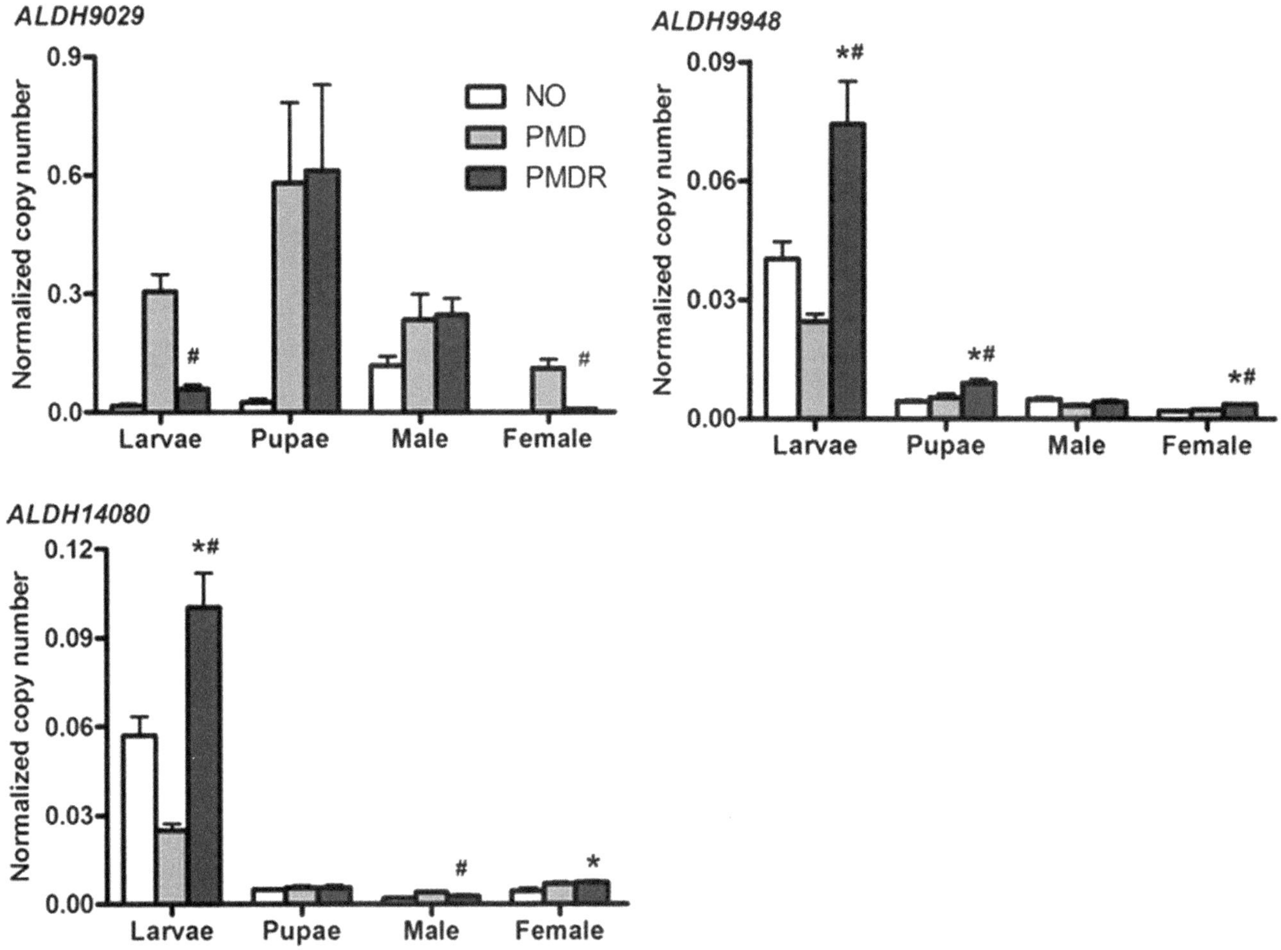

Figure 2. Transcription profiles of *ALDH9029, ALDH9948* and *ALDH14080* in three strains of *Ae. aegypti*. Complementary DNA from three different biological replicates (ten mosquitoes each) was used as templates. Four life-stages were analysed: larvae (L), pupae (P), adult male (M), and adult female (F). Each sample was analysed in duplicate in each experiment, and the results were averaged from three independent experiments. The mRNA copy numbers were determined by comparison with known concentrations of standard plasmids and normalised against the copy number of the ribosomal S7 transcript. Error bars indicate standard error of the mean. Statistically significant differences were evaluated with ANOVA followed by Tukey's multiple comparison test (*$p<0.05$ versus New Orleans strain; $^{\#}p<0.05$ versus PMD strain).

Western Blot analysis

To confirm the expression of ALDHs at the protein level, western blots were performed using specific polyclonal antibodies against ALDH9948 and ALDH14080. To validate the specificity of these polyclonal antibodies, immuno-cross-reactivity between ALDH9948 and ALDH14080 was investigated. The polyclonal antibody for ALDH9948 exhibited low-level cross-reactivity with ALDH14080 (Figure 3B), whereas the anti-ALDH14080 antibody was observed to have high specificity. Protein expression profiles of ALDH were investigated in crude homogenates of four developmental stages of three *Ae. aegypti* strains. Expression levels of ALDH9948 and ALDH14080 were increased in the PMD-R strain in almost all developmental stages (pupae and adult males and females), except for larvae when compared to the NO and PMD strains (Figure 3A). In all three strains, no visible bands of ALDHs were detected in the larval stage, whereas strong bands were presented in pupae and adult males and females. Meanwhile, crude homogenates from the larval stage gave no smearing bands when stained with Coomassie blue, indicating no protein degradation. The expression of rat ALDH has been reported to increase with age [26]. This might indicate that early stages express ALDH proteins at low levels that could not be detected in the small number of larvae used in this study.

Recombinant protein expression

To determine whether the *Ae. aegypti* ALDHs contribute to permethrin metabolism, recombinant ALDHs were produced, and the ability of these proteins to metabolise permethrin was determined. The full-length sequences of two ALDHs, ALDH9948 and ALDH14080, were amplified by PCR using cDNA templates from the PMD-R strain and subcloned into the *E. coli* expression vector pET 100-D/TOPO. Expression of His_6-tagged ALDHs in *E. coli* BL21 Star (DE3) yielded soluble recombinant proteins at the 37°C expression temperature. The purity of His_6-tagged recombinant ALDHs was verified in 12.5% SDS-PAGE and corresponded to the predicted size of approximately 65 kDa (Figure S2).

Biochemical characterisation of ALDHs

Both His_6-tagged recombinant ALDHs possess ALDH activity to catalyse the oxidation of intermediate aldehyde of permethrin, PBald. The ALDH activity was measured by spectrophotometry,

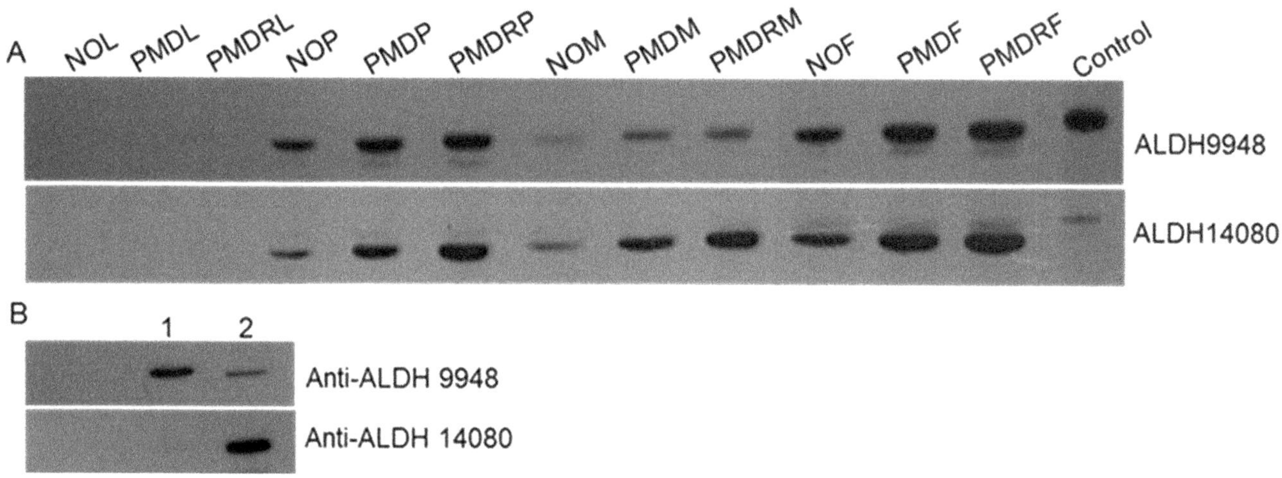

Figure 3. Western blot analysis of ALDH9948 and ALDH14080. (A) Elevated protein of ALDH9948 and ALDH14080 in PMD-R strain. Fifty micrograms of protein from New Orleans (NO), PMD and PMD-R strains in four life stages; larva (L), pupae (P), adult male (M) and adult female (F) including purified recombinant His-tagged ALDH9948 and 14080 (25 ng each) were resolved by SDS-PAGE. Proteins were transferred to a nitrocellulose membrane and probed with anti-ALDH9948 and anti-ALDH14080. Peroxidase labelled anti-rabbit antibody was used as a secondary antibody. Proteins were visualised by enhancing the chemiluminescence using ECL Advanced Blotting Detection Kit (Amersham Biosciences). Equal protein loading was confirmed by staining the membrane with Ponceus S. (B) Determination of antibodies specificity by western blot. Fifty nanograms of non-fusion ALDH9948 and ALDH14080 (Lane1, 2 and 3, respectively) were resolved in SDS-PAGE. Western blotting was performed as described.

mediated by the formation of NAD(P)H as products of the reaction. The oxidation reactions of recombinant ALDH9948 and ALDH14080 required either NAD^+ or $NADP^+$ as a cofactor; however, these enzymes prefer NAD^+ to $NADP^+$ (Table 1). It has been noted that most ALDHs prefer to use NAD^+ over $NADP^+$ as a cofactor [27]. Generally, ALDHs exhibit esterase activity in vitro [28]. In this study, recombinant ALDHs also have esterase activities that catalyse the hydrolysis of p-nitrophenyl acetate to produce p-nitrophenol and acetate (Table 1). The highest esterase activity belongs to recombinant ALDH14080, with a specific activity of 13.11±0.98 μmole/min/mg proteins.

Kinetic parameters of purified ALDHs were determined using PBald and NAD^+ as substrate and cofactor, respectively. Michaelis-Menten constants (K_m) for ALDH9948 and ALDH14080 were 153.8±30.0 and 34.4±6.8 nM, respectively, in respect to PBald (Table 2). The ($V_{max}/K_{m\ PBald}$) value of ALDH14080 was higher than that of ALDH9948, indicating the catalytic efficiency of this enzyme at low substrate concentrations.

To determine whether recombinant ALDHs readily oxidised PBald, HPLC was performed to identify the product of PBacid. The metabolite profile of *trans/cis*-permethrin is shown in Figure S3. Pyrene was spiked as an internal control, given the extraction recovery range of 81–97%. The HPLC results indicated that PBald was oxidised by recombinant ALDH9948 and ALDH14080 with specificities of 1192±55 and 1119±14 nmole PBacid formed/min/mg protein, respectively (Table 3). Because recombinant ALDHs exhibit esterase activity, the ability of these enzymes to catalyse the hydrolysis of the parent permethrin was investigated. The incubation of recombinant ALDHs with *trans/cis*-permethrin did not produce PBalc, suggesting that ALDHs are not associated with permethrin hydrolysis (data not shown). The incubation of denatured recombinant ALDHs with PBald in the presence of NAD^+ did not produce PBacid, indicating that the oxidation of PBald was mediated by recombinant ALDHs.

Discussion

Overexpression of detoxification genes has been well documented in association with insecticide resistance of many insect species. P450s, GSTs and CEs are primarily implicated in the detoxification of insecticides in insects. It has been reported that P450s contribute to resistance in all classes of insecticides [29]. The upregulation of several P450s, particularly those belonging to the CYP6Z, CYP6M or CYP9J subfamilies, has been reported to be involved in resistance to pyrethroids in mosquitoes [30,31,32,33]. Some species, including *Ae. aegypti* CYP9J32, *An. gambiae* CYP6M2 and *An. gambiae* CYP6Z8, have the ability to metabolise pyrethroids [34,35,36]. GSTs, especially GSTE2, GSTE4 and GSTE7, were also observed to be overexpressed in resistant populations [32,33,37]. Recombinant GSTE2-2 showed

Table 1. Substrate specificity of *Ae. aegypti* recombinant ALDH isoforms.

	Substrate/cofactor	ALDH 9948	ALDH 14080
Esterase activity (μmole/min/mg)	pNPA	13.11±0.98	0.14±0.02
ALDH activity	PBald/NAD^+	483±9	254±24
(nmole/min/mg)	PBald/$NADP^+$	58±4	19±2

ALDH activity was performed in the presence 4 mM PBald and 2.5 mM $NAD(P)^+$. The oxidation of PBald was monitored by the formation of NAD(P)H.

Table 2. Kinetic parameter of *Ae. aegypti* recombinant ALDH isoforms.

Enzyme	V_{max} (nmole NADH/min/mg)	PBald		NAD$^+$	
		K_m (nM)	V_{max}/K_m	K_m (nM)	V_{max}/K_m
ALDH 9948	627.4±34.0	153.8±30.8	4.1	139.1±27.9	3.2
ALDH 14080	208.2±9.8	34.4±6.8	6.1	193.8±34.5	1.3

Kinetic studies were performed by varying the concentration of PBald and cofactor NAD$^+$ at fixed saturated concentrations of NAD$^+$ and PBald, respectively. The oxidation of PBald to PBacid was monitored by the formation of NADH in the reaction at 37°C for 4 min. Three independent assays were performed. The results are shown as the mean ± SE.

DDT dehydrochlorinase activity to metabolise DDT, but the recombinant GSTE7-7 did not appear to metabolise DDT. Therefore, the role of GSTE7 in insecticide resistance remains unclear [21]. A recent study suggested that a single point mutation of *GSTe2* (L119F) associated with metabolic resistance to DDT and permethrin in mosquito *An. funetus* [38]. Many genes encoding CE enzymes were identified to be upregulated in organophosphate-, carbamate- and pyrethroid-resistant insects [39].

However, other genes that are responsible for insecticide resistance cannot be excluded. To date, microarray technology has been utilised to expand the number of detoxification genes and has identified new relevant genes that might be involved in metabolic resistance [19,32,33,40,41,42]. Aside from P450s, GSTs and CEs, microarray data also identified secondary detoxification genes that may confer insecticide resistance. For example, aldo-ketoreductase, an NAD(P)(H) oxidoreductases that catalyse the reduction of aldehydes to alcohols, was over-transcribed in temephos- and permethrin- selected strain of *Ae. aegypti* [43,44]. UDP-glucuronosyltransferases (UGTs), phase II detoxification enzymes involved in the conjugation of xenobiotics, were also identified as upregulated after permethrin exposure and in response to carbamate, respectively. ALDHs were also found to be upregulated in insecticide resistance in insects [19,40]. However, the functions of these enzymes in insecticide detoxification require further investigation. In mammals, the oxidation of pyrethroids was catalysed by ALDH [16]. A study in insecticide metabolism revealed the important role of ALDH in the detoxification of pyrethroid in mosquito [20]. Multiple detoxification enzymes were identified as a target of pyrethroid activity-based probes in rat proteome, including P450s, UDP-glucuronosyltransferases, Flavin-containing monooxygenase and ALDH [45].

Aldehyde dehydrogenases are a family of enzymes that oxidise a broad range of endogenous, xenobiotic and lipid peroxidation products that contain the highly reactive aldehyde to their corresponding carboxylic acid [25]. In mammals, ALDHs are involved in both the detoxification of aldehydes and the biosynthesis of pheromones [27]. However, few studies of ALDHs have been reported in insects. In *Drosophila*, ALDHs play a vital role in ethanol metabolism by mediating the oxidation of acetaldehyde to acetate, which is involved in ethanol resistance [27,28,46,47].

In this study, transcript levels for three of the *Ae. aegypti* ALDH genes were quantified. *ALDH9948* was significantly overexpressed in the insecticide-resistant PMD-R strain in almost all developmental stages, except adult males, when compared to the susceptible PMD line. In contrast, *ALDH14080* was upregulated relative to the PMD strain only in the larval stage. Quantitative PCR results revealed that insecticide selection increased the expression of these ALDHs, although the overexpression was not observed in all life stages. The altered expression of ALDH9948 and ALDH14080 was confirmed at the protein level, indicating that the increase in these proteins is strongly associated with resistance to permethrin. Inconsistencies between the mRNA and protein levels of the same gene may be caused by differences in post-translational regulation between the different developmental stages. Although high levels of ALDH mRNA were found in the larval stage, there was no protein detected by western blot, suggesting that the protein may be expressed at a level below the detection limit in early stages. However, low-abundance ALDH was detected by 2D-gel electrophoresis from a large sample of larvae used in combination with the sub-proteome approach for the enrichment of low-abundance proteins. The recombinant ALDH isoforms exhibited oxidase activity to catalyse the oxidation of aldehyde moiety of pyrethroids, but subcellular localisation of individual ALDHs was not investigated further in this study. These experiments suggested that ALDH9948 and ALDH14080 may play a role in insecticide resistance to permethrin in the PMD-R strain of *Ae. aegypti*.

Collectively, in *Ae. aegypti*, it has been reported that parental permethrin can be hydrolysed in vitro. Our previous study demonstrated that the formation of PBacid was decreased in the presence of an esterase inhibitor, BNPP, suggesting the function of

Table 3. Specific activity of *Ae.aegypti* recombinant ALDH isoforms to oxidise PBald.

Enzyme	Specific activity (nmole PBacid formed/min/mg protein)
ALDH 9948	1192±55
ALDH 14080	1119±14

Recombinant ALDH (5 µg) was incubated with 2 mM PBald in the presence of 3 mM NAD+ in 0.1 M Tris-Cl buffer, pH 7.4 at 37°C for 10 min. PBacid formation was determined by HPLC as described. Three independent assays were performed. The results are shown as the mean ± SE.

esterases in permethrin metabolism [20]. The importance of particular CEs in pyrethroid detoxification has not yet been studied. However, it has been proposed that non-specific esterases may be involved in pyrethroid hydrolysis in insects [48]. A recent study demonstrated that both PBalc and PBald were oxidised by *Ae. aegypti* CYP6Z8 [34]. In addition, our finding also clearly revealed that recombinant ALDH9948 and ALDH14080 have the ability to catalyse the oxidation of PBald. The results of this study indicate the role of *Ae. aegypti* ALDHs in pyrethroid degradation pathway and this knowledge will improve our ability to manage insecticide resistance in the field such as the use of synergists to increase the efficacy of certain insecticides.

In conclusion, we identified two ALDHs that are upregulated in permethrin-resistant *Ae. aegypti* mosquitoes in Thailand. Functional characterisation of recombinant ALDHs clearly demonstrates that these enzymes are capable of metabolising PBald. This report indicates the importance of *Ae. aegypti* ALDHs in permethrin degradation.

Supporting Information

Figure S1 Phylogenetic relationship of ALDHs in *Ae. aegypti* (AAEL) with *An. gambiae* ALDHs (AGAP).

Figure S2 SDS-PAGE analysis of His-tagged recombinant ALDHs in *Ae. aegypti* produced in *E. coli* BL21 Star (DE3).

Figure S3 Chromatograms of PBald oxidation to PBacid by recombinant aldehyde dehydrogenase.

Table S1 Sequences of oligonucleotide primers used to amplify the cDNA full-length of ALDHs for in vitro protein expression.

Table S2 Sequences of oligonucleotide primers used to amplify the fragment of *Ae. aegypti* ALDHs for quantitative PCR.

Table S3 Quantitative PCR results of *Ae. aegypti* ALDH.

Author Contributions

Conceived and designed the experiments: NL PS WC. Performed the experiments: NL JW PL. Analyzed the data: NL. Contributed reagents/materials/analysis tools: NL. Wrote the paper: NL. Interpreted the results: NL PS.

References

1. Chareonviriyaphap T, Bangs MJ, Suwonkerd W, Kongmee M, Corbel V, et al. (2013) Review of insecticide resistance and behavioral avoidance of vectors of human diseases in Thailand. Parasit Vectors 6: 280.
2. Hemingway J, Field L, Vontas J (2002) An overview of insecticide resistance. Science 298: 96–97.
3. Maharaj R (2011) Global trends in insecticide resistance and impact on disease vector control measures. Insect Physiol 3: 27–33.
4. Soderlund DM, Clark JM, Sheets LP, Mullin LS, Piccirillo VJ, et al. (2002) Mechanisms of pyrethroid neurotoxicity: implications for cumulative risk assessment. Toxicology 171: 3–59.
5. Brengues C, Hawkes NJ, Chandre F, McCarroll L, Duchon S, et al. (2003) Pyrethroid and DDT cross-resistance in *Aedes aegypti* is correlated with novel mutations in the voltage-gated sodium channel gene. Med Vet Entomol 17: 87–94.
6. Martins AJ, Lima JB, Peixoto AA, Valle D (2009) Frequency of Val1016Ile mutation in the voltage-gated sodium channel gene of *Aedes aegypti* Brazilian populations. Trop Med Int Health 14: 1351–1355.
7. Rajatileka S, Black WCt, Saavedra-Rodriguez K, Trongtokit Y, Apiwathnasorn C, et al. (2008) Development and application of a simple colorimetric assay reveals widespread distribution of sodium channel mutations in Thai populations of *Aedes aegypti*. Acta Trop 108: 54–57.
8. Saavedra-Rodriguez K, Urdaneta-Marquez L, Rajatileka S, Moulton M, Flores AE, et al. (2007) A mutation in the voltage-gated sodium channel gene associated with pyrethroid resistance in Latin American *Aedes aegypti*. Insect Mol Biol 16: 785–798.
9. Yanola J, Somboon P, Walton C, Nachaiwieng W, Somwang P, et al. (2011) High-throughput assays for detection of the F1534C mutation in the voltage-gated sodium channel gene in permethrin-resistant *Aedes aegypti* and the distribution of this mutation throughout Thailand. Trop Med Int Health 16: 501–509.
10. Feyereisen R (2005) Insect Cytochrome P450. In: Gilbert LI, Iatrou K, Gill SS, editors. Comprehensive Molecular Insect Science Elsevier pp. 1–77.
11. Oakeshott JG, Devonshire AL, Claudianos C, Sutherland TD, Horne I, et al. (2005) Comparing the organophosphorus and carbamate insecticide resistance mutations in cholin- and carboxyl-esterases. Chem Biol Interact 157–158: 269–275.
12. Ranson H, Hemingway J (2005) Glutathione transferases In: Gilbert LI, Iatrou K, Gill SS, editors. Comprehensive Molecular Insect Science - Pharmacology: Elsevier. pp. 383–402.
13. Kaneko H (2011) Pyrethroids: mammalian metabolism and toxicity. J Agric Food Chem 59: 2786–2791.
14. Nishi K, Huang H, Kamita SG, Kim IH, Morisseau C, et al. (2006) Characterization of pyrethroid hydrolysis by the human liver carboxylesterases hCE-1 and hCE-2. Arch Biochem Biophys 445: 115–123.
15. Ross MK, Borazjani A, Edwards CC, Potter PM (2006) Hydrolytic metabolism of pyrethroids by human and other mammalian carboxylesterases. Biochem Pharmacol 71: 657–669.
16. Choi J, Rose RL, Hodgson E (2002) In vitro human metabolism of permethrin: the role of human alcohol and aldehyde dehydrogenase. Pest Biochem Physiol 22: 249–261.
17. Nakamura Y, Sugihara K, Sone T, Isobe M, Ohta S, et al. (2007) The in vitro metabolism of a pyrethroid insecticide, permethrin, and its hydrolysis products in rats. Toxicology 235: 176–184.
18. Hodgson E (2003) In vitro human phase I metabolism of xenobiotics I: pesticides and related compounds used in agriculture and public health J Biochem Mol Toxicol 17: 201–206.
19. Vontas J, Blass C, Koutsos AC, David JP, Kafatos FC, et al. (2005) Gene expression in insecticide resistant and susceptible *Anopheles gambiae* strains constitutively or after insecticide exposure. Insect Mol Biol 14: 509–521.
20. Somwang P, Yanola J, Suwan W, Walton C, Lumjuan N, et al. (2011) Enzymes-based resistant mechanism in pyrethroid resistant and susceptible *Aedes aegypti* strains from northern Thailand. Parasitol Res 109: 531–537.
21. Lumjuan N, McCarroll L, Prapanthadara LA, Hemingway J, Ranson H (2005) Elevated activity of an Epsilon class glutathione transferase confers DDT resistance in the dengue vector, *Aedes aegypti*. Insect Biochem Mol Biol 35: 861–871.
22. Thompson JD, Higgins DG, Gibson TJ (1994) CLUSTAL W: improving the sensitivity of progressive multiple sequence alignment through sequence weighting, position-specific gap penalties and weight matrix choice. Nucleic Acids Res 22: 4673–4680.
23. Lumjuan N, Rajatileka S, Changsom D, Wicheer J, Leelapat P, et al. (2011) The role of the *Aedes aegypti* Epsilon glutathione transferases in conferring resistance to DDT and pyrethroid insecticides. Insect Biochem Mol Biol 41: 203–209.
24. Bradford MM (1976) A rapid and sensitive method for the quantitation of microgram quantities of protein utilizing the principle of protein-dye binding. Anal Biochem 72: 248–254.
25. Marchitti SA, Brocker C, Stagos D, Vasiliou V (2008) Non-P450 aldehyde oxidizing enzymes: the aldehyde dehydrogenase superfamily. Expert Opin Drug Metab Toxicol 4: 697–720.
26. Yoon M, Madden MC, Barton HA (2006) Developmental expression of aldehyde dehydrogenase in rat: a comparison of liver and lung development. Toxicol Sci 89: 386–398.
27. Morse D, Meighen E (1984) Aldehyde Pheromones in Lepidoptera: Evidence for an Acetate Ester Precursor in *Choristoneura fumiferana*. Science 226: 1434–1436.
28. Fry JD, Bahnck CM, Mikucki M, Phadnis N, Slattery WC (2004) Dietary Ethanol Mediates Selection on Aldehyde Dehydrogenase Activity in *Drosophila melanogaster*. Integr Comp Biol 44: 275–283.
29. Li X, Schuler MA, Berenbaum MR (2007) Molecular mechanisms of metabolic resistance to synthetic and natural xenobiotics. Annu Rev Entomol 52: 231–253.
30. Bariami V, Jones CM, Poupardin R, Vontas J, Ranson H (2012) Gene amplification, ABC transporters and cytochrome P450s: unraveling the molecular basis of pyrethroid resistance in the dengue vector, *Aedes aegypti*. PLoS Negl Trop Dis 6: e1692.

31. Marcombe S, Mathieu RB, Pocquet N, Riaz MA, Poupardin R, et al. (2012) Insecticide resistance in the dengue vector *Aedes aegypti* from Martinique: distribution, mechanisms and relations with environmental factors. PLoS One 7: e30989.
32. Marcombe S, Poupardin R, Darriet F, Reynaud S, Bonnet J, et al. (2009) Exploring the molecular basis of insecticide resistance in the dengue vector *Aedes aegypti*: a case study in Martinique Island (French West Indies). BMC Genomics 10: 494.
33. Strode C, Wondji CS, David JP, Hawkes NJ, Lumjuan N, et al. (2008) Genomic analysis of detoxification genes in the mosquito *Aedes aegypti*. Insect Biochem Mol Biol 38: 113–123.
34. Chandor-Proust A, Bibby J, Regent-Kloeckner M, Roux J, Guittard-Crilat E, et al. (2013) The central role of mosquito cytochrome P450 CYP6Zs in insecticide detoxification revealed by functional expression and structural modelling. Biochem J 455: 75–85.
35. Stevenson BJ, Bibby J, Pignatelli P, Muangnoicharoen S, O'Neill PM, et al. (2011) Cytochrome P450 6M2 from the malaria vector *Anopheles gambiae* metabolizes pyrethroids: Sequential metabolism of deltamethrin revealed. Insect Biochem Mol Biol 41: 492–502.
36. Stevenson BJ, Pignatelli P, Nikou D, Paine MJ (2012) Pinpointing P450s associated with pyrethroid metabolism in the dengue vector, *Aedes aegypti*: developing new tools to combat insecticide resistance. PLoS Negl Trop Dis 6: e1595.
37. Ding Y, Hawkes N, Meredith J, Eggleston P, Hemingway J, et al. (2005) Characterization of the promoters of Epsilon glutathione transferases in the mosquito *Anopheles gambiae* and their response to oxidative stress. Biochemical Journal 387: 879–888.
38. Riveron JM, Yunta C, Ibrahim SS, Djouaka R, Irving H, et al. (2014) A single mutation in the GSTe2 gene allows tracking of metabolically-based insecticide resistance in a major malaria vector. Genome Biol 15: R27.
39. Montella IR, Schama R, Valle D (2012) The classification of esterases: an important gene family involved in insecticide resistance - A review. Memórias do Instituto Oswaldo Cruz 107: 437–449.
40. Silva AX, Jander G, Samaniego H, Ramsey JS, Figueroa CC (2012) Insecticide resistance mechanisms in the green peach aphid *Myzus persicae* (Hemiptera: Aphididae) I: A transcriptomic survey. PLoS One 7: e36366.
41. Saavedra-Rodriguez K, Suarez AF, Salas IF, Strode C, Ranson H, et al. (2012) Transcription of detoxification genes after permethrin selection in the mosquito *Aedes aegypti*. Insect Mol Biol 21: 61–77.
42. Grisales N, Poupardin R, Gomez S, Fonseca-Gonzalez I, Ranson H, et al. (2013) Temephos resistance in *Aedes aegypti* in Colombia compromises dengue vector control. PLoS Negl Trop Dis 7: e2438.
43. David JP, Faucon F, Chandor-Proust A, Poupardin R, Riaz MA, et al. (2014) Comparative analysis of response to selection with three insecticides in the dengue mosquito *Aedes aegypti* using mRNA sequencing. BMC Genomics 15: 174.
44. Strode C, de Melo-Santos M, Magalhaes T, Araujo A, Ayres C (2012) Expression profile of genes during resistance reversal in a temephos selected strain of the dengue vector, *Aedes aegypti*. PLoS One 7: e39439.
45. Ismail HM, O'Neill PM, Hong DW, Finn RD, Henderson CJ, et al. (2013) Pyrethroid activity-based probes for profiling cytochrome P450 activities associated with insecticide interactions. Proc Natl Acad Sci U S A 110: 19766–19771.
46. Fry JD, Saweikis M (2006) Aldehyde dehydrogenase is essential for both adult and larval ethanol resistance in *Drosophila melanogaster*. Genet Res 87: 87–92.
47. Fry JD, Donlon K, Saweikis M (2008) A worldwide polymorphism in aldehyde dehydrogenase in *Drosophila melanogaster*: evidence for selection mediated by dietary ethanol. Evolution 62: 66–75.
48. Hemingway J, Hawkes NJ, McCarroll L, Ranson H (2004) The molecular basis of insecticide resistance in mosquitoes. Insect Biochem Mol Biol 34: 653–665.

2

Elucidating Turnover Pathways of Bioactive Small Molecules by Isotopomer Analysis: The Persistent Organic Pollutant DDT

Ina Ehlers[1], Tatiana R. Betson[1], Walter Vetter[2], Jürgen Schleucher[1]*

1 Department of Medical Biochemistry & Biophysics, Umeå University, Umeå, Sweden, 2 Department of Food Chemistry, University of Hohenheim, Stuttgart, Germany

Abstract

The persistent organic pollutant DDT (1,1,1-trichloro-2,2-bis(4-chlorophenyl)ethane) is still indispensable in the fight against malaria, although DDT and related compounds pose toxicological hazards. Technical DDT contains the dichloro congener DDD (1-chloro-4-[2,2-dichloro-1-(4-chlorophenyl)ethyl]benzene) as by-product, but DDD is also formed by reductive degradation of DDT in the environment. To differentiate between DDD formation pathways, we applied deuterium NMR spectroscopy to measure intramolecular deuterium distributions (^{2}H isotopomer abundances) of DDT and DDD. DDD formed in the technical DDT synthesis was strongly deuterium-enriched at one intramolecular position, which we traced back to ^{2}H/^{1}H fractionation of a chlorination step in the technical synthesis. In contrast, DDD formed by reductive degradation was strongly depleted at the same position, which was due to the incorporation of ^{2}H-depleted hydride equivalents during reductive degradation. Thus, intramolecular isotope distributions give mechanistic information on reaction pathways, and explain a puzzling difference in the whole-molecule ^{2}H/^{1}H ratio between DDT and DDD. In general, our results highlight that intramolecular isotope distributions are essential to interpret whole-molecule isotope ratios. Intramolecular isotope information allows distinguishing pathways of DDD formation, which is important to identify polluters or to assess DDT turnover in the environment. Because intramolecular isotope data directly reflect isotope fractionation of individual chemical reactions, they are broadly applicable to elucidate transformation pathways of small bioactive molecules in chemistry, physiology and environmental science.

Editor: Dariush Hinderberger, Martin-Luther-Universität Halle-Wittenberg, Germany

Funding: This work was funded by The Swedish Research Council for Environment, Agricultural Sciences and Spatial Planning; JS; grant 216-2008-764 (http://www.formas.se/en/). The funders had no role in study design, data collection and analysis, decision to publish, or preparation of the manuscript.

Competing Interests: The authors have declared that no competing interests exist.

* Email: jurgen.schleucher@chem.umu.se

Introduction

The chloropesticide DDT (1,1,1-trichloro-2,2-bis(4-chlorophenyl)ethane) is one of the most controversial chemicals developed in the twentieth century. Its application as the first effective pesticide to combat malaria has saved the lives of millions of people. This success triggered its general use as a pesticide on a million-ton scale in agriculture and by households worldwide. Due to its detrimental properties (lipophilicity and persistency), the use of DDT has led to its ubiquitous occurrence in the environment even in remote areas where it has never been used [1,2]. In addition, serious toxicological effects such as thinning of egg shells of birds in the Baltic region have been linked to DDT [3]. As a consequence, DDT has strongly influenced the public perception and regulation of man-made chemicals. Although the use of DDT has been drastically reduced for decades, it is still abundant in the environment and is still having detrimental effects. For this reason, DDT has been classified as a persistent organic pollutant (POP) by the Stockholm Convention, with the purpose of phasing out its use globally. However, it is still used against vector-borne diseases [4], primarily malaria transmitted by *Anopheles* mosquitos.

Since the discovery of the persistence of DDT, numerous studies have addressed environmental and human health hazards of DDT and its congeners [5,6]. Recent studies have detected *p, p'*-DDE (1,1-bis-(4-chlorophenyl)-2,2-dichloroethene) in human blood serum [6], even in countries that stopped DDT usage decades ago, and link high *p, p'*-DDE blood serum levels with prevalent hypertension [7] and an increased risk of Alzheimer disease [8]. Technical DDT is a mixture of compounds, the major components being the *p, p'* and *o, p'* isomers of DDT and some of the minor components being the *p, p'* and *o, p'* isomers of DDD (1-chloro-4-[2,2-dichloro-1-(4-chlorophenyl)ethyl]benzene) [9] (Fig. 1). DDD not only occurs as a by-product of technical DDT production but has itself been used as a pesticide, and is formed in the environment by hydrodechlorination of DDT [9–11]. DDD is of particular ecotoxicological relevance because it accumulates in the environment and has been implicated as a possible endocrine disruptor [5,12]. As for POPs in general [13], tracing DDT in the environment is critical to understand its transformation processes, to judge environmental and human health risks, and to develop remediation strategies.

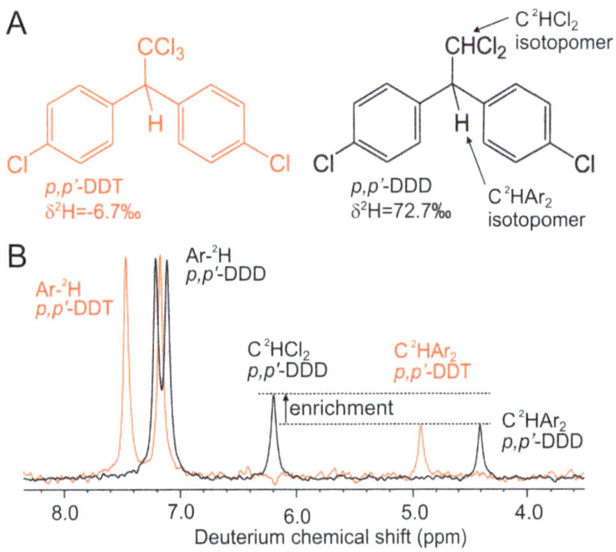

Figure 1. Comparison of *p, p'*-DDT with its congener *p, p'*-DDD. (A) Chemical structures and whole-molecule δ^2H values. (B) Overlay of deuterium NMR spectra of reference samples of *p, p'*-DDD (black, reference 1) and *p, p'*-DDT (red). Integrals of the signals are proportional to the abundance of the respective ^2H isotopomers. Ar-^2H denotes isotopomers carrying ^2H in the aromatic moieties of the respective compound. Positions of signals ("chemical shifts") differ between DDT and DDD because they reflect the stereoelectronic properties of each molecule; these chemical shift differences do not influence the integration of signals.

Increasingly, stable heavy isotopes, which occur naturally in the environment with varying abundance, are used to trace compounds [14]. In most applications, the abundance ratio between a heavy isotope and a light isotope of a chemical element is measured by isotope-ratio mass spectrometry (IRMS). Variation observed in isotope ratios is usually expressed as δ value (in ‰), the abundance deviation of the respective heavy isotope relative to a standard with known isotope ratio [15]. However, IRMS only allows measurement of the average isotope ratio of a whole molecule, because any molecule to be studied must be converted into defined gases for IRMS analysis. It is known, though, that the abundance of heavy isotopes varies among intramolecular groups of non-symmetric molecules [16], but the consequences of this intramolecular variation cannot be assessed by whole-molecule δ measurements. A molecule carrying a specific isotope at a specified intramolecular position is called an isotopomer. Intramolecular isotope variation therefore means that the isotopomers of non-symmetric molecules differ in their abundances. This variation in abundance is mainly caused by kinetic isotope effects during chemical reactions. Thus, isotopomer abundances are directly related to reaction mechanisms and turnover along chemical pathways [17]. As has been observed for several types of compounds, kinetic protium/deuterium ($^1H/^2H$) isotope effects can easily double the abundance of particular ^2H isotopomers [18–20]. Relative abundances of ^2H isotopomers can be measured by deuterium NMR, because each ^2H isotopomer creates one signal in the deuterium NMR spectrum. If the signals can be resolved and using suitable experimental conditions, the integrals of the deuterium NMR spectrum directly reflect relative isotopomer abundances [18,21,22]. Furthermore, when isotopomer abundance can be linked to defined chemical steps, isotopomer variation yields mechanistic information on reaction pathways. Here we use deuterium NMR to identify ^2H fractionations due to individual reactions during DDD formation, as basis to distinguish DDD formed in technical DDT production from DDD formed by hydrodechlorination of DDT in the environment.

Results & Discussion

Reference samples of *p, p'*-DDT and *p, p'*-DDD have been found to differ by as much as 79‰ in δ^2H [23] (Fig. 1A). Variation in this range would normally rule out a common source of the compounds. We measured deuterium NMR spectra of the same *p, p'*-DDT and *p, p'*-DDD reference samples, which show indistinguishable abundance patterns for the C^2HAr_2 and Ar-^2H isotopomers (Fig. 1B). In contrast, the C^2HCl_2 isotopomer in DDD was nearly twice as abundant as the C^2HAr_2 isotopomer (Table 1, *p, p'*-DDD reference sample 1). By combining relative isotopomer abundances from NMR spectra with the δ^2H values of the compounds [23], we calculated how much the high abundance of the C^2HCl_2 isotopomer influenced the δ^2H of DDD. This isotope balance calculation (see Materials and Methods) shows that the high abundance of the C^2HCl_2 isotopomer ($\delta^2H = 739‰$) completely explains the 79‰ difference in whole-molecule δ^2H between *p, p'*-DDT and *p, p'*-DDD. Thus, the molecular part that is common to DDD and DDT ($CHAr_2$ fragment) showed indistinguishable δ^2H values (Table 1). Consequently, the isotopomer analysis demonstrates that both compounds can have a common source.

Measurements on two further independent samples of *p, p'*-DDD (reference 2 and 3 in Table 1) gave indistinguishable results. Furthermore, analogous experiments on reference material of the *o, p'* isomer of DDD showed that the C^2HCl_2 isotopomer was again almost twice as abundant as the C^2HAr_2 isotopomer (Table 1). These results suggest that the C^2HCl_2 isotopomer of synthetic DDD generally has a very high abundance, which significantly shifts the whole-moleculeδ^2H value.

To apply isotopomer analysis to DDT as used in practice, we recorded deuterium NMR spectra of technical DDT (Fig. 2B) to analyze the by-product *p, p'*-DDD. Technical DDT is synthesized by reacting CCl_3CHO with chlorobenzene (Fig. 2A). The technical product contains only 4% DDD, and the amount of DDD in the technical DDT sample was only approximately 15 µmol. Deconvolution allowed quantifying the signals originating from *p, p'*-DDD, although the signal of the C^2HCl_2 isotopomer partly overlapped with the signal of an unknown impurity. Measurements on two replicate NMR samples of the same technical DDT gave similar results (Table 1), which showed that its C^2HCl_2 isotopomer was about twice as abundant as the C^2HAr_2 isotopomer. This indicates that acceptable signal-to-noise ratios can be obtained using sample amounts of 15 µmol, using a cryogenically cooled probe [24]. The synthesis of CCl_3CHO proceeds via chlorination of CH_3CHO. Incomplete chlorination gives rise to residual $CHCl_2CHO$ from which the corresponding DDD isomers are formed. Chlorination of carbonyl compounds is associated with a kinetic $^1H/^2H$ isotope effect of up to seven [25], which is close to the maximum classical kinetic $^1H/^2H$ isotope effect for C–H bond breakage [26]. According to the theory of isotope effects [26], chlorination of CH_3CHO will - with increasing turnover - progressively enrich deuterium in residual $CHCl_2$ groups. Supporting this explanation, a commercial $CHCl_2CHO$ product showed a 1.27-fold enrichment of the C^2HCl_2 isotopomer compared to the C^2HO isotopomer (Table 1). Thus, we conclude that the high abundance of the C^2HCl_2 isotopomer of DDD is the result of the $^1H/^2H$ isotope effect

Table 1. Deuterium abundance data of DDT and related compounds.

Sample	C^2HCl_2 isotopomer abundance[a]	δ^2H, ‰	
		Whole molecule[b]	$CHAr_2$ fragment[c]
p, p'-DDT reference	–	−6.7±3.4	−6.7±3.4
p, p'-DDD reference 1	1.77±0.10	72.7±7.0	−1.3±11.4
p, p'-DDD reference 2	1.80±0.04	–	–
p, p'-DDD reference 3	1.76±0.11	–	–
o, p'-DDD reference	1.89±0.09	74.8±7.6	−6.8±10.2
Commercial $CHCl_2CHO$	1.27±0.02	–	–
p, p'-DDD in technical DDT		–	–
replicate 1	2.17±0.44	–	–
replicate 2	1.78±0.26	–	–
p, p'-DDD from DDT hydrodechlorination:			
Experiment 1	0.69±0.05	–	–
Experiment 2	0.62±0.07	–	–
Experiment 3	0.58±0.09	–	–

[a]Relative to $C^2HR_2 = 1$.
[b]from reference [23].
[c]See Materials and Methods for calculations.

during the chlorination of CH_3CHO. The original enigma of how *p, p'*-DDD and *p, p'*-DDT can have a difference in δ^2H of 79‰ can be attributed to the isotope effect during synthesis of CCl_3CHO.

DDT is known to be degraded into DDD in the environment through hydrodechlorination under anaerobic conditions [10]. In this reaction, a chlorine atom in the CCl_3 moiety of DDT is replaced by a hydride equivalent to form the $CHCl_2$ group of DDD. The hydride equivalents are thought to be derived from the biological cofactors $FADH_2$ or reduced Fe porphyrins [10]. An analogous hydrodechlorination reaction using Fe(0) has been proposed for remediation of DDT contamination [11]. Thus, we experimentally produced DDD by hydrodechlorination of *p, p'*-DDT with Fe(0). The deuterium NMR spectrum of the resulting reaction mixture (Fig. 2C) shows that the *p, p'*-DDD formed by hydrodechlorination was approximately 30% depleted in the C^2HCl_2 isotopomer. To test the reproducibility of the method, we carried out triplicate experiments (see Material S1), which gave statistically indistinguishable results with an average abundance of the C^2HCl_2 isotopomer of 0.63±0.05, relative to the C^2HPh_2 isotopomer. The strong ^2H depletion of the $CHCl_2$ group formed by hydrodechlorination is for a fundamental biophysical reason. Because $^1H^2HO$ is a weaker acid than H_2O, the hydronium ions in water are ^2H-depleted by approximately −700 ‰ [27,28]. When these ions enter reduction reactions leading to abiogenic or biogenic hydride equivalents, the resulting hydride equivalents become accordingly ^2H-depleted. The resulting depletion, of the order of −300‰, has been observed in hydride-derived isotopomers of biochemical metabolites [17]. Thus, a strong ^2H depletion of the $CHCl_2$ group, similar to our Fe(0) reduction experiment, can be expected for DDD formed by hydrodechlorination of DDT in the environment. This depletion can serve to identify DDD formed in situ, either as result of natural degradation, or of remediation efforts, and can therefore be used to monitor breakdown of DDT.

In summary, DDD from technical synthesis is characterized by an increased abundance of the C^2HCl_2 isotopomer, while DDD

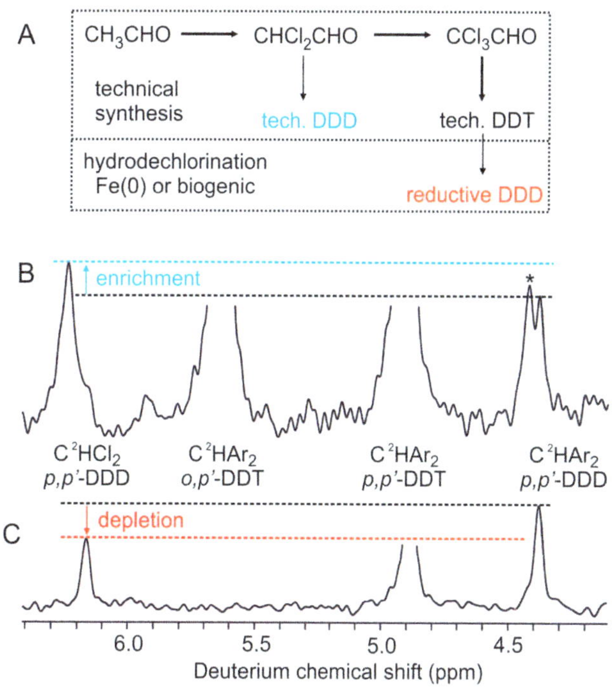

Figure 2. Pathways of DDD formation and their ^2H fractionation. (A) Technical synthesis of DDT, and pathways of formation of DDD. (B) Deuterium NMR spectrum of technical DDT (replicate 1). (C) Deuterium NMR spectrum of a reaction mixture containing DDD formed by hydrodechlorination (experiment 1). Blue and red lines qualitatively illustrate enrichment or depletion of the C^2HCl_2 isotopomer in technical and reductive DDD, respectively. To take differing linewidths and overlap into account, numerical values for isotopomer abundances were obtained by deconvolution of the signals. Note that signal positions differ slightly between samples, because of differing analyte concentrations. The asterisk denotes an unknown impurity in technical DDT.

formed by reductive dechlorination is depleted in the C^2HCl_2 isotopomer, and both isotopomer fractionations are inherent properties of the respective pathway of DDD generation. In light of the opposite deuterium fractionations, technical DDD and DDD formed by hydrodechlorination should generally differ by approximately 1,000‰ in the abundance of the C^2HCl_2 isotopomer (Table 1). This isotopomer difference can be used to estimate fractional contributions of both pathways to environmental samples of DDD. Unfortunately, the sample amount currently required for ^2H isotopomer measurements by NMR (approximately 10 mg per congener, see above) restricts the application of this approach to highly contaminated sites where large quantities of DDT congeners can be isolated. However, the C^2HCl_2 isotopomer signatures of technical and environmental DDD formation respectively induce more positive or negative whole-molecule δ^2H values of DDD, as compared to DDT. Compound-specific IRMS methods for δ^2H measurements for polyhalogenated compounds [29,30] have recently been introduced; therefore comparison of δ^2H values of DDD and DDT may allow tracing sources of DDD to technical synthesis or breakdown of DDT in the environment.

Conclusion

Our D isotopomer measurements represent a striking demonstration that individual ^2H isotopomer abundances can have a decisive effect on whole-molecule δ^2H values. Hence isotopomer abundances give essential insights into causes of variation in δ^2H, and must be taken into account in interpretations of whole-molecule isotope ratios, even when structural differences between related compounds appear to be minor. For DDT, sources of DDD may be traced based on isotopomer results. In general, isotopomer analysis of POPs can be applied: (1) to identify the responsible polluter; (2) to understand POP transformation processes in the environment; and (3) to gauge the efficacy of remediation approaches.

Isotope fractionation of chemical reactions directly affects isotopomer abundances, which in turn affects whole-molecule isotope ratios. However, without prior knowledge a change in a whole-molecule isotope ratio cannot be related to fractionation by a chemical reaction. Therefore isotopomer measurements are particularly valuable for deducing mechanistic detail of reaction pathways. The availability of cryogenically cooled probes has reduced the sample amount needed for isotopomer analysis to only micromoles, which enables applications in fields ranging from organic and pharmaceutical chemistry, to metabolic studies, and environmental science.

Materials and Methods

Sources of compounds

See Material S1.

Deuterium NMR

Several conditions must be fulfilled for isotopomer measurements by NMR. First, line widths must be minimized to resolve the signals and to optimize sensitivity. To achieve this goal, we used low-viscosity solvents C_6F_6 or hexane/C_6F_6 and elevated measurement temperatures around 50°C [21]. Second, the NMR signals must have pure Lorentzian lineshapes, so that their integrals can be determined by lineshape fitting. This goal was achieved by careful homogenizing the magnetic field of the NMR spectrometer ("shimming"). Third, because many individual spectra ("scans") are added to obtain each NMR spectrum, the influence of the previous scan on the deuterium nuclei must have decayed when the next scan is acquired. To achieve this goal, we determined deuterium relaxation times (T_1) of all compounds using an inversion-recovery pulse sequence, and acquired spectra with appropriate recycle times between scans (6 T_1) to ensure complete relaxation. A DRX600 spectrometer (Bruker, Karlsruhe, Germany) equipped with a 5-mm broadband probe and a ^{19}F lock device was used to record ^2H NMR spectra of DDT and DDD reference materials; C_6F_6 was used for field-frequency locking. Spectra of dichloroacetaldehyde diethyl acetal were acquired on a DRX500 spectrometer in unlocked mode using a 10-mm broadband probe. Spectra of technical DDT and of the hydrodechlorination reaction mixture were obtained on an Avance III 850 spectrometer equipped with a cryogenic probe optimized for deuterium detection and equipped with ^{19}F lock. Parameter settings for all samples are given in Material S1. All spectra were recorded with proton decoupling. Spectra were processed with exponential line-broadening, and signal integrals were obtained by deconvolution in the program topspin (version 3.0; Bruker). Typically, five spectra were recorded. The standard deviation of the C^2HCl_2/C^2HR_2 abundance ratio among replicate spectra was used to estimate the overall precision of our measurements. For three independent hydrodechlorination experiments, the standard deviations of replicate spectra were 0.05, 0.07 and 0.09, respectively. The experiments led to results which did not significantly differ from each other (ANOVA), and the standard deviation among the experiments was 0.05, indicating that the standard deviation among spectra is a realistic estimate of the overall precision (Table 1).

DDT hydrodechlorination

p, p'-DDT was reduced with Fe(0) under anaerobic conditions following Sayles et al. [11]. A 1-L flask was filled with 1 L 20 mM MOPS buffer (pH 7, deoxygenated by purging with N_2); 15 g Fe(0) powder and 60 mg *p, p'*-DDT (dissolved in 20 mL acetone) were added. The flask was sealed and mixed by rotation. After 18 h at room temperature, the reaction mixture was extracted consecutively with *n*-hexane and chloroform. The organic phases were dried with Na_2SO_4, the solvent was removed under vacuum, and the resulting product was used for NMR measurements. Analysis of the product using ^1H NMR showed that the main components of the product were un-reacted *p, p'*-DDT and approximately 30% *p, p'*-DDD.

Isotope balance calculations

Deuterium NMR spectra yield relative abundances of ^2H isotopomers, but position-specific $^2H/^1H$ ratios cannot be derived from NMR alone. When the whole-molecule $^2H/^1H$ ratio is available and if all the signals in the ^2H NMR spectrum can be integrated, position-specific $^2H/^1H$ ratios can be calculated according to the equation 1,

$$\left(\frac{^2H}{^1H}\right)_i = \frac{f_i}{F_i}\left(\frac{^2H}{^1H}\right)_G \qquad (1)$$

where $(^2H/^1H)_i$ and $(^2H/^1H)_G$ are the isotope ratios of isotopomer i and of the whole molecule, respectively. f_i is the experimentally determined molar fraction of isotopomer i, that is, the integral of its NMR signal compared to the integral of all isotopomers of the molecule. F_i is the statistical molar fraction [31], that is, the stoichiometric fraction hydrogen in position i according to the molecular formula, and the term f_i/F_i represents the abundance of the ^2H isotopomer i, relative to the $^2H/^1H$ ratio of the whole molecule. Note that the isotope ratios $(^2H/^1H)_i$ and $(^2H/^1H)_G$ must be expressed as true ratios, i.e. $\delta^2H = 72.7‰$ for *p, p'*-DDD (Table 1) must be converted to $(^2H/^1H)_G = 1.0727$, i.e. the DDD sample has a 1.0727-fold higher $^2H/^1H$ ratio than VSMOW. For example, the C^2HCl_2 isotopomer of *p, p'*-DDD contributed fraction $f_{C2HCl2} = 0.1621$ to the integral of the ^2H NMR spectrum, while F_{C2HCl2} is 0.1 (1 hydrogen out of 10 in the molecule). Therefore the $^2H/^1H$ ratio of the CHCl2 group is $(^2H/^1H)_{C2HCl2} = 1.621 \times 1.0727 = 1.739$ and means that this isotopomer has an abundance corresponding to a position-specific δ^2H of 739‰. An analogous calculation for the rest of the molecule-- the $CHAr_2$ fragment ($f_{rest} = 1 - f_{C2HCl2} = 0.8379$, $F_{rest} = 0.9$) which is common to *p, p'*-DDT and *p, p'*-DDD-- yields a $^2H/^1H$ ratio of 0.9987, corresponding to a δ^2H of $-1.3‰$ for this fragment. The 74‰ difference between this number and the whole-molecule δ^2H shows how much the whole-molecule δ^2H is influenced by the high abundance of the C^2HCl_2 isotopomer. In the isotope balance results in Table 1, error estimates reflect the combined uncertainties of the whole-molecule δ^2H measurements and of NMR-derived ^2H isotopomer distributions. For the DDD component of technical DDT, this calculation is not possible because the aromatic signals of all congeners overlap.

Acknowledgments

We thank several colleagues for valuable advice on the manuscript.

Author Contributions

Conceived and designed the experiments: IE TRB WV JS. Performed the experiments: IE TRB JS. Analyzed the data: IE TRB JS. Contributed reagents/materials/analysis tools: TRB WV. Contributed to the writing of the manuscript: IE WV JS.

References

1. Sladen WJL, Menzie CM, Reichel WL (1966) DDT residues in Adelie penguins and a crabeater seal from Antarctica. Nature 210: 670–673.
2. Sheng JS, Wang X, Gong P, Joswiak DR, Tian L, et al. (2013) Monsoon-driven transport of organochlorine pesticides and polychlorinated biphenyls to the Tibetan Plateau: Three year atmospheric monitoring study. Environ. Sci. Technol. 47: 3199–3208.
3. Vos JG, Dybing E, Greim HA, Ladefoged O, Lambré C, et al. (2000) Health effects of endocrine-disrupting chemicals on wildlife, with special reference to the European situation. Crit. Rev. Toxicol. 30: 71–133.
4. World Health Organization (2013) The use of DDT in malaria vector control.
5. Turusov V, Rakitsky V, Tomatis L (2002) Dichlorodiphenyltrichloroethane (DDT): Ubiquity, persistence, and risks. Environ. Health Persp. 110: 125–128.
6. Eskenazi B, Chevrier J, Goldman Rosas L, Anderson HA, Bornman MS, et al. (2009) The Pine River Statement: Human health consequences of DDT Use. Environ. Health Persp. 117: 1359–1367.
7. Lind PM, Penell J, Salihovic S, van Bavel B, Lind L (2014) Circuling levels of *p, p'*-DDE are related to prevalent hypertension in the elderly. Environ. Res. 129: 27–31.
8. Richardson JR, Roy A, Shalat SL, von Stein RT, Hossain MM, et al. (2014) Elevated Serum Pesticide Levels and Risk for Alzheimer Disease. JAMA Neurol. do i:10.1001/jamaneurol.2013.6030.
9. Fishbein L (1974) Chromatographic and biological aspects of DDT and its metabolites. J. Chromatogr. 98: 177–251.
10. Aislabie JM, Richards NK, Boul HL (1997) Microbial degradation of DDT and its residues – a review. New Zeal. J. Agr. Res. 40: 269–282.
11. Sayles GD, You G, Wang M, Kupferle MJ (1997) DDT, DDD, and DDE dechlorination by zero-valent iron. Environ. Sci. Technol. 31: 3448–3454.
12. Guillette LJ Jr, Pickford DB, Crain DA, Rooney AA, Percival HF (1996) Reduction in penis size and plasma testosterone concentrations in juvenile alligators living in a contaminated environment. Gen. Comp. Endocr. 101: 32–42.
13. Fenner K, Canonica S, Wackett LP, Elsner M (2013) Evaluating Pesticide Degradation in the Environment: Blind Spots and Emerging Opportunities. Science 341: 752–758.
14. Elsner M (2010) Stable isotope fractionation to investigate natural transformation mechanisms of organic contaminants: principles, prospects and limitations. J. Environ. Monitor. 12: 2005–2031.
15. Wolfsberg M, Van Hook WA, Paneth P, Rebelo LPN (2010) Isotope effects in the chemical, geological, and bio sciences. Dordrecht, London: Springer. 466 p.
16. Schmidt HL (2003) Fundamentals and systematics of the non-statistical distributions of isotopes in natural compounds. Naturwissenschaften 90: 537–552.
17. Schmidt HL, Werner RA, Eisenreich W (2003) Systematics of ^2H patterns in natural compounds and its importance for the elucidation of biosynthetic pathways. Phytochem. Rev. 2: 61–85.
18. Remaud GS, Martin YL, Martin GG, Martin GJ (1997) Detection of sophisticated adulterations of natural vanilla flavors and extracts: Application of the SNIF-NMR method to vanillin and *p*-hydroxybenzaldehyde. J. Agric. Food Chem. 45: 859–866.
19. Schleucher J, Vanderveer P, Markley JL, Sharkey TD (1999) Intramolecular deuterium distributions reveal disequilibrium of chloroplast phosphoglucose isomerase. Plant Cell Environ. 22: 525–533.
20. Markai S, Marchand PA, Mabon F, Baguet E, Billault I, et al. (2002) Natural deuterium distribution in branched-chain medium-length fatty acids is nonstatistical: A site-specific study by quantitative ^2H NMR spectroscopy of the fatty acids of capsaicinoids. ChemBioChem 3: 212–218.
21. Betson TR, Augusti A, Schleucher J (2006) Quantification of deuterium isotopomers of tree-ring cellulose using nuclear magnetic resonance. Anal. Chem. 78: 8406–8411.
22. Lesot P, Courtieu L (2009) Natural abundance deuterium NMR spectroscopy: Developments and analytical applications in liquids, liquid crystals and solid phases. Prog. Nucl. Mag. Res. Sp. 55: 128–159.
23. Armbruster W, Lehnert K, Vetter W (2006) Establishing a chromium-reactor design for measuring δ^2H values of solid polyhalogenated compounds using direct elemental analysis and stable isotope ratio mass spectrometry. Anal. Bioanal. Chem. 384: 237–243.
24. Kovacs H, Moskau D, Spraul M (2005) Cryogenically cooled probes – a leap in NMR technology. Prog. Nucl. Magn. Reson. Spectrosc. 46: 131–155.
25. Wiberg KB (1955) The deuterium isotope effect. Chem. Rev. 55: 713–743.
26. Melander L, Saunders WH Jr (1980) Reaction Rates of Isotopic Molecules. Wiley, New York.
27. Luo Y-H, Sternberg L, Suda S, Kumazawa S, Mitsui A (1991) Extremely low D/H ratios of photoproduced hydrogen by cyanobacteria. Plant Cell Physiol. 32: 897–900.
28. Walter S, Laukenmann S, Stams AJM, Vollmer MK, Gleixner G, et al. (2012) The stable isotopic signature of biologically produced molecular hydrogen (H_2). Biogeosciences 9: 4115–4123.
29. Kuder T, Philp P (2013) Demonstration of compound-specific isotope analysis of hydrogen isotope ratios in chlorinated ethenes. Environ. Sci. Technol. 47: 1461−1467.
30. Shouakar-Stash O, Drimmie RJ (2013) Online methodology for determining compound-specific hydrogen stable isotope ratios of trichloroethene and 1,2-*cis*-dichloroethene by continuous-flow isotope ratio mass spectrometry. Rapid Commun. Mass. Spectrom. 27: 1335–1344.
31. Martin GJ, Martin ML, Mabon F, Michon MJ (1982) Identification of the origin of natural alcohols by natural abundance hydrogen-2 nuclear magnetic resonance. Anal. Chem. 54: 2380–2382.

3

Monocrotophos Pesticide Decreases the Plasma Levels of Total 3,3′,5-Triiodo-L-Thyronine and Alters the Expression of Genes Associated with the Thyroidal Axis in Female Goldfish (*Carassius auratus*)

Xiaona Zhang, Hua Tian, Wei Wang, Shaoguo Ru*

College of Marine Life Sciences, Ocean University of China, Qingdao, China

Abstract

Our recent study showed that monocrotophos (MCP) pesticide disrupted the hypothalamic-pituitary-thyroid (HPT) axis in male goldfish (*Carassius auratus*); however, the effects of MCP on the thyroid system in female goldfish are remain unclear. In the present study, plasma thyroid hormone (TH) and thyroid-stimulating hormone (TSH) levels were evaluated in female goldfish exposed to 0.01, 0.10, and 1.00 mg/L of 40% MCP-based pesticide for 21 days in a semi-static exposure system. Expression profiles of HPT axis-responsive genes, including transthyretin (*ttr*), deiodinases (*d1*, *d2*, and *d3*), *tshβ*, thyrotropin-releasing hormone (*trh*), and corticotrophin-releasing hormone (*crh*), were determined. The results indicated that MCP decreased the plasma levels of total 3,3′,5-triiodo-L-thyronine (TT_3) and the ratio of TT_3 to total 3,3′,5,5′-L-thyroxine (TT_4), and induced alternative expression of TH-related genes. Exposure to 0.01 and 0.10 mg/L MCP pesticide resulted in the up-regulation of *ttr* mRNA. The reduction of plasma TT_3 levels was partly attributed to an increase in the metabolism of T_3 in the liver, as revealed by the highly elevated hepatic *d1* and *d3* mRNA levels in the MCP treatment groups, and the expression of hepatic *d3* showed a negative correlation with the plasma TT_3/TT_4 levels in females. Moreover, the plasma TSH levels were lower in females exposed to 0.01 and 0.10 mg/L MCP pesticide, whereas the up-regulation of *tshβ* mRNA levels was compensated by the decreased plasma TT_3 levels. These results indicated that MCP had the potential to influence several pathways of HPT axis homeostasis in female goldfish.

Editor: Vincent Laudet, Ecole Normale Supérieure de Lyon, France

Funding: This work was supported by the National Natural Science Foundation of China (31202001), www.nsfc.gov.cn, the Natural Science Foundation of Shandong Province (ZR2012CQ010), www.sdnsf.gov.cn, and the Innovation and Research Foundation of Ocean University of China, http://www.moe.edu.cn/. The funders had no role in study design, data collection and analysis, decision to publish, or preparation of the manuscript.

Competing Interests: The authors have declared that no competing interests exist.

* Email: rusg@ouc.edu.cn

Introduction

In fish, growth and reproduction are, at least partly, under the control of thyroid hormones (THs), 3,3′,5,5′-L-thyroxine (T_4) and 3,3′,5-triiodo-L-thyronine (T_3) [1,2]. Considering that reproduction in females from the early development of follicles to oocyte maturation and ovulation involves a significant investment of energy to support the related physiological functions, the reproductive-related allocation of energy is particularly important during ovarian development in oviparous species, including fish. Since THs have profound effects on energy metabolism, for example, they are the major regulators of oxidative energy metabolism at the level of the mitochondria in teleost fish [3], they should invariably be involved in the multifactorial regulation of metabolism associated with reproduction. They might also act as direct modulators of the reproductive cycle. A previous study on goldfish (*Carassius auratus*) has shown that T_4 could act synergistically with gonadotropin to influence ovarian development by increasing ovarian sensitivity to gonadotropic stimulation [4]. Moreover, the expression of steroidogenic enzymes and steroid receptors in goldfish could be modulated by THs [5].

Monocrotophos (MCP; CAS number, 6923-22-4), an organophosphorus pesticide, has been banned in developed countries due to its high toxicity; however, it is still used extensively in agricultural practices in developing countries such as Pakistan, China, and India, where high pesticide residue levels have led to excessive MCP levels in the environment. For example, the concentrations of MCP detected in water sources in China and rain water in India were 0.165 and 4 μg/L, respectively [6,7]. Anjum and Malik [8] reported the presence of organophosphorus pesticide in the industrial wastewater around Lucknow, India, and the concentration of MCP was determined to be 8.32±3.9 ng/mL. Our recent study showed that the MCP pesticide exhibited thyroid-disrupting effects by interfering with the thyroidal axis, resulting in decreased plasma total T_3 (TT_3) levels in male goldfish [9]. Further, a gender difference in the thyroid system in response to metolachlor has been reported by Jin et al. [10], and the authors considered that endogenous sex hormones might modify the response of the thyroid system to the environmental chemical. In our previous study, MCP was found to interfere with the reproductive axis *via* several pathways, thereby inducing increases in plasma 17β-estradiol (E_2) levels and decreases in testosterone (T) levels in both male and female goldfish [11,12]. Balanced plasma TH levels are known to be crucial for normal reproductive function; in mammals, both hyperthyroidism and hypothyroidism were found to result in reproductive impairment and lower fertility [13]. Considering gender differences in the response of the thyroid endocrine system in fish exposed to environmental pollutants, the thyroid disruption effects of MCP in females might be speculated to be similar to those found in males. The effects of MCP on the thyroid endocrine system in females, however, remain unclear.

THs are synthesized and secreted by the thyroid follicles under the control of the HPT axis. In teleosts, the thyrotropin-releasing hormone (TRH) and/or corticotrophin-releasing hormone (CRH), released from the hypothalamus, coordinate the HPT axis function by controlling the release of thyrotropin (TSH) from the pituitary, which could stimulate TH synthesis and release [1,14]. Most of the plasma THs in fish are bound to transthyretin (TTR), a specific TH transport protein in teleosts [15], and only free hormones can enter target cells to elicit a response. In the liver and some other peripheral tissues, three types of deiodinases (type I, D1; type II, D2; and type III, D3) are known to control the conversion of T_4 to the more physiologically active T_3 or the production of metabolically inactive counterparts [16]. Such complex regulatory pathways are involved in thyroid homeostasis. Therefore, environmental chemicals can act at multiple stages in the HPT axis.

In the present study, following administration of 40% MCP pesticide to female goldfish, plasma TH levels, including TT_3, total T_4 (TT_4), and the physiologically relevant free T_3 (FT_3) and free T_4 (FT_4) levels, were determined, and the changes in endocrine-mediated responses along the HPT axis were evaluated, including the regulation of TRH and CRH, synthesis and secretion of TSH, and expression of deiodinases (*d1*, *d2*, and *d3*) and transthyretin (*ttr*) genes.

Materials and Methods

Pesticide

MCP pesticide (3-hydroxyl-*N*-methyl-*cis*-crotonamidedimethyl phosphate, 40% water-soluble preparation) was purchased from the Qingdao pesticide factory in China. The concentration on the label was 40%, which was consistent with the actual concentration determined by gas chromatography (40±0.1%) [17]. The half-life of MCP is approximately 66 days at pH 7.0 and 20°C [18].

Fish exposure and sample protocols

Sexually mature goldfish (*C. auratus*) with mature and fully developed gonads (8.7±0.8 cm standard length; 21.2±4.1 g wet weight; 6.8±0.3% gonadal somatic index) were obtained from a local dealer in Qingdao, PR China, and sampled in late spring. Fish were handled according to the National Institute of Health Guidelines for Handling and Care of Experimental Animals. The animal utilization protocol was approved by the Institutional Animal Care and Use Committee of the Ocean University of China. Fish were maintained in a 70-L aquarium containing 50-L dechlorinated tap water at ambient temperature (18±2°C) with dissolved oxygen content of 7.0±0.1 mg/L and were fed small shrimp daily.

After the female goldfish were acclimated to laboratory conditions for 2 weeks, they were exposed to nominal 0.01, 0.10, and 1.00 mg/L 40% MCP pesticide (4, 40, and 400 µg/L equivalent levels of pure MCP), the concentrations of which were 1/10,000; 1/1,000; and 1/100 of the 96-h LC_{50} [11], respectively. The experiments were conducted in the same 70-L aquarium containing 50-L dechlorinated tap water by using a semi-static toxicity test (20-L water renewal daily to constantly maintain the MCP concentration). Each group of fish was exposed in three aquaria (seven fish/tank) and a control (dechlorinated tap water) was included in the exposure design. Less than 10% mortality was observed in all the treatments during the experiment, and abnormal behaviors such as unusual swimming pattern or bumping against the tank were observed in the group exposed to the highest dose.

After 21 days of exposure, all fish were anesthetized with 75 mg/L MS-222 (Sigma, St. Louis, MO, USA) and sampled between 9:00 and 11:00 h to avoid the possible influences of diurnal fluctuations in hormone levels on the results [19]. Fish from each group were divided into two subgroups at the time of sampling. One subgroup of 9–10 fish was sampled to collect 0.6–0.8 mL blood for TH measurements, whereas the remaining fish were used to collect blood with the same volume for TSH and cortisol measurements. Blood was collected from the caudal vein by using chilled heparinized syringes and maintained on ice. Plasma samples were obtained after centrifugation at 1000×*g* for 10 min and stored at −80°C until the hormone assay could be performed. The liver, pituitary, and hypothalamus tissues were dissected (n = 9), frozen in liquid nitrogen, and then stored at −80°C for the quantification of gene expression by real-time polymerase chain reaction (PCR).

Hormone assay

The plasma levels of TT_3, TT_4, FT_3, FT_4, TSH, and cortisol were measured using radioimmunoassay (RIA) by using commercially available kits following the protocols provided by the manufacturers (Beijing North Institute of Biological Technology, Beijing, China). The RIA kits for human TT_3, TT_4, FT_3, FT_4, TSH, and cortisol were validated for use with goldfish samples by showing parallelism between a series of diluted and spiked samples in relation to the standard curve included with the assay kits. Standards and samples were added to test tubes in duplicate. The assay detection limits were 0.05 ng/mL for T_3, 2 ng/mL for T_4, 0.06 pg/mL for FT_3, 0.23 pg/mL for FT_4, 0.1 µIU/mL for TSH, and 1 ng/mL for cortisol. The inter- and intra-assay coefficients of variation for all the above-mentioned hormones were <10% and <15%, respectively.

Gene expression analysis

Total RNA from each tissue was isolated using the phenolic reagent TRIzol (Invitrogen, Carlsbad, CA, USA) according to the manufacturer's protocol. The extracted RNA was measured by spectrometry at $OD_{260/280}$ before treatment with DNase I (Promega, Madison, WI, USA). Equal amounts of RNA (1 µg) were reverse-transcribed into cDNA in 20-µL reactions containing 10 pmol oligo$(dT)_{20}$, 4 µL 5× RT Buffer, 2 µL dNTP mixture, 10 U RNase inhibitor, and 1 µL ReverTra Ace (Toyobo, Tokyo, Japan). Reverse transcription reactions were conducted in a Bio-Rad DNA Thermal Cycler (Hercules, CA, USA) at 42°C for 20 min and terminated for 5 min at 85°C.

Oligonucleotide primers were designed for the specific amplification of *tshβ*, *trh*, *crh*, *ttr*, *d1*, *d2*, *d3*, *β*-actin, and 18S rRNA by using the Primer Premier 5.0 software (PREMIER Biosoft Int., Palo Alto, USA) and the sequences for goldfish available from the GenBank database (Table 1). The amplifications were performed using an Eppendorf MasterCycler ep *RealPlex*[4] (Eppendorf, Wesseling-Berzdorf, Germany). Parallel PCR analyses were conducted to amplify the cDNA of the target and reference genes. Real-time PCR was performed in 20-µL reaction mixtures containing 1× SYBR Premix Ex *Taq*, 0.4 µM of each primer,

0.4 μL of ROX reference Dye (Takara Bio Inc., Shiga, Japan), and 4 μL of first-strand cDNA (template). The thermal profile was 95°C for 30 s, followed by 40 cycles of 95°C for 5 s and 60°C for 30 s. Amplification of a single product was ensured by performing a melting curve analysis by using the PCR products obtained at the end of each PCR run. In addition, 2% agarose gel electrophoresis of the PCR products was performed to confirm the presence of single amplicons having the predicted sizes (data not shown). The *β*-actin and 18S rRNA transcripts were used as housekeeping genes to standardize the results and eliminate variations in mRNA and cDNA quantity and quality. Neither the *β*-actin nor 18S rRNA levels were affected by any of the experimental conditions used in the study. For each reaction, the relative target gene mRNA expression levels were normalized to the geometric mean of *β*-actin and 18S rRNA expression levels by using the formula $2^{-\Delta\Delta Ct}$ and plotted on a logarithmic scale [20].

Statistical analysis

The experimental data were presented as the means ± standard deviations, and the differences between the control and each exposure group were evaluated using one-way analysis of variance (ANOVA) followed by Tukey's test. Before parametric analysis, assumptions of normality and homogeneity of variances were assessed using probability plots and normality tests. Pearson's correlation coefficient was used to calculate the relationship between the expression of certain genes involved in the HPT axis and the plasma TH levels. Values were considered significant when $0.01<P<0.05$ and highly significant when $P<0.01$.

Results

Effects of MCP pesticide on plasma TT_3 and TT_4 levels and the ratio of TT_3 to TT_4

The TT_3 content in the plasma of the control female goldfish was 1.31±0.29 ng/mL, whereas that in the fish exposed to 0.10 and 1.00 mg/L MCP pesticide was significantly decreased by 30% to 0.92±0.23 ng/mL ($0.01<P<0.05$) and 51% to 0.64±0.21 ng/mL ($P<0.01$), respectively (Fig. 1A). At the end of the MCP treatment, the TT_4 content was 7.21±1.55 ng/mL in the control group, and there was no significant change in the TT_4 content of any of the MCP pesticide-treated groups (Fig. 1B). There was a dose-related decrease in the ratio of TT_3 to TT_4 after MCP treatment ($P<0.01$; Fig. 1C).

Effects of MCP pesticide on plasma FT_3 and FT_4 levels and hepatic *ttr* mRNA expression

The basal FT_3 levels in the plasma of the control female goldfish were 2.39±0.39 pg/mL; they were significantly reduced by 32%, 32%, and 50% in the 0.01, 0.10, and 1.00 mg/L MCP pesticide-treated groups, respectively ($P<0.01$; Fig. 2A). Plasma FT_4 levels in the control females were 2.10±0.80 pg/mL; they were significantly increased to 4.71±0.85 pg/mL in the 0.01 mg/L MCP pesticide-treated group ($P<0.01$) but decreased to 1.26 pg/mL in the 1.00 mg/L MCP pesticide-treated group ($P<0.01$; Fig. 2B). Furthermore, the mRNA expression of hepatic *ttr* was significantly higher after treatment with 0.01 and 0.10 mg/L of MCP pesticide ($P<0.01$), whereas there was no significant difference in the hepatic *ttr* mRNA expression between the group treated with the highest dose and the control group (Fig. 2C).

Table 1. Nucleotide sequences of the primers used for real-time polymerase chain reaction and product sizes.

Gene	GenBank accession no.	Primer sequence (5′-3′)	Amplicon size (bp)
tshβ	AB003584	F: TGGCTGTCAACACCACCATCTG	94
		R: CCCCTTTGAACCAGGAAACGAG	
trh	AB179819	F: GAACAAGAGTCGCTGTAAGGGAG	102
		R: GGATGCTGTGGAAAACAATGAAC	
crh	AF098629	F: CCAGAATTATCTCCGATCCCCAG	109
		R: CCCAACTTTCCCCTCCAACAG	
ttr	EU313781	F: TTCAGATGCTCACTGCCCTTTG	103
		R: CCCACCTTGTTCTTGGCGATAG	
d1	EU313785	F: GGCAGATTTCCTTGTTGTTTACC	112
	EF190702	R: CGCAGCCAGTCTCTCCTCC	
d2	EU313786	F: CAAACTCCAAGGTGGTGAAGGT	86
	EF190703	R: GTCCAGCAGATGGCACTCGT	
d3	EF190704	F: GCTGTTCCTGGATTTCCTGTGC	163
	EU313787	R: GAAGAAGTCCAGCTTGTGTCCGT	
β-actin	AB039726	F: GAAACTGGAAAGGGAGGTAGC	115
		R: CTGTGAGGGCAGAGTGGTAGA	
18S rRNA	AF047349	F: AGAAACGGCTACCACATCCAAG	169
		R: GCACCAGATTTGCCCTCCAG	

Goldfish contain duplicate genes encoding for D1, D2, and D3, and the primer pairs amplify the two genes.

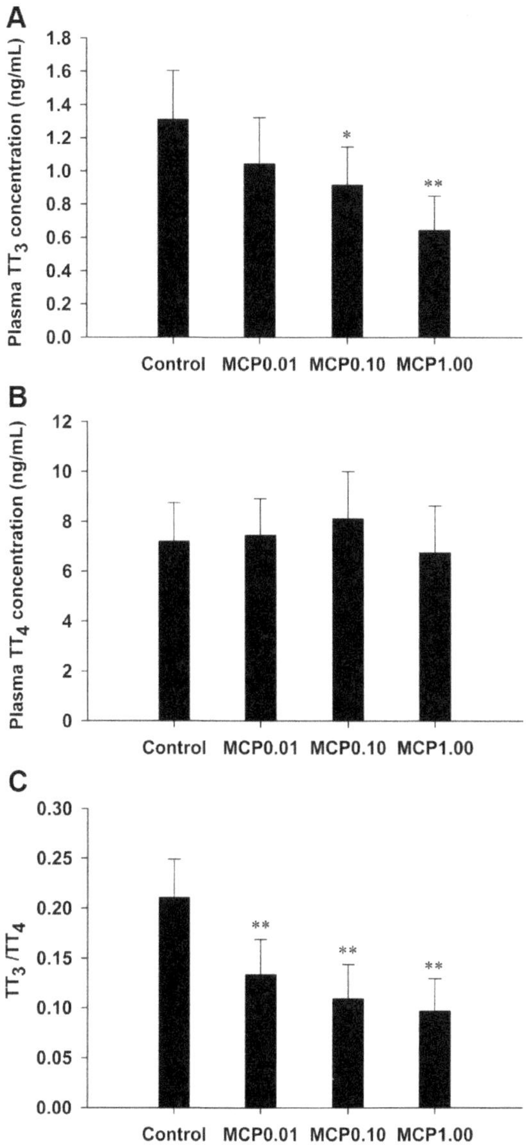

Figure 1. Quantification of plasma total 3,3′,5-triiodo-L-thyronine (T_3) and total 3,3′,5,5′-L-thyroxine (T_4) levels in female goldfish exposed to 0, 0.01, 0.10, and 1.00 mg/L 40% monocrotophos (MCP) pesticide. (designated C, MCP0.01, MCP0.10, and MCP1.00, respectively). The data are presented as the means ± standard deviations (n = 9). Asterisks indicate statistically significant differences from the control group (*0.01 < P < 0.05, ** P < 0.01).

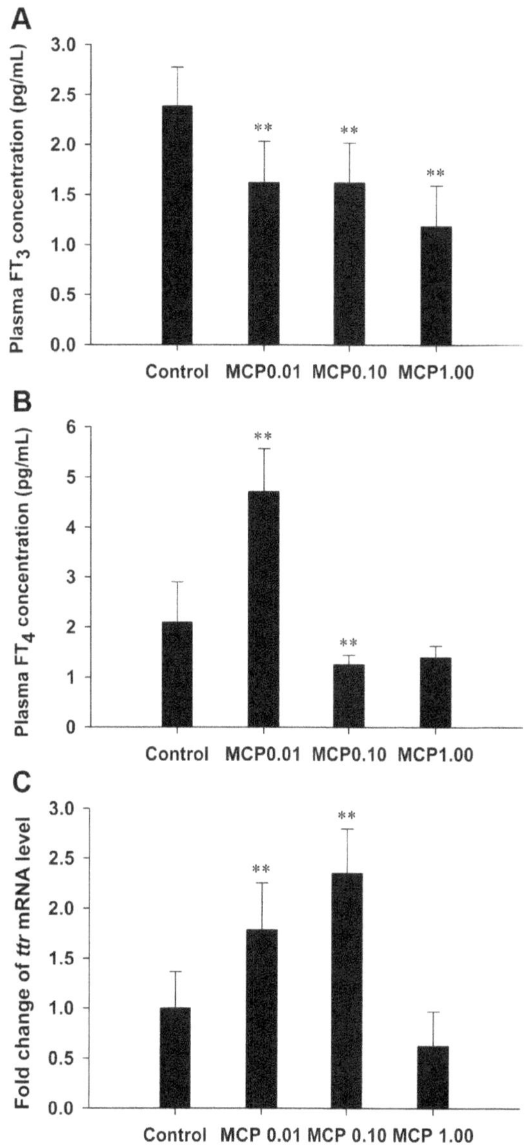

Figure 2. Quantification of plasma free 3,3′,5-triiodo-L-thyronine (T_3) and free 3,3′,5,5′-L-thyroxine (T_4) contents and the relative mRNA expression levels of hepatic transthyretin (*ttr*) in female goldfish exposed to 0, 0.01, 0.10, and 1.00 mg/L 40% monocrotophos (MCP) pesticide. (designated C, MCP0.01, MCP0.10, and MCP1.00, respectively). For panel C, fold change (Y axis) represents the expression of the target gene mRNA relative to that of the control group (equals 1 by definition). The data are presented as the means ± standard deviations (n = 9). Asterisks indicate statistically significant differences from the control group (*0.01 < P < 0.05, ** P < 0.01).

Effects of MCP pesticide on *d1*, *d2*, and *d3* mRNA expression levels in the liver, brain, and kidney

The *d1* mRNA levels in the liver were significantly increased by 0.72- and 2.15-fold and the hepatic *d2* mRNA expression was significantly up-regulated by 1.30- and 0.84-fold after treatment with 0.01 and 0.10 mg/L MCP pesticide, respectively (P < 0.01), relative to those in the control (Fig. 3A and 3B). Hepatic *d3* mRNA levels were significantly higher in all the MCP-treated groups, especially in the 0.01 and 0.10 mg/L MCP pesticide-treated groups (2.66- and 4.50-fold higher, respectively), than those in the control (P < 0.01; Fig. 3C).

With regard to the gene transcription level of deiodinase in the brain (Fig. 3D–3F), there was no significant difference in the *d1* gene transcription in any of the MCP-exposed groups. MCP pesticide exposure significantly stimulated *d2* gene transcription in the 0.01 mg/L and 0.10 mg/L groups (P < 0.01) and *d3* gene transcription in the 0.10 mg/L group (P < 0.01). However, the transcription of both *d2* and *d3* genes was significantly inhibited after exposure to 1.00 mg/L MCP pesticide (P < 0.01).

Transcription of the *d1* gene in the kidney was significantly up-regulated after exposure to 0.10 mg/L MCP (P < 0.01; Fig. 3G). The transcription of *d2* was significantly down-regulated after exposure to 0.01 and 1.00 mg/L MCP pesticide, but was stimulated in the 0.10 mg/L group (0.01 < P < 0.05; Fig. 3H). The expression levels of *d3* mRNA in the kidney were significantly

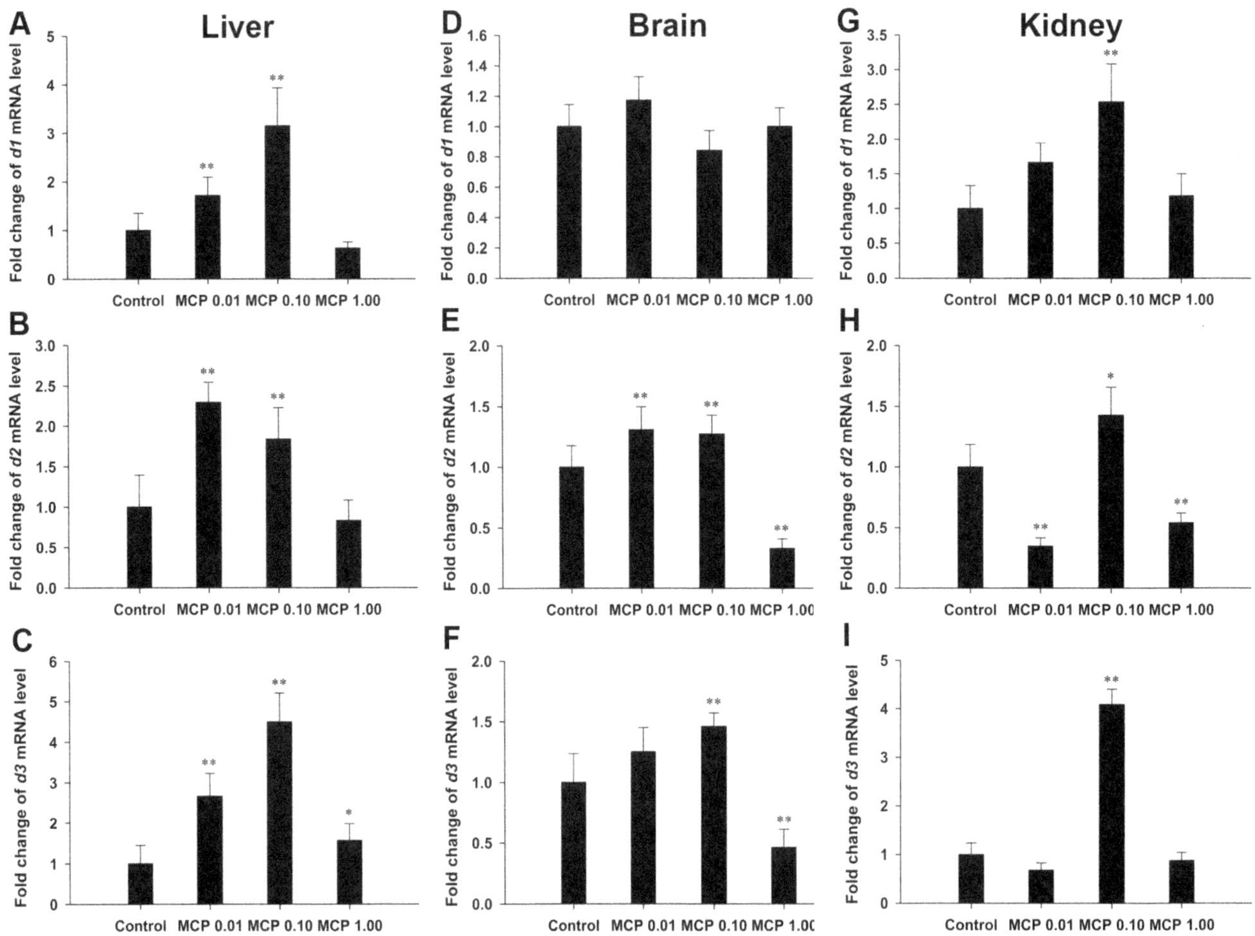

Figure 3. Relative mRNA expression levels of types I, II, and III deiodinases (*d1, d2,* and *d3*) in the liver, brain, and kidney of female goldfish exposed to 0, 0.01, 0.10, and 1.00 mg/L 40% monocrotophos (MCP) pesticide. (designated C, MCP0.01, MCP0.10, and MCP1.00, respectively). Fold change (Y axis) represents the expression of the target gene mRNA relative to that of the control group (equals 1 by definition). The data are presented as the means ± standard deviations (n = 9). Asterisks indicate statistically significant differences from the control group (*$0.01<P<0.05$, **$P<0.01$).

up-regulated only in the 0.10 mg/L MCP-exposed group (4.0-fold, $P<0.01$; Fig. 3I).

Effects of MCP pesticide on pituitary *tshβ* mRNA expression and plasma TSH levels

The mRNA expression of pituitary *tshβ* was significantly up-regulated in both the 0.10 and 1.00 mg/L MCP pesticide-treated groups ($P<0.01$; Fig. 4A). In contrast, the plasma TSH levels in the 0.10 and 1.00 mg/L MCP pesticide-treated groups were markedly decreased to 0.85±0.16 μIU/mL ($P<0.01$) and 0.91±0.14 μIU/mL ($0.01<P<0.05$), respectively, compared to those in the control group (1.24±0.30 μIU/mL).

Effects of MCP pesticide on hypothalamic *trh* and *crh* mRNA and plasma cortisol levels

The expression of hypothalamic *trh* was significantly down-regulated after treatment with the MCP pesticide (Fig. 5A). The transcription of the *crh* gene was significantly stimulated after treatment with 0.10 mg/L MCP pesticide ($P<0.01$), but inhibited in the 1.00 mg/L MCP pesticide-treated group ($0.01<P<0.05$; Fig. 5B). The measured plasma cortisol content was 142.01±24.11 ng/mL in the control female goldfish and was significantly decreased to 74.53±18.26 ng/mL and 89.88±22.05 ng/mL in the 0.10 and 1.00 mg/L MCP pesticide-treated groups, respectively ($P<0.01$; Fig. 5C).

The correlation coefficient for plasma THs levels and the expression of certain genes related to the HPT axis

The correlation coefficient for plasma THs levels and the expression of certain genes associated with the HPT axis are shown in Table 2. Significant correlations of the ratio of TT_3 to TT_4 with hepatic *d3* expression, TT_3 with pituitary *tshβ* expression, and FT_4 with hepatic *ttr* expression were found.

Discussion

An increasing number of studies have reported that groups of pesticides, including acetochlor, amitrole, and metolachlor, have the potential to influence several steps in HPT axis homeostasis and to induce THs disturbance in adult fish, particularly with respect to sex differences occurring in response to chemical-induced thyroid system disruption [10,21,22]. For example, Li et al. [21] showed that TH-related genes such as malic enzyme and sodium iodide symporter were significantly down-regulated in the brains of the rare minnow *Gobiocypris rarus*, and that the expression of these genes in females was more sensitive to

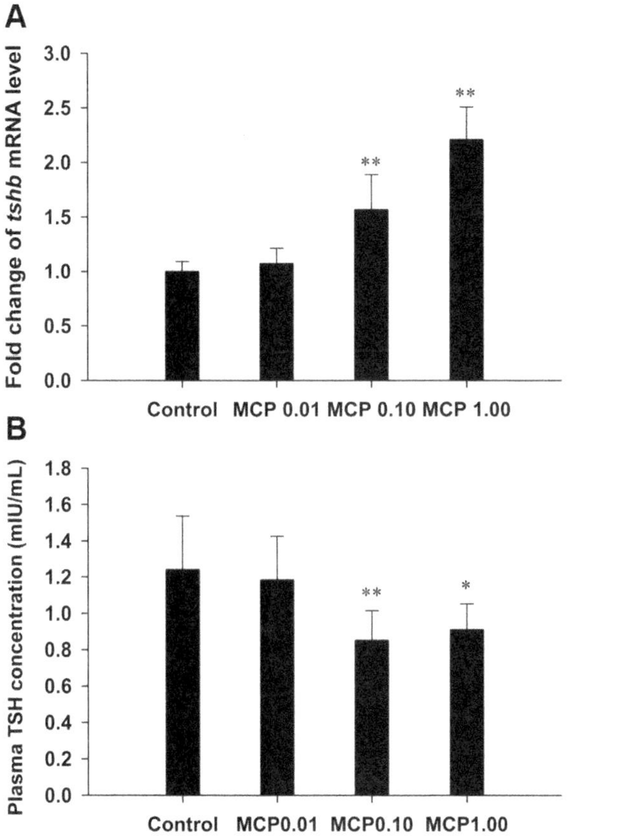

Figure 4. Relative mRNA expression levels of pituitary thyroid-stimulating hormone β subunit (*tshβ*) and quantification of plasma TSH content in female goldfish exposed to 0, 0.01, 0.10, and 1.00 mg/L 40% monocrotophos (MCP) pesticide. (designated C, MCP0.01, MCP0.10, and MCP1.00, respectively). For panel A, fold change (Y axis) represents the expression of the target gene mRNA relative to that of the control group (equals 1 by definition). The data are presented as the means ± standard deviations (n = 9). Asterisks indicate statistically significant differences from the control group (*$0.01<P<0.05$, **$P<0.01$).

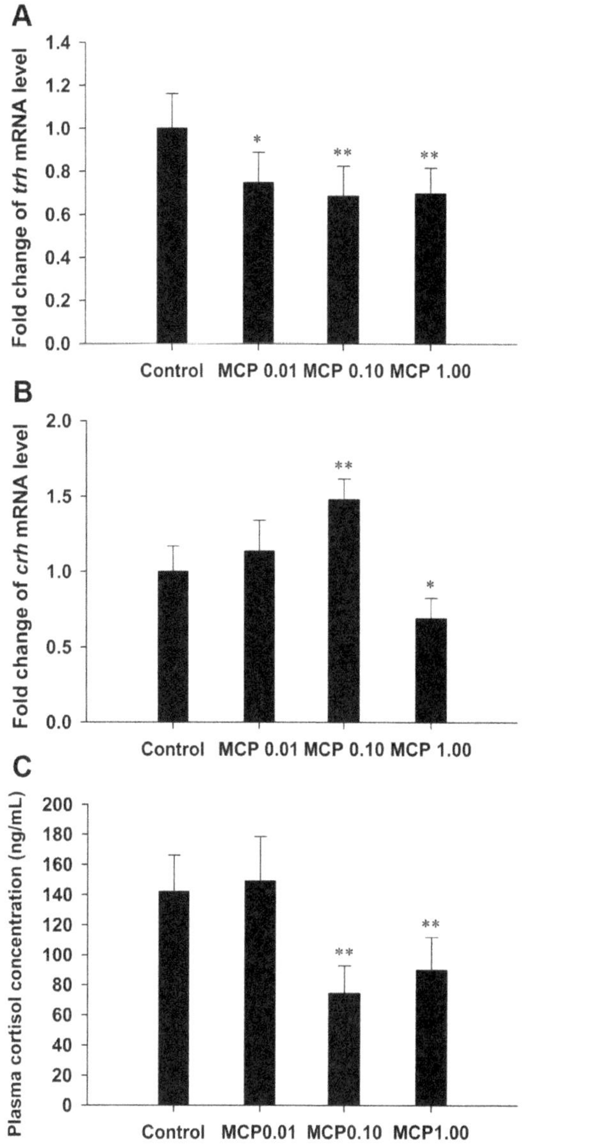

Figure 5. Relative mRNA expression levels of thyrotropin-releasing hormone (*trh*) and corticotrophin-releasing hormone (*crh*) in the hypothalamus glands and quantification of plasma cortisol content in female goldfish exposed to 0, 0.01, 0.10, and 1.00 mg/L 40% monocrotophos (MCP) pesticide. (designated C, MCP0.01, MCP0.10, and MCP1.00, respectively). For panels A and B, fold change (Y axis) represents the expression of the target gene mRNA relative to that of the control group (equals 1 by definition). The data are presented as the means ± standard deviations (n = 9). Asterisks indicate statistically significant differences from the control group (*$0.01<P<0.05$, **$P<0.01$).

acetochlor than that in males. Recently, we found that a 21-d exposure to MCP pesticide caused significant decreases in plasma TT_3 levels and TT_3-to-TT_4 ratios in male goldfish [9]; however, whether similar effects occur in females is not clear.

In the present study, although the plasma levels of TT_4 remained unchanged, those of TT_3, FT_4, and FT_3 in female goldfish were significantly altered after a 21-day exposure to MCP pesticide, suggesting a failed adaptation and auto-regulation of THs homeostasis. Almost all THs circulating in the plasma are bound to transporter proteins, and the equilibrium of TH binding to the plasma proteins determines the concentration of free THs within the plasma [23]. TTR, which is primarily a secretory product of the liver, has been proposed to be the major TH-carrier protein that binds THs and transports them to target tissues in fish [24–26]. In our study, TTR gene expression was up-regulated after exposure to 0.01 and 0.10 mg/L MCP pesticide. This up-regulation might have resulted in higher TTR mRNAs and thus TTR proteins, leading to decreases in plasma FT_3 and/or FT_4 levels. Moreover, the expression of *ttr* showed a positive correlation with plasma FT_4 levels in females. Notably, less than 1% of plasma TT_4 is free with 99% reversibly bound to plasma proteins [23]. Indeed, changes of such small amount of FT_4 contents may not represent a dynamic circulating TT_4 reservoir and vice verse. For example, in brown trout (*Salmo trutta*) fed diets enriched with β–Tetrabromoethylcyclohexane for 56 days, there was no significant difference among treatments in FT_4, but TT_4 was significantly reduced in the high dose group relative to the control [27]. In the 0.01 mg/L MCP group in particular, one possible explanation for the increased plasma FT_4 levels could be that feedback systems attempt to respond to the reduction in plasma TT_3 levels. However, the stimulatory effects of 0.01 mg/L MCP pesticide on plasma FT_4 levels in females were not observed in males [9], indicating that the thyroidal system in female goldfish

Table 2. Pearson correlation coefficients for the plasma thyroid hormone levels and the expression of certain genes related to the hypothalamus-pituitary-thyroid axis.

Paired samples	Correlation coefficient
TT_3/TT_4 vs. hepatic *d1*	−.189
TT_3/TT_4 vs. hepatic *d2*	−.232
TT_3/TT_4 vs. hepatic *d3*	−.384*
TT_3 vs. pituitary *tshβ*	−.469*
TT_3 vs. hypothalamus *trh*	.356
TT_3 vs. hypothalamus *crh*	.025
FT_3 vs. hepatic *ttr*	−.069
FT_4 vs. hepatic *ttr*	.396*

*Statistically significant at $0.01<P<0.05$.

is more sensitive to MCP than that of males. In the group treated with the highest dose of MCP pesticide, the total TH level was more readily responsive to the changes in free hormone levels, since TTR gene expression was not stimulated in this group.

Among the thyroid follicle secretions, T_4 is the predominant circulating hormone in the blood of fish, and T_3 appears to be produced largely by enzymatic deiodination of T_4 in the peripheral tissues [28]. Consequently, plasma T_3 levels mostly decline due to a drop in thyroidal T_4 production and secretion and/or changes in the peripheral TH metabolism [29]. Our finding that decreases in plasma TT_3 levels along with relatively stable plasma TT_4 levels suggested possible changes in peripheral TH deiodination or metabolism. Iodothyronine deiodinases play a crucial role in the mechanism of TH biotransformation in extra-thyroidal tissues. Conversion from T_4 to T_3 is mediated by outer-ring deiodination (ORD). Both inner-ring deiodination (IRD) and ORD are involved in the inactivation of T_4 to 3,3′,5′-triiodo-L-thyronine (rT_3) and T_3 and rT_3 to 3,3′-diiodo-L-thyronine (3,3′-T_2) [16,30]. Three types of deiodinases have been identified in teleosts: type I (D1) has both ORD and IRD activities, whereas types II (D2) and III (D3) have ORD and IRD activity, respectively [31–34]. At the end of the 21-day exposure to 0.01 and 0.10 mg/L MCP pesticide, the transcription of all three types of deiodinases was stimulated in the liver: changes in the expression of *d1* and *d3* mRNA were higher than those of *d2* mRNA. Further, in the 1.00 mg/L treatment, only hepatic *d3* mRNA levels were significantly up-regulated. Van der Geyten et al. [35] showed that the changes in hepatic deiodinases activities tend to be consistent with those of their mRNA levels, indicating pre-translated regulation of the hepatic deiodinases. Both D1 and D3 are involved in the metabolism of THs. D3 catalyzes the degradation of T_4 and T_3 to inactive metabolites and could protect tissues from an excess of biologically active TH, namely, T_3 [36]. Although D1 also has the ability to degrade T_3, it is less effective than D3, and its preferred substrate is rT_3 [37]. Therefore, the stimulated enzymatic metabolism of T_3 in the liver could cause a reduction in the levels of circulating TT_3. Consistently, a significant negative correlation between hepatic *d3* mRNA expression and the ratio of TT_3 to TT_4 was observed. It is also worth noting that, in the highest dose group, alterations of gene expression of the three types of deiodinases in females were not consistent with those in males in which hepatic *d1* and *d2* mRNA levels were significantly down-regulated, whereas *d3* mRNA levels were not affected [9].

Although liver is considered to be the main peripheral source of circulating T_3, TH deiodination also occurs in other extra-thyroidal tissues such as brain and kidney; however, the available T_3 derived from these tissues is primarily utilized by the same tissues itself [38]. Exposure to MCP pesticide had no effect on the *d1* mRNA expression in the brain, but stimulated the transcription of *d1* gene in the kidney of female goldfish. The functional roles of D1 might indicate why the patterns of *d1* mRNA regulation varied among tissues after exposure to MCP. Recent studies have suggested that the major role of D1 might be to clear rT_3 and sulfated iodothyronines from circulation. Indeed, it functions as a scavenger enzyme to remove inactive iodothyronines and recycle iodine within the organisms [39–41]. Thus, 0.10 mg/L MCP pesticide might enhance the metabolism of THs by up-regulating the expression of *d1* mRNA in the kidney. Further, the specific patterns of deiodinase-mediated gene regulation show tissue-specific variations, probably in accordance with the distinct thyroid status and tissue-specific requirements for available THs [42]. In the brain, the transcriptions of *d2* and *d3* gene were stimulated after treatment with 0.01 and 0.10 mg/L MCP but inhibited by 1.00 mg/L MCP pesticide, whereas, in the kidney, the transcriptions of *d2* and *d3* gene were stimulated after treatment with 0.10 mg/L MCP but inhibited by treatment with 0.01 and 1.00 mg/L MCP pesticide. Such expression profiles of *d2* and *d3* changing in a parallel way indicates that intracellular T_3 levels are tightly controlled [43].

THs are also regulated by TSH and TRH/CRH, but can interfere with the synthesis and release of TSH and TRH/CRH *via* feedback mechanisms. This regulatory system plays an important role in maintaining the homeostatic balance along the HPT axis [44,45]. In the present study, the *tshβ* mRNA levels were up-regulated and showed a negative correlation with plasma TT_3 levels. The T_3 feedback mechanism is known to control TSH expression in certain teleosts, and Sohn et al. [46] provided evidence that T_3 acted directly on the pituitary and inhibited *tshβ* gene expression in goldfish, probably *via* the negative T_3-responsive elements in the *tshβ* gene. Hence, the decreased plasma TT_3 levels would feedback to stimulate the transcription of *tshβ*, and similar negative feedback has also been observed in males [9]. TSH is a glycoprotein secreted by the pituitary thyrotrope cells. In our previous study, Bing [47] reported direct damage of the MCP pesticide to the structure of thyrotrope cells in adenohypophysis. The partly dissolved nuclear membrane, prominent dilated rough endoplasmic reticulum, and slightly dissolved mitochondrial cristae indicated a reduction on the hormone secretion yield of thyrotrope cells. Accordingly, in the current study, although pituitary *tshβ* mRNA levels were elevated, the reduced plasma TSH levels in the 0.10 and 1.00 mg/L MCP treatments were probably resulted from the diminished TSH protein synthesis and secretion of thyrotrope cells. In goldfish, both TRH and CRH might be involved in the regulation of the pituitary-thyroid axis, and TRH might act as a multifunctional hypophysiotropic factor. For instance, TRH is a potent stimulator of alpha melanocyte-stimulating hormone release from the pars intermedia cells in various species of teleost fish, including goldfish [48], and is also involved in the regulation of feeding and locomotor behaviors [49]. Therefore, the up-regulated *trh* mRNA expression levels caused by the MCP pesticide would not be the only response of the HPT axis. Instead, CRH might act as a TSH-releasing factor in lower vertebrates [14], and might also be mediated and/or intensified by a concomitant increase in corticosteroids [50]. It is also worth noting that CRH plays important roles in mediating the hypothalamic-pituitary-interrenal axis (HPI axis) to produce cortisol in response to stress, and that cortisol could also tightly control the release of CRH *via* a negative feedback loop [51,52]. The reductions in plasma cortisol levels in the 0.10 mg/L group would be the result of MCP interference with cortisol synthesis and metabolism at some level. A recent study in our lab revealed that 0.10 mg/L MCP pesticide decreased interrenal synthesis of cortisol and further promoted the metabolism of cortisol, resulting in lower overall cortisol levels and a diminished stress capacity in female zebrafish (*Danio rerio*) [53]. In goldfish, whether CRH acts on thyroid activity either directly or *via* adrenal steroids has yet to be determined; the decreased plasma TT_3 and cortisol levels would also signal the up-regulation of *crh* gene expression, as observed in the 0.10 mg/L group. On the other hand, in the 1.00 mg/L MCP pesticide group, both plasma cortisol levels and hypothalamic *crh* mRNA levels were decreased, indicating that the high concentration of MCP pesticide disrupted the feedback mechanisms along the HPI axis. Some investigators have suggested that hyperactivity of the cortisol-producing cells associated with severe stress might lead to an inter-renal exhaustion with very low levels of cortisol [54,55].

Our results showed that the MCP pesticide disturbed the THs homeostasis and interfered with the transport and conversion of THs, synthesis and secretion of pituitary TSHs, and regulation of hypothalamic TRH/CRH in female goldfish. Similar to the findings in males, MCP pesticide decreased plasma TT_3 and FT_3 levels with relatively stable plasma TT_4 levels, and profiles of the relative abundance of deiodinase transcripts were observed in the liver, brain, and kidneys in female goldfish; however, 0.01 mg/L MCP pesticide merely enhanced the plasma FT_4 levels in females, and gender differences existed in the hepatic deiodinase transcripts at the highest dose. Although the reproductive and thyroidal endocrine systems have been reported to be related to the neuroendocrine control of many complex functions, including growth, metabolism, and reproduction in teleosts [1,2,5], and many chemicals, including polychlorinated biphenyl and perfluor-ooctanoic acid, could simultaneously cause a variety of effects along both endocrine systems [56–59], further studies unraveling the interaction between the MCP-induced reproductive axis disruption and thyroidal axis disruption are necessary to explain gender differences observed in the thyroid system response to MCP exposure in goldfish.

Author Contributions

Conceived and designed the experiments: XZ HT WW SR. Performed the experiments: XZ HT. Analyzed the data: XZ HT. Wrote the paper: XZ SR.

References

1. Cyr DG, Eales JG (1996) Interrelationships between thyroidal and reproductive endocrine systems in fish. Rev Fish Biol Fish 6: 165–200.
2. Power DM, Llewellyn L, Faustino M, Nowell MA, Bjornsson BT, et al. (2001) Thyroid hormones in growth and development of fish. Comp Biochem Physiol C Toxicol Pharmacol 130: 447–459.
3. Oommen OV, Sreejith P, Beyo RS, Divya L, Vijayasree AS, et al. (2006) Thyroid hormones regulate mitochondrial respiration as well as antioxidant defense in teleosts too! J Endocrinol Reprod 102: 96–105.
4. Hurlburt ME (1977) Role of the thyroid gland in ovarian maturation of the goldfish, *Carassius auratus* L. Can J Zool 55: 1906–1913.
5. Habibi HR, Nelson ER, Allan ERO (2012) New insights into thyroid hormone function and modulation of reproduction in goldfish. Gen Comp Endocrinol 175: 19–26.
6. Kang Y, Zhang G (2000) Analysis of organophosphorous pesticides from source water using solid-phase extraction technique. China Environ Sci 20: 1–4.
7. Kumari B, Madan VK, Kathpal TS (2007) Pesticide residues in rain water from Hisar, India. Environ Monit Assess 133: 467–471.
8. Anjum R, Malik A (2013) Evaluation of mutagenicity of wastewater in the vicinity of pesticide industry. Environ Toxicol Pharmacol 35: 284–291.
9. Zhang X, Tian H, Wang W, Ru S (2013) Exposure to monocrotophos pesticide causes disruption of the hypothalamic–pituitary–thyroid axis in adult male goldfish (*Carassius auratus*). Gen Comp Endocrinol 193: 158–166.
10. Jin Y, Chen R, Wang L, Liu J, Yang Y, et al. (2011) Effects of metolachlor on transcription of thyroid system-related genes in juvenile and adult Japanese medaka (*Oryzias latipes*). Gen Comp Endocrinol 170: 487–493.
11. Tian H, Ru S, Bing X, Wang W (2010) Effects of monocrotophos on the reproductive axis in the male goldfish (*Carassius auratus*): potential mechanisms underlying vitellogenin induction. Aquat Toxicol 98: 67–73.
12. Tian H, Ru S, Wang W, Bing X (2010) Effects of monocrotophos on the reproductive axis in the female goldfish (*Carassius auratus*). Comp Biochem Physiol C Toxicol Pharmacol 152: 107–113.
13. Jahnke GD, Choksi NY, Moore JA, Shelby MD (2004) Thyroid toxicants: assessing reproductive health effects. Environ Health Perspect 112: 363–368.
14. De Groef B, Van der Geyten S, Darras VM, Kuhn ER (2006) Role of corticotrophin releasing hormone as a thyrotropin-releasing factor in non-mammalian vertebrates. Gen Comp Endocrinol 146: 62–68.
15. Morgado I, Hamers T, Van der Ven L, Power DM (2007) Disruption of thyroid hormone binding to sea bream recombinant transthyretin by ioxinyl and polybrominated diphenyl ethers. Chemosphere 69: 155–163.
16. Orozco A, Valverde-RC (2005) Thyroid hormone deiodination in fish. Thyroid 15: 799–813.
17. Ru S, Wei X, Jiang M, Li Y (2003) In vivo and in vitro inhibitions of red drum (*Sciaenops ocellatus*) brain acetylcholinesterase and liver carboxylesterase by monocrotophos at sublethal concentrations. Water Air Soil Pollut 149: 17–25.
18. Wang HM, Zhang ZM (1989) Monocrotophos. In: Wang, H.M., Zhang, Z.M. (Eds.), A New Pesticide Manual. China Agriculture Press Beijing p 19.
19. Leiner KA, Han GS, MacKenzie DS (2000) The effects of photoperiod and feeding on the diurnal rhythm of circulating thyroid hormones in the red drum, *Sciaenops ocellatus*. Gen Comp Endocrinol 120: 88–98.
20. Vandesompele J, De Preter K, Pattyn F, Poppe B, Van Roy N, et al. (2002) Accurate normalization of real-time quantitative RT-PCR data by geometric averaging of multiple internal control genes. Genome Biol 3, research0034.10–research0034.11.
21. Li W, Zhan J, Spear PA, Li Z, Yang L, et al. (2009) Changes of thyroid hormone levels and related gene expression in Chinese rare minnow (Gobiocypris rarus) during 3-amino-1,2,4-triazole exposure and recovery. Aquat Toxicol 92: 50–57.
22. Li W, Zha J, Li Z, Yang L, Wang Z (2009) Effects of exposure to acetochlor on the expression of thyroid hormone related genes in larval and adult rare minnow (*Gobiocypris rarus*). Aquat Toxicol 94: 87–93.
23. Brown SB, Adams BA, Cyr DG, Eales JG (2004) Contaminant effects on the teleost fish thyroid. Environ Toxicol Chem 23: 1680–1701.
24. Kawakami Y, Seoka M, Miyashita S, Kumai H, Ohta H (2006) Characterization of transthyretin in the Pacific bluefin tuna, *Thunnus orientalis*. Zoolog Sci 23: 443–448.
25. Power DM, Elias NP, Richardson SJ, Mendes J, Soares CM, et al. (2000) Evolution of the thyroid hormone-binding protein, transthyretin. Gen Comp Endocrinol 119: 241–255.
26. Schreiber G (2002) The evolutionary and integrative roles of transthyretin in thyroid hormone homeostasis. J Endocrinol 175: 61–73.
27. Park BJ, Palace V, Wautier K, Gemmill B, Tomy G (2011) Thyroid Axis Disruption in Juvenile Brown Trout (*Salmo trutta*) Exposed to the Flame Retardant *β*-Tetrabromoethylcyclohexane (*β*-TBECH) via the Diet. Environ Sci Technol 45: 7923–7927.

28. Van der Geyten S, Mol KA, Pluymers W, Kühn ER, Darras VM (1998) Changes in plasma T_3 during fasting/refeeding in tilapia (*Oreochromis niloticus*) are mainly regulated through changes in hepatic type II iodothyronine deiodinase. Fish Physiol Biochem 19: 135–143.
29. Li D, Xie P, Zhang X (2008) Changes in plasma thyroid hormones and cortisol levels in crucian carp (*Carassius auratus*) exposed to the extracted microcystins. Chemosphere 74: 13–18.
30. Van der Geyten S, Byamungu N, Reyns GE, Kühn ER, Darras VM (2005) Iodothyronine deiodinases and the control of plasma and tissue thyroid hormone levels in hyperthyroid tilapia (*Oreochromis niloticus*). J Endocrinol 184: 467–479.
31. Kalsbeek A, Buijs RM, van Schaik R, Kaptein E, Visser TJ, et al. (2005) Daily variations in type II iodothyronine deiodinase activity in the rat brain as controlled by the biological clock. Endocrinology 146: 1418–1427.
32. Köhrle J (1999) Local activation and inactivation of thyroid hormones: the deiodinase family. Mol Cell Endocrinol 151: 103–119.
33. Sanders JP, van der Geyten S, Kaptein E, Darras VM, Kühn ER, et al. (1997) Characterization of a propilthiouracil-insensitive type I iodothyronine deiodinase. Endocrinology 138: 5153–5160.
34. Valverde C, Croteau W, Lafleur GJ Jr, Orozco A, Germain DL (1997) Cloning and expression of a 5′-iodothyronine deiodinase from the liver of *Fundulus heteroclitus*. Endocrinology 138: 642–648.
35. Van der Geyten S, Toguyeni A, Baroiller JF, Fauconneau B, Fostier A, et al. (2001) Hypothyroidism induces type I iodothyronine deiodinase expression in tilapia liver. Gen Comp Endocrinol 124: 333–342.
36. Sutija M, Longhurst TJ, Joss JMP (2004) Deiodinase type III in the Australian lungfish, *Neoceratodus forsteri*. Gen Comp Endocrinol 136: 152–161.
37. Moreno M, Berry MJ, Horst C, Thoma R, Goglia F, et al. (1994) Activation and inactivation of thyroid hormone by type I iodothyronine deiodinase. FEBS Lett 344: 143–146.
38. Eales JG, Morin PP, Tsang P, Hara TJ (1993) Thyroid hormone deiodination in brain, liver, gill, heart and muscle of Atlantic salmon (*Salmo salar*) during photoperiodically-induced parr-smolt transformation. II. Outer- and inner-ring 3,5,3′-triiodo-L-thyronine and 3,3′,5′-triiodo-L-thyronine (reverse T_3) deiodination. Gen Comp Endocrinol 90: 157–167.
39. Bianco AC, Salvatore D, Gereben B, Berry MJ, Larsen PR (2002) Biochemistry, cellular and molecular biology, and physiological role of the iodothyronine selenodeiodinases. Endocr Rev 23: 38–89.
40. Maia AL, Goemann IM, Meyer EL, Wajner SM (2011) Deiodinases: the balance of thyroid hormone: type 1 iodothyronine deiodinase in human physiology and disease. J Endocrinol 209: 283–297.
41. St Germain DL, Galton VA, Hernandez A (2009) Minireview: defining the roles of the iodothyronine deiodinases: current concepts and challenges. Endocrinology 150: 1097–1107.
42. Darras VM, Van Herck SL (2012) Iodothyronine deiodinase structure and function: from ascidians to humans. J Endocrinol 215: 189–206.
43. Becker KB, Stephens KC, Davey JC, Schneider MJ, Galton VA (1997) The type 2 and type 3 iodothyronine deiodinases play important roles in coordinating development in Rana catesbeianatadpoles. Endocr inology138: 2989–2997.
44. Ng TB, Idler DR, Eales JG (1982) Pituitary hormones that stimulate the thyroidal system in teleost fishes. Gen Comp Endocrinol 48: 372–389.
45. Yoshiura Y, Sohn Y, Munakata A, Kobayashi M, Aida K (1999) Molecular cloning of the cDNA encoding the beta subunit of thyrotropin and regulation of its gene expression by thyroid hormones in the goldfish, *Carassius auratus*. Fish Physiol Biochem 21: 201–210.
46. Sohn YC, Yoshiura Y, Suetake H, Kobayashi M, Aida K (1999) Isolation and characterization of the goldfish thyrotropin β subunit gene including the 5′-flanking region. Gen Comp Endocrinol 115: 463–473.
47. Bing X (2003) The environmental estrogen effect of monocrotophos on male goldfish, *Carassius Auratus*. M.Sc. Thesis, Ocean University of China. Available: http://cdmd.cnki.com.cn/Article/CDMD-10423-2003093472.htm.
48. Tran TN, Fryer JN, Bennett HP, Tonon MC, Vaudry H (1989) TRH stimulates the release of POMC-derived peptides from goldfish melanotropes. Peptides 10: 835–841.
49. Abbott M, Volkoff H (2011) Thyrotropin Releasing Hormone (TRH) in goldfish (*Carassius auratus*): role in the regulation of feeding and locomotor behaviors and interactions with the orexin system and cocaine- and amphetamine regulated transcript (CART). Horm Behav 59: 236–245.
50. Decuypere E, Scanes CG, Kuhn ER (1983) Effects of glucocorticoids on circulating concentrations of thyroxine (T_4) and triiodothyronine (T_3) and on peripheral monodeiodination in pre and posthatching chickens. Horm Metab Res 15: 223–236.
51. Huising MO, Metz JR, van Schooten C, Taverne-Thiele AJ, Hermsen T, et al. (2004) Structural characterisation of a cyprinid (*Cyprinus carpio* L.) CRH, CRH-BP and CRH-R1, and the role of these proteins in the acute stress response. J Mol Endocrinol 32: 627–648.
52. Mommsen TP, Vijayan MM, Moon TW (1999) Cortisol in teleosts: dynamics, mechanisms of action, and metabolic regulation. Rev Fish Biol Fisheries 9: 211–268.
53. Zhong Y (2014) Effects of monocrotophos pesticide on the interrenal stress axis in zebrafish (*Danio rerio*). M.Sc. Thesis, Ocean University of China. In press.
54. Hontela A, Rasmussen JB, Audet C, Chevalier G (1992) Impaired cortisol stress response in fish from environments polluted by PAHs, PCBs, and mercury. Arch Environ Contam Toxicol 22: 278–283.
55. Thangavel P, Sumathiral K, Karthikeyan S, Ramaswamy M (2005) Endocrine response of the freshwater teleost, *Sarotherodon mossambicus* (Peters) to dimecron exposure. Chemosphere 61: 1083–1092.
56. Desaulniers D, Leingartner K, Wade M, Fintelman E, Yagminas A, et al. (1999) Effects of acute exposure to PCBs 126 and 153 on anterior pituitary and thyroid hormones and FSH isoforms in adult Sprague Dawley male rats. Toxicol Sci 47: 158–169.
57. Ulbrich B, Stahlmann R (2004) Developmental toxicity of polychlo-rinated biphenyls (PCBs): a systematic review of experimental data. Arch Toxicol 78: 252–268.
58. Lau C, Anitole K, Hodes C, Lai D, Pfahles-Hutchens A, et al. (2007). Perfluoroalkyl acids: A review of monitoring and toxicological findings. Toxicol Sci 99: 366–394.
59. DeWitt JC, Shnyra A, Badr MZ, Loveless SE, Hoban D, et al. (2009) Immunotoxicity of perfluorooctanoic acid and perfluorooctane sulfonate and the role of peroxisome proliferator-activated receptor alpha. Crit Rev Toxicol 39: 76–94.

4

Microemulsion Formulation of Carbendazim and Its *In Vitro* Antifungal Activities Evaluation

Pengfei Leng[1], Zhiming Zhang[1], Qian Li[2], Maojun Zhao[3]*, Guangtang Pan[1]*

1 Maize Research Institute, Sichuan Agricultural University, Ya'an, PR China, **2** Chongqing College of Humanities, Science and Technology, School of Foreign Languages, Hechuan District, Chongqing, PR China, **3** College of Life and Basic Sciences, Sichuan Agricultural University, Sichuan Ya'an, PR China

Abstract

The fungus *Rhizoctonia solani* Kuhn is a widespread and destructive plant pathogen with a very broad host range. Although various pathogens, including *R. solani*, have been traditionally controlled using chemical pesticides, their use faces drawbacks such as environmental pollution, development of pesticide resistance, and other negative effects. Carbendazim is a well-known antifungal agent capable of controlling a broad range of plant diseases, but its use is hampered by its poor aqueous solubility. In this study, we describe an environmentally friendly pharmaceutical microemulsion system using carbendazim as the active ingredient, chloroform and acetic acid as solvents, and the surfactants HSH and 0204 as emulsifiers. This system increased the solubility of carbendazim to 30 g/L. The optimal microemulsion formulation was determined based on a pseudo-ternary phase diagram; its physicochemical characteristics were also tested. The cloud point was greater than 90°C and it was resistant to freezing down to −18°C, both of which are improvements over the temperature range in which pure carbendazim can be used. This microemulsion meets the standard for pesticide microemulsions and demonstrated better activity against *R. solani* AG1-IA, relative to an aqueous solution of pure carbendazim (0.2 g/L). The mechanism of activity was reflected in the inhibition of against *R. solani* AG1-IA including mycelium growth, and sclerotia formation and germination were significantly better than that of 0.2 g/L carbendazim water solution according to the results of *t*-test done by SPSS 19.

Editor: Beom Seok Kim, Korea University, Republic of Korea

Funding: This work was funded by the Application Foundation of Science and Technology Bureau of Sichuan Province (Grant No. 2006J13-039) and Priority Projects of Sichuan Education Office (Grant No. 07ZA063). The funders had no role in study design, data collection and analysis, decision to publish, or preparation of the manuscript.

Competing Interests: The authors declare that no competing interests exist.

* Email: wlisa0920@163.com (MZ); pfleng@163.com (GP)

Introduction

In recent years, there has been an increase in the prevalence of fungal infections of crops worldwide. *R. solani* is a particularly widespread, soil-borne plant-pathogenic fungus which causes seed and hypocotyl rot, collar rot, root rot, damping off and wire stem in many plant species [1–4]. One of the most well-known symptoms of *R. solani* infection is black scurf on potato tubers, corresponding to the sclerotia of the fungus. The anastomosis group of *R. solani*, AG1-IA, is a major pathogen in Latin America, causing Sheath Blight (SB) on rice and Banded Leaf Sheath Blight (BLSB) on maize, particularly in Venezuela [5]. Soybean yield losses of up to 35% have been attributed to Rhizoctonia foliar blight (RFB), with potential yield reductions of up to 70% [6]. Chemical pesticides have contributed significantly towards controlling crop diseases, including those caused by *Rhizoctonia*. An evaluation of the inhibitory effects of 18 types of fungicides towards *R. solani* AG1-IA in vitro and in vivo revealed that mycelial growth was strongly inhibited and disease symptoms could be alleviated to a large extent [7]. However, pesticide resistance, environmental pollution and other negative effects are attributed to the overuse of pesticides. Worldwide, it is estimated that about 2.5 million tons of pesticides are used on crops annually, with the environmental costs associated with pesticide overuse approximating $100 billion annually [8]. The risk of fungicide resistance increases when a single chemical pesticide is used intensively for an extended period [9]. Consequently, the biological control of *R. solani* with antagonists has been examined in detail [10–12], although such methods are strongly influenced by environmental conditions. For example, a preparation of *Trichoderma harzianum* significantly reduced the incidence of bean diseases caused by both *Sclerotium rolfsii* and *R. solani*, but its biocontrol capacity was inversely correlated with temperature [13]. This supports the requirement to develop more efficient methods of pesticide delivery, thereby minimizing the application rates of the active pesticide ingredient. Microemulsions of pesticides thus represent a relatively environmentally friendly formulation that decrease the use of organic solvents and increase the wettability, dispersibility, and penetration ability of the pesticide droplets. A microemulsion is defined as "a colloidal dispersion composed of oil phase, aqueous phase, surfactant and co-surfactant at appropriate ratios, which is a single optically isotropic and thermodynamically stable liquid solution" [14,15]. Our previous study on a difenoconazole/propiconazole microemulsion indicated that the usage of xylene was reduced by 75%, while the use of toluene in a triadimefon microemulsion was

Table 1. Surfactant screening assay.

Surfactants	Appearance	Appearance after 24 h
Agricultural emulsifier 500	White emulsion	Evident destabilization
Agricultural emulsifier 700	White emulsion	Evident destabilization
Agricultural emulsifier 0201	White, turbid	Destabilization, crystallization
Agricultural emulsifier 0204	Semitransparent solution	Slight destabilization
Agricultural emulsifier 0207	White, turbid	Destabilization, crystallization
Agricultural emulsifier 1600	White, turbid	Destabilization, crystallization
Agricultural emulsifier 2201	Semitransparent solution	Destabilization, crystallization
By-140	White, turbid	Destabilization, crystallization
Emulsifier HSH	Transparent solution	Transparent
Emulsifier OP-10	White turbid	Destabilization, crystallization
Tween40	White turbid	Destabilization, crystallization

reduced by 70%, relative to commercial formulations [16,17]. Studies have indicated that the control efficiency of pesticide microemulsions is superior to those of ordinary emulsions, wettable powders, and other pesticide formulations [16,17]. It has been reported that a microemulsion of λ-cyhalothrin possesses better performance and stability than its commercial formulation [18]. Similarly, a microemulsion of triadimefon displayed significantly better antifungal activity against *R. solani* AG1-IA, when compared to a 20% triadimefon emulsion [17].

Carbendazim is a well-known antifungal agent that can be used to control a broad range of diseases on field crops, fruits, and vegetables, including sclerotinia rot of canola, wheat head blight, peanut leaf spot, and SB on rice [19]. Carbendazim is fairly easily degraded through chemical, physical and biological processes, such as UV, H_2O_2 and microorganisms [20–22]. It has been officially registered in several countries and is used extensively in controlling diseases in field and greenhouse crops, and in forests. A mixture of flusilazole and carbendazim has been recommended for soybean rust control [23]. Another study revealed that carbendazim was the most effective fungicide for inhibiting the mycelial growth of *Moniliophthora. pernicious* [24]. Unfortunately, its solubility in water is 6.11 μg/mL and it possesses a high melting point, making it difficult to apply [25].

The aim of the present study was to establish a pharmaceutical microemulsion system for the application of carbendazim, using chloroform and acetic acid as co-solvents, the surfactants HSH and '0204' as emulsifiers, and methanol as a co-surfactant. The optimal microemulsion formulation was selected based on testing the dilution stability, low and high temperature stabilities, freezing stability, and other physiochemical parameters. Finally, the antifungal activities of the optimal carbendazim microemulsion formulation against *R. solani* AG-IA were assessed.

Materials and Methods

Materials

Chemicals. Carbendazim was purchased from Shandong Zouping Agrochemical Co. Ltd (Zouping, Shandong, China). A pure carbendazim standard was purchased from the Chinese CRM/RM Information Center. The emulsifier HSH and the commercially used agricultural emulsifier 0204 (Hai'an Petrochemicals, Jiangsu, China) were of commercial grade. Dyes Sudan IV and methylene blue were of analytical grade and supplied by Sinopharm Group Chemical Reagent Co. Ltd. (Shanghai, China). Methanol, chloroform, and acetic acid were purchased from Sichuan Xilong Chemical Co. Ltd (Chengdu, Sichuan, China). Deionized water was used in all experiments.

Microorganism. *Rhizoctonia solani* AG1-IA and *Alternaria alternata* (Fr.) Keisskr were maintained on potato dextrose agar medium (PDA: potato, 200 g; dextrose, 20 g; agar, 20 g; deionized water, 1 L) at 28°C.

Solvent and surfactant screening

Solvent screening. Pure carbendazim was placed in different test tubes, and its solubility at various temperatures was assessed in 10 mL different solvents. The solvent that completely dissolved carbendazim was selected for subsequent experiments; the solution was maintained at 0°C to prevent the formation of a precipitate or phase inversion. The solution got from this is the "Oil" will be used the next steps.

Surfactant assay. In order to determine the most effective surfactant(s), microemulsion formulations were devised by maintaining the contents of various surfactants equal to that of the active ingredient (carbendazim). The microemulsion was then initially placed in a water bath at 20°C for 24 h. The quality of each surfactant was then determined according to the appearance of the samples, the temperature range over which transparency was maintained, and low temperature and freezing stability.

Pseudo-ternary phase diagram construction

Pseudo-ternary phase diagrams were constructed using the SAA (optimal formulation of emulsifier HSH, agricultural emulsifier 0204 and methanol got from the solvent and surfactant screening part) titration method; first, a carbendazim compound solution was prepared using a mixture of the emulsifier HSH, agricultural emulsifier 0204 and methanol (4: 2: 1, w/w/w) [26]. Nine ratios of the carbendazim compound solution and SAA (1: 9, 2: 8…9: 1, w/w) were dispersed in vials. To each vial, water was added dropwise until a critical change in transparency was observed. The amount of water required to induce this change was measured for each of the nine vials; this experiment was conducted in triplicate. The mean value was used for constructing a pseudo-ternary phase diagram using the software Origin 7.5. Based on the proportions of carbendazim, emulsifiers, water and other components, an optimal formulation was finally determined.

Table 2. Results of the surfactant screening assay.

Emulsifier HSH: agricultural emulsifier 0204	Appearance	Appearance after 24 h	Low temperature stability	Cloud point (°C)
1:4	Turbid	-	-	-
2:4	Turbid	-	-	-
3:4	Transparent	Transparent	Some precipitation	>54
4:4	Transparent	Transparent	Transparent	>54
4:3	Transparent	Transparent	Transparent	>54
4:2	Transparent	Transparent	Transparent	>54
4:1	Turbid	-	-	-

"-": not tested.

Physicochemical characteristics

Microemulsion type. The appearance, centrifugation properties, staining reactions, and dilution characteristics, as described in our previous work, were used to characterize the obtained microemulsion [16].

HPLC analysis. A Shimadzu (Kyoto, Japan) liquid chromatograph, equipped with an LC-10 AT VP solvent pump unit and an SPD-10A VP UV-Visible detector, was used to analyze the optimal carbendazim microemulsion indicated in the previous section. Samples were injected manually through a 10 μL loop with a Rheodyne injector. Separations were conducted at room temperature on a C18 VP-ODS, 250×4.6 mm column with a mobile phase of methanol/0.1% ammonium hydroxide (65/35, v/v), with a flow rate of 1.0 mL/min. Carbendazim was detected at 280 nm [27]. A standard solution of carbendazim was injected to construct a stable calibration curve. The microemulsion was injected in triplicate, after which the concentration of carbendazim was calculated according to the following formula:

$$\text{Concentration} = \frac{A_2 M_1 P}{A_1 M_2} \times 100\%$$

Where: A_1: peak area of standard carbendazim; A_2: peak area of the tested microemulsion sample; M_1: the weight of standard carbendazim (g); M_2: the weight of the tested microemulsion sample (g); P: purity of standard carbendazim (%).

Physiochemical stability. The following four aspects were evaluated: emulsion stability, low temperature stability, reaction to freezing, and turbidity point. Three types of microemulsion formulations were assayed, differing in the type of water used: distilled water, water containing known concentrations of Ca^{2+} and Mg^{2+} (182 mg/L each), or a standard hard water solution (342 mg/L total dissolved solids) [16,28].

Antifungal activity assay

Toxicity bioassay. Assay of the inhibition against *R. solani* AG1-IA was conducted based on a previous study, with some modifications [29]. A mixture of carbendazim microemulsion and PDA medium was prepared, with a 0.2 g/L solution of carbendazim in water as a positive control and PDA medium with no carbendazim as a negative control. Microemulsions with carbendazim concentrations of 0.05, 0.1, 0.2, 0.4 and 0.8 mL/L and aqueous solutions with carbendazim concentrations of 1, 2, 4, 8 and 16 mL/L were assayed in this experiment. All of the test plates were inoculated with healthy *R. solani* AG1-IA isolates with a colony diameter of 0.4 cm. The plates were incubated in the dark at 28°C and the diameters of the fungal colonies were measured every 12 h. Subsequently, the inhibition ratios were calculated, while regression analysis was used to determine the EC_{50} and EC_{90} values. Each treat was performed in three biological replicates. EC_{50} and EC_{90} were got according to regression equation. The inhibitory effects of the carbendazim microemulsion and solutions were calculated as indicated below.

$$\text{Inhibition} = \frac{D_0 - D_1}{D_0} \times 100\%$$

Where: D_0: colony diameter in the negative control; D_1: colony diameter in the treatment.

Another antifungal activity assay through mycelial growth inhibition with *A. alternata* was conducted as indicated above. Microemulsions with carbendazim concentrations of 0.4, 0.8, 1.6, 3.2 and 6.4 mL/L and aqueous solutions with carbendazim concentrations of 8, 16, 32, 64 and 128 mL/L were analyzed. The plates were incubated in the dark at 25°C and the diameters of the fungal colonies were measured 7 days later.

Sclerotia formation. Plates with carbendazim microemulsion, at concentrations of 0.05, 0.1, 0.2, 0.4 and 0.8 mL/L, were respectively compared with those with the positive control, at concentrations of 1, 2, 4, 8 and 16 mL/L (recorded as contrast groups A, B, C, D and E, respectively), following inoculation with *R. solani* AG1-IA. After incubation for 15 days at 28°C, sclerotia were collected and dried to a constant weight at 80°C. The inhibition of sclerotia formation by the carbendazim microemulsions and 0.2 g/L carbendazim aqueous solutions were calculated with respect to the negative control. Each experiment was conducted in triplicate.

Sclerotia germination. The mature sclerotia of *R. solani* AG1-IA were dipped in carbendazim microemulsions (0.05, 0.1, 0.2, 0.4 and 0.8 mL/L) or in aqueous carbendazim solutions of 0.2 g/L (1, 2, 4, 8 and 16 mL/L) (the respective comparison groups were recorded as A, B, C, D and E, respectively) for 30 seconds and then air-dried. Sclerotia were dipped in sterile water as a negative control. Treated sclerotia were inoculated onto untreated PDA plates and incubated at 28°C for 24 h. Each treatment consisted of 30 sclerotia and was repeated in triplicate.

Statistical analysis

All the regression functions were performed by Microsoft Excel (Redmond, WA, USA) and the other statistical analysis was conducted using SPSS 19.0 software (SPSS, Chicago, IL, USA); a *t*-test was used to analyze the inhibitory effects of the carbendazim microemulsion and solutions.

Table 3. Assay of carbendazim microemulsion formulations.

Oil/SAA/H_2O	Appearance	Freezing experiment	Low temperature stability	Heat storage stability	Cloud point (°C)
1/4/5	Transparent, Tyndall effect	Froze, recovered upon thawing	Transparent, no precipitate	Transparent	72
1/3/4	Bluish transparent, Tyndall effect	Did not freeze, low viscosity	Transparent, no precipitate	Transparent	>90
2/5/3	Bluish transparent, Tyndall effect	Did not freeze but became turbid	Precipitate	Transparent	>90
2/3/3	Transparent, Tyndall effect	Turbidity and crystallization observed	Precipitate	Transparent	>90

Results and Discussion

Components of carbendazim microemulsion

Carbendazim solubility was limited in all of the solitary solvents screened. However, our analysis suggested that a combination of acetic acid and chloroform could be used as an effective solvent. When chloroform and acetic acid were mixed in a volumetric ratio of 3: 2, the solubility of carbendazim attained 30 g/L, which was significantly ($P<0.001$) higher than its highest known solubility in solitary solvents. Buffers, co-solvents, surfactants, and complexants

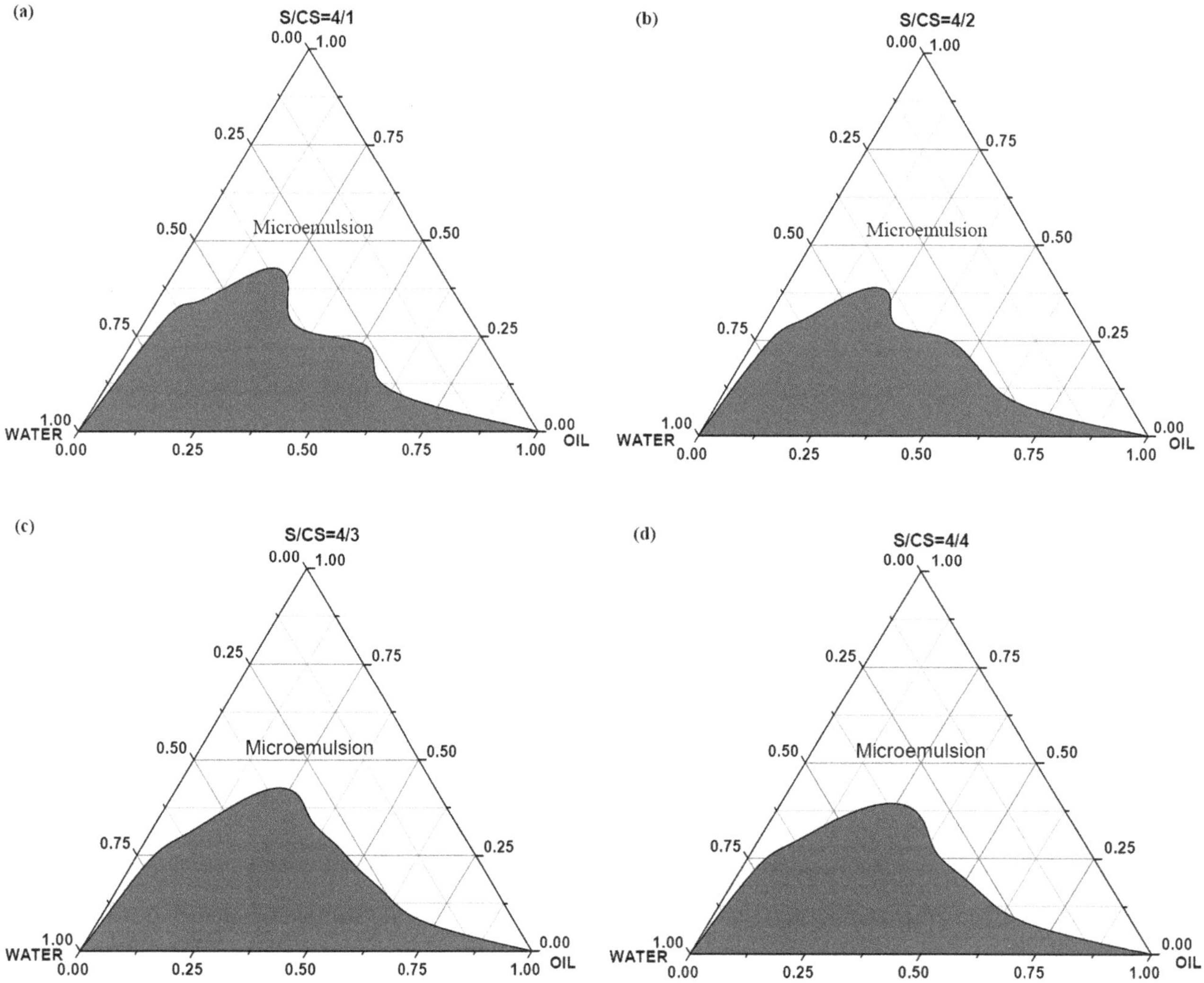

Figure 1. Pseudoternary phase diagrams of carbendazim microemulsion using different ratios of emulsifiers/methanol: (a) 4: 1, (b) 4: 2, (c) 4: 3, (d) 4: 4.

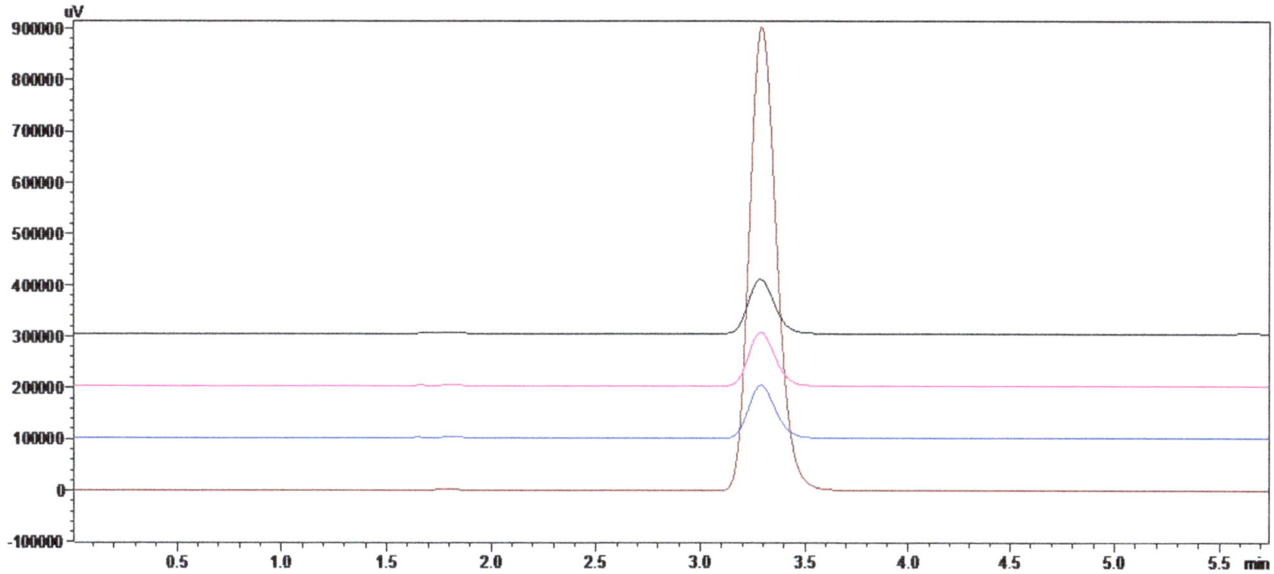

Figure 2. Chromatograms of a carbendazim standard and carbendazim microemulsion samples. HPLC conditions are described in the text. Note: Chromatography of the carbendazim standard (top) and microemulsion samples (bottom) were repeated in triplicate.

are the most commonly used excipients to improve the solubility of a nonpolar drug or pesticide [30]. Our results suggest the value of experimentally testing the solubility of such compounds in mixed solvents. For subsequent experiments, the concentration of carbendazim was maintained at 20 g/L.

As shown in Table 1, the emulsifier HSH and the agricultural emulsifier 0204 were best-suited for use in carbendazim microemulsion. Therefore, they were mixed in gravimetric ratios of 1: 4, 2: 4, 3: 4, 4: 4, 4: 3, 4: 2 and 4: 1 to determine the best formulation. The optimal ratio (by weight) of the emulsifier HSH to the agricultural emulsifier 0204 was 4: 2; this sample exhibited the best low temperature stability and thermal stability (Table 2). This phenomenon may be explained by the formation of novel polymers of these compounds. Many studies have indicated that a blend of two emulsifiers is more efficient than a single substance, given comparable hydrophilic–lipophilic balance (HLB) values [31,32]. In particular, one study indicated that the maximum stability of oil-in-water emulsions using a mixture of the commercial surfactants Tween and Span was achieved at a certain molar ratio of these two surfactants [31]. Others have indicated that block copolymers prefer to localize to the interface, which significantly reduces the interfacial tension [33,34]. Research on oil-in-water emulsion systems stabilized with non-ionic surfactant blends revealed the presence of thin aqueous films between the oil phases and oil droplets, coalescing against their homophase [35].

Pseudo-ternary phase diagrams and preparation of microemulsion

Pseudo-ternary phase diagrams are presented in Figure 1a–d. A smaller microemulsion region (39.26%) was formed when the surfactants/methanol ratio was 4: 1 (Figure 1a). As shown in Figure 2b, when this ratio was decreased to 4: 2, the proportion of the mono-phase region decreased (36.57%). When the ratio was reduced further to 4: 3 (Figure 1c), a larger mono-phase region was observed, accounting for 40.39% of the area in the phase diagram. Finally, decreasing the ratio to 4: 4 (Figure 1d) diminished the proportional area of this microemulsion to 37.01%. Therefore, a surfactant: methanol ratio of 4: 3 was selected for preparing subsequent microemulsion. Candidate microemulsion with certain proportions of Oil, SAA and water were designed according to the pseudo-ternary phase diagram indicated in Figure 1c. Their characteristics were assessed (Table 3) and the optimum formulation was determined to be as follows: Oil/SAA/water = 1: 3: 4; SAA consisted of HSH, the agricultural emulsifier 0204, and methanol, with a surfactant: methanol ratio of 4: 3.

Next, we assessed the formation of the microemulsion at room temperature. As indicated in Figure 1, the proportional area of the microemulsion in the corresponding phase diagrams first decreased and then increased as the ratio of methanol increased. These results are comparable with those of a previous study, the authors of which suggested that the increase in the isotropic phase

Table 4. Inhibition of *R. solani* AG1-IA mycelial growth at 36 h after treatment with carbendazim.

Treatment	Regression function	Relation coefficient	EC_{50}/(mL/L)	EC_{90}/(mL/L)
Carbendazim microemulsion	y = 1.9266x+6.3607	0.9585	0.20a	0.91a
0.2 g/L Aqueous carbendazim solution	y = 1.4972x+3.2813	0.952	14.06b	100.91b

Table 5. Inhibition of *A. alternata* mycelial growth at 7 days after treatment with carbendazim.

Treatment	Regression function	Relation coefficient	EC_{50}/(mL/L)	EC_{90}/(mL/L)
Carbendazim microemulsion	y = 1.7582x+5.1597	0.9433	0.81a	4.35a
0.2 g/L Aqueous carbendazim solution	y = 1.8527x+1.4769	0.9728	79.73b	392.01b

area indicates an increased incorporation of solvent (in their case, ethanol) into the surfactant film, thereby leading to a decrease in the bending modulus of this film [36]. The addition of short-chain alcohols to non-ionic oil/water microemulsion decreases the bending energy [37]. Methanol, as a co-surfactant, plays an important role in microemulsion formulations, as it reduces the interfacial tension and bending energy, adjusts the HLB value, and is easily inserted into the structure of the surfactant. As the relative proportion of methanol increased, the liquidity and appearance of the microemulsion improved noticeably. Our previous research on triadimefon microemulsion revealed that the addition of methanol, up to a certain limit, significantly increased the area occupied by the microemulsion in the corresponding phase diagram and also improved its appearance [17].

Physiochemical detection

Type of carbendazim microemulsion. The carbendazim microemulsion exhibited bluish transparent fluorescence and did not form any precipitate, which is typical of most microemulsions. In view of the colloidal properties of microemulsions, we tested the appearance and occurrence of the Tyndall effect in this microemulsion. Directing monochromatic light through it clearly produced a beam of light. Furthermore, no phase separation was observed after centrifugation at 8000 *g* for 20 min at room temperature. These results are consistent with the results of Zhong [38] and our previous studies [16,17]. These results may be additionally explained by the theory of mutual dissolution.

Carbendazim concentration in the microemulsion. As shown in Figure 2, the retention time of carbendazim was 3.29 min, with good peak symmetry at 280 nm and high peak purity. The mean concentration of carbendazim was 1.19%, with a standard deviation and coefficient of variation of 0.006 and 0.60%, respectively. The analysis method was precise, suggesting that it has potential in future assays of carbendazim content in microemulsion or other matrices.

Physiochemical stability assays. The physiochemical stability of the microemulsion was assessed in triplicate as follows.

Dilution stability: Carbendazim microemulsion was diluted with standard hard water to concentrations of 5%, 2%, 1% and 0.5%. All samples remained transparent and uniform; no floating oil or precipitate was observed.

Low temperature stability: After storage at 0°C for 14 days, there was no precipitate formation or phase separation; no significant change in the mobility or stability was observed.

Freezing experiment: The sample did not freeze or destabilize after storage for 7 days at −18°C. This is a particularly noteworthy characteristic of the microemulsion, since it permits the application of carbendazim at very low temperatures if necessary.

Cloud point: This was determined as the temperature at which the sample became observably turbid; this experiment was conducted by increasing the temperature of the water bath at 2.0°C/min with gentle stirring of the immersed sample. The turbidity point occurred at temperatures greater than 90°C.

The useful characteristics of the carbendazim microemulsion assessed in this study may be explained by the unique interaction between the co-solvents, resulting in enhanced solute stability as a function of the balance between thermodynamics and kinetics. A previous study reported that a λ-cyhalothrin microemulsion formulation exhibited acceptable transparency over a broad temperature range, and superior performance and stability compared to other commercial formulation [18].

When using ionic surfactants, phase transition occurs as the concentrations of electrolytes change. We prepared microemulsion using distilled water, naturally hard water, and a standard sample of hard water, using the same formula; physiochemical testing revealed that there were no differences among these methods. This is particularly surprising with respect to the turbidity point (> 90°C), a parameter that is easily affected by water hardness. This finding is different from what has been reported by Song [39] and in previous studies [16,17]. This may potentially be explained by the turbidity point of carbendazim being higher than 90°C, with the differences between the three types of water being too small to produce observable effects. Although there were no obvious changes in the physiochemical stabilities of the microemulsions, it is possible that slight reduction in these factors occurred when hard water was used. The concentrations of Ca^{2+} and Mg^{2+} often affect the HLB values of the surfactants in microemulsion systems [39,40]. These changes may include thinning of the interface film, an increase in the sizes of oil droplets a decrease in the uniformity of their distribution, and an increased rate of particle collisions, thereby leading to phase separation at high temperatures.

Antifungal activities

Inhibition of *R. solani* mycelial growth. 18.13% of mycelial growth was inhibited after the *R. solani* were cultivated for 36 h in the presence of 0.05 mL/L of carbendazim microemulsion, which was significantly higher than that of a 1 mL/L aqueous solution of carbendazim (6.23%) ($P<0.001$). When the concentration of the carbendazim microemulsion was increased to 0.2 mL/L, the inhibition ratio went to 44.63%, significantly higher than that of a 4 mL/L carbendazim aqueous solution (17.32%). About 91% inhibition ratio was tested when *R. solani* were treated with 0.8 mL/L microemulsion and was significantly higher than that of 16 mL/L aqueous solution of carbendazim (60%) ($P<0.001$). With these inhibition ratios (Table 4), regression functions were then formed. The EC_{50} and EC_{90} values of the carbendazim microemulsion after 36 h of treatment were 0.20 mL/L and 0.91 mL/L, while those of a 0.2 g/L aqueous solution of carbendazim were 14.06 mL/L and 100.91 mL/L, respectively.

Inhibition of *A. alternata* mycelial growth. Cultivation for 7 days with the presence of 0.4 mL/L of carbendazim microemulsion resulted in a mycelial inhibition ratio of 37.1%. This was significantly higher than that of a 8 mL/L aqueous solution of carbendazim (3.7%) ($P<0.001$). When the concentration of the carbendazim microemulsion was quadrupled to 1.6 mL/L, the inhibition ratio was 65.6%, which was significantly higher than

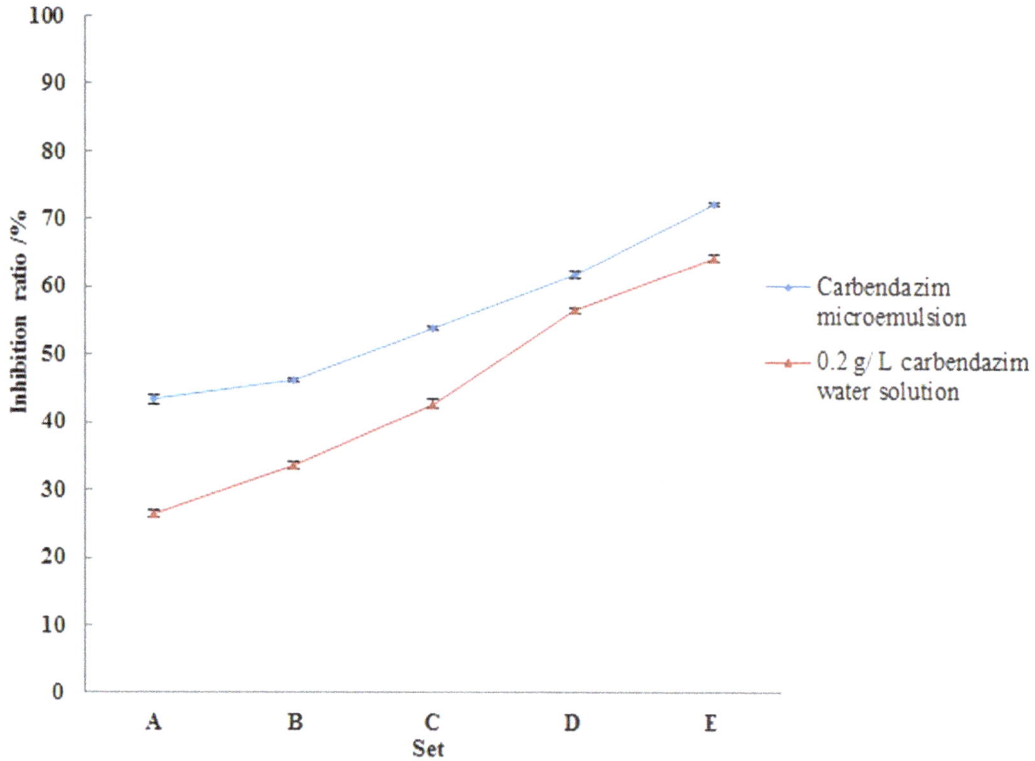

Figure 3. Comparisons of the inhibitory effects of the tested carbendazim microemulsions and the positive control (0.2 g/L aqueous solution of carbendazim): (a) 0.05 mL/L microemulsion and 1 mL/L water solution, (b) 0.1 mL/L microemulsion and 2 mL/L water solution, (c) 0.2 mL/L microemulsion and 4 mL/L water solution, (d) 0.4 mL/L microemulsion and 8 mL/L water solution, (e) 0.8 mL/L microemulsion and 16 mL/L water solution.

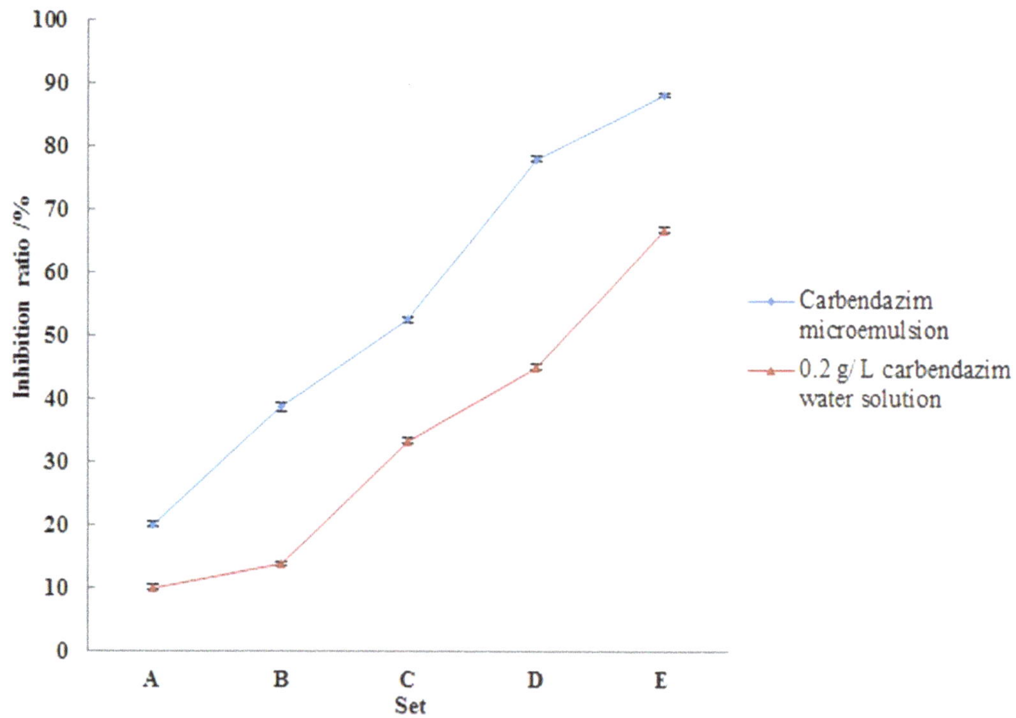

Figure 4. ***R. solani*** **AG1-IA sclerotia germination rate at 24 h after treatment with carbendazim microemulsions and aqueous solutions.**

that of a 32 mL/L aqueous solution of carbendazim (23.1%). Treatment with the microemulsion at a concentration of 6.4 mL/L revealed an inhibition ratio was 96.5%, which was significantly higher than that of a 16 mL/L aqueous solution of carbendazim (60.4%) ($P<0.001$). Regression equations were then obtained using these inhibition ratios (Table 5). The EC_{50} and EC_{90} values of the carbendazim microemulsion after 7 days of treatment were about 0.8 mL/L and 4.4 mL/L, respectively, while those of a 0.2 g/L aqueous solution of carbendazim were almost 80 mL/L and 392 mL/L, respectively.

The aforementioned results confirmed that the carbendazim microemulsion had a significantly stronger inhibitory effect on mycelial growth than did an aqueous solution of carbendazim. A reason for this observation may be that carbendazim in the form of a microemulsion could easily permeate the fungal tissues and inhibit the polymerization of free tubulin molecules by binding an arginin residue of the β-tubulin subunit and acts by disrupting cell division through linking to the nuclear spindle, which inhibits fungal growth. A microemulsion system allows the pesticide to easily overcome capillary resistance due to a decrease in the interfacial tension. Additionally, many traditional chemical and pharmaceutical products are more effective when mixtures of solvents are employed, as opposed to a single solvent [41].

Inhibition of *R. solani* sclerotia formation. After cultivation for 15 days, sclerotia of *R. solani* were collected, and dried to constant weight; subsequently, the inhibition ratios of different carbendazim formulations were assessed (Figure 3). When the concentration of the carbendazim microemulsion was 0.05 mL/L, the inhibition ratio was 43.33%, significantly higher than the 26.43% obtained using a 1 mL/L aqueous solution of carbendazim ($P<0.001$). Sclerotial formation was further inhibited (53.83%) when the concentration of the microemulsion was increased to 0.2 mL/L, significantly higher than that obtained using a 4 mL/L aqueous solution of carbendazim ($P<0.001$). When the concentration of the microemulsion was increased to 0.8 mL/L, the inhibition ratio was 72.30%, relative to 64.30% obtained using a 16 mL/L aqueous solution of carbendazim ($P<0.001$). This may be attributed to the fact that the improved solubility of carbendazim results in a steeper gradient of the pesticide between the treatment liquid and the fungal tissues, thereby increasing the velocity of its transport into the organism's cells. Regression equations were obtained based on the inhibition ratios (Table 4). The EC_{50} and EC_{90} values of the carbendazim microemulsion after 36 h were 0.12 mL/L and 12.03 mL/L, respectively, while those of a 0.2 g/L aqueous solution of carbendazim were 5.84 mL/L and 181.50 mL/L, respectively.

Inhibition of *R. solani* sclerotial germination. Inhibition of sclerotial germination is vital in preventing crop diseases; mature sclerotia are the main sources of diseases attributed to *Rhizoctonia*. The inhibition ratios of the microemulsion were calculated, as shown in Figure 4. The rate of sclerotial germination decreased as the concentration of carbendazim increased. When the sclerotia were incubated for 24 h, increases in the strengths of both the microemulsion and aqueous solutions of carbendazim resulted in an inhibition of sclerotial germination. As observed for the inhibition of mycelial growth and sclerotia formation, carbendazim in the form of a microemulsion was consistently more effective than that in the form of aqueous solutions. At the lowest application rates, the inhibition ratio of the microemulsion was 20.03%, compared to 10.03% obtained using an aqueous solution of carbendazim ($P<0.001$). The discrepancies between the two types of treatments increased as the concentration of carbendazim increased. The inhibition ratios was 88.23% when the concentration of the microemulsion was 0.8 mL/L, compared to 66.9% obtained using a 16 mL/L aqueous solution of carbendazim. The EC_{50} and EC_{90} values of the carbendazim microemulsion after 36 hours were 0.16 mL/L and 8.76 mL/L, respectively, while those of a 0.2 g/L aqueous solution of carbendazim were 0.89 mL/L and 65.73 mL/L, respectively.

The aforementioned results are similar to those reported by Wang, who suggested that tebuconazole is more effective at inhibiting sclerotial germination than mycelial growth [42]. Based on the results of the antifungal activity assays, the application of carbendazim in the form of microemulsions should be aimed at controlling diseases, as opposed to preventing them.

In this study, we successfully developed an effective and conveniently used carbendazim microemulsion system, with improved inhibitory activity against mycelial growth and sclerotia formation in *R. solani* AG1-IA. Notably, this microemulsion was resistant to storage at $-18°C$ with no solidification or disruption of stability occurring; similarly, it was physically stable at temperatures up to and including 90°C. This improved formulation may prove to be invaluable in controlling pathogens sensitive to carbendazim. However, further studies should be conducted to evaluate the efficacy of carbendazim microemulsion in controlling BLSB on maize, SB on rice, and other Rhizoctonia diseases at the field scale.

Author Contributions

Conceived and designed the experiments: PL ZZ MZ GP. Performed the experiments: PL. Analyzed the data: PL GP. Contributed reagents/materials/analysis tools: PL MZ. Contributed to the writing of the manuscript: PL QL.

References

1. Sneh B (1996) Rhizoctonia species: taxonomy, molecular biology, ecology, pathology and disease control. New York: Springer.
2. Ogoshi A (1987) Ecology and pathogenicity of anastomosis and intraspecific groups of *Rhizoctonia solani* Kuhn. Ann Rev of Phytopathology 25: 125–143.
3. Pascual C, Hyakumachi M (2000) Distribution of vegetatively compatible populations of *Rhizoctonia solani* AG-1 IA in a field planted with different host species. J Gen Plant Pathol 66: 206–209.
4. Anderson NA (1982) The genetics and pathology of *Rhizoctonia solani*. Ann Rev of Phytopathology 20: 329–347.
5. González-Vera A, Bernardes-de-Assis J, Zala M, McDonald B, Correa-Victoria F, et al. (2010) Divergence between sympatric rice-and maize-infecting populations of *Rhizoctonia solani* AG-1 IA from Latin America. Phytopathology 100: 172–182.
6. Sinclair JB (1982) Compendium of soybean diseases. American Phytopathological Society and University of Illinois.
7. Meyer MC, Bueno CJ, De Souza NL, Yorinori JT (2006) Effect of doses of fungicides and plant resistance activators on the control of Rhizoctonia foliar blight of soybean, and on *Rhizoctonia solani* AG1–IA in vitro development. Crop Pro 25: 848–854.
8. Mohan M, Haider S, Andola H, Purohit V (2011) Essential Oils as Green Pesticides: For Sustainable Agriculture. Res J Pharm Bio and Chem Sci 2: 102–106.
9. Cooke LR, Carlisle DJ, Donaghy C, Quinn M, Perez FM, et al. (2006) The Northern Ireland Phytophthora infestans population 1998–2002 characterised by genotypic and phenotypic markers. Plant Pathol 55: 320–330.
10. Weller DM, Cook RJ (1986) Suppression of root diseases of wheat by fluorescent pseudomonads and mechanisms of action. In Iron, Siderophores, and Plant Diseases, New York: Springer 99–107.
11. Papavizas G (1985) Trichoderma and Gliocladium: biology, ecology, and potential for biocontrol. Ann Rev Phytopathology 23: 23–54.
12. Wells H, Bell D, Jaworski C (1972) Efficacy of *Trichoderma harzianum* as a biological control for *Sclerotium rolfsii*. Phytopathology 62: 442–447.
13. Elad Y, Chet I, Katan J (1980) *Trichoderma harzianum*: A biocontrol agent effective against Sclerotium rolfsii and *Rhizoctonia solani*. Phytopathology 70: 119–121.

14. Changez M, Varshney M (2000) Aerosol-OT microemulsions as transdermal carriers of tetracaine hydrochloride. Drug Dev Ind Pharm 26: 507–512.
15. Tenjarla S (1999) Microemulsions: an overview and pharmaceutical applications. Crit Rev Ther Drug 16: 461–521.
16. Leng P, Zhang ZH, Li Q, Zhang Y, Zhao M, et al. (2012) Development of a difenoconazole/propiconazole microemulsion and its antifungal activities against *Rhizoctonia solani* AG1-IA. Die Pharmazie 67: 534–541.
17. Leng P, Liu L, Li Q, Zhao M, Pan G (2013) Preparation of Triadimefon Microemulsion and its Antifungal Activity against *Rhizoctonia solani* AG1-IA Analysis. Cur Pharm Anal 9: 396–403.
18. Chin CP, Lan CW, Wu HS (2012) Study on the Performance of Lambda Cyhalothrin Microemulsion with Biodiesel as an Alternative Solvent. Ind Eng Chem Res 51: 4710–4718.
19. Zhang LZ, Wei N, Wu QX, Ping ML (2007) Anti-oxidant response of *Cucumis sativus* L. to fungicide carbendazim. Pestic Biochem and Phys 89: 54–59.
20. Helweg A (1977) Degradation and adsorption of carbendazim and 2-aminobenzimidazole in soil. Pestic Sci 8: 71–78.
21. Mazellier P, Leroy E, De Laat J, Legube B (2003) Degradation of carbendazim by UV/H_2O_2 investigated by kinetic modelling. Environ Chem Lett 1: 68–72.
22. Lin X, Hou Z, Feng Y, Zhao S, Ye J (2011) Isolation and Characteristics of Efficient Carbendazim Degradation Bacterium. International Conference on Agricultural and Biosystems Engineering, Advances in Biomedical Engineering 1–2: 384–388.
23. Galloway J (2008) Effective management of soyabean rust and frogeye leaf spot using a mixture of flusilazole and carbendazim. Crop Prot 27: 566–571.
24. Gea FJ, Tello JC, Navarro MJ (2010) Efficacy and effects on yield of different fungicides for control of wet bubble disease of mushroom caused by the mycoparasite *Mycogone perniciosa*. Crop Prot 29: 1021–1025.
25. Ni N, Sanghvi T, Yalkowsky SH (2002) Solubilization and preformulation of carbendazim. Int J Pharm 244: 99–104.
26. Zhang H, Feng F, Li J, Zhan X, Wei H, et al. (2008) Formulation of food-grade microemulsions with glycerol monolaurate: effects of short-chain alcohols, polyols, salts and nonionic surfactants. Euro Food Res Technol 226: 613–619.
27. Zhu F, Dong P (1999) Determination of Active Ingredients in Carbendazim by High Performance Liquid Chromatography. Chem Ind Stand Qual Sup 8: 29–30.
28. Aboofazeli R, Patel N, Thomas M, Lawrence M (1995) Investigations into the formation and characterization of phospholipid microemulsions. IV. Pseudo-ternary phase diagrams of systems containing water-lecithin-alcohol and oil; the influence of oil. Int J Pharm 125: 107–116.
29. Zhang S (1999) Fungicidal bioassay methods. Plant Prot 3: 35–37.
30. Yalkowsky SH (1999) Solubility and solubilization in aqueous media. Washington, DC: American Chemical Society.
31. Boyd Jv, Parkinson C, Sherman P (1972) Factors affecting emulsion stability, and the HLB concept. J Colloid and Interf Sci 41: 359–370.
32. Shah DO (1971) Significance of the 1: 3 molecular ratio in mixed surfactant systems. J Colloid Interf Sci 37: 744–752.
33. Marques C, Joanny J, Leibler L (1988) Adsorption of block copolymers in selective solvents. Macromolecules 21: 1051–1059.
34. Noolandi J, Hong K (1982) Interfacial properties of immiscible homopolymer blends in the presence of block copolymers. Macromolecules 15: 482–492.
35. Velev O, Gurkov T, Chakarova SK, Dimitrova B, Ivanov I, et al. (1994) Experimental investigations on model emulsion systems stabilized with non-ionic surfactant blends. Colloids and Surfaces A 83: 43–55.
36. Garti N, Yaghmur A, Leser ME, Clement V, Watzke HJ (2001) Improved oil solubilization in oil/water food grade microemulsions in the presence of polyols and ethanol. J Agri Food Chem 49: 2552–2562.
37. Gradzielski M (1998) Effect of the cosurfactant structure on the bending elasticity in nonionic oil-in-water microemulsions. Langmuir 14: 6037–6044.
38. Zhong F, Yu M, Luo C, Shoemaker CF, Li Y, et al. (2009) Formation and characterisation of mint oil/S and CS/water microemulsions. Food Chem 115: 539–544.
39. Song F, Wang X, Jiang SR, Wu XM (2005) The Effect of Different Factors on the Cloud Point of Microemulsion. Chin J Pestic Sci 44: 460–462.
40. Kabalnov A, Olsson U, Wennerstroem H (1995) Salt effects on nonionic microemulsions are driven by adsorption/depletion at the surfactant monolayer. J Phy Chem 99: 6220–6230.
41. Martin D Abildskov, EJ John, P (2010) Molecular Thermodynamic Modeling of Mixed Solvent Solubility. Ind Eng Chem Res 49: 11620–11632.
42. Wang HC, Zhou MG, Zhang YJ, Chen CJ, Wang JX (2007) Fungicidal activity of tebuconazole against *Rhizoctonia solani* and its application to rice. Chin J Pes Sci 9(4): 357–362.

5

Area-Wide Ground Applications of *Bacillus thuringiensis* var. *israelensis* for the Control of *Aedes albopictus* in Residential Neighborhoods: From Optimization to Operation

Gregory M. Williams[1,2*], Ary Faraji[2,3], Isik Unlu[2,4], Sean P. Healy[2,5], Muhammad Farooq[6], Randy Gaugler[2], George Hamilton[2], Dina M. Fonseca[2]

1 Hudson Regional Health Commission, Secaucus, New Jersey, United States of America, **2** Center for Vector Biology, Rutgers University, New Brunswick, New Jersey, United States of America, **3** Salt Lake City Mosquito Abatement District, Salt Lake City, Utah, United States of America, **4** Mercer County Mosquito Control, West Trenton, New Jersey, United States of America, **5** Department of Pathobiological Sciences, Louisiana State University School of Veterinary Medicine, Baton Rouge, Louisiana, United States of America, **6** Navy Entomology Center of Excellence, Jacksonville, Florida, United States of America

Abstract

The increasing range of *Aedes albopictus*, the Asian tiger mosquito, in the USA and the threat of chikungunya and dengue outbreaks vectored by this species have necessitated novel approaches to control this peridomestic mosquito. Conventional methods such as adulticiding provide temporary relief, but fail to manage this pest on a sustained basis. We explored the use of cold aerosol foggers and misting machines for area-wide applications of *Bacillus thuringiensis* var. *israelensis* (VectoBac WDG) as a larvicide targeting *Aedes albopictus*. During 2010–2013 we performed initially open field trials and then 19 operational area-wide applications in urban and suburban residential areas in northeastern USA to test three truck-mounted sprayers at two application rates. Area-wide applications of WDG in open field conditions at 400 and 800 g/ha killed on average 87% of tested larvae. Once techniques were optimized in residential areas, applications with a Buffalo Turbine Mist Sprayer at a rate of 800 g/ha, the best combination, consistently provided over 90% mortality. Importantly, there was no significant decrease in efficacy with distance from the spray line even in blocks of row homes with trees and bushes in the backyards. Under laboratory conditions *Bti* deposition in bioassay cups during the operational trials resulted in over 6 weeks of residual control. Our results demonstrate that area-wide truck mounted applications of WDG can effectively suppress *Ae. albopictus* larvae and should be used in integrated mosquito management approaches to control this nuisance pest and disease vector.

Editor: Nigel Beebe, University of Queensland & CSIRO Biosecurity Flagship, Australia

Funding: This work was funded by Cooperative Agreement USDA-ARS-58-6615-8-105 between the US Department of Agriculture, Agricultural Research Service (USDA-ARS) and Rutgers University (PI at Rutgers: DMF) and Regional IPM Grant 2008-34103-18978 between USDA-ARS and Rutgers University (PI at Rutgers: GH). The funders had no role in study design, data collection and analysis, decision to publish, or preparation of the manuscript.

Competing Interests: The authors have declared that no competing interests exist.

* Email: gwilliams@hudsonregionalhealth.org

Introduction

The range of *Aedes albopictus*, the Asian tiger mosquito, is expected to expand in highly urbanized temperate regions in North America and Europe in response to climate change [1–3]. An aggressive human biter, this mosquito is often the primary pest species eliciting complaints from the public in areas where it occurs [4] and it also presents a significant health risk [1,5,6]. Unfortunately, conventional mosquito abatement methods do not control *Ae. albopictus* effectively. Local mosquito control agencies across the US mount aggressive control campaigns against salt marsh, floodwater, and many other rural-based mosquito pest species [2]. Since the onset of the WNV epidemic mosquito control agencies in urban and suburban areas have targeted *Culex* spp. in residential storm drains and other large stagnant water sources [2]. These programs rarely impact *Ae. albopictus*, which thrive in small pockets of water in artificial containers such as buckets, toys and trash primarily within private yards [7]. Therefore, efforts tend to be limited to public education, placing the responsibility for control on unqualified citizens [8,9]. Having individual homeowners bear considerable responsibility for source reduction has hampered control efforts primarily because residents are not usually able to coordinate their efforts with enough of their neighbors to eliminate the larval sources that contribute biting *Ae. albopictus* to a neighborhood.

For any significant suppression, this pest must be managed on an area-wide basis [9]. Intensive source reduction efforts which involve teams of professionals emptying, removing or treating all sources of standing water in thousands of private yards can be effective. However, they are extremely labor intensive leading to unsustainably high costs [9–11] and often also temporary since new containers are continuously being added by uninformed homeowners [7,12]. Area-wide efforts require relatively swift and repeatable applications of insecticides at a large scale and low volume usually from a ground vehicle or aircraft.

Although *Ae. albopictus* is a day-biting mosquito, adulticide applications at night when thermal inversions optimize likelihood of insecticide deposition, non-target species are mostly inactive and residents are less likely to be outside, can be effective [13]. Unfortunately, results are temporary [13] because such methods kill adults without addressing the large pool of immature stages that continuously emerge into adults.

Area-wide application of mosquito larvicides involves applying liquid or emulsified larvicides with a low-volume sprayer so the droplets drift onto backyards. The droplets of larvicide then settle into the myriad of containers of water where *Ae. albopictus* can be found, delivering a lethal dose. Using this approach hundreds of residential yards can be treated in a single application during one night, an action that can be repeated as needed, as opposed to intensive source reduction approaches that would take minimally weeks to treat the same number of yards [9].

Bacillus thuringiensis subspecies *israelensis* (*Bti*) is an excellent candidate for a low-volume area-wide strategy. The efficacy of *Bti* has been demonstrated for many mosquito species in a variety of habitats. Today, formulations of *Bti* are one of the primary products used for larval mosquito control in the U.S. and other countries [14]. After ingestion and activation in the mosquito larvae, which have a highly basic gut pH, proteinaceous toxins are released which cause lysis of the epithelial cells of the larval midgut, killing the larvae. *Bti* contains four different larvicidal proteins, each acting in slightly different ways making the development of resistance to *Bti* difficult [15]. The specificity of these proteins to organisms with catalytic gut pH results in very few non-target impacts. With the exception of several families of Nematocera (many of them also pest species such as blackflies and biting midges), *Bti* has no known effects on other insects, invertebrates, or vertebrates [16].

Although *Bti* has often been applied as a solid directly to larval sources, recent studies have demonstrated that *Bti* liquid can be applied as ULV or low volume (LV) sprays to control mosquitoes. Several such studies have successfully used the VectoBac 12AS formulation (Valent BioSciences Corporation, Libertyville, IL, USA) [17–19], however 12AS can cause persistent spotting on automotive paint [20,21] and is therefore unsuitable for use in residential areas in the US.

Until now, studies that used the VectoBac WDG formulation (Valent BioSciences Corporation, Libertyville, IL, USA) focused on backpack mist sprayers that are insufficient for area-wide control efforts because they hold small volumes of product and require the operator to walk into backyards during the application [17,22,23]. These manual applications are problematic because it takes too much time to treat a neighborhood and access to yards is often restricted by fences, dogs, or uncooperative homeowners. Most studies were conducted in Southeast Asia where differences in the strains of *Ae. albopictus*, housing construction and condition, and accepted pest management practices limit the value of those data in the US. The present study describes the development of an area-wide larviciding strategy using VectoBac WDG applied with truck-mounted equipment. We evaluated several sprayers and application rates under both staged open-field as well as operational trials in urban and suburban settings, and used larval bioassays to examine the efficacy of this method over distance (from the road into the backyards) and over time (residual effects). Our ultimate goal is to develop a novel strategy for the control of *Ae. albopictus* larvae to be used in an integrated mosquito management program in hopes of reducing the nuisance and disease risk associated with this species.

Materials and Methods

Ethics Statement

No specific permits were required for the described field studies. All pesticide applications were made by county mosquito control agencies under the authority of Title 26:9 of the New Jersey Administrative Code. These studies did not involve endangered or protected species.

Pesticide

VectoBac WDG is a water-dispersible granule formulation containing 3,000 International Toxic Unit/mg of *Bti* (strain AM 65-52). Containing only *Bti* and food-grade inert ingredients, it has been approved for use in organic crops and sensitive habitats around the world. We chose it for this study because of the favorable environmental profile, safety for non-target organisms, and fast results; all mortality usually occurs within 12–72 hrs. The product is mixed by weight with tap water. The US product label allows for application rates of 123–988 g/ha for ground spraying equipment.

Equipment

We tested three sprayers: a Cougar ULV cold aerosol generator (Clarke, Roselle, IL), a CSM2 Mist Sprayer (Buffalo Turbine, Springville, NY), and an Ag-Mister LV-8 low volume sprayer (Curtis Dyna-Fog, Westfield, IN). We chose the Cougar to examine the feasibility of using conventional ULV equipment with WDG. The standard pump was replaced with an optional QB3CSC 1/2 inch pump (Fluid Metering, Inc., Syosset, NY) to increase flow rate. However, despite being rated for a flow rate of over 2 L/min, we were unable to achieve more than 1 L/min. That flow rate was too low to reach our lowest application rate of 400 g/ha and the Cougar was not evaluated further.

The Buffalo Turbine Mist Sprayer (CSM2) has a 189 L stainless spray tank. A mechanical twin piston pump sends fluid to a spray head contained within an air chute. A gas-powered wind turbine propels the fluid from the spray head. The standard spray head consists of four flat-fan spray nozzles. For this study, we replaced the standard head with an optional Monsoon gyratory atomizing spray head (Buffalo Turbine, Springville, NY) for increased flow rate and more precise control over the droplet spectra. With the atomizing head, fan blades cause the head to rotate in relation to the wind speed. Fluid released from the center of the head is forced through stainless mesh screens; larger mesh producing larger droplets. We evaluated 0.33, 0.51, and 0.91 mm screens. The smallest screen was too fragile to be used alone and was supported with the 0.51 mm screen. Pump pressure was set to 1,034 kPa and a ball valve on the pump outlet was adjusted to a flow rate of 8.3 L/min. Wind speed was controlled with the engine throttle and set at maximum (177 km/h) for all tests for greatest swath width.

The LV-8 is a low volume sprayer designed for orchard spraying. An electric pump pushes liquid to eight nozzles arranged in two vertical rows of four. A gas-powered roots type blower generates air pressure to atomize the pesticide. A ball valve adjusts the flow rate on the fluid supply line and was set to 8.3 L/min. Air pressure was adjusted by changing engine speed. Pump pressure was maintained at 68.9 kPa. As equipped from the factory, the nozzles point out to the sides of the spray truck and may not release droplets high enough to make it over buildings and other obstacles. Given the narrow streets in many urban neighborhoods, the LV-8 may end up applying material directly to cars parked along the streets. To minimize this occurrence, we cut and welded the booms containing the nozzles to angle at 45 degrees so that the spray would be released over parked cars.

Droplet Characterization

We characterized the droplet spectra at the Navy Entomology Center of Excellence (Jacksonville, FL) with a 2-D Phase Doppler Particle Analyzer system (TSI Inc., Shoreview, MN). The transmitter and receiver were outfitted with 500 and 300 mm lenses, resulting in a measurable droplet size range of 0.6–211 µm respectively. Equipment was positioned 1.2 m from the lens array. We measured the droplet size at the horizontal center of the spray scanning the whole plume vertically in a continuous mode and characterized the droplet spectra under a variety of flow rates and pressures with three replicated measurements recorded for each test.

We set all equipment for a flow rate of 8.3 L/min. The fluid pump of the LV-8 was therefore set at 68.9 kPa for each test and we adjusted the engine speed to change the air pressure through the nozzles, thereby varying the droplet size. With the CSM2 the fluid pump pressure was maintained at 1,034.2 kPa for all tests. Finally, we kept the Buffalo turbine wind speed at a maximum of 177 km/h for greatest swath width and varied droplet size by changing screen sizes on the atomizing spray head. VectoBac WDG was mixed with tap water at a concentration of 59 or 118 g/L with an electric paint mixer in 19 L plastic buckets. Target application rates were 400 and 800 g/ha (representing mid and high label rates on the USA pesticide label) assuming a swath width of 91.4 m and an application speed of 8 km/h.

Open-field Trials

We conducted open-field trials to determine efficacy, swath width and droplet deposition under ideal conditions. The LV-8 was tested on private land over an open grass field at Arnold Groves Ranch (Clermont, FL) in September 2010 with permission from the property owner, Robert Arnold. The CSM2 was tested over a blacktop tarmac at the Trenton-Mercer Airport (Ewing, NJ) in October 2010 with permission from the Airport Manager, Melinda Montgomery. We set up stations in a 3×8 grid out to 76.2 m for the LV-8 and a 3×9 grid out to 91.4 m for the CSM2. The grids were set up parallel to the prevailing wind. Each station contained an empty 500 ml polyethylene container for larval bioassays and one Kromekote C1S card (Mohawk Fine Papers, Inc., Cohoes, NY) on a plastic compact disc case (for ballast) at ground level to determine droplet deposition and characteristics.

Sprayers were mounted on pickup trucks and driven at 8 km/h perpendicular to the wind during applications. We mixed the WDG with tap water at 59 and 118 g/L for the 400 and 800 g/ha rates respectively. A 2% solution FD&C Red 40 granular dye (Glanbia Nutritionals, Carlsbad, CA) was added to the mixture for droplet analysis. We operated the LV-8 at 62 kPa and 8.3 L/min with a volume mean diameter (VMD) of 107 µm from the nozzles. The CSM2 was operated with a 0.51 mm screen at 1,034 kPa and 8.3 L/min with a VMD of 233 µm from the spray head with the CSM2 head positioned at a 60° angle from horizontal. We made control applications with dye and water (CSM2) or water only (LV-8). Applications were made when wind speeds were between 3.2 and 8 km/h and stations were moved as needed to remain parallel to wind direction. We retrieved cards and cups 15 min after the applications allowing time for droplets to settle. Lids were placed on the cups to prevent contamination or loss of product. We replicated each application three times and performed one control application with each machine. Cups were returned to the laboratory for analysis where we added 400 ml water and 10 (CSM2) or 20 (LV-8) 3rd instar *Ae. albopictus* larvae to each cup. Mortality was recorded at 24, 48 and 72 hours and final mortality rates were corrected for control mortality using the Schneider-Orelli formula [24]. We analyzed the Kromekote cards with the DropVision AG system (Leading Edge Associates, Inc., Waynesville, NC) a system that uses a portable scanner to digitize droplet cards. Stains on the card were counted and measured and the software provided droplet and volume data using user-defined application and mix rates.

Operational Trials

We developed operational field trials in Trenton (Mercer County), New Jersey, USA (40° 12′ N, 74° 44′ W) and Belmar (Monmouth County), NJ USA (40° 13′ N, 74° 44′ W). The two Trenton sites were urban residential neighborhoods: S. Olden (40° 22′ N, 74° 73′ W) was 48.6 ha consisting of 24 residential blocks, each containing a residential street on all four sides, sometimes divided lengthwise by a drivable alley. St. Olden included 1,250 parcels (i.e. house with surrounding yard) most often built as adjoining row homes or duplexes [25]. Parcel sizes were relatively constant at approximately 200 m^2. The second site, Cummings (40° 21′N, 74° 74′W), was 30 ha with 23 residential blocks, occasionally divided lengthwise by a drivable alley. Cummins included 1,122 parcels many adjoining (i.e. row houses). Almost all adjoining parcels contained a sheltered alcove area between homes, where vegetation and trash proliferated creating mosquito resting areas. Socioeconomic conditions in both St. Olden and Cummins have led to a large number of abandoned homes with neglected yards that accumulate containers such as buckets, cans and other trash that can hold water [26].

The field site in Belmar was a seaside resort community located along the Atlantic coast and consisting mainly of traditional suburban single- family home parcels with yard space on all four sides, interspersed with seasonal rental properties. A small section of the plot contained commercial storefront and several large condominiums and motels. This 121.9 ha site consisted of 59 residential blocks and 2,356 parcels averaging 0.1 ha in size. The socioeconomic conditions of the site can best be described as upper-middleclass. The majority of the homes are well maintained and container mosquito habitat, such as planters and flexible downspout extensions, tend to be cryptic but abundant. Rental properties often had more traditional container habitats such as buckets, toys, and recyclable containers.

We used either a LV-8 or a CSM2 mist sprayer for each application. We set up the equipment as previously described for a flow rate of 8.3 L/min. We mixed VectoBac WDG with tap water at a rate of 59 or 118 g/L and drove the vehicles at an average speed of 8 km/h for a final application rate of 400 or 800 g/ha. Meteorological data was recorded for wind speed, direction, humidity, and temperature at 1 m and 10 m heights for monitoring thermal inversion. A Vantage Pro2 wireless weather station (Davis Instruments, Hayward, CA) was set up within the treatment site 14 h prior to application and maintained until 8 h post application. We made applications from April through September 2011, June 2012, and July 2013 between 01:00 and 06:00 hours. All pesticide applications were made by county mosquito control agencies under the authority of Title 26:9 of the New Jersey Administrative Code.

We selected 30 residential parcels within our treatment sites and 10 parcels within a control site in either Mercer or Monmouth for placement of bioassay cups. Within each parcel, dry bioassay cups were placed in front, alongside, and in the backyard of each home. A total of 210 polyethylene 500 ml bioassay cups per replicate were placed in the field 2 hrs prior to application and removed 1 hr post application to allow time for droplets to settle into cups. Bioassay cups were used in lieu of sampling natural larval populations to get consistent and comparable data points. Existing larval habitats were too transient to be of statistical value and sampling of varied containers could not be standardized.

All cups were transported to the laboratory and loaded with 400 ml of distilled water and 40 mg of a 50:50 mix of yeast:lactalbumin (Sigma-Aldrich, St. Louis, MO) added as an emulsion in 1 ml of water as a food source for larvae. Then we added 10–15 late 2nd to early 3rd instar *Ae. albopictus*. Mortality was recorded at 24, 48, and 72 hrs post-application. Mortality rates were corrected for control mortality using the Schneider-Orelli formula [24].

Based on the outcome of the applications, we selected three of the operational trials to determine the residual efficacy of the treatments. For these experiments, we retained any cups which achieved 100% mortality after 72 hours. Dead larvae were removed from the cups and ten early 3rd instar *Ae. albopictus* were added. After 72 hours mortality was recorded, dead larvae were removed, and ten fresh larvae were added. We repeated this process until larval mortality dropped to or below 20%.

Statistical analyses

All mortality data were reported as proportions and arcsin square root transformed to normalize the distributions [27]. After the open field trials we performed least squares repeated measures analyses of variance in JMP (SAS Institute, Cary, NC) to examine the effects of equipment type and application rates on mortality rates. As mentioned, there were three spatial replicates at increasing distances from the application site and each trial was replicated 3 times. To assess the reliance of the results on expectations of homoscedasticity of the data (hard to assess with a feasible number of replicates), we also performed the repeated measures analysis with ranked data.

After the operational field trials, we examined the effects of equipment type, application rate and location of bioassay containers (front, middle, and back) with a least squares analysis of variance after confirming normality of the distributions using a goodness of fit test in JMP (SAS Institute, Cary, NC).

Results

Droplet Analysis

Unlike ULV sprayers that are optimized to produce droplets below 30 µm, the LV-8 is designed to produce larger droplets for agricultural use. Under laboratory conditions, median droplet sizes ranged from 74–141 µm (Table 1). In general, increasing air pressure led to larger droplet sizes with the LV-8. Beyond 34.5 kPa there was a direct relationship between air pressure and droplet size. With the CSM2, VMD's ranged from 214–233 µm.

Open-Field Trials

Weather conditions during the LV-8 tests averaged 23.9°C, 85% RH and 4.8 km/h wind speed. Average conditions during the CSM2 runs were 14.4°C, 55% RH and 3.2 km/h wind speed. The LV-8 resulted in an average mortality of 89.4±6.4 and 76.3±9.2% out to 80 m (maximum distance tested) for the 400 and 800 g/ha rates, respectively. The CSM2 averaged 98.9±0.9% mortality out to 100 m at the 800 g/ha rate and 85.1±5.1% at the 400 g/ha rate (Fig. 1). There was no significant difference in the overall average mortality at each distance from deployment between rates or machines. However, when only data for the CMS2 is examined, the average mortality was greater at the 800 g/ha rate than at the 400 g/ha rate ($P=0.04$) (Fig. 1). Repeating the analysis with ranked data yielded the same results.

We conducted a droplet analysis to determine droplet size, swath width and deposition rates (Fig. 2). The LV-8 had greater deposition at the 10 m stations than the CSM2. At distances beyond 10 m, there was little difference in the deposition rates regardless of equipment or application rate. Minimal deposition at the farther distances still resulted in significant mortality (Fig. 1). Application rates were determined by the mix ratio alone. The 800 g/ha rate resulted in greater volume deposition for the LV-8 but lower deposition for the CSM2. Generally, larger droplets settled out closer to the spray line and decreased swath width (Fig. 2). There was a clear correlation between VMD and distance for the CSM2 (average $r^2=0.9$) while the droplets from the LV-8 appear to be relatively uniform across the measured swath width.

Operational Trials

The efficacy of control in urban and suburban neighborhoods in 2011 were mixed with an average efficacy of 72.0±1.3%. In contrast, by the end of 2012 we were able to consistently achieve high mortalities in urban settings (over 91%), which we further increased in 2013 to 94.5±1% (Fig. 3). However, operational results varied greatly for the LV-8. Of the ten applications made with the LV-8, we were only able to exceed 80% mortality on three occasions. In contrast, the CSM2 averaged 89.4±1.0% mortality over all nine applications and exceeded 90% mortality for five of those applications.

Although we had slightly greater success in urban (Mercer) than in suburban areas (Monmouth) (Fig. 4), we only performed two applications in suburban Monmouth and both were performed with the CSM2 at the 400 g/ha rate. Since these were the only operational trials with the CSM2 at the lower rate we did not include these two suburban trials in the statistical analysis. Despite a slight decline in mortality as distance from the spray line increased, operational mortality rates were similar across all cup locations and the differences were not significant ($F_{2,2}=1.31$, $P=0.28$). We did however see significantly greater efficacy at the higher application rate of 800 g/ha ($F_{1,1}=4.26$, $P=0.04$), and the CSM2 outperformed the LV-8 ($F_{1,1}=8.64$, $P=0.005$). Under laboratory conditions, the residual activity of the applications achieved over 80% mortality for 6 weeks and lasted nearly 10 weeks (Fig. 5).

Discussion

Unlike most pressurized spray equipment where air pressure has an inverse relationship with droplet size, the LV-8 exhibited a direct relationship between air pressure and VMD. It is unclear if this trend was a result of nozzle design or was an artifact of the properties of the liquid WDG, which is an emulsion rather than a solution. However, the largest droplets did occur at the lowest pressure setting indicating that the nozzles were not operating within their optimal range when spraying the viscous mixture of WDG. Also of note, screen size did not seem to have a predictable effect on droplet size for the CSM2. It is plausible that running the head at maximum rpm canceled out any effect from the screens by creating additional wind shear. Reducing turbine speed may result in different droplet sizes but will reduce swath width. We only tested the 59 g/L mix ratio for the particle analyzer tests since existing literature [28] demonstrated that there was little advantage to application rates above 400 g/ha.

Based on equipment design, the droplets from these mist blowers were significantly larger than the droplets created by a ULV machine applying the same material [29]. Similar studies

Table 1. Droplet spectra for VectoBac WDG determined by 2-D phase doppler particle analyzer ± SEM.*

Equipment	Setting	$DV_{0.1}$ (μm)	$DV_{0.5}$ (μm)	$DV_{0.9}$ (μm)
LV-8	34.5 kPa	35.7±2.2	140.7±22.4	276.0±13.0
	41.4 kPa	31.5±0.5	73.5±1.5	202.5±0.5
	42.7 kPa	34.0±0.6	81.0±6.2	217.0±26.0
	63.4 kPa	28	92	250
	65.5 kPa	34.7±0.9	109.7±9.7	262.0±6.4
CSM2	0.33 & 0.51 mm screen	71.3±8.2	214.0±11.5	313.7±6.6
	0.51 mm screen	82.0±3.1	232.7±2.4	324.7±2.4
	0.91 mm screen	64.3±9.2	218.0±10.1	313.3±16.2

*VectoBac WDG mixed with tap water at 59 g/l. Equipment set for final application rate of 400 g/ha.

with back pack mist blowers produced droplets comparable in size to the LV-8 [22]. The insufficient flow rates we experienced with the modified Clarke Cougar suggest that most ULV sprayer will be unsuitable for WDG application without extensive modification.

In open field trials we observed the expected decrease in mortality with increasing distance from the spray line regardless of equipment or rate; however the CSM2 outperformed the LV-8 in both swath width and efficacy. The LV-8 had greater deposition at the 10 m stations than the CSM2 indicating that larger drops settled out earlier likely because the LV-8 is a passive sprayer (i.e. it does not induce a strong directional force to the droplets the way the CSM2 does). The wind speed generated by the turbine of the CSM2 propelled the large droplets further, resulting in a more uniform deposition rate. The performance of the LV-8 was more dependent on wind speed than the CSM2: for example, with the LV-8, the 800 g/ha rate did not perform as well as the 400 g/ha rate probably because of environmental conditions as the wind averaged 4.8–9.6 km/h for the 400 g/ha trials and only 0–3.2 km/h for the 800 g/ha trials. While the CSM2 was less dependent on environmental conditions, the wind from the turbine pushed droplets beyond the closest stations, resulting in lower deposition rates near the spray vehicle. Importantly, we observed very high toxicity of WDG to mosquito larvae such that the lower deposition rates did not result in lower mortality.

Mortality in the controls was low for the LV-8 but significant mortality was observed in the CSM2 controls. This is likely because the LV-8 is an easier machine to clean and load and had been used for orchard spraying of agricultural pests with agricultural pesticides, not mosquito larvicides. In contrast, the CSM2 had routinely been used in a mosquito control program for several years, exclusively with mosquito larvicides. Further, on the LV-8 the removable spray tank is translucent and the fluid lines are transparent, making it easy to see when the unit has been sufficiently cleansed. The spray tank on the CSM2, is stainless steel and is welded into the machine. The tank is filled through a 51 mm diameter neck and emptied through a 19 mm drain plug. Flushing the tank is difficult and there is no way to see inside the tank to determine when the tank is clean. The fluid lines also have metal fittings that corrode and provide recesses for pesticide residue to collect. Despite repeated efforts to clean the CSM2 tank, residual material may have persisted during the control applications. We have since also conducted experiments to confirm that the food dye used in the controls was not toxic to *Ae. albopictus* at the rates applied [30].

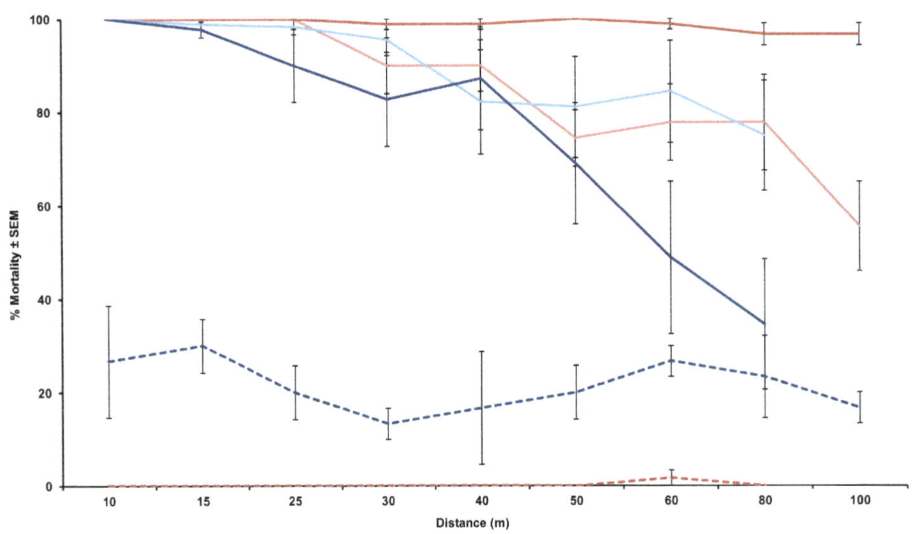

Figure 1. ***Aedes albopictus*** **72 hr mortality rates for VectoBac WDG applied with a CSM2 and LV-8 at 400 and 800 g/ha in an open-field.** Red = CSM2, blue = LV-8, light colors = 400 g/ha, dark colors = 800 g/ha, dashed line = control.

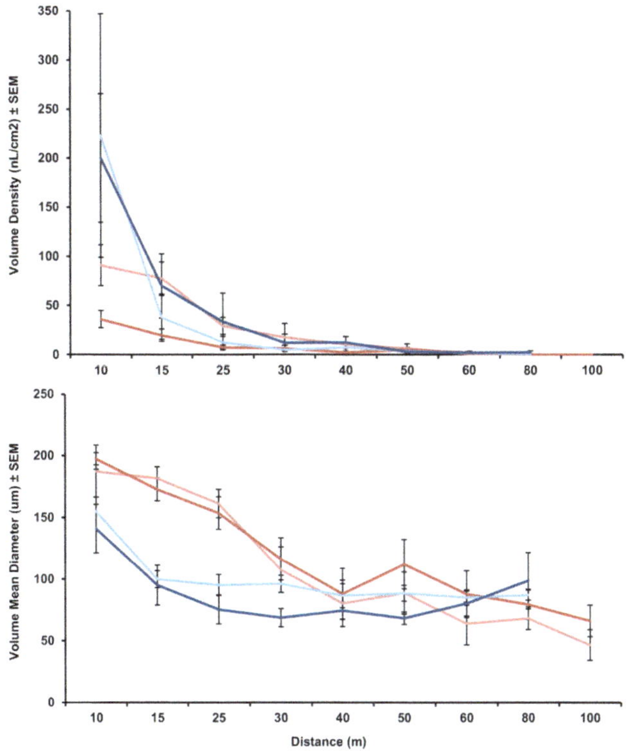

Figure 2. Volume density and droplet sizes of VectoBac WDG applied with a CSM2 and LV-8 at 400 and 800 g/ha in an open-field. Red = CSM2, blue = LV-8, light colors = 400 g/ha, dark colors = 800 g/ha.

The VMD values at the 10 m stations suggest that the droplet data from the spray cards are reliable as they are close to the VMD values from the nozzle as determined by the laser analysis. While it is logical to presume that the droplets from the Buffalo Turbine are too large (VMD = 233 µm) to travel 100 m, the 177 km/h wind created by the turbine was sufficient to push the drops, compensating for the extra mass of the large drops. The smaller drops from the LV-8 (VMD = 107 µm) were still small enough to drift out to 80 m from wind currents alone. We did not find significant differences in droplet size based on the mix ratio of the product which contradicts previous studies that found that higher mix ratios resulted in larger droplets [22]. This is likely because of the low power of the backpack mist blower used in that study. The more powerful motors of the truck-mounted machines we used appear better equipped to handle the more viscous mix. The problem with many water-based formulations is the evaporation of droplets. Personal communication from Valent technical staff (Peter DeChant, 21 July 2010) indicated that WDG will not evaporate beyond a terminal droplet size of around 80 µm. Our data support this assertion. Regardless of the initial size, beyond 40 m droplet sizes converged to an average VMD of 80.7 µm.

Smaller droplets are lighter than larger droplets and therefore should drift farther yielding greater swath width. Other open-field trials with ULV sprayers produced small droplets of 20–50 µm yet a swath of only 30 m [31]. These sprayers are normally used for mosquito adulticides where small drops are expected to drift up to 90 m without settling. It is possible that many of the droplets created by the ULV sprayer in Lee et al. [31] drifted beyond the sample area. For the purposes of applying mosquito larvicides, larger droplets are preferable so that they settle into containers before drifting past the target area. Based on our results we recommend that equipment should be optimized for larger droplets (≥80 µm).

We expected a positive correlation between mortality and volume density, higher volume density resulting in greater mortality. However, the LD_{50} values for WDG against *Ae. albopictus* are so low, 3 ppb as determined against our laboratory strain, that even a single droplet in the bioassay cups was sufficient to cause 100% mortality. Any deposition over 18 nL/cm^2 resulted in 100% mortality and in many cases rates below 1 nL/cm^2 were still able to provide 100% mortality. While there is no correlation between volume density and mortality, the data demonstrate that equipment can be calibrated without bioassays. As long as droplets are visible on spray cards, mortality should be acceptable at that distance.

The open-field trials tested area-wide larval control from truck-mounted equipment under ideal conditions in the absence of any obstructions. Operational conditions present many obstacles to any type of space spray. Houses, automobiles, and vegetation all serve to potentially impede the drift of pesticide sprays. To evaluate the operational utility of this technique, we conducted operational applications under operational conditions. These data

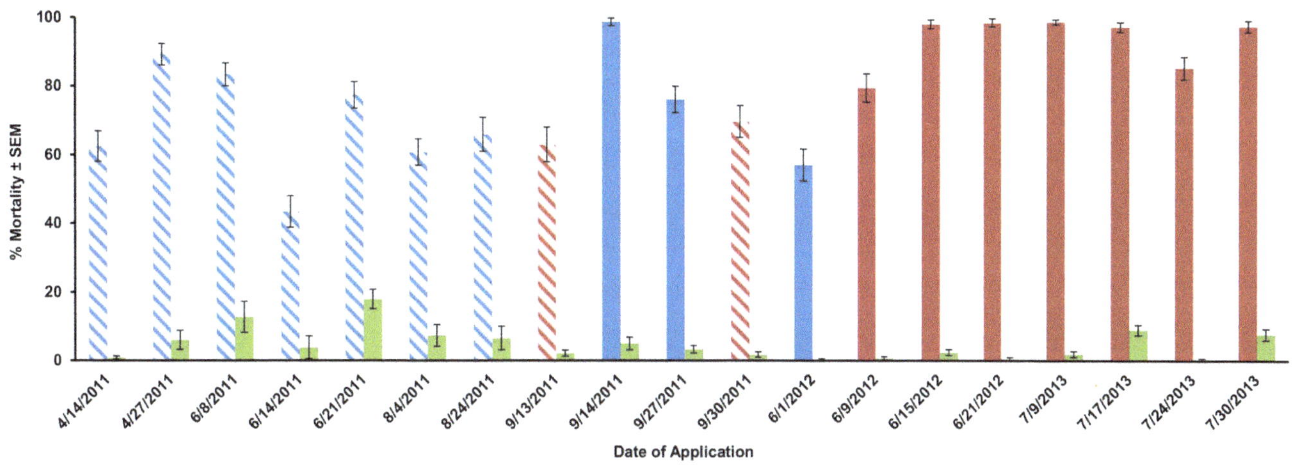

Figure 3. Efficacy summary of area-wide VectoBac WDG applications by date. Red = CSM2, blue = LV-8, green = control, dashed lines = 400 g/ha, solid lines = 800 g/ha.

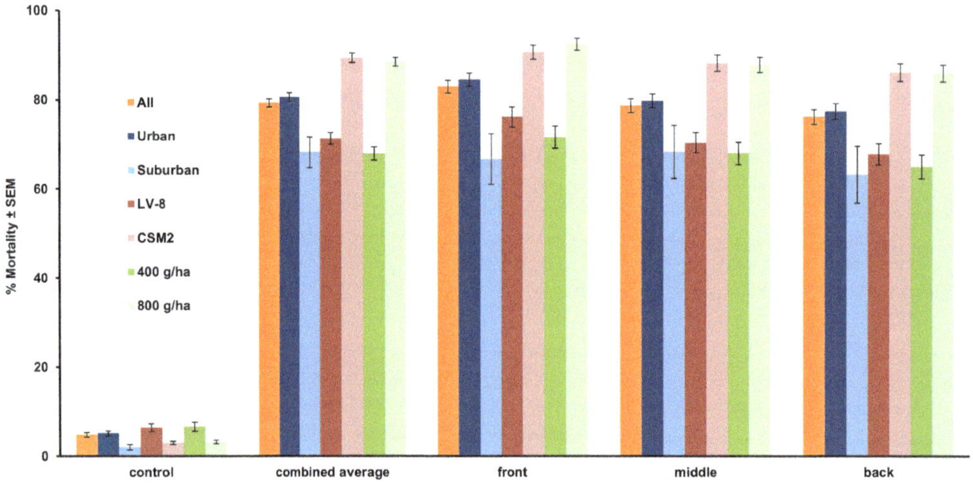

Figure 4. Efficacy by location of VectoBac WDG applied with a CSM2 and LV-8 at 400 and 800 g/ha in residential areas. Orange = average efficacy, dark blue = urban plots, light blue = suburban plots, red = LV-8, pink = CSM2, dark green = 400 g/ha, light green = 800 g/ha.

show a clear improvement in our area-wide technique over the three seasons of the study. Early in the study we focused on a 400 g/ha rate from the LV-8. Equipment problems such as pump failure, solenoid malfunctions, and short circuits of the electrical system led to poor results in several of the early trials. Further, contrary to what we found in the open field trials, bioassay results indicated that the 400 g/ha rate was not delivering enough material into distant containers. In late 2011 we increased the rate to 800 g/ha and found that this change yielded improved yet inconsistent results from the LV-8. In 2012 and 2013 all applications were made at the 800 g/ha rate and all but one application was made with the CSM2. This combination of higher application rate and better machine led to high mortality rates in all trials.

Within each parcel, the consistent results of the operational trials demonstrate that this technique can be used effectively in any type of residential environment with different types of equipment. The WDG spray effectively traveled around obstructions and settled in containers. In total, larval mortality was observed in 90% of the treatment cups despite efforts to place cups in inaccessible areas such as under decks and in dense vegetation. Our results are limited to bioassay cups but we expect similar results in other habitats such as roof gutters although these areas were not monitored. Additionally, the observed persistence of WDG indicates that rainfall following WDG applications would likely wash *Bti* to non-target areas such as catch basins and sump pumps leading to increased control.

As a result of the thorough coverage, there was no significant difference in mortality based on the location of the cups. Bioassay cups placed in the backyards of homes did just as well as cups placed in front of homes nearest the spray line. This is similar to Jacups et al. [22] using a backpack mist blower over various distances. However our truck mounted strategy has the advantage of being more consistent and keeps the operators inside the truck cab away from the *Bti* spray.

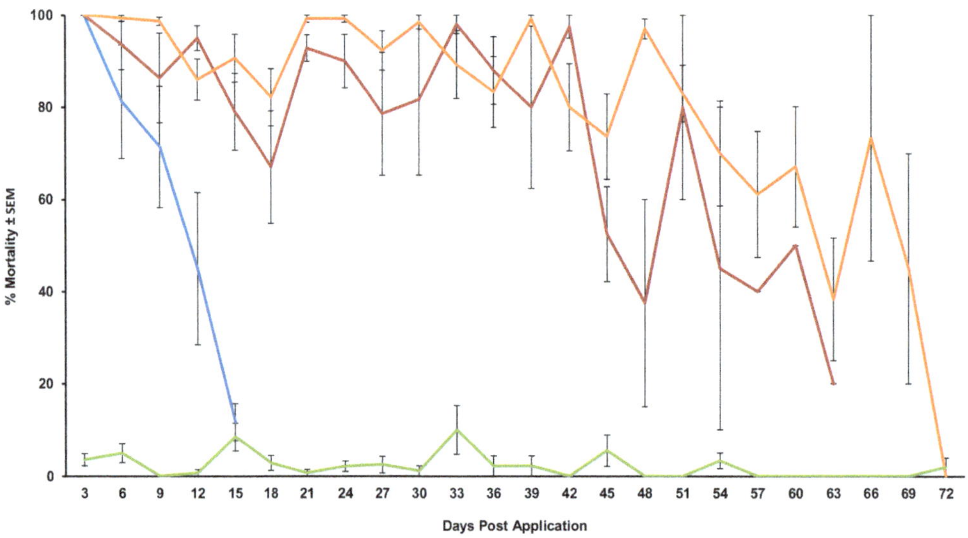

Figure 5. Residual efficacy of VectoBac WDG applied with a CSM2 and LV-8 at 400 and 800 g/ha in residential areas. Orange and red = CSM2 at 800 g/ha, blue = LV-8 at 400 g/ha, green = control.

The greater efficacy we observed in the urban areas was not a result of the smaller parcel sizes or easier access in those plots, as there was no difference in efficacy across distances even in the suburban parcels. Instead, it is more likely due to the fact that the suburban trials were only run at the 400 g/ha rate at a time when we thought based on the results of the open-field trials, and the results of Jacups et al. [22] in a bushland setting, that the higher application rate was unnecessary. Obstacles present during operational trials likely resulted in lower deposition rates which would explain the increased mortality at the higher application rate. We were unable to measure deposition during operational trials because we could not apply dyed material in a residential setting. High environmental humidity precluded the use of water sensitive paper and ultraviolet dye did not mix well with the WDG solution. However, it is logical to conclude that deposition rates would decrease given the physical obstacles present under operational conditions. With fewer droplets reaching the cups, the amount of *Bti* in each droplet becomes increasingly critical especially at the farthest distances. Significantly, our application rate was below the maximum of 988 g/ha specified on the product label. Higher application rates could provide even greater mortality, although as discussed mixing and spraying at high rates has limitations.

The CSM2 performed significantly better than the LV-8 overall most likely because the LV-8 is a passive machine relying on wind to carry the spray cloud. Wind conditions were less than optimal for most applications, rarely exceeding 4.8 km/h, putting the LV-8 at a disadvantage. By contrast the turbine of the CSM2 generates its own wind and is therefore less reliant on environmental conditions. This is further evidenced by the consistency of the CSM2 across all distances. It is possible, however, that under more favorable wind conditions, the efficacy of the LV-8 would be higher.

Although not considered a residual product, *Bti* has been shown to give extended control in containers [32–35]. Our applications achieved over 80% mortality for 6 weeks and lasted nearly 10 weeks. It is important to note that the residual trials were conducted in the laboratory. While environmental conditions could reduce the residual activity of *Bti*, trials conducted under semi-field conditions have provided residual control for up to 11 weeks [22]. In that study, WDG was applied closer to the containers with a backpack mist blower leading to droplet densities up to 300 drops/cm^2. In our open-field trials, droplet densities averaged 10 drops/cm^2 and never exceeded 81 drops/cm^2 and Jacups et al. [22] found that higher application rates resulted in longer residual control. Given that there were significant differences in mortality between machines and rates, it is likely that the reduced residual efficacy of the LV-8 application at 400 g/ha was due to fewer *Bti* particles reaching the containers.

Based on our open-field and operational trials, both sprayers are suitable for area wide larval control operations. The CSM2 offers the advantages of higher potential flow rates, adjustable nozzle angle, and increased swath width due to the 177 km/h wind drift created by the turbine. However, capacity is limited to 189 L (about 25 min of spray time), and the hurricane force wind from the turbine may present problems in a residential areas rattling windows, blowing over potted plants, scattering debris, etc. The LV-8 has the advantage of an optional 568 L spray tank with an onboard mixer, which increases the potential run time and makes loading the WDG easier. However, as discussed the LV-8 is a passive sprayer that relies on the wind to carry the pesticide to the target areas. Wind speeds often averaged less than 4.8 km/h during our operational trials, a standard condition in very urban areas at night, leading to inconsistent efficacy with LV8-applications. With the reconfigured nozzles, we found that WDG deposition to parked cars was minimal and similar to what was left by the CSM2. Observations by the team indicate that applications left a fine dust on automobiles not easily distinguished from dirt or pollen. The WDG rinsed easily from vehicles and caused no damage to automotive paint. Although citizens were in regular contact with the mosquito control programs, we received no negative comments from the public regarding visible residues from any of the 19 area-wide applications over 3 years.

The present study indicates that area-wide applications of larvicides with truck-mounted equipment have the clear potential to reduce *Ae. albopictus* populations in urban and suburban residential areas by killing immature mosquitoes in open containers. Both larvicide doses effected high larval mortality that could be sustained for over one week and high dose applications exhibited up to 90 days of residual control. Street-level applications were able to drift past obstructions and settle into containers located up to 100 m away with no loss of efficacy at greater operational distances. In conclusion, we provide evidence that a nighttime LV application of VectoBac WDG can be efficacious in killing larvae of *Ae. albopictus* in urban and suburban neighborhoods.

It is important to note, however, that effective larval control does not necessarily translate into a reduction in the adult mosquito population or human landing rates. In fact, a reduction in larval population can actually result in an increase in adult emergence for some species [36], but this does not appear to be the case with *Ae. albopictus* [37]. We deployed a high-density of BG Sentinel (Biogents AG, Regensburg, Germany) mosquito traps throughout each operational research plot for the duration of the study and have measures of adult populations of *Ae. albopictus*. These data were not presented here because we were performing multiple control strategies simultaneously, as needed operationally to safeguard the well being of the residents. Those data are being developed into a separate publication. In brief, we found that when area-wide applications of WDG were coordinated with a degree-day model and traditional adulticide spray missions, there was a synergistic effect resulting in a significant decrease in the adult *Ae. albopictus* populations.

This study was conducted with readily available equipment adapted for larvicide applications. Purpose-built equipment could provide even better efficacy and easier operational use. New formulations of *Bti* developed for this technique that mix with water more easily or come pre-mixed would greatly reduce the labor involved in preparing and loading the product. Future studies are needed to determine the optimal timing of applications so as to have maximum impact on adult mosquito populations. Based on our findings, we recommend that nighttime applications of LV larvicides in areas with large *Ae. albopictus* populations be considered as part of an integrated approach for public health protection.

Acknowledgments

We acknowledge Peter DeChant and Jim Andrews from Valent BioSciences for providing product and technical assistance. We thank Linda McCuiston and Rafael Valentin for laboratory assistance, and multiple students and postdocs from the Center for Vector Biology, Rutgers University, and personnel from Mercer, Monmouth, and Hudson Counties for laboratory and field work. In addition, we thank LCDR Jeff Stancil for access to NECE facilities and staff, Robert Arnold, the owner of Arnold Groves Ranch, and Melinda Montgomery, the Mercer Airport Manager, for permission to develop open field trials on their lands.

Author Contributions

Conceived and designed the experiments: GMW AF SPH MF RG GH DMF. Performed the experiments: GMW AF SPH MF DMF. Analyzed the data: GMW AF IU SPH MF DMF. Contributed reagents/materials/analysis tools: GMW AF SPH MF DMF. Contributed to the writing of the manuscript: GMW AF IU SPH RG DMF.

References

1. Benedict MQ, Levine RS, Hawley WA, Lounibos LP (2007) Spread of the tiger: global risk of invasion by the mosquito *Aedes albopictus*. Vector Borne Zoonotic Dis 7: 76–85.
2. Rochlin I, Ninivaggi DV, Hutchinson ML, Farajollahi A (2013) Climate change and range expansion of the Asian tiger mosquito (*Aedes albopictus*) in northeastern USA: implications for public health practitioners. PLOS ONE 8: e60874.
3. Unlu I, Farajollahi A (2012) Vectors without borders: Imminent arrival, establishment and public health implications of the Asian bush (*Aedes japonicus*) and the Asian tiger (*Aedes albopictus*) mosquitoes in Turkey. Hacettepe J Biol Chem, 40: 23–36.
4. Farajollahi A, Nelder MP (2009) Changes in *Aedes albopictus* (Diptera: Culicidae) populations in New Jersey and implications for arbovirus transmission. J Med Entomol 46: 1220–1224.
5. Armstrong P, Anderson JK, Farajollahi A, Healy SP, Unlu I, et al. (2013) Isolations of Cache Valley virus from *Aedes albopictus* (Diptera: Culicidae) in New Jersey and evaluation of its role as a regional arbovirus vector. J Med Entomol 50: 1310–1314.
6. Gratz NG (2004) Critical review of the vector status of *Aedes albopictus*. Med Vet Entomol 18: 215–27.
7. Unlu I, Farajollahi A, Strickman D, Fonseca DM (2013) Crouching tiger hidden trouble: urban sources of *Aedes albopictus* (Diptera: Culicidae) refractory to source reduction. PLoS One 8: e77999.
8. Bartlett-Healy K, Unlu I, Obenauer P, Hughes T, Healy S, et al. (2012) Larval mosquito habitat utilization and community dynamics of *Aedes albopictus* and *Aedes japonicus* (Diptera: Culicidae). J Med Entomol 49: 813–824.
9. Fonseca DM, Unlu I, Crepeau T, Farajollahi A, Healy SP, et al. (2013) Area-wide management of *Aedes albopictus*: II. Gauging the efficacy of traditional integrated pest control measures against urban container mosquitoes. Pest Manag Sci 69: 1351–1361.
10. Halasa YA, Shepard DS, Fonseca DM, Farajollahi A, Healy S, et al. (2014) Quantifying the impact of mosquitoes on quality of life and enjoyment of yard and porch activities in New Jersey PLOS ONE 9: e89221. doi:10.1371/journal.pone.0089221.
11. Zhou YB, Zhao TY, Leng PE (2009) Evaluation on the control efficacy of source reduction to *Aedes albopictus* in Shanghai, China. Chin J Vector Biol Control 20: 3–6.
12. Bartlett-Healy K, Hamilton G, Healy SP, Crepeau T, Unlu I, et al. (2011) Source reduction behavior as an independent measurement of the impact of a public health education campaign in an integrated vector management program for the Asian tiger mosquito. Int J Environ Res Public Health 8: 1358–1367.
13. Farajollahi A, Healy SP, Unlu I, Gaugler R, Fonseca DM (2013) Effectiveness of ultra-low volume nighttime applications of an adulticide against diurnal Aedes albopictus, a critical vector of dengue and chikungunya viruses PLoS One 7: 11. e49181 doi:10.1371/journal.pone.0049181.
14. Lacey LA (2007) *Bacillus thuringiensis* serovariety *israelensis* and *Bacillus sphaericus* for mosquito control. J Am Mosq Control Assoc 23: 133–163.
15. Regis L, Nielson-LeRoux C (2000) Management of resistance to bacterial vector control. In: Charles J-F, Delécluse A, Nielsen-LeRoux C, editors. Entomopathogenic Bacteria: From laboratory to field application. Dordecht: Kluwer Academic Publishers. 419–438.
16. Lacey LA, Merritt RW (2003) The safety of bacterial microbial agents used for black fly and mosquito control in aquatic environments. In: Hokkanen HMT, Hajek AE, editors. Environmental Impacts of Microbial Insecticides: Need and Methods for Risk Assessment. Dordecht: Kluwer Academic Publishers. 151–168.
17. Lee HL, Chen CD, Mohd Masri S, Chiang YF, Chooi KH, et al. (2007) Impact of larviciding with a *Bacillus thuringiensis israelensis* formulation of VectoBac WDG, on dengue mosquito vector population in a dengue endemic site in Selangor State, Malaysia. SE Asian J Trop Med Pub Health 39: 601–609.
18. Seleena P, Lee HL (1998) Field trials to determine the effectiveness of *Bacillus thuringiensis* subsp. *israelensis* application using an ultra low volume generator for the control of *Aedes* mosquitoes. Israel J Entomol 32: 25–31.
19. Yap HH, Chong ASC, Adanan CR, Chong NL, Rohaizat B, et al. (1997) Performance of ULV formulations (Pesguard 102/VectoBac 12AS) against three mosquito species. J Am Mosq Control Assoc 13: 384–388.
20. Valent BioSciences Corporation (2003) Technical Use Bulletin for VectoBac 12AS Mosquito and Black Fly Larvicide. Valent Biosciences Corporation Public Health Products website. Available: http://publichealth.valentbiosciences.com/docs/public-health-resources/vectobac-12as-technical-use-bulletin_mosquito-and-black-fly. Accessed 25 April 2014.
21. Valent BioSciences Corporation (2012) VectoBac 12AS specimen label. Valent Biosciences Corporation Public Health Products website. Available: http://publichealth.valentbiosciences.com/docs/public-health-resources/vectobac-sup-sup-12as-specimen-label. Accessed 25 April 2014.
22. Jacups SP, Rapley LP, Johnson PH, Benjamin S, Ritchie SA (2013) *Bacillus thuringiensis* var. *israelensis* misting for control of *Aedes* in cryptic ground containers in North Queensland, Australia. Am J Trop Med Hyg 88: 490–496.
23. Sun D, Williges E, Unlu S, Healy S, Williams GM, et al. (2014) Taming a tiger in the city: Comparison of motorized backpack mist applications and source reduction against the Asian tiger mosquito, *Aedes albopictus*. J Am Mosq Control Assoc 30: 99–105.
24. Punter W (1981) Manual for field trials in plant protection. Second edition. Basle: Agricultural Division, Ciba-Geigy Limited.
25. Unlu I, Farajollahi A, Healy SP, Crepeau T, Bartlett-Healy K, et al. (2011) Area-wide management of *Aedes albopictus*: choice of study sites based on geospatial charateristics, socioeconomic factors and mosquito populations. Pest Manag Sci 8: 965–974.
26. Unlu I, Farajollahi A (2012) To catch a tiger in a concrete jungle: operational challenges for trapping *Aedes albopictus* in an urban environment. J Am Mosq Control Assoc 28: 334–337.
27. Sokal RR, Rohlf FJ (1995) Biometry: the principles and practice of statistics in biological research, 3rd ed. Freeman: New York. 887.
28. World Health Organization (2004) Report of the seventh WHOPES working group meeting: Review of VectoBac WG.
29. Lam PH, Boon CS, Yng NY, Benjamin S (2010) *Aedes albopictus* control with spray application of *Bacillus thuringiensis israelensis* strain AM 65–52. SE Asian J Trop Med Pub Health 41: 1071–1081.
30. Dow GW, Leisnham P, Williams GM, Kirchoff N, Jin S, et al. (2014) Effects of a red marker dye on *Aedes* and *Culex* mosquitoes: Implications for operational mosquito control. J Am Mosq Control Assoc. In press.
31. Lee HL, Gregorio ER Jr, Khadri MS, Seleena P (1996) Ultralow volume applications of *Bacillus thuringiensis* ssp. *israelensis* for the control of mosquitoes. J Am Mosq Control Assoc 12: 651–655.
32. Batra CP, Mittal PK, Adak T (2000) Control of Aedes aegypti breeding in desert coolers and tires by use of *Bacillus thuringiensis* var. *israelensis* formulation. J Am Mosq Control Assoc 16: 321–323.
33. Mulla MS, Tharavara U, Tawatsin A, Chompoosri J (2004) Procedures for the evaluation of field efficacy of slow-release formulations of larvicides against *Aedes aegypti* in water-storage containers. J Am Mosq Control Assoc 20: 64–73.
34. Vilarinhos PTR, Monerat R (2004) Larvicidal persistence of formulations of *Bacillus thuringiensis* var. *israelensis* to control larval *Aedes aegypti*. J Am Mosq Control Assoc 20: 311–314.
35. Farajollahi A, Williams GM, Condon G, Kesavaraju B, Unlu I, et al. (2013) Assesment of a direct application of two *Bacillus thuringiensis israelensis* formulations for immediate and residual control of *Aedes albopictus*. J Am Mosq Control Assoc 29: 385–388.
36. Agnew P, Haussy C, Michalakis Y (2000) Effects of density and larval competition on selected life history traits of *Culex pipiens quinquefasciatus* (Diptera: Culicidae). J Med Entomol 37: 732–735.
37. Reiskind MH, Lounibos LP (2009) Effects of intraspecific larval competition on adult longevity in the mosquitoes *Aedes aegypti* and *Aedes albopictus*. Med Vet Entomol 23: 62–68.

6

Use of Insecticide-Treated School Uniforms for Prevention of Dengue in Schoolchildren: A Cost-Effectiveness Analysis

Yesim Tozan[1,2]*, Pitcha Ratanawong[1], Valérie R. Louis[1], Pattamaporn Kittayapong[3], Annelies Wilder-Smith[1,4,5]

1 Institute of Public Health, Ruprecht-Karls-University, Heidelberg, Germany, **2** Steinhardt School of Culture, Education and Human Development, New York University, New York, New York, United States of America, **3** Center of Excellence for Vectors and Vector-Borne Diseases, Faculty of Science, Mahidol University at Salaya, Nakhon Phatom, Thailand, **4** Lee Kong Chian School of Medicine, Nanyang Technological University, Singapore, Singapore, **5** University of Umea, Umea, Sweden

Abstract

Background: Dengue-related illness is a leading cause of hospitalization and death, particularly among children. Practical, acceptable and affordable measures are urgently needed to protect this age group. Schools where children spend most of their day is proposed as an ideal setting to implement preventive strategies against day-biting *Aedes* mosquitoes. The use of insecticide-treated school uniforms is a promising strategy currently under investigation.

Methods: Using a decision-analytic model, we evaluated the cost-effectiveness of the use of insecticide-treated school uniforms for prevention of dengue, compared with a "do-nothing" alternative, in schoolchildren from the societal perspective. We explored how the potential economic value of the intervention varied under various scenarios of intervention effectiveness and cost, as well as dengue infection risk in school-aged children, using data specific to Thailand.

Results: At an average dengue incidence rate of 5.8% per year in school-aged children, the intervention was cost-effective (ICER≤$16,440) in a variety of scenarios when the intervention cost per child was $5.3 or less and the intervention effectiveness was 50% or higher. In fact, the intervention was cost saving (ICER<0) in all scenarios in which the intervention cost per child was $2.9 or less per year and the intervention effectiveness was 50% or higher. The results suggested that this intervention would be of no interest to Thai policy makers when the intervention cost per child was $10.6 or higher per year regardless of intervention effectiveness (ICER>$16,440).

Conclusions: Our results present the potential economic value of the use of insecticide-treated uniforms for prevention of dengue in schoolchildren in a typical dengue endemic setting and highlight the urgent need for additional research on this intervention.

Editor: Thomas A. Smith, Swiss Tropical & Public Health Institute, Switzerland

Funding: DengueTools is funded under the Seventh Framework Programme of the European Community, Grant Agreement Number: 282589 Dengue Tools. The funders had no role in study design, data collection and analysis, decision to publish, or preparation of the manuscript.

Competing Interests: The authors have declared that no competing interests exist.

* Email: tozan@nyu.edu

Introduction

Recent estimates put the global public health burden of dengue infections in 2010 at 390 million infections per year with 96 million symptomatic cases, affecting Southeast Asian countries disproportionately [1]. Among symptomatic cases, disease severity varies from mild, self-limiting febrile illness to severe to fatal hemorrhagic disease–the latter more commonly experienced by children and adolescences under the age of fifteen [2,3]. There are currently no vaccines or specific antiviral drugs. Dengue-related illness is a leading cause of hospitalization, particularly among children [2–4], with case fatality rates of 1-5% among patients with dengue shock syndrome [5], placing heavy socio-economic burden on households and putting enormous pressure on strained health systems in endemic countries [4,6,7], particularly during outbreaks [8–11].

In the absence of vaccines and antiviral therapies, dengue prevention and control have relied heavily on vector control interventions that aim to reduce the population of dengue-carrying *Aedes* mosquitoes through the application of larvicides and adulticidal insecticide space sprays and management of breeding sites [12]. These community-based vector control efforts have, however, had limited impact on the increasing incidence and the geographic expansion of dengue in endemic countries and beyond [13–15]. Practical, acceptable and affordable measures are urgently needed, particularly to protect vulnerable children at risk of dengue infection. *Aedes* mosquitoes mainly bite during the day [16]. Because children spend most of their day at school, it has been suggested that preventive strategies should target schools and school activities [17]. Schoolchildren in most endemic countries wear school uniforms as a social norm [18]. A recent review on the safety and effectiveness of the use of insecticide-treated clothing indicated that it is a promising intervention, depending on the targeted vector and the pathogen transmission potential, and that studies demonstrated a wide range of effectiveness from nil to 79% in reducing disease incidence [19]. A randomized controlled trial is underway in Thailand to establish the effectiveness of insecticide-treated school uniforms for prevention of dengue in schoolchildren [20]. Further, permethrin-treated school uniforms are currently being tested under laboratory conditions to measure their knock-down efficacy under different types of treatments [19,20]. A recent mathematical modeling study showed that the use of insecticide-treated school uniforms could potentially reduce the incidence of dengue infection up to 55% in schoolchildren, depending on a number of factors such as the proportion of mosquito bites received during school time, the probability that mosquitoes will come into contact with the insecticide, and the level of compliance among schoolchildren with the intervention [21].

Economic evaluation using decision analytical modeling makes it possible to assess the potential health and economic value of new health technologies in advance of randomized controlled trials. As a vehicle for economic evaluation, decision models can synthesize the available epidemiological, clinical, and economic evidence, examine multiple sources and consequences of uncertainty in the available evidence, and identify the parameters that have the greatest effect on the cost-effectiveness of new technologies. Results can guide further research on new technologies during development, or inform policy decisions about their adoption and use, particularly in resource-limited health care settings. Cost-effectiveness frameworks can be updated as new evidence on the safety, efficacy and effectiveness of new technologies and their costs becomes available from randomized trials, observational studies, and systematic reviews.

Using a decision-analytic model, we evaluated the cost-effectiveness of the use of insecticide-treated school uniforms for prevention of dengue in schoolchildren from the societal perspective. We explored how the potential economic value of the intervention varied under various scenarios of intervention effectiveness and cost, as well as dengue infection risk in school-aged children, using data specific to Thailand.

Methods

Using TreeAge Pro 2014 (TreeAge Software Inc., Williamstown, MA, USA), we developed a decision-analytic model to simulate the decision on using insecticide-treated school uniforms for prevention of dengue in schoolchildren in an endemic setting. The decision model evaluated the expected costs and health outcomes of the intervention in a hypothetical cohort of schoolchildren. The intervention was then compared with a "do-nothing" alternative, following the standard guidelines of economic analyses [22]. The time horizon of the analysis is one year under a conservative assumption that children would require a new set of school uniforms each year. The outcome of the cost-effectiveness analysis was expressed as a ratio of incremental costs to incremental health outcomes–that is, incremental cost-effectiveness ratios (ICERs). Incremental health outcomes were estimated in terms of DF cases (non-hospitalized and hospitalized), DHF cases, and disability-adjusted life years (DALYs) averted. DALYs combine years of life lost because of premature death and years of life lived with disability in a single health outcome measure. ICERs were calculated as the cost per DALY averted. We reported ICERs when an alternative strategy was not ruled out of the decision analysis by simple dominance (i.e. less costly and more effective). Incremental costs and ICERs were calculated in United States (US) dollars ($) for the year 2012. All input parameters, their distributions, and data sources are listed in Table 1.

Following the recommendations of the World Health Organization (WHO) Commission on Macroeconomics and Health, we used the 2012 per head gross domestic product (GDP) for Thailand of $5,480 as a benchmark for intervention cost-effectiveness. The intervention was classified as highly cost-effective if the cost per DALY averted was less than the GDP per head ($5,480), and as cost-effective if this cost was less than one-to-three times the GDP per head ($5,480–$16,440). A summary measure of gross value added by all resident producers in a given country, the GDP per head reflects the fair share of residents to national economic output that can be devoted to health care. Given these cost-effectiveness benchmarks, the intervention would be unlikely to be considered by Thai policy-makers for implementation when the cost per DALY averted is higher than $16,440.

Estimating health outcomes

The effectiveness of the intervention has not been established in real-life settings yet, but is expected to be around 50% [18]. This intervention cannot be 100% effective because school uniforms do not cover the entire body and are not worn all day long, during the weekends and school holidays. Children are not being bitten by *Aedes* mosquitoes only during school time. The knock-down effect of the insecticide on school uniforms does not reach 100% even under ideal laboratory conditions with currently used impregnation methods [21]. Lastly, there might also be compliance issues with the intervention in an everyday context [19]. In the absence of evidence on intervention effectiveness, we considered a very broad range of possible estimates of effectiveness from 5% to 100% to determine the threshold price points at which the intervention would be considered cost-effective given the cost-effectiveness benchmarks stated above. But we reported the results of the probabilistic cost-effectiveness analysis for intervention effectiveness 50% or higher. A public health intervention for prevention with less than 50% effectiveness (i.e. averting less than 50% of a disease) would be of no policy interest in resource-limited settings [23,24], which set the effectiveness level at the low end for the analysis. Drawing on the range of possible effectiveness reported for insecticide-treated clothing in the literature [19], we assumed 75% effectiveness at the high end. The full effectiveness (100%) scenario was included to show the maximum potential benefit of the intervention for illustrative purposes. In the decision model, schoolchildren wearing insecticide-treated uniforms had a decreased risk of acquiring dengue infection by 1-intervention effectiveness. Acquiring dengue infection could result in asymptomatic infection or symptomatic infection, the latter requiring ambulatory care or hospitalization, followed by full recovery or death.

The clinical outcome probabilities used in the analysis were derived primarily from a multi-year prospective study on the epidemiology of acute dengue infection in a well-defined schoolchildren population in Thailand [25]. Over a period of three years, the overall incidence of dengue infection was 5.8%, and 54% of the infections were asymptomatic [25]. Of the symptomatic cases, the proportion of non-hospitalized dengue fever (DF), hospitalized DF, and hospitalized dengue hemorrhagic

Table 1. Decision model parameters.

Parameter	Base value (SD or range)	Distribution	Source
Annual dengue incidence rate (%)	5.8	Beta (331, 5429)	[25]
Proportion of asymptomatic cases (%)	53.4	Beta (177, 154)	[25]
Proportion of non-hospitalized DF cases (%)	81.5	Point estimate	[25]
Proportion of hospitalized DF cases (%)	7.4	Point estimate	[25]
Proportion of hospitalized DHF cases (%)	11.1	Point estimate	[25]
Case fatality rate for DF (%)	0.0027	Point estimate	[31]
Case fatality rate for DHF (%)	0.15 (0.0002)	Beta	[32]
Duration of illness for non-hospitalized DF (days)	4.4 (1–25)	LogNormal (1.48, 0.09)	[26]
Duration of illness for hospitalized DF (days)	6.4 (2–17)	LogNormal (1.85, 0.11)	[26]
Duration of illness for DHF (days)	8.4 (3–25)	LogNormal (2.12, 0.09)	[26]
Duration of hospitalization for dengue (days)	4.9 (3.3)	LogNormal (1.59, 0.05)	[6]
Disability weight for symptomatic DF	0.197	Point estimate	[34]
Disability weight for DHF	0.545	Point estimate	[34]
Social discount rate	0.03	Point estimate	[35]
Effectiveness of insecticide-treated school uniforms (%)	5–100	Threshold analysis	–
	50 (low) 75 (high) 100 (full)	Scenario analysis	
Number of ambulatory visits for non-hospitalized dengue[a]	4.2 (2.7)	LogNormal (1.43, 0.02)	[6]
Number of ambulatory visits for hospitalized dengue	4.2 (2.0)	LogNormal (1.43, 0.04)	[6]
Number of school days lost for non-hospitalized dengue[a]	4.2 (3.2)	LogNormal (1.43, 0.02)	[6]
Number of school days lost for hospitalized dengue	5.5 (3.4)	LogNormal (1.70, 0.05)	[6]
Number of work days lost for non-hospitalized dengue[a]	4.0 (5.6)	LogNormal (1.39, 0.05)	[6]
Number of work days lost for hospitalized dengue[a]	3.9 (5.0)	LogNormal (1.36, 0.10)	[6]
Cost per ambulatory care visit[b]	15.87	Point estimate	[6]
Minimum daily wage[c] (300 TBH)	9.81	Point estimate	[36]
Daily cost of providing education per student[b]	2.34	Point estimate	[6]
Cost of food for an attendant family member per day of hospitalization[c] (50 THB)	1.63	Point estimate	[26]
Cost of transportation for an attendant family member per clinical visit or day of hospitalization[c] (10 THB)	0.33	Point estimate	[26]
Mark-up cost of impregnation per child per year	0–10	Threshold analysis	–
	2.5 (low) 5 (moderate) 10 (high)	Scenario analysis	

(all costs are in US dollars for the year 2012).
DF = Dengue Fever; DHF = Dengue Hemorrhagic Fever; THB: Thai Baht.
[a] Values correspond to mean values reported for 8 dengue endemic countries [6].
[b] Costs reported for 2005 were adjusted to 2012, using an inflation rate of 3%.
[c] Using an average exchange rate of 1 USD = THB 30.60 for the year 2012 [36].

fever (DHF) were 81.5%, 7.4% and 11.1%, respectively [25]. These clinical outcome probabilities were used for the base case analysis.

The dengue disease burden (incidence, symptomatic cases, hospitalizations, deaths) varies from year to year with the circulating dengue virus strains. The prospective study in Thai schoolchildren reported that there were marked variations in the incidence of dengue infection and the spectrum of illness over the study period. This form of inter-annual variability in the clinical outcomes of dengue infections are unavoidable and may partially explain differences in parameter estimates across different endemic settings. It is not appropriate to use this type of variability, as opposed to other types of uncertainty, in the decision model to draw conclusions on the cost-effectiveness of the intervention. However, it is important to recognize the effect of inter-annual variability of dengue disease on the results. Therefore we examined the potential economic value of the intervention for a high incidence year and a low incidence year, with a dengue incidence rate of 7.9% and 2.2% per year, respectively [25]. During the high incidence year, 54% of the dengue infections were asymptomatic, and of the symptomatic cases, the proportion of non-hospitalized DF, hospitalized DF, and hospitalized DHF were 80.6%, 8.3% and 11.1%, respectively [25]. During the low incidence year, 63.6% of the infections were asymptomatic, and of the symptomatic cases, the proportion of non-hospitalized DF, hospitalized DF, and hospitalized DHF were 87.5%, 0% and 12.5%, respectively [25]. Lastly, the prospective study in Thai schoolchildren looked at the burden of symptomatic dengue infection and reported a mean duration of illness for non-hospitalized DF, hospitalized DF and hospitalized DHF of 4.4, 6.4 and 8.4 days, respectively [26].

The reported case fatality rates for the WHO Southeast Asia Region are around 1% [27,28]. Focal outbreaks have, however, resulted in case fatality rates as high as 3–5% in some countries in the region [28]. It is possible to keep case fatality rates below 1% with early diagnosis and detection of shock [28–30]. Despite a steep increase in incidence and severity of dengue cases, Thailand has steadily reduced the case fatality rate of dengue to less than 0.2% over the past decade and pioneered the clinical management guidelines for DF and DHF [6,29]. Based on the published literature, we used a case fatality rate of 0.0027% for DF patients (non-hospitalized or hospitalized) [31] and 0.15% for DHF patients [32]. DALY calculations considered an average life expectancy of 65.08 years based on the life tables for Thai men and women [33] and an average age of death of 10 years in this age group (5–15 years). We used a disability weight of 0.197 for DF and 0.545 for DHF [34]. DALYs were discounted at annual rate of 3%, as recommended by the World Bank [35]. Valuing a year of health life equally at all ages, we did not apply age weighting.

Estimation of costs

Taking the societal perspective, we considered the direct cost of the intervention and the direct (medical and non-medical) and indirect costs of dengue-related illness over one year. The direct cost of the intervention included the cost of treating pre-fabricated school uniforms with a long-lasting insecticide (permethrin) formulation. Home- and factory-based impregnation methods are commonly used to treat clothing and nets used against vector-born diseases. New microencapsulation technologies, which may improve the residual activity of insecticides on clothing and reduce human exposure to insecticides, are on the horizon [19]. At the time of this analysis, the mark-up cost of insecticide treatment of pre-fabricated school uniforms was unknown in dengue endemic settings. In the randomized control trial in Thailand, the insecticide treatment cost of pre-fabricated school uniforms at an overseas factory using a factory proprietary method was about $5.50 per child, excluding the costs of international shipment and of collection of uniforms from and their distribution to families (Pattamaporn Kittayapong, personal communication). Therefore we ranged the mark-up cost of insecticide treatment of school uniforms between $0–10 per child per year in the threshold analysis and estimated the price points at which the intervention would be cost-effective across different levels of intervention effectiveness using the base values of the other parameters as presented in Table 1.

The published literature provides a detailed account of the economic burden of dengue-related illness on households and health systems. Most of the economic data used in this analysis were derived from a dengue cost-of-illness study in Asian and Latin American countries [6]. We considered the direct medical and non-medical costs of treatment for both ambulatory and hospitalized dengue patients. The cost-of-illness study reported that ambulatory and hospitalized patients averaged 4.2 ambulatory care visits during an illness episode, at an average cost of $15.87 per visit [6]. The costs of hospitalization were estimated by multiplying the mean duration of hospitalization with the cost per inpatient bed day. The mean duration of hospitalization for dengue was 4.9 days at a mean cost of $105.77 per inpatient bed day in Thailand [6]. We also estimated the direct non-medical costs associated with seeking care for dengue-related illness, such as food and transportation costs, and made the same unit cost assumptions with the prospective study in Thai schoolchildren [26]. We assumed that transport would be required for all ambulatory care visits and hospitalizations, and food costs would accrue for each day of hospitalization.

The published literature indicates that dengue-related illness negatively affects school attendance and productivity of household members. We estimated the indirect costs of ambulatory and hospitalized dengue cases. The cost-of-illness study reported that non-hospitalized and hospitalized dengue patients lost on average 4.2 and 5.5 school days, respectively [6]. The economic value of school absences was estimated as the product of the number of school days lost because of dengue-related illness and the daily per capita cost of providing education, which was estimated at $2.34 in Thailand [6]. Household members lost on average 4.0 and 3.9 workdays to care for non-hospitalized and hospitalized dengue patients, respectively. The economic value of lost earnings was conservatively estimated as the product of the average number of workdays lost because of dengue-related illness by the minimum daily wage of $9.81 in Thailand [36]. We excluded the economic value of premature deaths due to inherent difficulty of placing an economic value on death.

Sensitivity analysis

Using scenario analysis, we examined the effects of two most critical parameters, namely intervention effectiveness and intervention cost per child per year, on the potential economic value of the intervention and determined threshold price points at which the intervention would cease to be cost-saving, highly cost-effective or cost-effective, using the base case values presented in Table 1. We also examined how the cost-effectiveness of the intervention varied with another critical parameter, annual dengue incidence rate, to recognize the effect of inter-annual variability in the clinical outcomes of dengue infections on the model outputs. All parameter estimates used to populate the model were obtained from published secondary sources. We performed univariate sensitivity analyses to examine the effect on the ICER of each parameter within its reported range (table 1). To represent uncertainty in parameter estimates obtained from secondary sources, distributions were chosen and fitted to reflect sampling uncertainty associated with their estimation by following standard approaches to distributional assumptions in health economic evaluation [37]. To test the robustness of the cost-effectiveness results, we performed probabilistic sensitivity analyses using a Monte Carlo simulation method where parameters were randomly sampled from their distributions (table 1). We considered three scenarios for intervention effectiveness– low effectiveness at 50%; high effectiveness at 75%; and full effectiveness at 100%– and three scenarios for the mark-up cost of insecticide treatment of pre-fabricated school uniforms per child– a low cost of $2.5, a moderate cost of $5, and a high cost of $10 per year. The results of the probabilistic cost-effectiveness analysis are reported under each of these scenarios.

Results

In the base case analysis, the cost-effectiveness of the intervention was assessed from the societal perspective over one year across varying levels of intervention effectiveness and costs at an average dengue incidence rate of 5.8% per year in school-aged children. Figure 1 presents the threshold price points for the intervention. We found the intervention to be cost-effective (ICER≤$16,440) in all scenarios in which intervention cost per child was $5.3 or less per year and intervention effectiveness was 50% or higher. In fact, the intervention was cost saving (ICER<0) in all scenarios in which intervention cost per child was $2.9 or less per year at intervention effectiveness 50% or higher. The intervention proved to be not cost-effective (ICER>$16,440) in scenarios in which intervention cost per child was more than $5.3 per year and intervention effectiveness was less than 50%. In

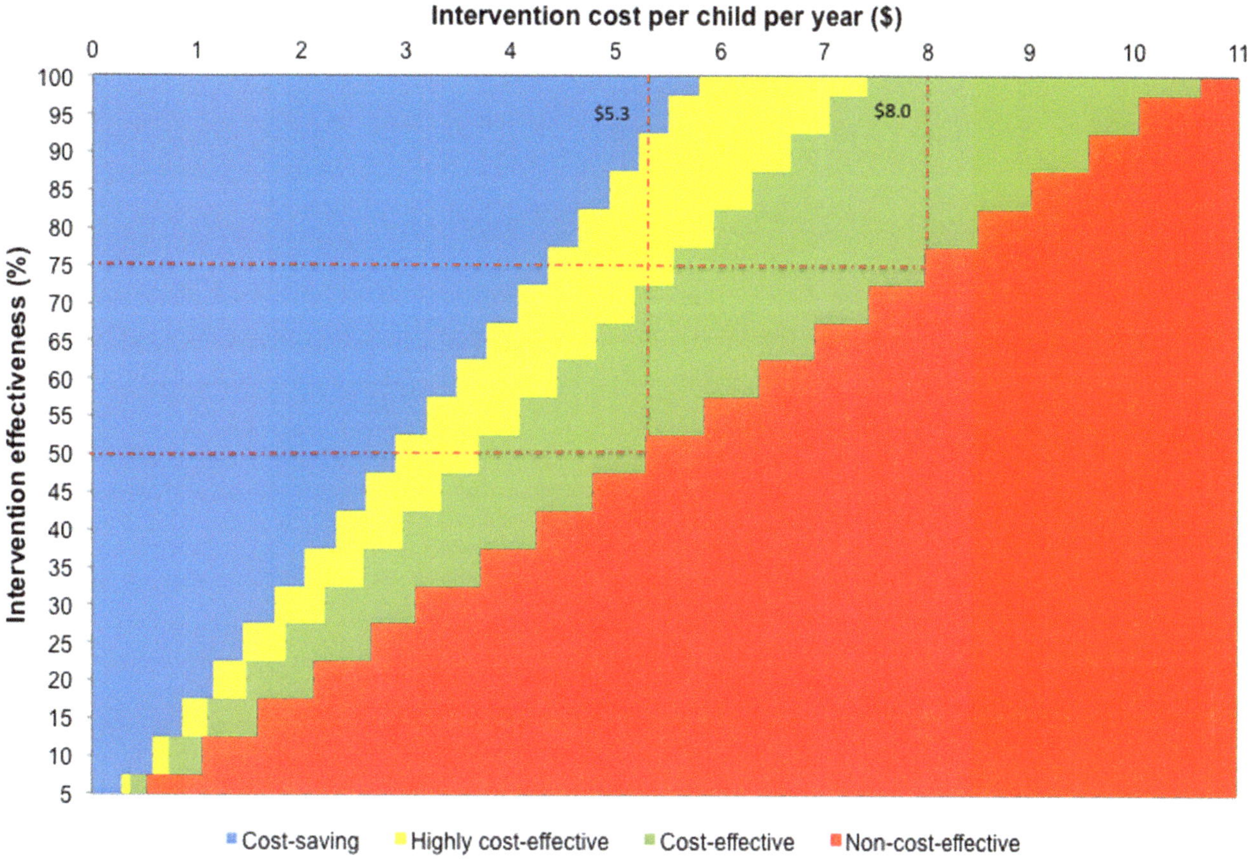

Figure 1. Threshold price points for the use of insecticide-treated uniforms for prevention of dengue in schoolchildren[a] (all costs are in US dollars for the year 2012). [a] Using Thailand's GDP per capita of $5,480 as a threshold value, cost-saving, ICER <0; highly cost-effective, ICER <$5,480; cost-effective, $5,480 ≤ ICER ≤ $16,440; non-cost-effective, ICER> $16,440.

general, the intervention would be of no interest to policy makers if intervention cost per child was $10.6 or higher per year.

Table 2 presents the results of the probabilistic cost-effectiveness analysis for the base case scenario. This table shows how the number of DF (non-hospitalized and hospitalized) and DHF cases averted varied by intervention effectiveness over one year. Under the full (100%) effectiveness scenario, the intervention would have averted on average 22 (95% CI, 19-25) cases of non-hospitalized DF, 2 (95% CI, 2-2) cases of hospitalized DF, and 3 (95% CI, 3-3) cases of DHF in a cohort of 1,000 schoolchildren over one year. At 50% effectiveness, 11 (95% CI, 9-13) cases of non-hospitalized DF, 1 (95% CI, 1-1) hospitalized DF, and 2 (95% CI, 1-2) cases of DHF would have been averted in the same cohort over the same period. Table 2 also shows that the cost of averting a DF or DHF case increased with increasing intervention cost at any given intervention effectiveness, and decreased with increasing intervention effectiveness at any given intervention cost. Cost-offsets included costs of prevented health care utilization related to treatment of dengue illness and costs of prevented productivity losses and school absences, and increased with increasing intervention effectiveness. In scenarios in which intervention effectiveness was 50% or higher and intervention cost is $2.9 or less per child, cost savings were associated with averting a case of DF or DHF, and the intervention was economically dominant over no intervention alternative. At a moderate intervention cost of $5 per child per year, per-case costs of averting non-hospitalized DF, hospitalized DF, and DHF could be as high as $198 (95% CI, 130-208), $2,182 (95% CI, 1,422-3,064), and $1,442 (95% CI, 986-2,018), respectively, while the intervention remained cost-effective for intervention effectiveness between 50%–100%.

The results of the univariate sensitivity analyses are presented in a tornedo diagram in Figure 2. In the tornedo diagram, each bar represents the impact of variation in an individual parameter on the ICER and parameters are arranged from top to bottom in the order of their impact so that comparisons can be done visually. The solid vertical line on the diagram marks the ICER at an intervention cost of $5 and intervention effectiveness of 50% for the base case analysis. Horizontal bars to the left of this line indicate when the intervention is cost-effective, and bars to the right when it is not cost-effective. The results show that ICERs are most sensitive to dengue incidence rate, intervention effectiveness, and intervention cost, followed by duration of hospitalization for dengue and proportion of asymptomatic infections. This is because these parameters directly affect the avertable disease burden of dengue, modulating the incremental costs.

Table 3 presents the results of the probabilistic cost-effectiveness analysis for a low and a high dengue incidence year. The results are presented across varying levels of intervention effectiveness and costs similar to the base case analysis. In a low incidence year, the intervention was non-cost-effective (ICER>$16,440) in all scenarios except when the intervention cost per child was $2.5 or less per year and the intervention effectiveness was 60% or higher. In a high incidence year, the intervention was the economically dominant strategy (ICER<0) in most scenarios in which the

Table 2. Results of the probabilistic cost-effectiveness analysis for the base case scenario with an annual dengue incidence rate of 5.8%.

Intervention effectiveness	Number of non-hospitalized DF cases averted[a] (95% CI)	Cost per non-hospitalized DF cases averted[c] (95% CI)	Number of hospitalized DF cases averted[a] (95% CI)	Cost per hospitalized DF cases averted[c] (95% CI)	Number of DHF cases averted[a] (95% CI)	Cost per DHF case averted[b] (95% CI)	Number of DALYs averted[a] (95% CI)	Cost per DALY averted[c] (ICER) (95% CI)
Intervention cost: $2.5 per child per year								
50%	11 (9−13)	−33 (−69−9)	1 (1−1)	−364 (−766 - 86)	2 (1−2)	−275 (−507−65)	0.14 (0.12−0.18)	***Dominant***
75%	16 (14−19)	−110 (−136−−81)	2 (1−2)	−1,212 (−1,487−−896)	2 (2−3)	−808 (−1,005−−601)	0.22 (0.18−0.27)	***Dominant***
100%	22 (19−25)	−148 (−170−−125)	2 (2−2)	−1,634 (−1,872−−1,388)	3 (3−3)	−1,089 (−1,248−−930)	0.29 (0.24−0.35)	***Dominant***
Intervention cost: $5 per child per year								
50%	11 (9−13)	198 (130−278)	1 (1−1)	2,182 (1,422 - 3,064)	2 (1 - 2)	1,442 (986 - 2,018)	0.14 (0.12 - 0.18)	***14,934 (9,662 – 21,498)***
75%	16 (14 - 19)	44 (−3 - 97)	2 (1 - 2)	484 (−22 - 1,068)	2 (2−3)	319 (−20−695)	0.22 (0.17−0.27)	***3,383 (−319 – 7,520)***
100%	22 (19−25)	−33 (−69−9)	2 (2−3)	−364 (−756 - 88)	3 (3−3)	−243 (−501−31)	0.29 (0.24−0.35)	***Dominant***
Intervention cost: $10 per child per year								
50%	11 (9−13)	659 (524−815)	1 (1−1)	7,285 (5,812 - 9,027)	2 (1−2)	4,879 (3,874 - 6,012)	0.15 (0.12−0.21)	49,480 (37,759 – 63,970)
75%	16 (14−19)	353 (262−457)	2 (1−2)	3,877 (2,887 - 5,029)	2 (2−3)	2,591 (2,919 - 3,335)	0.22 (0.17−0.27)	26,782 (19,342 – 35,487)
100%	22 (19−25)	198 (129−276)	2 (2−2)	2,188 (1,428 - 3,047)	3 (3−3)	1,455 (981 - 2,032)	0.29 (0.23 - 0.36)	***14,937 (9.278 – 21,100)***

(all costs are in US dollars for the year 2012).
DF = Dengue Fever; DHF = Dengue Hemorrhagic Fever; DALY = Disability-Adjusted Life Years; ICER = Incremental Cost-Effectiveness Ratio.
[a] Incremental health outcomes are presented per 1,000 children over one year.
[b] Negative cost values indicate cost-savings.
[c] Boldface font indicates scenarios that were cost-effective (ICER≤$16,440). Underlined entries in boldface font are scenarios in which the intervention was the dominant strategy (i.e. less costly and more effective than no intervention alternative).

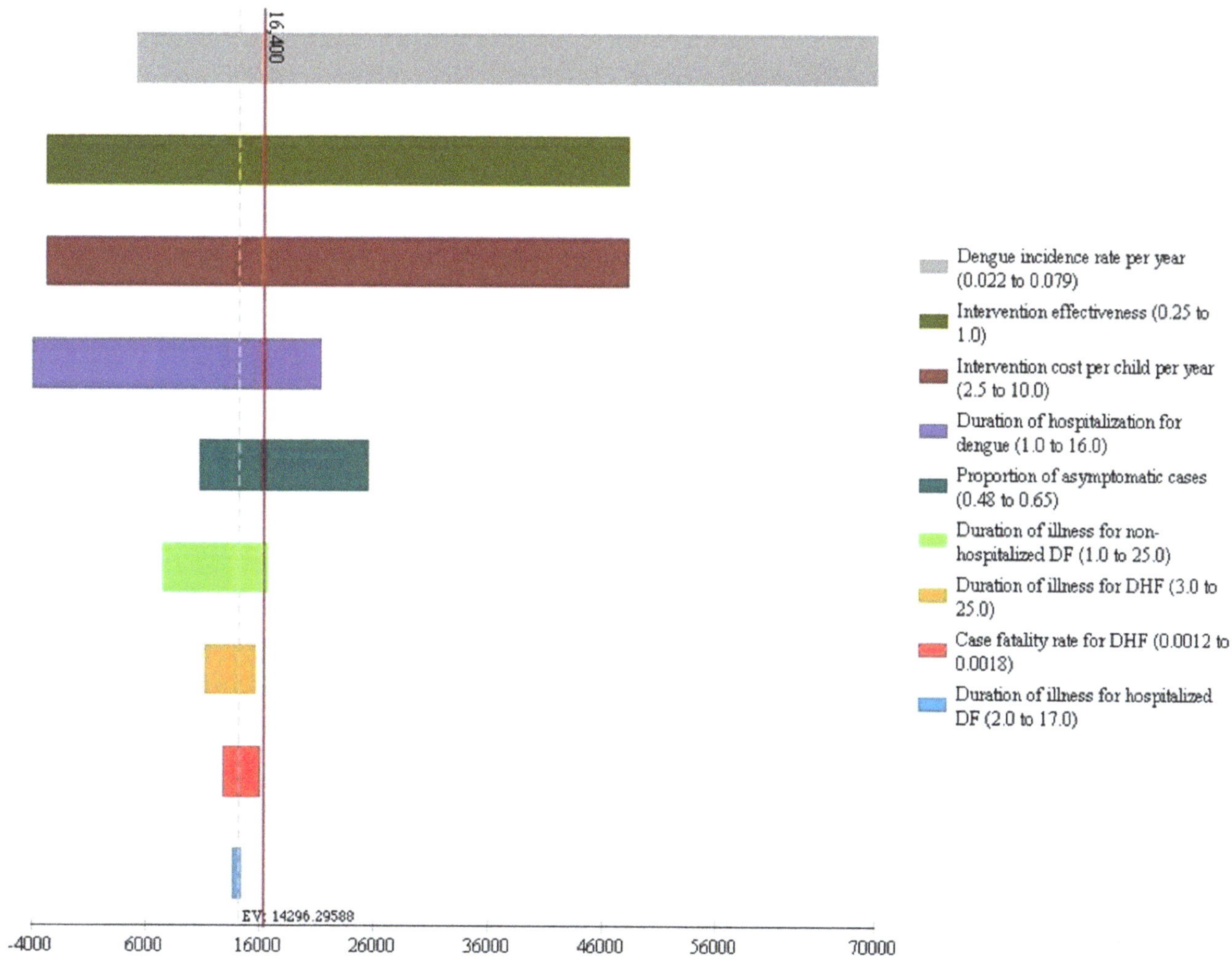

Figure 2. Tornado diagram of univariate sensitivity analyses. DF = Dengue Fever; DHF = Dengue Hemorrhagic Fever.

intervention cost per child was $5 or less per year regardless of intervention effectiveness. The intervention proved to be cost-effective (ICER≤$16.440) even at a high intervention cost of $10 per year when intervention effectiveness was 70% or higher.

Discussion

The use of insecticide-treated school uniforms for prevention of dengue in schoolchildren is a promising intervention given the

Table 3. Incremental cost-effectiveness ratios (ICERs) for low and high dengue incidence years[a].

	Intervention cost per child per year					
	Annual dengue incidence rate: 2.2%			Annual dengue incidence rate: 7.9%		
Intervention effectiveness	$2.5	$5	$10	$2.5	$5	$10
50%	25,301 (21,620–29,207)	70,638 (60,934–81,071)	161,522 (140,176–185,879)	***Dominant***	***5,248 (4,039 - 6,506)***	30,587 (26,415 - 35,843)
75%	***10,190 (8,565–11,995)***	40,550 (34,818–47,129)	101,056 (86,536–116,880)	***Dominant***	***Dominant***	***13,637 (11,721 - 15,744)***
100%	***2,639 (1,620–3,664)***	25,281 (21,913–29,394)	70,742 (60,712–81,514)	***Dominant***	***Dominant***	***5,251 (3,941 - 6,573)***

(all ICERs are in US $ per DALY averted for the year 2012).
[a] Values represent ICERs expressed as cost per DALY averted. Boldface font indicates scenarios that were cost-effective (ICER≤$16,440). Underlined entries in boldface font are scenarios in which the intervention was the dominant strategy (i.e. less costly and more effective than no intervention alternative).

existing body of evidence on the effectiveness of insecticide-treated clothing against vector-born diseases [19]. This modeling-based cost-effectiveness analysis examined the economic value of the intervention in a typical dengue endemic setting from the societal perspective. Cost-effectiveness is only one criterion to assess the merit of an intervention. Although insecticide treated clothing has been used by the military and can be found in recreational markets for personal protection for many years [19], the success of the intervention depends ultimately on the acceptability of insecticide-treated school uniforms by parents and the user compliance rate with the intervention, which would require extensive community mobilization efforts. A limited number of intervention trials reported greater acceptability and higher compliance rates if treated clothing or materials were personal items [19,38,39]. A recent study that took place in the context of the randomized clinical trial showed that the acceptability of the insecticide-treated school uniforms was high among parents, teachers, school principals, reflected by the lack of concern about and willingness to pay for and recommend them [40]. Further studies are needed to monitor the residual activity of insecticides on clothing over time to ensure its operational practicality in real world settings and to establish its long-term safety [19].

Using scenario analysis, we determined the threshold price points for the mark-up cost of insecticide treatment on school uniforms across different levels of intervention effectiveness at which their use would be cost-effective for prevention of dengue in schoolchildren (figure 1). At 50% effectiveness, the intervention cost per child should not exceed $5.3 per year. The threshold price point went up to $8 per child per year at 75% effectiveness. A major source of uncertainty in this analysis is the inter-annual variability of dengue disease patterns. While we showed its impact on the ICERs (table 3), our base case analysis using an average dengue incidence rate of 5.8% per year showed that the intervention could be cost-effective, even cost-saving, under a variety of scenarios (figure 1 and table 2).

Our cost-effectiveness framework can be easily updated as new evidence, particularly on the efficacy, safety, acceptability and cost of the intervention, becomes available from randomized trials, observational studies and systematic reviews. The framework can also be adapted to other dengue endemic settings to inform local decision-making; but policymakers should always contextualize costs and cost-effectiveness benchmarks and assess local disease and health care utilization patterns and other key parameters in their own settings to arrive at more locally representative ICERs. Nevertheless, cost-effectiveness does not necessarily indicate affordability, particularly in developing country settings where health budgets are constrained. The budget impact of the adoption of a new intervention in a specific setting is an important decision criterion for priority-setting in health care along with effectiveness, cost-effectiveness, and severity of illness. Commissioned by the government agency which manages the largest health plan in Thailand, a recent study described the evaluation process of a set of interventions to be included in the universal health coverage benefit package and defined a low budget impact intervention as one that would cost THB 200 million per annum or less to implement [24]. There are currently about 8 million schoolchildren (6–14 years) in Thailand [41]. Based on this budget impact criterion, the use of insecticide-treated school uniforms for prevention of dengue among schoolchildren would not be a candidate for public reimbursement if the mark-up cost of insecticide treatment of school uniforms is more than $1 per child per year.

Increasing severity and frequency of dengue outbreaks and the geographic expansion of dengue transmission within endemic countries pose a serious challenge to vulnerable populations. Children carry the brunt of the disease burden of dengue with high case fatality rates. Previous research has been inconclusive on whether children are more frequently contracting dengue at school or home [25,42–44]. Nevertheless, current prevention programs that mainly target residential areas have had limited impact on the increasing burden of dengue illness. Strategies to protect children against dengue infection at schools remain a neglected research area [43]. There have been calls for practical, acceptable and affordable preventive strategies to protect children living in dengue endemic areas. As we wait for the development of an effective vaccine in the fight against dengue, this model-based economic evaluation study indicates the potential economic value of a newly proposed preventive strategy and highlights the need for additional research on the intervention to policy makers, manufacturers, researchers and other key stakeholders.

Conclusions

Current dengue control efforts failed to address the increasing disease and economic burden of dengue in endemic countries. Practical, acceptable and affordable preventive strategies are needed to protect children at risk of dengue infection. Our results present the potential economic value of the use of insecticide-treated uniforms for prevention of dengue in school children in a typical dengue endemic setting and highlight the urgent need for additional research on this intervention.

Author Contributions

Conceived and designed the experiments: YT. Performed the experiments: YT PR. Analyzed the data: YT. Contributed reagents/materials/analysis tools: YT PR VL PK AWS. Wrote the paper: YT. Edited and commented on the manuscript: ASW PR VL PK.

References

1. Bhatt S, Gething PW, Brady OJ, Messina JP, Farlow AW, et al. (2013) The global distribution and burden of dengue. Nature 496: 504–507.
2. Guzman MG, Halstead SB, Artsob H, Buchy P, Farrar J, et al. (2010) Dengue: a continuing global threat. Nat Rev Micro.
3. Gubler D (2002) The global emergence/resurgence of arboviral diseases as public health problems. Arch Med Res 33: 330–342.
4. Torres JR, Castro J (2007) The health and economic impact of dengue in Latin America. Cadernos de Saúde Pública 23: S23–S31.
5. Gubler DJ (2004) The changing epidemiology of yellow fever and dengue, 1900 to 2003: full circle? Comp Immunol Microbiol Infect Dis 27: 319–330.
6. Suaya JA, Shepard DS, Siqueira JB, Martelli CT, Lum LC, et al. (2009) Cost of dengue cases in eight countries in the Americas and Asia: a prospective study. Am J Trop Med Hyg 80: 846–855.
7. Shepard DS, Coudeville L, Halasa YA, Zambrano B, Dayan GH (2011) Economic Impact of Dengue Illness in the Americas. The American Journal of Tropical Medicine and Hygiene 84: 200–207.
8. Siqueira JJ, Martelli CMT, Coelho GE, da Rocha Simplício AC, Hatch DL (2005) Dengue and Dengue Hemorrhagic Fever, Brazil, 1981–2002 Emerg Infect Dis [serial on the Internet].
9. Tomashek KM, Gregory CJ, Rivera Sánchez A, Bartek MA, Garcia Rivera EJ, et al. (2012) Dengue Deaths in Puerto Rico: Lessons Learned from the 2007 Epidemic. PLoS Negl Trop Dis 6: e1614.
10. Guzmán MG, Kourí G, Valdés L, Bravo J, Vázquez S, et al. (2002) Enhanced severity of secondary dengue-2 infections: death rates in 1981 and 1997 Cuban outbreaks. Pan Am J Public Health 11: 223–227.
11. Stahl H-C, Butenschoen V, Tran H, Gozzer E, Skewes R, et al. (2013) Cost of dengue outbreaks: literature review and country case studies. BMC Public Health 13: 1048.
12. WHO (1995) Global Strategy for Prevention and Control of Dengue Fever and Dengue Hemorrhagic Fever: report of the consultation on key issues in dengue vector control towards the operationalization of a global strategy. Geneva.

13. Horstick O, Runge-Ranzinger S, Nathan MB, Kroeger A (2010) Dengue vector-control services: how do they work? A systematic literature review and country case studies. Transactions of The Royal Society of Tropical Medicine and Hygiene 104: 379–386.
14. Dick OB, San Martín JL, Montoya RH, del Diego J, Zambrano B, et al. (2012) The History of Dengue Outbreaks in the Americas. The American Journal of Tropical Medicine and Hygiene 87: 584–593.
15. Guzman A, Istúriz RE (2010) Update on the global spread of dengue. International Journal of Antimicrobial Agents 36, Supplement 1: S40–S42.
16. Hawley WA (1988) The biology of Aedes albopictus. J Am Mosq Control Assoc Suppl 1: 1–39.
17. Wilder-Smith A, Lover A, Kittayapong P, Burnham G (2011) Hypothesis: Impregnated school uniforms reduce the incidence of dengue infections in school children. Med Hypotheses 76: 861–862.
18. Wilder-Smith A, Byass P, Olanratmanee P, Maskhao P, Sringernyuang L, et al. (2012) The impact of insecticide-treated school uniforms on dengue infections in school-aged children: study protocol for a randomised controlled trial in Thailand. Trials 13: 212.
19. Banks SD, Murray N, Wilder-Smith A, Logan JG (2014) Insecticide-treated clothes for the control of vector-borne diseases: a review on effectiveness and safety. Med Vet Entomol.
20. Wilder-Smith A, Renhorn KE, Tissera H, Abu Bakar S, Alphey L, et al. (2012) DengueTools: innovative tools and strategies for the surveillance and control of dengue. Glob Health Action 5.
21. Massad E, Amaku M, Coutinho FA, Kittayapong P, Wilder-Smith A (2013) Theoretical impact of insecticide-impregnated school uniforms on dengue incidence in Thai children. Glob Health Action 6: 20473.
22. Drummond M, Stoddart G, Torrance G (2005) Methods for the Economic Evaluation of Health Programs. New York: Oxford University Press.
23. Wilder-Smith A, Byass P, Olanratmanee P, Maskhao P, Sringernyuang L, et al. (2012) The impact of insecticide-treated school uniforms on dengue infections in school-aged children: study protocol for a randomised controlled trial in Thailand. Trials 13: 212.
24. Youngkong S, Baltussen R, Tantivess S, Mohara A, Teerawattananon Y (2012) Multicriteria Decision Analysis for Including Health Interventions in the Universal Health Coverage Benefit Package in Thailand. Value in Health 15: 961–970.
25. Endy TP, Chunsuttiwat S, Nisalak A, Libraty DH, Green S, et al. (2002) Epidemiology of inapparent and symptomatic acute dengue virus infection: a prospective study of primary school children in Kamphaeng Phet, Thailand. Am J Epidemiol 156: 40–51.
26. Anderson KB, Chunsuttiwat S, Nisalak A, Mammen MP, Libraty DH, et al. (2007) Burden of symptomatic dengue infection in children at primary school in Thailand: a prospective study. Lancet 369: 1452–1459.
27. WHO Regional Office for South-East Asia (2011) Comprehensive Guidelines for Prevention and Control of Dengue and Dengue Haemorrhagic Fever. Geneva.
28. WHO (2009) Dengue - Guidelines for diagnosis, treatment, prevention and control. Geneva: World Health Organization.
29. Kalayanarooj S (1999) Standardized Clinical Management: Evidence of Reduction of Dengue Haemorrhagic Fever Case-Fatality Rate in Thailand. Dengue Bulletin 23: 10–17.
30. Monath TP (1994) Dengue: the risk to developed and developing countries. Proc Natl Acad Sci U S A 91: 2395–2400.
31. Lee BY, Connor DL, Kitchen SB, Bacon KM, Shah M, et al. (2011) Economic value of dengue vaccine in Thailand. Am J Trop Med Hyg 84: 764–772.
32. WHO SEARO.
33. WHO Global Health Observatory Data Repository: Life expectancy.
34. Murray CJL, Lopez AD (1996) The Global Burden of Disease, vol. 1.; Health HSoP, editor. Cambridge, MA: Harvard University Press.
35. World Bank (1993) World Development Report: Investing in Health. Washington, DC.
36. Office of the Prime Minister In: Thailand, editor. Thailand.
37. Briggs A, Claxton K, Sculpher M (2006) Decision Modelling for Health Economic Evaluation. New York: Oxford University Press Inc.
38. Kimani E, Vulule J, Kuria I, Mugisha F (2006) Use of insecticide-treated clothes for personal protection against malaria: a community trial. Malaria Journal 5: 63.
39. Rowland M, Durrani N, Hewitt S, Mohammed N, Bouma M, et al. (1999) Permethrin-treated chaddars and top-sheets: appropriate technology for protection against malaria in Afghanistan and other complex emergencies. Trans R Soc Trop Med Hyg 93: 465–472.
40. Murray N, Jansarikij S, Olanratmanee P, Maskhao P, Soures A, et al. (Forthcoming 2014) Acceptability of impregnated school uniforms for dengue control on Thailand: a mixed methods approach. Global Health Action.
41. UNESCO Institute for Statistics.
42. Endy TP, Nisalak A, Chunsuttiwat S, Libraty DH, Green S, et al. (2002) Spatial and Temporal Circulation of Dengue Virus Serotypes: A Prospective Study of Primary School Children in Kamphaeng Phet, Thailand. American Journal of Epidemiology 156: 52–59.
43. García-Rejón JE, Loroño-Pino MA, Farfán-Ale JA, Flores-Flores LF, López-Uribe MP, et al. (2011) Mosquito Infestation and Dengue Virus Infection in Aedes aegypti Females in Schools in Mérida, México. The American Journal of Tropical Medicine and Hygiene 84: 489–496.
44. Mammen MP Jr, Pimgate C, Koenraadt CJM, Rothman AL, Aldstadt J, et al. (2008) Spatial and Temporal Clustering of Dengue Virus Transmission in Thai Villages. PLoS Med 5: e205.

7

Discovery and Characterization of a Potent and Selective Inhibitor of *Aedes aegypti* Inward Rectifier Potassium Channels

Rene Raphemot[1,2], Matthew F. Rouhier[3], Daniel R. Swale[1], Emily Days[4], C. David Weaver[2,4], Kimberly M. Lovell[2,6], Leah C. Konkel[2,6], Darren W. Engers[2,6], Sean F. Bollinger[2,6], Corey Hopkins[2,5,6], Peter M. Piermarini[3]*, Jerod S. Denton[1,2,4,5]*

1 Department of Anesthesiology, Vanderbilt University Medical Center, Nashville, TN, United States of America, **2** Department of Pharmacology, Vanderbilt University School of Medicine, Nashville, TN, United States of America, **3** Department of Entomology, Ohio Agricultural Research and Development Center, The Ohio State University, Wooster, OH, United States of America, **4** Institute of Chemical Biology, Vanderbilt University School of Medicine, Nashville, TN, United States of America, **5** Institute for Global Health, Vanderbilt University, Nashville, TN, United States of America, **6** Department of Chemistry, Vanderbilt University School of Medicine, Nashville TN, United States of America

Abstract

Vector-borne diseases such as dengue fever and malaria, which are transmitted by infected female mosquitoes, affect nearly half of the world's population. The emergence of insecticide-resistant mosquito populations is reducing the effectiveness of conventional insecticides and threatening current vector control strategies, which has created an urgent need to identify new molecular targets against which novel classes of insecticides can be developed. We previously demonstrated that small molecule inhibitors of mammalian Kir channels represent promising chemicals for new mosquitocide development. In this study, high-throughput screening of approximately 30,000 chemically diverse small-molecules was employed to discover potent and selective inhibitors of *Aedes aegypti* Kir1 (*Ae*Kir1) channels heterologously expressed in HEK293 cells. Of 283 confirmed screening 'hits', the small-molecule inhibitor VU625 was selected for lead optimization and in vivo studies based on its potency and selectivity toward *Ae*Kir1, and tractability for medicinal chemistry. In patch clamp electrophysiology experiments of HEK293 cells, VU625 inhibits *Ae*Kir1 with an IC_{50} value of 96.8 nM, making VU625 the most potent inhibitor of *Ae*Kir1 described to date. Furthermore, electrophysiology experiments in *Xenopus* oocytes revealed that VU625 is a weak inhibitor of *Ae*Kir2B. Surprisingly, injection of VU625 failed to elicit significant effects on mosquito behavior, urine excretion, or survival. However, when co-injected with probenecid, VU625 inhibited the excretory capacity of mosquitoes and was toxic, suggesting that the compound is a substrate of organic anion and/or ATP-binding cassette (ABC) transporters. The dose-toxicity relationship of VU625 (when co-injected with probenecid) is biphasic, which is consistent with the molecule inhibiting both *Ae*Kir1 and *Ae*Kir2B with different potencies. This study demonstrates proof-of-concept that potent and highly selective inhibitors of mosquito Kir channels can be developed using conventional drug discovery approaches. Furthermore, it reinforces the notion that the physical and chemical properties that determine a compound's bioavailability in vivo will be critical in determining the efficacy of Kir channel inhibitors as insecticides.

Editor: Joseph Clifton Dickens, United States Department of Agriculture, Beltsville Agricultural Research Center, United States of America

Funding: This work was funded by a grant from the Foundation for the National Institutes of Health through the Vector-Based Transmission of Control: Discovery Research (VCTR) program of the Grand Challenges in Global Health initiative. This work was also supported in part by a grant from the National Institute of Diabetes and Digestive and Kidney Diseases (1R01DK082884; JSD). The funders had no role in study design, data collection and analysis, decision to publish, or preparation of the manuscript.

Competing Interests: The authors have declared that no competing interests exist.

* Email: piermarini.1@osu.edu (PMP); jerod.s.denton@vanderbilt.edu (JSD)

Introduction

Mosquitoes are vectors of protozoan, filarial nematode, and viral pathogens that cause numerous human diseases, including malaria, lymphatic filariasis, and dengue fever. These diseases impose an enormous burden on global health and profoundly impair socioeconomic advancement in developing countries [1]. The overuse of a limited number of insecticides has led to the emergence of insecticide-resistant populations of mosquitoes, which is hampering the effectiveness of vector control efforts [2,3,4]. Consequently, there is a need to identify new molecular targets against which insecticides can be developed and deployed.

An emerging body of evidence from our group supports the idea that inward rectifier potassium (Kir) channels represent viable targets for insecticide development [5,6,7]. Kir channels are tetrameric proteins that conduct K^+ ions across the cell membrane and thereby generate an ionic current that underlies various cellular functions. Dipteran insects possess three major Kir channel subtypes, denoted Kir1, Kir2 and Kir3. In *Drosophila melanogaster*, there are three genes that encode Kir channels

(*Dr*Kir1, *Dr*Kir2, *Dr*Kir3), which play important roles in osmoregulation, immunity, and development [8,9,10,11]. In *Aedes aegypti*, there are five Kir channel genes (*AeKir1, AeKir2A, AeKir2B, AeKir2B'* and *AeKir3*), which are expressed in various body segments and tissues such as the carcass (thorax and abdomen), head, Malpighian tubules, midgut, and hindgut [6,12]. We showed previously in vitro that the *A. aegypti* Kir1 (*Ae*Kir1) channel mediates strong inward rectifying K^+ currents that are blocked by barium and the small molecule inhibitors, VU573 and VU590 [7,12,13]. Moreover, a hemolymph injection of either VU573 or VU590 inhibits the excretion of urine by adult female mosquitoes, leads to abdominal bloating, and incapacitates mosquitoes within 24 h [5].

Taken together, the above studies indicate that Kir channels represent promising molecular targets for insecticides that have a novel mechanism of action by disrupting the renal-dependent regulation of extracellular fluid homeostasis (i.e., renal failure). However, in mammals, Kir channels regulate the electrical excitability of neurons and cardiac cells, hormone secretion, and transport of K^+ ions across epithelial tissues of the kidney and gut [14]. Missense mutations that perturb the activity of Kir channels cause human diseases of the heart, nervous system, pancreas, and kidney [15,16,17]. Thus, efforts aimed at developing insecticides to target Kir channels must verify that lead compounds do not perturb the functions of mammalian Kir channels.

As such, the above 'tool' compounds VU573 and VU590 allowed us to establish proof-of-concept, but are not suitable for insecticide development, in part, because they inhibit mammalian Kir channels with greater potency than *Ae*Kir1 [18,19]. Here, we aim to discover new chemical probes of *Ae*Kir1 channels that exhibit improved potency and selectivity compared to the tool compounds by optimizing and validating an existing fluorescent thallium (Tl^+) flux-based assay of *Ae*Kir1 function [5] for high-throughput screening (HTS) of small molecule libraries. Screening approximately 30,000 small molecules from the chemical library of the Vanderbilt Institute of Chemical Biology (VICB) resulted in the identification of 283 compounds with activity against *Ae*Kir1 channels. We focus on the in vitro and in vivo activity of one of these compounds, N-(3-methoxyphenyl)-2-methyl-1-propionylindoline-5-sulfonamide (VU625), which exhibits nanomolar affinity and is highly selective for *Ae*Kir1 over mammalian Kir channels.

Materials and Methods

Tl^+ flux assays

Tl^+ flux assays were performed essentially as described previously [13,18,19]. Briefly, stably transfected T-Rex-HEK-293 cells expressing *Ae*Kir1 channels were cultured overnight in 384-well plates (20,000 cells/20 µL/well black-walled, clear-bottomed BD PureCoat amine-coated plates (BD, Bedford, MA) with a plating media containing DMEM, 10% dialyzed FBS and 1 µg/mL tetracycline. The next day, the cell culture medium was replaced with a dye-loading solution containing assay buffer (Hanks Balanced Salt Solution with 20 mM HEPES, pH 7.3), 0.01% (*w/v*) Pluronic F-127 (Life Technologies, Carlsbad, CA), and 1.2 µM of the thallium-sensitive dye Thallos-AM (TEFlabs, Austin, TX). Following 1 hr incubation at room temperature, the dye-loading solution was washed from the plates and replaced with 20 µL/well of assay buffer.

The plates were transferred to a Hamamatsu Functional Drug Screening System 6000 (FDSS6000; Hamamatsu, Hamamatsu (or Bridgewater, NJ), Japan) where 20 µL/well of test compounds in assay buffer (as prepared below) were added and allowed to incubate with the cells for 20 min. After the incubation period, a baseline recording was collected at 1 Hz for 10 s (excitation 470±20 nm, emission 540±30 nm) followed by a Tl^+ stimulus buffer addition (10 µL/well) and data collection for an additional 4 min. The Tl^+ stimulus buffer contains in (mM) 125 $NaHCO_3$, 1.8 $CaSO_4$, 1 $MgSO_4$, 5 glucose, 12 Tl_2SO_4, 10 HEPES, pH 7.4. For Tl^+ flux assays on the mammalian channels Kir2.x, Kir4.1 and Kir6.2/SUR1 expressing cells, the Tl^+ stimulus buffer contained 1.8 mM Tl_2SO_4. Also, Tl^+ flux assays on Kir3.1/3.2/mGlu8 expressing cell, required addition of an EC_{80} concentration of glutamate (Sigma-Aldrich, St. Louis, MO) with the Tl^+ stimulus buffer [19].

The test compounds were transferred to daughter polypropylene 384-well plates (Greiner Bio-One, Monroe, NC) using an Echo555 liquid handler (Labcyte, Sunnyvale, CA), and then diluted into assay buffer to generate a 2X stock in 0.6% DMSO (0.3% final). For Tl^+ flux assays on Kir6.2/SUR1 expressing cells, test compounds were diluted in assay buffer containing diazoxide (250 µM final) to induce channel activation [20]. Concentration-response curves (CRCs) were generated by screening compounds at 3-fold dilution series in 4-point (1 µM–30 µM) or 11-point (1 nM–30 µM) CRCs.

Tl^+ flux data were analyzed as previously described [19,21,22] using a combination of Excel (Microsoft Corp, Redmond, WA) with XLfit add-in (IDBS, Guildford, Surrey, UK) and OriginPro (OriginLab, Northampton, MA) software. Raw data were opened in Excel and each data point in a given trace was divided by the first data point from that trace (static ratio) followed by subtraction of data points from control traces generated in the presence of vehicle controls. The slope of the fluorescence increase beginning 5 s after Tl^+ addition and ending 15 s after Tl^+ addition was calculated.

Compound synthesis

2,2,2-trifluoro-1-(2-methylindolin-1-yl)ethan-1-one. The reagents and conditions are illustrated in Figure S1. To a round bottom flask equipped with a magnetic stir bar, 2-methylindoline (4.8 mL, 37 mmol, 1 eq.) and pyridine (46 mL) were added. The reaction mixture was cooled to 0°C and trifluoroacetic anhydride (6.3 mL, 44 mmol, 1.2 eq.) was added dropwise. The reaction mixture was allowed to warm to room temperature and was stirred an additional 2 hours. The reaction was quenched with water (50 mL) and diluted with DCM (100 mL). The organic layer was separated and washed subsequently with water (50 ml) and brine (50 mL), dried over Na_2SO_4, and concentrated under reduced pressure. The crude material (8.33g, 98%) was used without purification. LCMS: $R_T = 0.785$ min, $[M+H]^+ = 229.6$; >98%.

2-methyl-1-(2,2,2-trifluoroacetyl)indoline-5-sulfonyl chloride: Chlorosulfonic acid (22 mL, 330 mmol, 9 equiv.) was added to a 100 mL round bottom flask equipped with a reflux condensor, and cooled to 0°C. To this, 2,2,2-trifluoro-1-(2-methylindolin-1-yl)ethan-1-one, (8.5 g, 37 mmol, 1 eq.) was added dropwise. The reaction mixture was removed from the ice bath. The vial was heated to 40°C for 1 hour. The reaction was subsequently cooled to room temperature and PCl_5 (7.7 g, 37 mmol, 1 equiv.) was added slowly. After gas evolution ceased, the reaction mixture was heated to 80°C for 1 hour. The reaction mixture was cooled to room temperature and then placed in an ice bath. Water was added very slowly to the reaction mixture. Subsequently, DCM was added and the reaction was filtered through a phase separator. The organic layer was concentrated under reduced pressure and used without subsequent purification (6.46 g, 53%).

N-(3-methoxyphenyl**)-2-methylindoline-5-sulfonamide**: 2-methyl-1-(2,2,2-trifluoroacetyl)indoline-5-sulfonyl chloride (2.5 g,

7.6 mmol, 1 eq.) was diluted with DCM (10 mL). 3-methoxyaniline (1.71 mL, 15.2 mmol, 2 eq.) followed by *N,N*-Diisopropylethylamine (5.3 mL, 31 mmol, 4 eq.) was added to the reaction. Reaction progress was monitored by LCMS. Once the reaction was deemed complete, it was diluted with DCM (40 mL) and washed with water (2x, 50 mL) and brine (50 mL). The organic layer was dried over Na_2SO_4 and concentrated under reduced pressure. Purification by flash chromatography (0%–100% EtOAc in Hexanes) afforded the desired product (2.66 g, 85%). LCMS: $R_T = 0.800$ min., $[M+H]^+ = 414.7$; >98% @ 220 and @ 254 nm. The trifluoroacetate was removed by stirring in a 1:1:1 mixture of MeOH, THF, and 10% NaOH affording the title compound (782 mg, 38%). LCMS: $R_T = 0.665$ min., $[M+H]^+ = 318.8$; >98% @ 220 and @ 254 nm.

N-(3-methoxyphenyl)**-2-methyl-1-propionylindoline-5-sulfonamide (VU0077625)**: N-(3-methoxyphenyl)-2-methylindoline-5-sulfonamide (11 mg, 0.035 mmol, 1 eq.) was diluted with DCM (0.3 mL). To this reaction, pyridine was added (0.011 mL, 0.14 mmol, 4 eq.) followed by propionyl chloride (0.003 mL, 0.05 mmol, 1.5 eq.). Reaction progress was monitored by LCMS. Once the reaction was deemed complete it was concentrated under forced air and heat and was subsequently purified on preparative HPLC (3 mg, 26%). ^1H NMR (400.1 MHz, $CDCl_3$) δ ppm): 8.17 (bs, 1 H); 7.67 (dd, *J* = 1.69, 8.72 Hz, 1 H); 7.58 (s, 1 H); 7.16 (t, *J* = 8.25 Hz, 1 H); 6.69–6.57 (m, 4 H); 4.58 (bs, 1 H); 3.74 (s, 3 H); 3.38–3.32 (m, 1 H); 2.66–2.47 (m, 4 H); 1.29–1.22 (m, 5 H). HRMS (TOF, ES^+) $C_{19}H_{23}N_2O_4S$ $[M+H]^+$ calc'd for 375.1379, found 375.1381.

Patch clamp electrophysiology

T-REx-HEK293-*Ae*Kir1 cells were voltage clamped in the whole-cell configuration of the patch clamp technique after overnight induction with tetracycline (1 µg/ml) essentially as described earlier [5]. Briefly, patch electrodes were pulled from silanized 1.5 mm outer diameter borosilicate microhematocrit tubes using a Narishige PC-10 two-stage puller. Electrode resistance ranged from 3.5 to 5.5 MΩ when filled with the following intracellular solution (in mM): 135 KCl, 2 $MgCl_2$, 1 EGTA, 10 HEPES free acid, 2 Na_2ATP (Roche, Indianapolis, IN), pH 7.3, 275 mOsm. The standard bath solution contained (in mM): 135 NaCl, 5 KCl, 2 $CaCl_2$, 1 $MgCl_2$, 5 glucose, 10 HEPES free acid, pH 7.4, 290 mOsm. Whole-cell currents were recorded under voltage-clamp conditions using an Axopatch 200B amplifier (Molecular Devices, Sunnyvale, CA). Electrical connections to the amplifier were made using Ag/AgCl wires and 3 M KCl/agar bridges. Electrophysiological data were collected at 5 kHz and filtered at 1 kHz. Data acquisition and analysis were performed using pClamp 9.2 software (Axon Instruments). All recordings were made at room temperature (20–23°C).

Heterologous expression of *Ae*Kir1 and *Ae*Kir2B in Xenopus oocytes

*Ae*Kir1 and *Ae*Kir2B channels were expressed heterologously in *Xenopus laevis* oocytes as described previously [6]. In brief, defolliculated *Xenopus* oocytes (purchased from Ecocyte Bioscience, Austin, TX) were injected with 10 ng (0.35 ng/nL) of either *Ae*Kir1 or *Ae*Kir2B cRNA and cultured for 3–7 days in OR3 media at 18°C. Oocytes injected with 28 nl of nuclease-free H_2O served as controls.

Electrophysiology of Xenopus oocytes

All electrophysiological experiments on *Xenopus* oocytes were performed at room temperature. The compositions of the solutions used in these experiments are shown in Table 1. When present, VU625 was dissolved in solution *III* or solution *V* to a final concentration of 0.1, 1, 5, 15, or 50 µM (0.05% DMSO). All solutions were delivered by gravity to a RC-3Z oocyte chamber (Warner Instruments, Hamden, CT) via polyethylene tubing at a flow rate of ~2 ml/min. Solution changes were made with a Rheodyne Teflon 8-way Rotary valve (Model 5012, Rheodyne, Rohnert Park, CA).

Electrophysiological recordings from oocytes were conducted as described previously [6] In brief, each oocyte was transferred to the holding chamber under superfusion with solution *I* and impaled with two conventional-glass microelectrodes backfilled with 3 M KCl (resistances of 0.5–1.5 MΩ) to measure membrane potential (V_m) and whole-cell membrane current (I_m), respectively. Current-voltage (I–V) relationships of oocytes were acquired as described previously [6]. In brief, the oocytes were subjected to a voltage-stepping protocol consisting of 20 mV steps from −140 mV to +40 mV (100 ms each). After the conclusion of the voltage-stepping protocol, the clamp was turned off and a new solution was superfused through the chamber for ~90 s before acquiring another I–V relationship. All V_m and I_m values were recorded by a Digidata 1440A Data Acquisition System (Molecular Devices) and the Clampex module of pCLAMP. The I–V plots were generated using the Clampfit module of pCLAMP.

To evaluate the inhibition of *Ae*Kir1 and *Ae*Kir2B activity by VU625, we focused on the maximal inward currents elicited by the voltage-stepping protocol, which occur at a voltage of −140 mV. For *Ae*Kir1 oocytes, the background, inward currents in solution *II* (i.e., 0.5 mM K^+) were subtracted from those in 1) solution *III* (i.e., 10 mM K^+) to calculate the total inward current for an oocyte before exposure to VU625 (I_A), and 2) solution *III* with VU625 to calculate the inward current after exposure to the small molecule (I_B). The percent inhibition of the inward current was calculated by subtracting I_B from I_A and then dividing by I_A. For *Ae*Kir2B oocytes, a similar protocol was followed and similar calculations were made, except solution *IV* replaced solution *II* and solution *V* replaced solution *III*.

Mosquito colony

The *Aedes aegypti* mosquito colony used in the present study is identical to that described previously [6]. As before, only adult female mosquitoes 3–10 days post emergence were utilized for experiments.

Mosquito toxicology experiments

Adult female mosquitoes for injection were anesthetized on ice and impaled through the metapleuron using a pulled-glass capillary attached to a nanoliter injector (Nanoject II, Drummond Scientific Company, Broomall, PA). Each mosquito received a single hemolymph injection of 69 nL of solution. The injection solution consisted of a potassium-rich phosphate buffered saline (K^+-PBS), 15% DMSO, 1% β-cyclodextrin, 0.1% Solutol, and a concentration of VU625 to deliver the doses indicated. In experiments where probenecid was used, water-soluble probenecid (Biotium, Hayward CA) was included in the injection solution at 50 mM, thereby providing a dose of 3.4 nmol per mosquito.

The K^+-PBS solution consisted of the following in mM: 92.2 NaCl, 47.5 KCl, 10 Na_2HPO_4, and 2 KH_2PO_4 (pH 7.5). A total of 10 mosquitoes were injected for a given treatment or dose, and then were placed into small cages within a rearing chamber (28°C, 80% relative humidity, 12:12 light:dark) and allowed free access to a solution of 10% sucrose. The mosquitoes were observed at 24 hr after injection. For each treatment, 3–7 replicates of 10 mosquitoes each were performed.

Table 1. Compositions (in mM) of solutions used in *Xenopus* oocyte electrophysiology.

Solution #	I	II	III	IV	V
NaCl	96	88.5	88.5	73.5	73.5
NMDG-Cl	0	9.5	0	24.5	0
KCl	2	0.5	10	0.5	25
$MgCl_2$	1.0	1.0	1.0	1.0	1.0
$CaCl_2$	1.8	1.8	1.8	1.8	1.8
HEPES	5	5	5	5	5

The pH of all solutions was adjusted to 7.5 with NMDG-OH.
The osmolality of each solution was verified to be 190 mOsm kg^{-1} H_2O (±5 mOsm kg^{-1} H_2O) by vapor pressure osmometry.
NMDG = N-methyl-D-glucamine.

Mosquito excretion experiments

The excretory capacity of mosquitoes was measured as described [6]. In brief, after anesthetizing mosquitoes on ice, their hemolymph was injected as described above with 900 nL of a K^+-PBS vehicle containing 1.15% DMSO, 0.077% β-cyclodextrin, and 0.008% Solutol, or the vehicle containing VU625 (0.77 mM) to deliver a dose of 690 pmol of VU625 per mosquito. In experiments where probenecid was used, the vehicle was supplemented with water- soluble probenecid (3.08 mM) to deliver a dose of 3.4 nmol of probenecid per mosquito. After injection, the mosquitoes were placed immediately in a graduated, packed-cell volume tube (MidSci, St. Louis, MO; 5 mosquitoes per tube) and held at 28°C. The volume of urine excreted at 60 min post injection was measured as described previously [6], and all mosquitoes were confirmed to be alive at the end of 60 min period. For each treatment, 6–18 independent trials of 5 mosquitoes per treatment were performed.

Statistical analyses

Tl+ flux assay. The Z′ value was calculated as described earlier [21], using the following formula:

$$Z' = 1 - (3SD_p + 3SD_n)/|Mean_p + Mean_n|$$

where SD is standard deviation, p and n are vehicle control and compound inhibited flux values respectively.

To compare the effect of DMSO on *Ae*Kir1-mediated Tl^+ flux, a one-way ANOVA was performed with a Tukey's multiple comparison test. Prism software (GraphPad Software) was used to generate CRC from Tl^+ flux. Half-inhibition concentration (IC_{50}) values were calculated from fits using a four parameter logistic equation.

Mosquito toxicology and urine excretion. Prism (GraphPad Software) was used to generate a dose-response curve for the toxicity of VU625; the doses (x-axis) were first log transformed and then the mortality data was normalized using Prism, where the smallest value and largest values in a data set equal '0%' and '100%', respectively. The data were then fitted using a 'biphasic' algorithm (<100 constraint) to calculate potencies (ED_{25} and ED_{75} values). To compare 1) the toxic effects among the vehicle, probenecid, VU625, and VU625 + probenecid treatments, and 2) the excretory capacity among the vehicle, probenecid, VU625, and VU625 + probenecid treatments, one-way ANOVAs were performed with Newman-Keuls posttests.

Results

Discovery of novel *Ae*Kir1 inhibitors via HTS

In an effort to discover mosquito-specific inhibitors of *Ae*Kir1, we optimized a Tl^+ flux assay for HTS of large libraries of chemically diverse small molecules. The assay utilizes a monoclonal T-REx-HEK293 cell line that expresses *Ae*Kir1 from a tetracycline-inducible promoter [5]. The fluorescent dye, Thallos, is used to report the inward flux of Tl^+ through the *Ae*Kir1 channel pore in a population of cells plated in individual wells of a 384-well plate. As shown in Figure 1A, overnight induction of *Ae*Kir1 expression with tetracycline leads to a robust Tl^+ flux compared to control cells that were not treated with tetracycline. This assay enables more than 300 compounds to be tested simultaneously in a single plate, and thousands of compounds to be tested daily, for effects on *Ae*Kir1 activity.

The assay was validated for HTS by meeting a series of performance benchmarks. First, the assay was tested for its tolerance to the small-molecule vehicle DMSO at concentrations up to 10% v/v. As shown in Fig. 1B, the Tl^+-flux mediated by *Ae*Kir1 is unaffected by DMSO concentrations up to 1.3% v/v as compared to the 0% DMSO control (one-way ANOVA, $P < 0.0001$). Next, the assay was tested for uniformity and reproducibility of HTS performance. As shown in Fig. 1C, the average Z′ statistic for these experiments was 0.69±0.05 ($Z' \geq 0.5$ is suitable for HTS), indicating that the assay is robust and will enable modulators of *Ae*Kir1 to be identified in HTS with a low false-positive rate.

Approximately 30,000 compounds from the VICB library were screened at a nominal concentration of 10 μM for inhibition of *Ae*Kir1. From this primary screen and following confirmation testing in tetracycline-induced and uninduced T-REx-HEK293-*Ae*Kir1 cells (see Methods), 283 authentic channel-dependent modulators were selected for further study. Because our ultimate goal is to develop Kir channel inhibitors that are active against mosquitoes and not humans, these 'hits' were subsequently tested for dose-dependent activity against a panel of mammalian Kir channels, which included Kir1.1, Kir2.1, Kir2.2, Kir2.3, Kir3.1/3.2, Kir4.1, Kir7.1(M125R), and Kir6.2/SUR1 [18,19,21]. Four-point concentration response curves (CRCs) were generated for the 283 compounds, resulting in 17 inhibitors with 11 unique chemical scaffolds that exhibited dose-dependent inhibition of *Ae*Kir1 with IC_{50} values below 5 μM and little to no activity ($IC_{50} \geq 30$ μM) against mammalian Kir channels (data not shown). These compounds were subsequently purchased from commercial vendors, freshly dissolved in DMSO, and assayed in 11-point CRCs against *Ae*Kir1 via the Tl^+-flux assay.

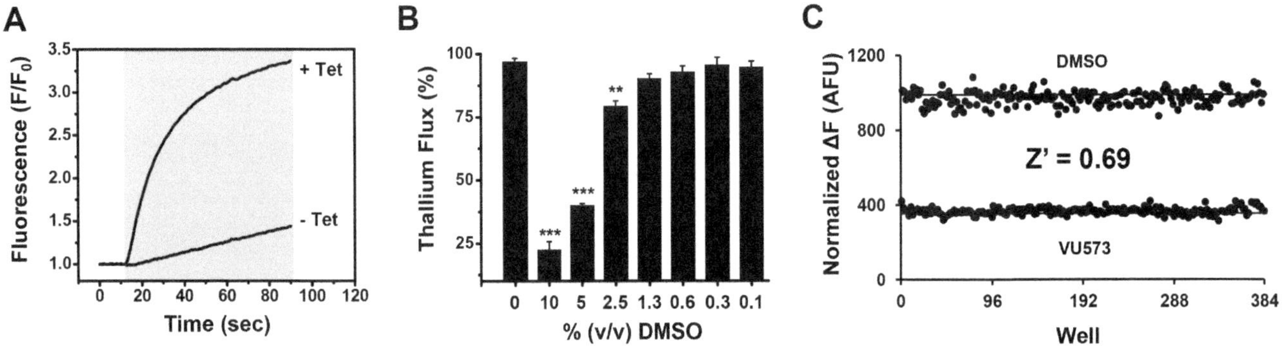

Figure 1. Tl$^+$ flux assay of *Ae*Kir1 channel activity for high-throughput screening. (A) Representative Tl$^+$-induced changes in Thallos fluorescence in T-Rex-HEK293-*Ae*Kir1 cells cultured overnight with (+Tet) or without (-Tet) tetracycline. The shaded box indicates the cell exposure to Tl$^+$. (B) DMSO concentrations up to 1.3% v/v DMSO have no effect on Tl$^+$ flux through *Ae*Kir1. Data are means ±SEM ($n = 3$). One-way ANOVA $P < 0.0001$, and asterisks (**, ***) indicate $P < 0.01$ or $P < 0.001$ respectively, when compared to 0% DMSO (Tukey's test). (C) Representative checkerboard analysis using 100 μM VU573 or 0.1% v/v DMSO as the vehicle control. The mean peak fluorescence amplitude of each sample population is indicated with a solid line and alternating samples for DMSO (top) and VU573 (bottom) are graphed as individual points. The mean ±SD Z′ calculated over 6 plates on 3 separate days was 0.69±0.05.

VU625 is a potent and preferential inhibitor of *Ae*Kir1 vs. mammalian Kir and *Ae*Kir2B channels

From the aforementioned Tl$^+$ flux assays, one compound—N-(3-methoxyphenyl)-2-methyl-1-propionylindoline-5-sulfonamide, termed VU625—was found to inhibit *Ae*Kir1 in 11-point CRCs with an IC_{50} of 0.32 μM (95% CI: 0.25–0.39 μM) and a Hill coefficient value of 0.98 (95% CI: 0.8–1.2) (Fig. 2A–C). VU625 also had no significant effects on the mammalian Kir channels assayed via Tl$^+$ flux with the exception of G-protein coupled Kir channels comprised of Kir3.1/3.2 subunits ($IC_{50} = 8.6$ μM; Table S1). Furthermore, in radioligand displacement assays against 68 mammalian GPCR's, ion channels, and transporters, 10 μM VU625 was active (defined as>50% ligand displacement) against only three targets: adenosine A1 receptor (76% displacement), melatonin MT1 receptor (56% displacement) and 5-HT$_{2B}$ receptor (69% displacement) (Table S2).

To further confirm the activity of VU625 obtained from Tl$^+$-flux assays, we used patch-clamp electrophysiology to assay the inhibition of *Ae*Kir1 expressed in T-REx-HEK293 cells. In whole-cell patch clamp recordings, VU625 inhibited *Ae*Kir1 channel activity with an IC_{50} of 96.8 nM (95% CI: 75.4–124.2 nM) and a Hill coefficient value of 1.02 (95% CI: 0.8–1.3) (Fig. 3A, B).

In a previous paper, we demonstrated that other small molecule inhibitors of *Ae*Kir1 (i.e., VU573 and VU590) can have different pharmacological effects on *Ae*Kir2B [6]. Thus, we sought to determine the effects of VU625 on *Ae*Kir2B channel activity, utilizing *Xenopus* oocytes heterologously expressing *Ae*Kir2B. *Ae*Kir1 expressing oocytes served as positive controls. Figure 3C shows that VU625 inhibits *Ae*Kir1- and *Ae*Kir2B-mediated K$^+$ currents with IC_{50} values of 3.8 μM (95% CI: 2.3–6.3 μM) and 45.1 μM (95% CI: 31.7–64.2 μM), respectively. Thus, VU625 inhibits both *Ae*Kir1 and *Ae*Kir2B channels, albeit with greater affinity for *Ae*Kir1. It should be noted that the reduction in VU625 potency observed in *Xenopus* oocytes compared to HEK cells is typical for a small-molecule inhibitor of Kir channels and has been observed for structurally diverse compounds and Kir channels [5,6,19,23].

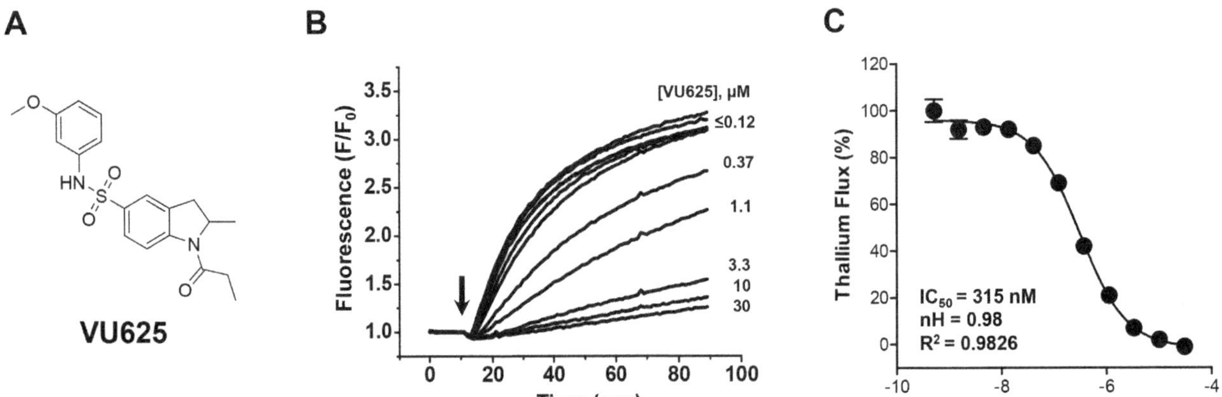

Figure 2. VU625 is a potent inhibitor of *Ae*Kir1 in Tl$^+$ flux assays. (A) Chemical structure of VU625. (B) Dose-dependent inhibition of the *Ae*Kir1-mediated Tl$^+$ flux by VU625 with concentrations ranging from ≤0.12 to 30 μM. The arrow indicates when Tl$^+$ was added to the extracellular bath. (C) Concentration-response curves of VU625 derived from Tl$^+$ flux assays. The IC_{50} and Hill-coefficient (nH) values are 315 nM (95% CI: 254.4–390.2 nM) and 0.98 respectively. Data are mean ±SEM. $n = 4$ independent experiments performed in triplicate.

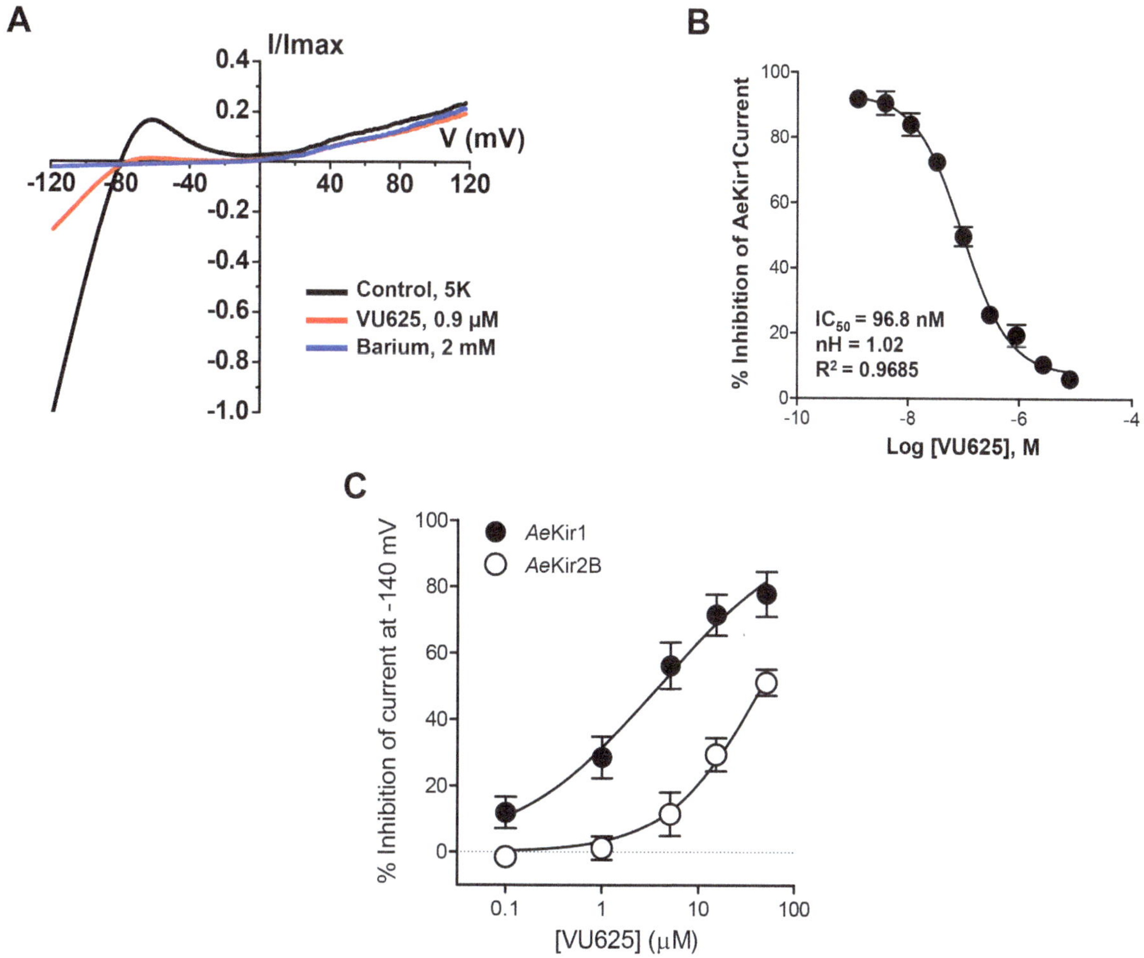

Figure 3. VU625 is a potent and preferential inhibitor of *Ae*Kir1 over *Ae*Kir2B in whole-cell electrophysiology. (A) Normalized *Ae*Kir1 current-voltage relationships obtained from heterologous expression in T-Rex-HEK293 cells, illustrating VU625-dependent inhibition before (control) and after addition of 0.9 μM VU625. Residual *Ae*Kir1 currents were inhibited with 2 mM barium. Cells were voltage clamped at −75 mV and ramped between −120 mV and +60 mV. (B) Concentration-response curve of VU625 derived from patch clamp experiments ($n = 4$–6). The IC_{50} of VU625 is 96.8 nM (95% CI: 75.4–124.2 nM). (C) Concentration-response curves of current inhibition mediated by heterologous expression in *Xenopus* oocytes of *Ae*Kir1 (filled circles) and *Ae*Kir2B (open circles) channels after bath application of VU625. $n = 4$–5 oocytes per concentration. The calculated IC_{50} values of VU625 for *Ae*Kir1 and *Ae*Kir2B current inhibition are 3.8 μM (95% CI: 2.3–6.3 μM) and 45.1 μM (95% CI: 31.7–64.2 μM), respectively.

Chemical lead optimization and structure-activity relationships

Because of its potency, clean ancillary pharmacology and chemical tractability (Figures 2–3, Tables S1–S2), VU625 was selected for lead optimization (**3a**, Table 2). We partitioned the compound into three areas for structure-activity relationship (SAR) exploration denoted as the sulfonamide, central core, and southern amide portions (Fig. 4A). The first generation libraries held the sulfonamide and the central core sections constant and diversified the southern amide portion (Table 2). The synthetic scheme (Fig. 4B) for this portion was straightforward and started with protection of the amine with trifluoracetamide (TFAA, pyridine) followed by sulfonyl chloride formation ($ClSO_3$, PCl_5). Next, the sulfonamide was formed, the protecting group was removed, and either the amide or sulfonamide was formed (see Methods for details). Little tolerance for steric bulk was seen in this portion of the molecule. That is, the trifluoroacetamide (VU0477197, **3b**, Table 2) retained potency (0.58 μM), however, larger aromatic amides were much less active (**3c–g**, Table 2). The same trend was observed for the sulfonamide compounds, with smaller sulfonamides retaining nanomolar activity (VU0477691, **3k**, 0.76 μM; VU0477692, **3l**, 0.82 μM) and the larger aromatic group leading to less activity (**3h**, Table 2).

Next, we evaluated the left-hand sulfonamide portion of the molecule, however, all efforts to change the 3-methoxyaryl moiety led to significant reductions in potency (see Table S3). Finally, we explored the central core with the intent of establishing the minimal pharmacophore needed for activity against *Ae*Kir1. To this end, the indoline core was replaced with simple aryl, heteroaryl or biaryl groups which all led to compounds with much reduced activity (>10-fold loss of potency). However, an interesting SAR was seen with very closely related 6,6- or 6,5-indole or dihydroquinolinone-like structures (**4a–f**, Table 3). The simple *N*-methyl indole (VU0481807, **4a**, 0.55 μM, Table 3) retained most of the activity as VU625 and addition of a 2-methyl (VU0486620, **4b**, 0.97 μM, Table 3) led to a further minor reduction in activity. Expanding the ring system and addition of a lactam (**4c–e**, Table 3) was not productive. Lastly, removal of the methyl group in the indoline system of VU625, led to a ~3-fold loss of potency (VU0483404, **4f**, 1.15 μM, Table 3).

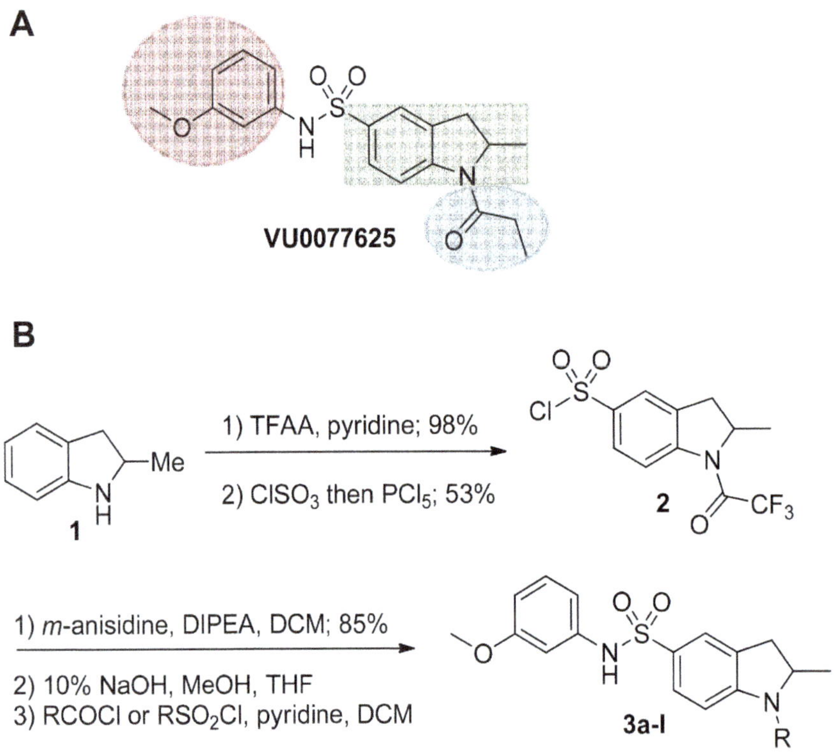

Figure 4. Design and chemical lead optimization strategy for VU625. (A) Modular approach to assess three areas of diversification of VU625: sulfonamide (red shading), central core (green shading), and southern amide (blue shading) portions. (B) General synthetic approach to access VU625 and analogs around the amide and sulfonamide portions.

Figure 5 summarizes the SAR observed for the VU625 scaffold. The left-hand sulfonamide portion offered the least amount of SAR traction as only the 3-methoxyphenyl (and weaker 2-methoxyphenyl) sulfonamide provided activity. Replacement with an amide, or substituting the 3-methoxyphenyl for other aryl groups all led to less potent compounds. Replacement of the proponamide in the southern fraction was tolerated as long as the substituent was small and aliphatic. Sulfonamides could be exchanged, although there was an observed ~2-3-fold loss of activity. Lastly, the central core was also important for potency. Only very similar compounds such as indole and des-methyl indoline were tolerated.

VU625-induced toxicity is increased by probenecid

Injection of the *Ae*Kir1 channel inhibitors VU573 or VU590 into the hemolymph of adult female *A. aegypti* mosquitoes leads to their incapacitation and/or death within 24 h [5,6]. Surprisingly, injection of a high dose (i.e. 690 pmol per mosquito) of VU625, which is a more potent inhibitor of *Ae*Kir1 than VU573 and VU590, into the hemolymph of mosquitoes had no significant effects on mosquito behavior or survival within 24 h (Fig. 6). Thus, we hypothesized that the lack of in vivo effects could be due to poor bioavailability of VU625 as a result of metabolic detoxification and/or excretion by the mosquito.

We therefore tested whether probenecid, which is a broad-spectrum inhibitor of organic anion transporters (OATs) and ATP-binding cassette (ABC) transporters [24,25,26], would improve the efficacy of VU625. Interestingly, both probenecid and VU625 have a sulfonamide moiety in their chemical structure (Fig. S2). As shown in Fig. 6, the injection of probenecid (3.4 nmol per mosquito) along with VU625 (690 pmol per mosquito) significantly increases the toxicity of VU625 within 24 h compared to injection of VU625 or probenecid alone. The abdomens of these mosquitoes were not severely bloated and obvious sub-lethal effects (e.g., loss of flight) were not apparent.

We next sought to characterize the dose-response relationship of VU625 in mosquitoes when co-injected with a constant dose of probenecid (3.4 nmol per mosquito). As shown in Fig. 7, co-injection of VU625 with probenecid induces mortality in mosquitoes within 24 h in a biphasic manner with 25% and 75% efficacious doses (ED_{25} and ED_{75}) of 9.96 pmol and 502 pmol, respectively. This biphasic dose-response relationship suggests the inhibition of at least two distinct molecular targets for which VU625 has different affinities, which is consistent with the inhibition of both *Ae*Kir1 and *Ae*Kir2B channels by VU625 (Fig. 3C).

VU625-induced reduction of urine excretion is enhanced by probenecid

We showed previously that pharmacological inhibition of *Ae*Kir1 with the small molecule inhibitors VU573 and VU590 leads to a decrease in the excretory capacity of *A. aegypti* mosquitoes after loading their hemolymph with 900 nl of a PBS vehicle [5] [6]. Therefore, we sought to similarly determine the effects of VU625 on mosquito excretory capacity. As shown in Fig. 8, mosquitoes injected with the PBS vehicle excreted 644±24.18 nL of urine within the next hour. Consistent with the toxicity studies, we found that adding VU625 (0.77 mM) to the vehicle, which delivers 690 pmol of VU625 per mosquito, did not significantly decrease the excretory capacity (583.3±29.52 nL/female) compared to the vehicle controls (Fig. 8). Interestingly, adding probenecid (50 mM) to the vehicle, which delivers

Table 2. Structure-activity relationships and lead optimization summary of VU0077625 scaffold.

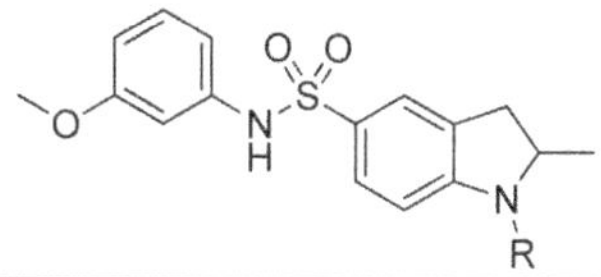

Cmpd	R	VU#	$IC_{50}\pm SEM$ (μM)
3a	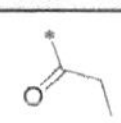	VU0077625	0.36±0.02
3b	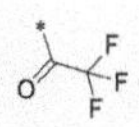	VU0477197	0.58±0.04
3c	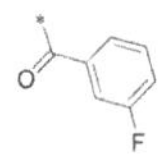	VU0477684	5.20±0.40
3d	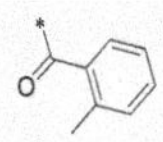	VU0477693	4.41±1.11
3e	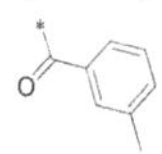	VU0477694	3.87±1.97
3f	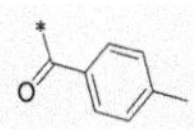	VU0477688	>30
3g	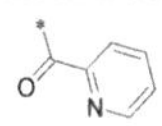	VU0477685	4.40±0.50
3h	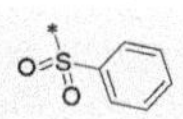	VU0477686	3.29±0.89
3i	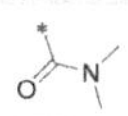	VU0477687	2.09±0.49
3j	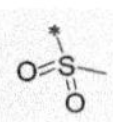	VU0477690	2.82±0.48
3k	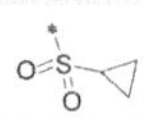	VU0477691	0.76±0.00
3l	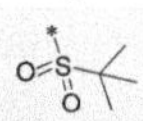	VU0477692	0.82±0.48

IC_{50} values were derived from 11-point CRCs on *Ae*Kir1 in Tl^{+} flux experiments performed in triplicate on two separate days.

3.4 nmol of probenecid per mosquito, causes a small but significant reduction in excretory capacity to 467.8±33.53 nL/female, suggesting a potential role of probenecid-sensitive transporters in urine excretion. However, adding both VU625 (0.77 mM) and probenecid (50 mM) to the vehicle significantly decreases the excretory capacity the furthest to 236.7±24.53 nL/female.

Discussion

Here, we report the discovery of VU625, the first submicromolar inhibitor of a mosquito Kir channel. VU625 is one of 283 confirmed *Ae*Kir1 inhibitors identified in a HTS of approximately 30,000 compounds from the VICB library. It was chosen for lead optimization based on its potency (IC_{50} = 96.8 nM), greater than 80-fold selectivity for the *Ae*Kir1 channel over 8 mammalian Kir channels, and clean ancillary pharmacology among a panel of 68 critical mammalian off-targets comprised of voltage-gated ion channels, ion transporters, and receptors (i.e., neurotransmitter, peptide, and G-protein coupled). VU625 is the most potent and selective mosquito Kir channel inhibitor reported to date.

This study provides proof-of-concept that conventional drug discovery approaches can be employed successfully to identify small-molecule tools for probing the physiology of insect Kir channels and potential lead compounds for insecticide development. A similar approach has been used recently in insecticide discovery efforts targeting mosquito G-protein coupled receptors [27].

VU625 exhibits inhibitory activity against both *Ae*Kir1 and *Ae*Kir2B, albeit with greater affinity for *Ae*Kir1. To date, we have reported the activity of two other small-molecule inhibitors of mosquito Kir channels that exhibit differential pharmacology. VU590 is a selective inhibitor of *Ae*Kir1 over *Ae*Kir2B, whereas VU573 inhibits *Ae*Kir1 and activates *Ae*Kir2B [6]. Thus, VU625 potentially represents a broad-spectrum, small-molecule blocker of mosquito Kir channels, pending the characterization of its effects on the other mosquito Kir channels (*Ae*Kir2A, *Ae*Kir2B' and *Ae*Kir3 channels), which to date have not yet been expressed functionally in a heterologous system ([12]; Denton and Piermarini, personal observations). Once the distinguishing pharmacological properties of each of these Kir channel inhibitors are fully characterized, they can potentially be employed to determine the relative contributions of Kir channel subtypes in the physiology of various mosquito tissues. This would provide an important

Table 3. Structure-activity relationships and lead optimization summary for the central core portion of VU0077625 scaffold.

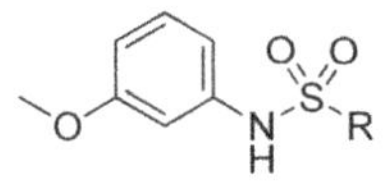

Cmpd	R	VU#	IC_{50}±SEM (μM)
4a	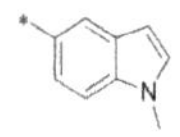	VU0481807	0.55±0.08
4b	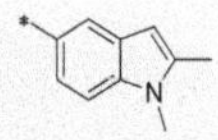	VU0486620	0.97±0.10
4c	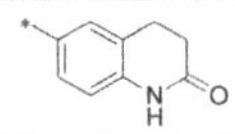	VU0481811	>30
4d	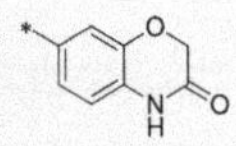	VU0483082	>30
4e	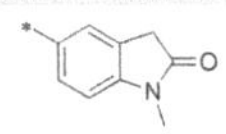	VU0483402	>30
4f	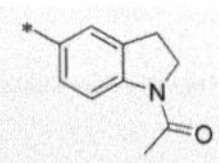	VU0483404	1.15±0.05

IC_{50} values were derived from 11-point CRCs on *Ae*Kir1 in Tl^+ flux experiments performed in triplicate on two separate days.

chemical tool set to validate and complement studies of mosquito Kir channels that employ functional genetic approaches (e.g., RNA interference).

Given the superior in vitro potency of VU625 compared to the *Ae*Kir1 inhibitors VU573 [5] and VU590 [6], we expected VU625 to elicit superior in vivo toxicity. Thus, we were surprised when high doses of VU625 elicited no observable effects on mosquito survival or excretory capacity when injected directly into the hemolymph. Since mosquitoes have evolved robust protective mechanisms for detoxifying and excreting xenobiotics that would harm them otherwise [28,29], we investigated whether the molecule may be detoxified and/or excreted.

Preliminary experiments with PBO did not improve the efficacy of VU625, suggesting that detoxification of the compound by cytochrome P450s is unlikely to contribute to its poor in vivo efficacy. The co-injection of VU625 with probenecid rescued not only its toxicity, but also its effects on excretory capacity, which suggests that VU625 is likely a substrate of OATs and/or ABC transporters in the mosquitoes and may be rendered ineffective in vivo through excretion. The potent toxicity of VU625 when co-injected with probenecid may be due to the ability of VU625 to inhibit at least two Kir channels, some of which are expressed in the central and peripheral nervous systems, such as Kir1 and Kir2B' [8,12,30,31], and/or a synergistic effect of probenecid that maintains high circulating concentrations of VU625 by preventing its renal excretion. Indeed, it is conceivable that the sulfonamide moiety in the structures of VU625 and probenecid causes them to be substrates for OATs and/or ABC transporters. Overall, these findings highlight efficient xenobiotic transport mechanisms in mosquitoes that render a nanomolar inhibitor of *Ae*Kir1 (VU625) ineffective in vivo, even when introduced directly to the hemolymph. The tissues that contribute to the excretion of VU625 remain to be determined, but presumably involve the Malpighian tubules and/or gut [32,33].

Lastly, the medicinal chemistry efforts put forth in the present study may be a valuable first step in determining which structural moieties are important for the excretion of VU625 by xenobiotic transporters and/or its in vivo activity in mosquitoes. Future

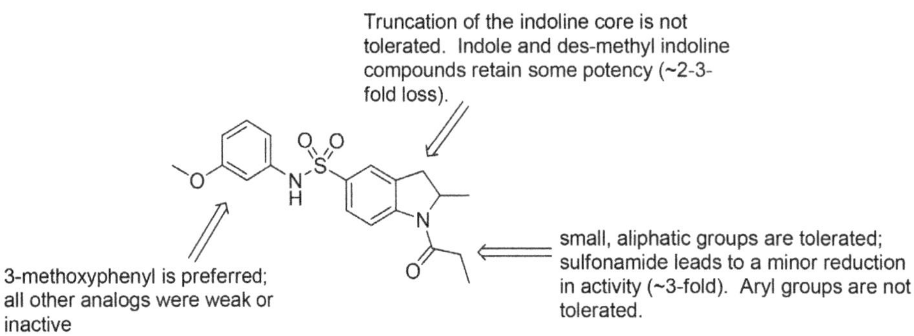

Figure 5. Summary of structure-activity relationship (SAR). Summary of observed SAR of over 100 analogs synthesized exploring all three regions of VU625.

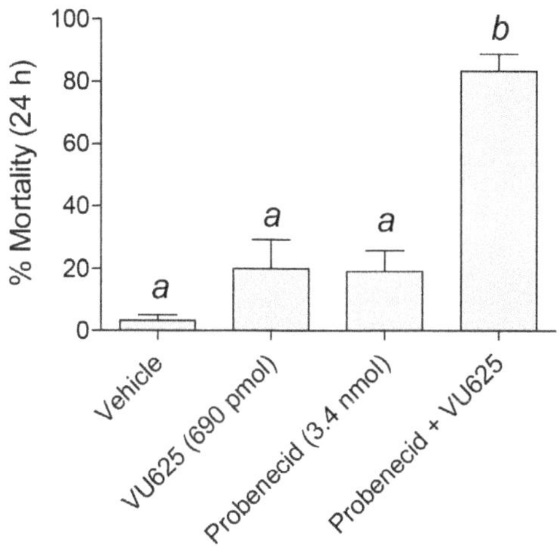

Figure 6. Effects of probenecid and VU625 on survival of adult female mosquitoes (*A. aegypti*). Percent mortality of mosquitoes at 24 h post-injection. Each mosquito was injected with 69 nl of the vehicle containing VU625 (10 mM), probenecid (50 mM), or both, to deliver the desired doses: 690 pmol of VU625, 3.4 nmol probenecid. $n = 6$–7 trials of 10 mosquitoes each per treatment. Lower-case letters indicate statistical categorization of the means as determined by a one-way ANOVA with a Newman-Keuls post-test ($P < 0.05$).

studies should assess the in vivo efficacy and probenecid-mediated clearance of the VU625 analog series we generated to determine if any of these compounds exhibit potent toxicity in mosquitoes without probenecid.

Perspectives

Here, we show a direct relationship between in vitro pharmacology and in vivo toxicity of VU625, which is consistent with our previous studies [5,6] suggesting that Kir channel inhibitors are promising chemicals for insecticide development. To date, none of the Kir channel inhibitors we have reported (i.e. VU573, VU590, VU625) exhibit toxicity when applied to the cuticle (Piermarini, unpublished observations), which is a waxy, lipophilic structure that creates a physical barrier to insecticide permeation into the hemocoel of mosquitoes. This lack of topical activity severely limits the potential use of the present Kir channel inhibitors as active compounds for incorporation into insecticide-treated bed nets and indoor-residual sprays. The efficacy of common insecticides, such as permethrin, is dependent in part on their lipophilic nature [34,35]. Thus, future chemistry efforts will focus on lipophilic inhibitors of Kir channels. Furthermore, prioritizing initial HTS 'hits' according to their hydrophobicity may facilitate the discovery of more suitable small-molecules compounds for insecticide development.

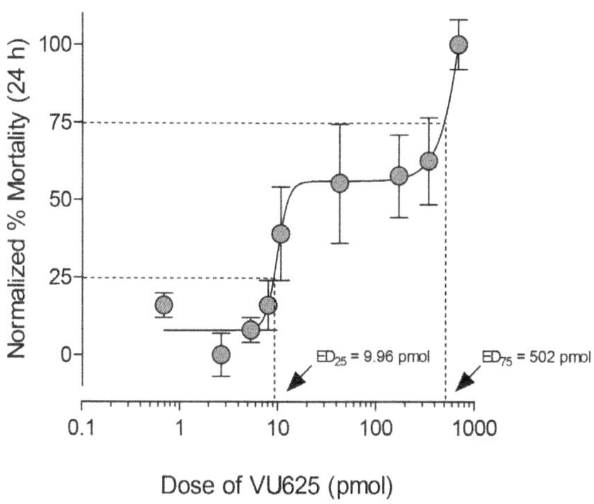

Figure 7. The dose-response curve of the toxic effects of VU625 on adult female mosquitoes (*A. aegypti*) is biphasic. Normalized percent mortality of mosquitoes at 24 h post-injection. Each mosquito was injected with 69 nL of the vehicle containing probenecid (50 mM) and an appropriate concentration of VU625 to deliver the doses of VU625 indicated and 3.4 nmol of probenecid. The ED_{25} and ED_{75} were determined by fitting a non-linear biphasic curve to the data. $n = 3$–4 trials of 10 mosquitoes each per dose.

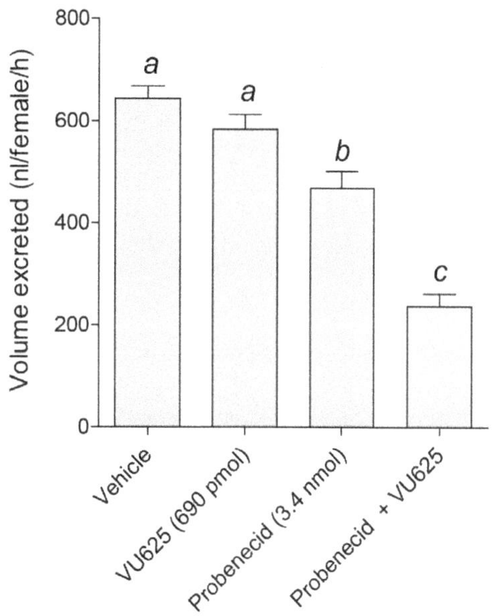

Figure 8. Effects of probenecid and VU625 on the in vivo excretory capacity of adult female mosquitoes (*A. aegypti*). Amount of urine excreted by mosquitoes 1 h after injection with 900 nL of the vehicle (K^+-PBS_{50} containing 1.8% DMSO, 0.077% β-cyclodextrin, and 0.008% Solutol), or the vehicle containing VU625 (0.77 mM), probenecid (3.85 mM), or both, to deliver the desired doses: 690 pmol of VU625, 3.4 nmol probenecid. Values are means ±SEM; $n = 6$–18 trials of 5 mosquitoes per treatment. Lower-case letters indicate statistical categorization of the means as determined by a one-way ANOVA with a Newman-Keuls posttest ($P < 0.05$).

Supporting Information

Figure S1 Reagents and conditions: (A) TFAA, pyridine, 0°C; (B) $ClSO_3H$, 40°C, 1 h; PCl_5, rt; (C) 3-methoxyaniline, DIEA, rt; MeOH:10% NaOH (1:1:1); (D) pyridine, CH_2Cl_2, $ClCOCH_2CH_3$.

Figure S2 VU625 and probenecid share a sulfonamide moiety. The sulfonamide moiety contained in the chemical structure of VU625 and probenecid is shaded in blue.

Table S1 Selectivity of VU625 against human Kir channels assessed in Tl^+ flux assays. $n = 2$ independent experiments in triplicate.

Table S2 Summary of results obtained from the activity of the VU625 compound in radioligand binding assays. The significant results are highlighted in grey.

Table S3 SAR around the left-hand sulfonamide. IC_{50} values were derived from 11-point CRCs on *Ae*Kir1 in Tl^+ flux experiments performed in triplicate on two separate days.

Acknowledgments

The authors thank Nuris Acosta (The Ohio State University), and the Vanderbilt High-Throughput Screening Center for technical assistance.

Author Contributions

Conceived and designed the experiments: RR MFR DRS ED CDW KML LCK DWE SFB CH PMP JSD. Performed the experiments: RR MFR DRS ED KML LCK DWE SFB PMP. Analyzed the data: RR MFR DRS ED CDW KML LCK DWE SFB CH PMP JSD. Contributed reagents/ materials/analysis tools: CDW KML LCK DWE SFB. Wrote the paper: RR MFR DRS CDW CH PMP JSD.

References

1. WHO (2013) World Malaria Report WHO; World Health Organization.
2. Ranson H, N'Guessan R, Lines J, Moiroux N, Nkuni Z, et al. (2011) Pyrethroid resistance in African anopheline mosquitoes: what are the implications for malaria control? Trends Parasitol 27: 91–98.
3. Asidi A, N'Guessan R, Akogbeto M, Curtis C, Rowland M (2012) Loss of household protection from use of insecticide-treated nets against pyrethroid-resistant mosquitoes, benin. Emerg Infect Dis 18: 1101–1106.
4. Maharaj R (2011) Global trends in insecticide resistance and impact on disease vector control measures. Open Access Insect Physiology: 27.
5. Raphemot R, Rouhier MF, Hopkins CR, Gogliotti RD, Lovell KM, et al. (2013) Eliciting renal failure in mosquitoes with a small-molecule inhibitor of inward-rectifying potassium channels. PLoS One 8: e64905.
6. Rouhier MF, Raphemot R, Denton JS, Piermarini PM (2014) Pharmacological Validation of an Inward-Rectifier Potassium (Kir) Channel as an Insecticide Target in the Yellow Fever Mosquito *Aedes aegypti*. PLoS One 9:e100700.
7. Raphemot R, Estevez-Lao TY, Rouhier MF, Piermarini PM, Denton JS, et al. (2014) Molecular and functional characterization of Anopheles gambiae inward rectifier potassium (Kir1) channels: A novel role in egg production. Insect Biochem Mol Biol 51: 10–19.
8. Döring F, Wischmeyer E, Kühnlein RP, Jäckle H, Karschin A (2002) Inwardly Rectifying K (Kir) Channels in *Drosophila*. J Biol Chem 277: 25554–25561.
9. Evans JM, Allan AK, Davies SA, Dow JA (2005) Sulphonylurea sensitivity and enriched expression implicate inward rectifier K^+ channels in *Drosophila* melanogaster renal function. J Exp Biol 208: 3771–3783.
10. Eleftherianos I, Won S, Chtarbanova S, Squiban B, Ocorr K, et al. (2011) ATP-sensitive potassium channel (K_{ATP})-dependent regulation of cardiotropic viral infections. Proc Natl Acad Sci 108: 12024–12029.
11. Dahal GR, Rawson J, Gassaway B, Kwok B, Tong Y, et al. (2012) An inwardly rectifying K^+ channel is required for patterning. Development 139: 3653–3664.
12. Piermarini PM, Rouhier MF, Schepel M, Kosse C, Beyenbach KW (2013) Cloning and functional characterization of inward-rectifying potassium (Kir) channels from Malpighian tubules of the mosquito *Aedes aegypti*. Insect Biochem Mol Biol 43: 75–90.
13. Raphemot R, Kadakia RJ, Olsen ML, Banerjee S, Days E, et al. (2013) Development and validation of fluorescence-based and automated patch clamp-based functional assays for the inward rectifier potassium channel kir4.1. Assay Drug Dev Technol 11: 532–543.
14. Hibino H, Inanobe A, Furutani K, Murakami S, Findlay I, et al. (2010) Inwardly Rectifying Potassium Channels: Their Structure, Function, and Physiological Roles. Physiol Rev 90: 291–366.
15. Denton JS, Pao AC, Maduke M (2013) Novel diuretic targets. Am J Physiol Renal Physiol 305: F931–942.
16. Pattnaik BR, Asuma MP, Spott R, Pillers DA (2012) Genetic defects in the hotspot of inwardly rectifying K^+ (Kir) channels and their metabolic consequences: a review. Mol Genet Metab 105: 64–72.
17. Denton JS, Jacobson DA (2012) Channeling dysglycemia: ion-channel variations perturbing glucose homeostasis. Trends Endocrinol Metab 23: 41–48.
18. Lewis LM, Bhave G, Chauder BA, Banerjee S, Lornsen KA, et al. (2009) High-throughput screening reveals a small-molecule inhibitor of the renal outer medullary potassium channel and Kir7.1. Mol Pharmacol 76: 1094–1103.
19. Raphemot R, Lonergan DF, Nguyen TT, Utley T, Lewis LM, et al. (2011) Discovery, characterization, and structure-activity relationships of an inhibitor of inward rectifier potassium (Kir) channels with preference for Kir2.3, Kir3.x, and Kir7.1. Front Pharmacol 2: 75.
20. Raphemot R, Swale DR, Dadi PK, Jacobson DA, Cooper P, et al. (2014) Direct Activation of beta-cell K_{ATP} Channels with a Novel Xanthine Derivative.Mol Pharmacol. 85: 858–65.
21. Raphemot R, Weaver CD, Denton JS (2013) High-throughput screening for small-molecule modulators of inward rectifier potassium channels. J Vis Exp. 71: 4209.
22. Niswender CM, Johnson KA, Luo Q, Ayala JE, Kim C, et al. (2008) A novel assay of Gi/o-linked G protein-coupled receptor coupling to potassium channels provides new insights into the pharmacology of the group III metabotropic glutamate receptors. Mol Pharmacol 73: 1213–1224.
23. Bhave G, Chauder BA, Liu W, Dawson ES, Kadakia R, et al. (2011) Development of a selective small-molecule inhibitor of Kir1.1, the renal outer medullary potassium channel. Mol Pharmacol 79: 42–50.
24. Feller N, Broxterman HJ, Währer DC, Pinedo HM (1995) ATP-dependent efflux of calcein by the multidrug resistance protein (MRP): no inhibition by intracellular glutathione depletion. FEBS Lett 368: 385–388.
25. Jaehde U, Sörgel F, Reiter A, Sigl G, Naber KG, et al. (1995) Effect of probenecid on the distribution and elimination of ciprofloxacin in humans. Clin Pharmacol Ther 58: 532–541.
26. Hill G, Cihlar T, Oo C, Ho ES, Prior K, et al. (2002) The anti-influenza drug oseltamivir exhibits low potential to induce pharmacokinetic drug interactions via renal secretion-correlation of in vivo and in vitro studies. Drug Metab Dispos 30: 13–19.
27. Meyer JM, Ejendal KF, Avramova LV, Garland-Kuntz EE, Giraldo-Calderon GI, et al. (2012) A "genome-to-lead" approach for insecticide discovery: pharmacological characterization and screening of *Aedes aegypti* D_1-like dopamine receptors. PLoS Negl Trop Dis 6: e1478.
28. Li X, Schuler MA, Berenbaum MR (2007) Molecular mechanisms of metabolic resistance to synthetic and natural xenobiotics. Annu Rev Entomol 52: 231–253.
29. Dermauw W, Van Leeuwen T (2014) The ABC gene family in arthropods: Comparative genomics and role in insecticide transport and resistance. Insect Biochem Mol Biol 45C: 89–110.
30. Baker DA, Nolan T, Fischer B, Pinder A, Crisanti A, et al. (2011) A comprehensive gene expression atlas of sex- and tissue-specificity in the malaria vector, *Anopheles gambiae*. BMC Genomics 12: 296.
31. Rouhier MF, Piermarini PM (2014) Identification of life-stage and tissue-specific splice variants of an inward rectifying potassium (Kir) channel in the yellow fever mosquito *Aedes aegypti*. Insect Biochem Mol Biol 48C: 91–99.
32. O'Donnell MJ (2009) Too much of a good thing: how insects cope with excess ions or toxins in the diet. J Exp Biol 212: 363–372.
33. O'Donnell MJ (2008) Insect Excretory Mechanisms. In: Simpson SJ, editor.Advances in Insect Physiology.London: Academic Press. pp.1–122.
34. Tice CM (2001) Selecting the right compounds for screening- does Lipinski's Rule of 5 for pharmaceuticals apply to agrochemicals? Pest Manag Sci 57: 3–16.
35. Akamatsu M (2011) Importance of physicochemical properties for the design of new pesticides. J Agric Food Chem 59: 2909–2917.

8

Weevil x Insecticide: Does 'Personality' Matter?

Juliana A. Morales[1], Danúbia G. Cardoso[1], Terezinha Maria C. Della Lucia[2], Raul Narciso C. Guedes[1]*

1 Departamento de Entomologia, Universidade Federal de Viçosa, Viçosa, Minas Gerais, Brazil, **2** Departamento de Biologia Animal, Universidade Federal de Viçosa, Viçosa, Minas Gerais, Brazil

Abstract

An insect's behavior is the expression of its integrated physiology in response to external and internal stimuli, turning insect behavior into a potential determinant of insecticide exposure. Behavioral traits may therefore influence insecticide efficacy against insects, compromising the validity of standard bioassays of insecticide activity, which are fundamentally based on lethality alone. By extension, insect 'personality' (i.e., an individual's integrated set of behavioral tendencies that is inferred from multiple empirical measures) may also be an important determinant of insecticide exposure and activity. This has yet to be considered because the behavioral studies involving insects and insecticides focus on populations rather than on individuals. Even among studies of animal 'personality', the relative contributions of individual and population variation are usually neglected. Here, we assessed behavioral traits (within the categories: activity, boldness/shyness, and exploration/avoidance) of individuals from 15 populations of the maize weevil (*Sitophilus zeamais*), an important stored-grain pest with serious problems of insecticide resistance, and correlated the behavioral responses with the activity of the insecticide deltamethrin. This analysis was performed at both the population and individual levels. There was significant variation in weevil 'personality' among individuals and populations, but variation among individuals within populations accounted for most of the observed variation (92.57%). This result emphasizes the importance of individual variation in behavioral and 'personality' studies. When the behavioral traits assessed were correlated with median lethal time (LT_{50}) at the population level and with the survival time under insecticide exposure, activity traits, particularly the distance walked, significantly increased survival time. Therefore, behavioral traits are important components of insecticide efficacy, and individual variation should be considered in such studies. This is so because population differences provided only crude approximation of the individual personality in a restrained experimental setting likely to restrict individual behavior favoring the transposition of the individual variation to the population.

Editor: Mark Briffa, University of Plymouth, United Kingdom

Funding: Funding was provided by the following Brazilian agencies: Minas Gerais State Foundation of Research Aid (FAPEMIG; www.fapemig.br), the CAPES Foundation (Brazilian Ministry of Education; www.capes.gov.br), and the National Council for Scientific and Technological Development (CNPq; Brazilian Ministry of Science and Technology; www.cnpq.br). The funders had no role in study design, data collection and analysis, decision to publish, or preparation of the manuscript.

Competing Interests: The authors have declared that no competing interests exist.

* E-mail: guedes@ufv.br

Introduction

"Curiosity killed the cat" is an animal aphorism that is widely used as a cautionary note to people who are too inquisitive, but the recent burgeoning interest in animal 'personality' most likely favors a more literal meaning of such statements. In fact, there has been a growing body of compelling evidence regarding the existence of 'personality' among animals and its eco-evolutionary importance [1–10]. Such growing interest naturally leads to conceptual misunderstandings, especially considering the long-standing interest and importance of human personality within the field of psychology and the more recent interest in animal 'personality' within biology, particularly in ecology and evolution [1,2,3,5,11]. Here, our working concept of 'personality' refers to an individual's integrated set of behavioral tendencies inferred from multiple empirical measures, which is more commonly used in psychology than in animal biology, in which personality is frequently treated as a synonym of behavioral syndromes and related concepts [5,11]. Furthermore, animal 'personality' as used in biology tend to focus on population variation rather than on individual variation [5,7,8,11], and is prone to jingle-jangle-fallacies (i.e., same term referring to different concepts and different terms referring to the same concept) [11]. Our stated working concept circumvents both shortcomings.

The study of personality differences in humans has proven useful in predicting the manifestation of certain behaviors, such as job satisfaction, risk taking, and social stress reactions, among others [1,12–15]. The expansion of personality studies to domestic animals led to improvements in animal production, welfare and conservation and thus drew attention from the general public [16–24]. By contrast, the study of animal 'personality' in arthropods lags behind, with studies of spiders and social insects as a main focus of attention [25–29], with very recent contributions regarding other species, such as the pea aphid *Acyrthosiphon pisum* and the confused flour beetle *Tribolium confusum* [30–33].

Theoretical considerations and frameworks attempting to address the roles of animal 'personality' variation within ecology and evolution have been receiving considerable attention, but adequate empirical tests of such hypotheses are lacking, especially those that consider testing the adaptive value of 'personality' [1–9]. In psychology, for instance, child personality has been associated with unintentional injury risk, and more specifically, with poison ingestion [12,15,34–37]. In applied entomology, the likelihood of insecticide exposure is potentially associated with insect behavioral traits. In fact, pesticide resistance mediated by insect behavior (i.e., avoidance) has been documented elsewhere [38–49]. However, the potential influence of 'personality' on insect mortality by insecticides has adaptive and applied consequences that have been neglected.

Bioassays of insecticidal activity usually consider mortality to be the assessment endpoint, particularly when insect pest species are considered, although insect behavior is an important determinant of insecticide exposure and, consequently, of its activity [42,43,50,51]. There is no denying the importance of studying the lethal effects of insecticides; nevertheless, it is unwise to neglect the influence of insect behavior on insecticide efficacy, or for that matter, insect 'personality'. Even when insect behavior is considered in insecticide studies, almost exclusively only the variation among insect populations is considered, as opposed to the variation within populations (i.e., individual variation), thus neglecting the importance of the latter [38–52]. Among studies of animal 'personality', the relative contribution of individual and population variation is also usually overlooked, as well as the multidimensional nature of 'personality'. By contrast, the tendency has been to focus on a category of behavior and to explore its correlations across different contexts, commonly referred to as 'behavioral syndromes', which is a term that is frequently used as a synonym of 'personality' in animal studies [2,4–11].

In this study, we assessed behavioral traits (within the following categories: activity, boldness/shyness, and exploration/avoidance) of the maize weevil *Sitophilus zeamais* Motschulsky (Coleoptera: Curculionidae) from 15 populations. The behavioral responses were tested for correlations with the efficacy of the pyrethroid insecticide deltamethrin. This analysis was performed at both the population and individual levels to determine the relative contribution of individual and population variation in the make-up of the insect 'personality', if indeed it exists, and to determine how individual and population variation in behavioral traits may relate to survival to insecticide exposure. As both behavior and insecticide susceptibility are conceptually attributes of the individual, we believe that individual variations in behavioral traits are fundamental contributors to 'personality' and its consequences, even if such 'personality' is expressed in a given population. However, insect 'personality', with its complexities, may not be necessary or even significant to explain insecticide susceptibility, which we also attempted to test in the present study.

Materials and Methods

Ethics Statement

No specific permits were required for the described studies, which were performed in the laboratory. Each of the insect colonies was initially established from over 200 individuals collected from storage units and maintained under mass rearing conditions at the Federal University of Viçosa. The insect species used here is a cosmopolitan pest species of cereal grains and is very common in Neotropical America and Africa. The study therefore did not involve any endangered or protected species.

Insect Populations

Fourteen populations of the maize weevil that were collected from storage units across Brazil, and one from Paraguay, were used in the experiments. These populations were mostly collected within the last five years, except the populations from Jacarezinho and Juiz de Fora, which were collected in the late 1980s and late 1990s, respectively. These populations varied in their resistance to pyrethroid insecticides, with some susceptible populations (e.g., Sete Lagoas) and some insecticide-resistant ones (e.g., Jacarezinho and Juiz de Fora). The prevailing resistance mechanism is altered target site sensitivity with secondary involvement of enhanced detoxification by glutathione-*S*-tranferases and esterases [52–55]. The specimens of each population were reared on maize kernels free of insecticide residues in 1.5 L glass jars maintained under controlled environmental conditions of 27±2°C, 70±10% relative humidity and a 12 h photoperiod.

Behavioral Bioassays

Adult sexed insects between one and three days old were used for both the behavioral and the insecticide bioassays, which were sequentially performed during one week with the same individuals. The behavioral bioassays were subjected to retest in subsequent days to ascertain the relative stability of the behavioral responses obtained before the (final) insecticide bioassay. The individual emerging insects were isolated, sexed using their pattern of rostrum texture and punctuation [56], and maintained in 20 mL transparent plastic vials containing maize kernels. Twenty-one individuals of each sex and from each population were subjected to the bioassays. The behavioral bioassays were performed under controlled laboratory conditions, as previously described for rearing insect populations, between 9:00 am and 6:00 pm. Six behavioral bioassays were performed focusing on measurable behavioral (or performance) traits demonstrating an individual's ability to perform a task, exploring three of the five categories of behavioral propensity (or personality traits) proposed by Reále et al. [5]: activity, boldness/shyness, and exploration/avoidance. Insect activity was assessed through a walking bioassay, two flight bioassays (flight mill and free-fall), and a body righting bioassay. Boldness/shyness was assessed through a death-feigning bioassay, and exploration/avoidance was assessed through bioassays of intra- and interspecific interactions using maize kernels infested with adults of either the maize weevil (intraspecific interaction) or the lesser grain borer (*Rhyzopertha dominica*) (interspecific interaction).

Walking activity bioassay. Walking activity was recorded for 10 min in Petri dish arenas (9 cm diameter) whose inner walls were coated with Teflon PTFE (DuPont, Wilmington, DE, USA) to prevent insect escape, as described elsewhere [46–49,57]. A single insect was released in the center of the arena, and its movement was recorded and digitally transferred to a computer using an automated video tracking system equipped with a CCD camera (ViewPoint Life Sciences, Montreal, Canada). The following characteristics were evaluated: distance walked (cm), walking velocity (cm/s), and resting time (s).

Free-fall flight bioassay. A hand-made square box (44 cm wide, 44 cm deep, 88 cm high) was used for the free-fall flight bioassay. The frame was made of wood, and the top was covered with thin transparent plastic film with a hole 5 cm in diameter located in the center, whereas the lateral sides were covered by organza tissue. The box was placed on a sheet of paper marked with a series of concentric circles 3 cm apart from one another. Each individual adult weevil was dropped through the central hole in the top of the box, and the landing distance from the center after wing fluttering was recorded following the methods adapted from a study on the cowpea beetle *Callosobruchus chinensis* [58]. The test was replicated three times, and the mean score was used as the flying activity.

Flight-mill bioassay. The flight-mill bioassay method used was that described by Riley et al. [59]. Each individual insect had its thorax attached by a thread to a rod in the mill, and the insect movement was recorded for 10 min, registering the wing beat and

Table 1. Summary results of the univariate analyses of variance (ANOVA) and repeatability (test-retest) results of the behavioral traits of both sexes from 15 populations of the maize weevil (*Sitophilus zeamais*).

Sources of variation (ANOVA)	Degrees of freedom	Walking activity						Flight activity						Conspecific interaction latency (min)		Heterospecific interaction latency (min)		Duration of death-feigning (s)		Length of time to body righting (s)	
		Distance walked (cm)		Resting time (s)		Walking velocity (cm/s)		Horizontal dislocation upon fall (cm)		Wing beat (no.)		Distance flown (m)									
		F	*P*	F	*P*	F	*P*	F	*P*	F	*P*	F	*P*	F	*P*	F	*P*	F	*P*	F	*P*
Model	29	3.78	<0.001	2.07	<0.001	2.71	<0.001	2.68	<0.001	3.47	<0.001	2.73	<0.001	1.47	0.06	1.93	0.002	2.50	<0.001	5.46	<0.001
Error	600	–	–	–	–	–	–	–	–	–	–	–	–	–	–	–	–	–	–	–	–
Population	14	4.78	<0.001	2.86	<0.001	3.94	<0.001	1.97	0.02	4.47	<0.001	3.00	<0.001	1.49	0.11	3.08	<0.001	4.11	<0.001	10.23	<0.001
Sex	1	0.79	0.37	0.54	0.46	3.08	0.08	0.08	0.77	6.76	0.009	10.76	0.001	1.56	0.21	1.62	0.20	0.99	0.32	0.01	0.93
Population x sex	14	2.99	<0.001	1.39	0.15	1.45	0.12	3.97	<0.001	2.24	0.006	1.88	0.02	1.45	0.13	0.79	0.68	0.99	0.46	1.09	0.36
Repeatability (test-retest) correlations [n = 630]	r	0.57		0.49		0.51		0.37		0.43		0.35		0.28		0.31		0.48		0.50	
	P	<0.001		<0.001		<0.001		<0.001		<0.001		<0.001		<0.001		<0.001		<0.001		<0.001	

Table 2. Behavioral traits (± EPM) of both sexes from 15 populations of the maize weevil (*Sitophilus zeamais*).

Population	Sex	Walking activity			Flight activity			Conspecific interaction latency (min)	Heterospecific interaction latency (min)	Duration of death-feigning (s)	Length of time to body righting (s)
		Distance walked (cm)	Resting time (s)	Walking velocity (cm/s)	Horizontal dislocation upon fall (cm)	Wing beat (no.)	Distance flown (m)				
Abre Campo	Female	168.17±13.55	189.45±10.13	0.68±0.02	6.73±0.99	0.58±0.08	0.37±0.20	13.43±0.93	9.13±1.39	9.41±1.92	1.45±0.17
	Male	290.49±17.14	176.61±15.71	0.68±0.02	5.13±0.75	0.11±0.04	0.18±0.10	9.91±1.35	11.86±1.18	12.44±3.30	2.81±0.70
Amambay	Female	278.65±13.82	191.88±14.77	0.68±0.01	6.35±0.92	0.49±0.09	0.85±0.28	9.17±1.35	10.26±1.27	5.03±1.59	1.69±0.23
	Male	267.30±13.18	199.96±14.35	0.66±0.01	5.46±0.60	0.27±0.07	0.22±0.15	9.65±1.25	12.55±1.01	4.28±1.01	1.21±0.13
Esp. Sant. Pinhal	Female	253.69±11.37	189.54±12.03	0.61±0.01	4.67±0.28	0.34±0.09	0.34±0.19	7.93±1.16	9.79±1.15	10.73±3.47	4.69±1.16
	Male	265.72±12.91	181.21±12.82	0.63±0.01	5.11±0.36	0.31±0.08	0.25±0.17	8.03±1.26	11.39±1.05	18.77±7.45	4.76±1.67
Guarapuava	Female	253.42±11.44	213.77±13.15	0.65±0.01	4.98±0.72	0.22±0.07	0.22±0.15	6.14±1.15	5.67±1.40	10.89±2.94	1.85±0.39
	Male	246.84±15.49	216.02±14.86	0.64±0.02	4.38±0.23	0.31±0.07	0.00±0.00	9,79±1.25	5.48±1.15	7.38±1.28	2.04±0.45
Guaxupé	Female	300.56±12.79	174.22±10.37	0.71±0.02	5.89±0.58	0.22±0.02	0.09±0.04	9.21±0.11	9.90±1.28	5.47±1.06	0.99±0.09
	Male	251.87±10.39	208.99±0.01	0.64±0.21	4.90±0.15	0.61±0.09	0.82±0.29	9.54±1.36	12.69±1.07	4.91±1.15	0.96±0.15
Jacarezinho	Female	281.53±10.12	174.87±9.82	0.66±0.01	5.00±0.40	0.50±0.08	0.91±0.31	8.65±1.26	8.44±1.47	7.36±1.36	1.24±0.16
	Male	429.52±8.30	169.93±8.20	0.70±0.01	4.48±0.23	0.54±0.09	0.61±0.28	12.31±1.05	8.70±1.43	7.71±1.95	1.69±0.30
Jacuí	Female	249.13±10.45	181.29±12.35	0.59±0.01	4.38±0.30	0.56±0.09	0.31±0.21	8.63±1.37	7.93±1.46	28.87±9.54	7.17±1.70
	Male	264.19±9.55	174.69±10.02	0.62±0.01	8.33±0.77	0.47±0.08	0.12±0.08	9.80±1.40	9.54±1.38	12.56±4.50	4.80±1.21
Juiz de Fora	Female	345.00±9.05	136.60±7.20	0.74±0.01	6.27±0.99	0.65±0.10	1.13±0.33	9.08±1.38	9.12±1.38	10.06±3.77	1.77±0.21
	Male	315.26±13.48	157.29±11.35	0.71±0.01	4.14±0.28	0.41±0.10	0.25±0.14	11.14±1.15	9.94±1.40	7.96±0.69	2.81±0.41
Piracicaba	Female	288.05±17.81	168.17±13.55	0.65±0.02	4.24±0.29	0.35±0.09	0.32±0.23	7.22±1.19	10.72±1.33	7.30±2.89	1.03±0.12
	Male	348.21±87.86	190.42±14.13	0.79±0.14	4.90±0.26	0.44±0.08	0.00±0.00	10.03±1.05	10.74±1.27	3.40±0.75	1.14±0.13
Proterito	Female	256.82±11.56	202.89±10.84	0.64±0.01	5.60±0.58	0.81±0.10	0.82±0.27	9.47±1.31	9.34±1.34	10.44±6.10	1.69±0.36
	Male	247.08±9.77	202.44±9.70	0.62±0.01	4.47±0.23	0.63±0.09	0.00±0.00	11.45±1.05	10.57±1.32	7.59±2.34	1.48±0.24
Rio Verde	Female	262.96±13.34	187.99±13.19	0.63±0.01	5.25±0.52	0.54±0.09	0.65±0.23	10.10±1.36	8.86±1.51	7.59±2.78	1.19±0.13
	Male	241.13±14.09	207.28±13.94	0.61±0.01	4.57±0.22	0.23±0.09	0.00±0.00	8.23±1.37	10.67±1.40	3.65±1.28	1.70±0.21
Sete Lagoas	Female	302.27±13.09	164.06±10.23	0.69±0.01	4.33±0.28	0.60±0.10	0.30±0.17	9.74±1.34	9.61±1.17	24.03±7.55	1.31±0.22
	Male	302.07±17.21	171.35±13.27	0.70±0.02	4.84±0.62	0.61±0.11	0.76±0.27	9.95±1.32	9.27±1.42	21.62±6.73	1.97±0.31
Unaí	Female	293.03±18.79	168.37±17.99	0.67±0.01	4.86±0.17	0.38±0.09	0.07±0.07	11.09±1.33	12.86±1.26	8.38±4.75	2.48±0.45
	Male	235.90±13.94	228.42±12.63	0.64±0.02	6.46±0.98	0.63±0.09	0.07±0.07	11.81±1.20	11.43±1.34	7.75±2.72	1.45±0.16
Viçosa	Female	268.94±13.64	174.51±13.27	0.64±0.02	5.24±0.23	0.39±0.09	0.21±0.15	10.98±1.30	12.10±1.05	7.48±1.74	2.10±0.30
	Male	311.85±13.22	149.02±11.11	0.69±0.01	5.71±0.86	0.19±0.06	0.37±0.20	8.53±1.14	10.27±1.31	7.02±1.27	2.13±0.34
Votuporanga	Female	270.52±15.26	187.43±13.81	0.65±0.02	4.81±0.32	0.47±0.09	0.98±0.23	10.10±1.33	9.78±1.27	9.16±3.45	2.97±0.45
	Male	291.70±12.75	190.58±10.53	0.71±0.03	4.92±0.29	0.49±0.08	0.69±0.25	9.75±1.08	7.37±1.21	14.44±2.66	2.39±0.26

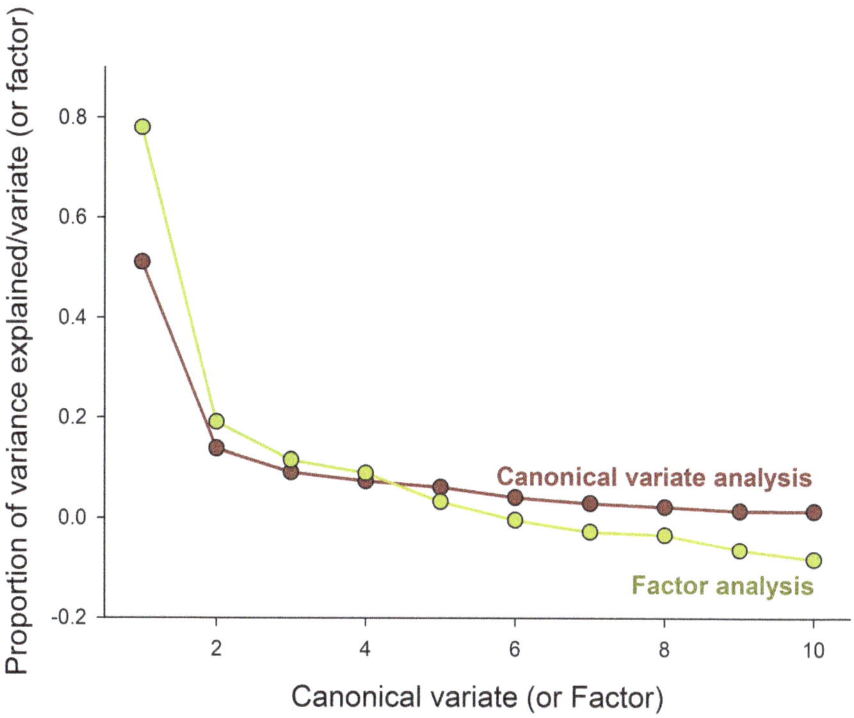

Figure 1. Proportion of the behavioral data variance explained by each canonical variate or factor generated from CVA or R-factor analysis, respectively.

the number of turns of each insect in the mill, allowing us to estimate the distance flown (m).

Body righting bioassay. Each insect was placed on its dorsum in an arena, and the time that was taken to recover its regular ventral posture was recorded. The procedure was replicated three times, and the mean score was used as the body righting activity, adapting a bioassay developed for the red flour beetle (*Tribolium castaneum*) [60].

Death-feigning bioassay. Death-feigning was induced by dorsally prodding the adult insect with a fine-haired brush and

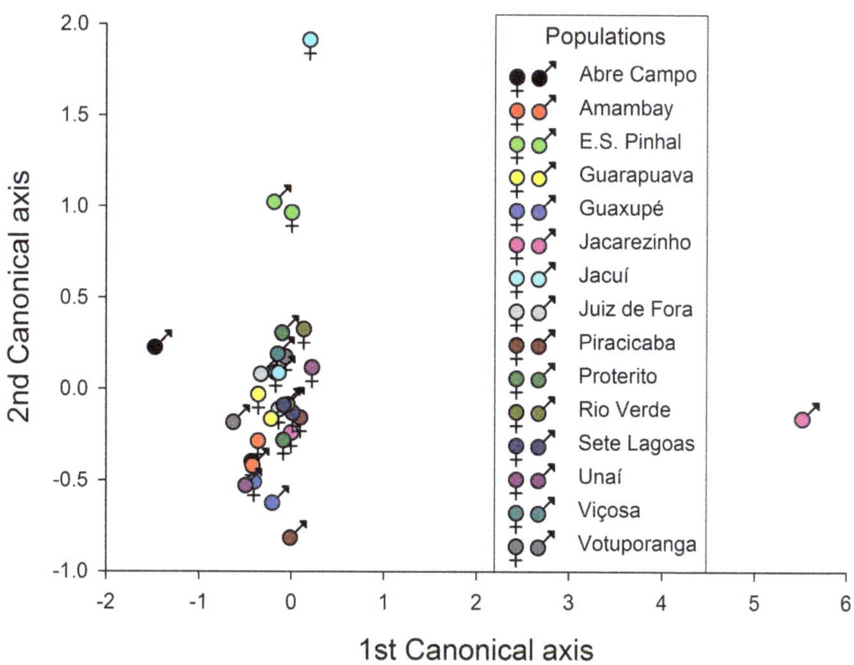

Figure 2. Ordination (CVA) diagram showing the behavioral divergence among males and females of populations of the maize weevil *Sitophilus zeamais* (see Table 3). The symbols are centroids of treatments representing the class mean canonical variates.

Table 3. Canonical (for CVA) and factor (for principal R-factor analysis) loadings of the significant ($p<0.05$) canonical axes (CVA) and factors for the behavioral traits of both sexes from 15 populations of the maize weevil (*Sitophilus zeamais*).

Variables	Canonical axes						Principal main factors (orthogonally rotated)	
	1	2	3	4	5	6	1	2
Distance walked	**0.68**	−0.20	0.06	−0.15	**0.52**	0.06	**0.97**	−0.06
Resting time	−0.15	−0.23	−0.12	0.28	**−0.86**	−0.00	**−0.63**	−0.11
Walking velocity	0.09	**−0.49**	0.11	−0.16	**0.51**	0.11	**0.89**	−0.03
Horizontal dislocation upon fall	−0.20	0.02	0.11	**0.75**	0.28	**−0.48**	0.05	−0.23
Wing beat	0.17	−0.26	**0.83**	0.36	0.02	0.15	0.01	−0.26
Distance flown	0.10	−0.16	**0.68**	−0.31	0.25	**−0.49**	0.06	−0.32
Conspecific interaction latency	0.25	−0.30	0.13	**0.53**	0.21	**0.33**	−0.01	−0.03
Heterospecific interaction latency	−0.16	−0.12	−0.37	0.20	**0.50**	**−0.60**	−0.13	−0.17
Duration of death−feigning	−0.04	**0.64**	0.49	−0.16	−0.03	**0.37**	−0.02	0.35
Length of time to upturn	−0.04	**0.97**	0.06	0.07	−0.06	0.02	−0.02	0.40
$F_{appr.}$	4.20	2.67	2.25	1.98	1.73	1.44	–	–
p	<0.001	<0.001	<0.001	<0.001	<0.001	0.001	–	–
Eigenvalue	1.21	0.33	0.22	0.17	0.15	0.10	2.17	0.53

Bold type indicates the main contributors of each axis.

subsequently recording the time taken for the insect to start moving after reaching its characteristic death-feigning (or thanatosis) posture. This posture in the maize weevil involves leg contraction and subsequent immobilization upon prodding. The procedure was replicated three times, and the mean score was used as the duration of death-feigning behavior, again adapting a method developed for the red flour beetle [60].

Intra- and interspecific interaction bioassays. The exploration/avoidance behavioral category was assessed through bioassays of intra- and interspecific interactions using maize kernels infested by adults. A Petri dish arena (18 cm diameter) whose inner walls were coated with Teflon PTFE to prevent insect escape was used in this bioassay. The bottom of the arena was covered with millimeter paper on which the center was marked. A

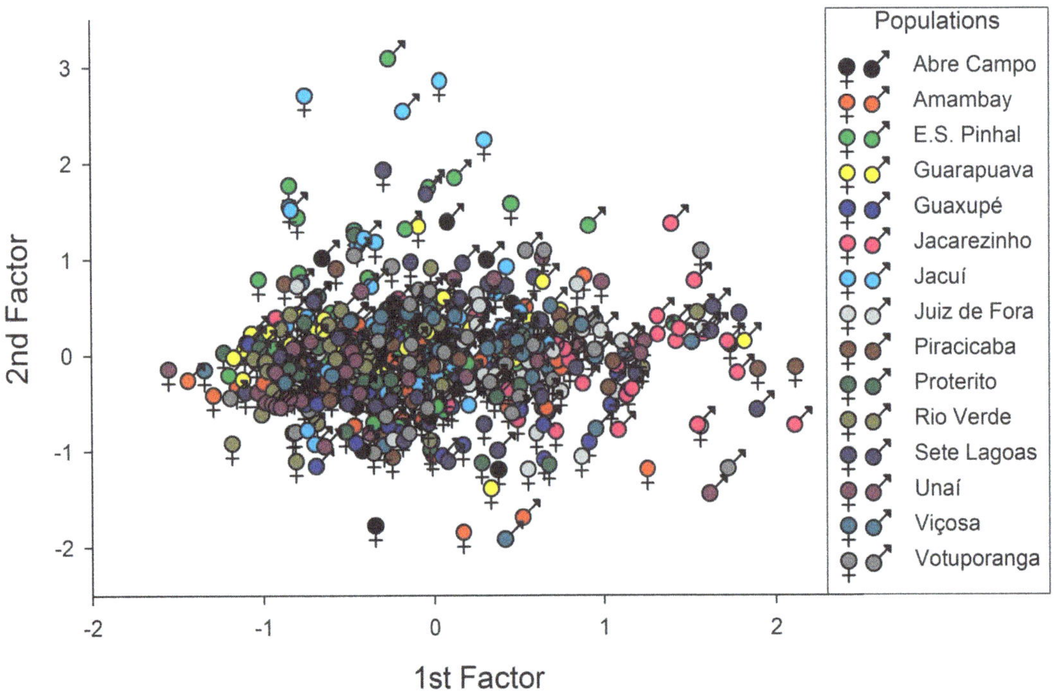

Figure 3. Ordination (principal R-factor) diagram showing the behavioral divergence among adult (male and female) individuals of the maize weevil *Sitophilus zeamais* belonging to 15 populations of this species (Table 2).

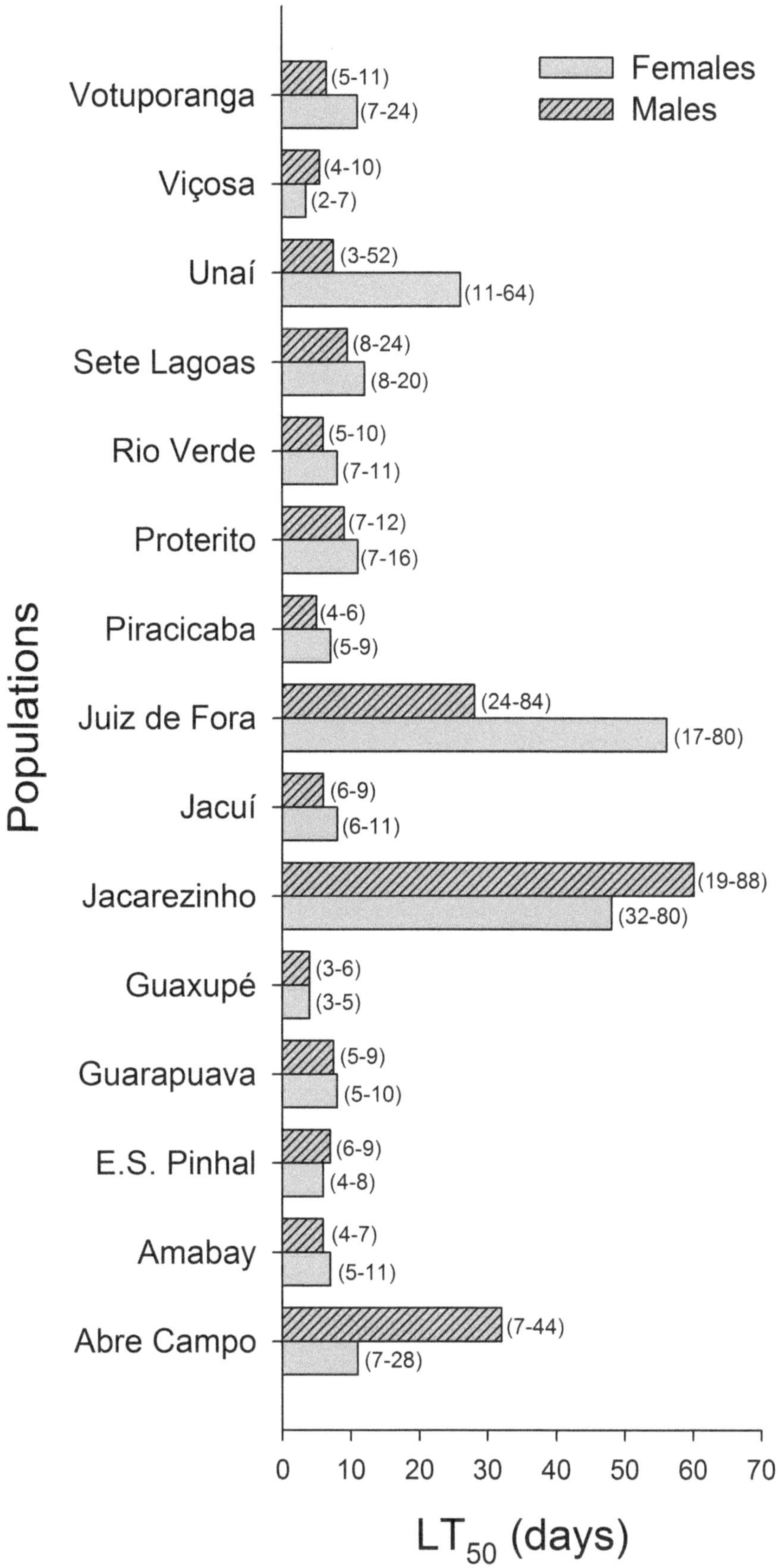

Figure 4. Median survival time (TL_{50}) to deltamethrin exposure of both sexes from 15 populations of the maize weevil *Sitophilus zeamais*. The histogram bars indicate the estimated TL_{50}'s, and their respective 95% CIs are between parentheses.

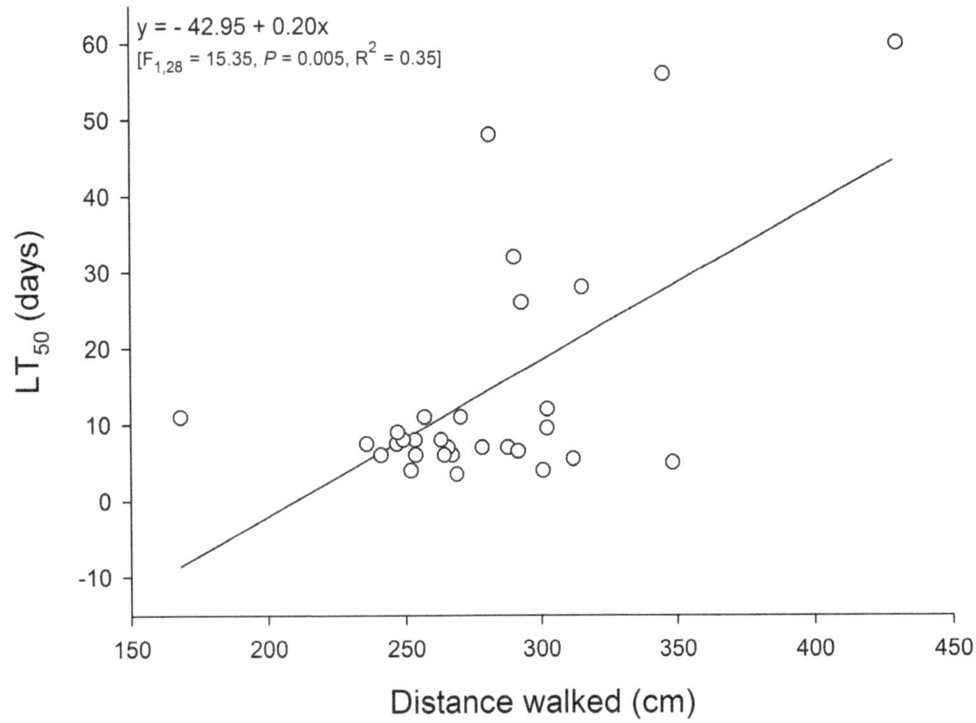

Figure 5. Effect of distance walked on the median lethal time (TL_{50}) to deltamethrin exposure of adults from 15 populations of the maize weevil *Sitophilus zeamais*. The symbols indicate the observed data.

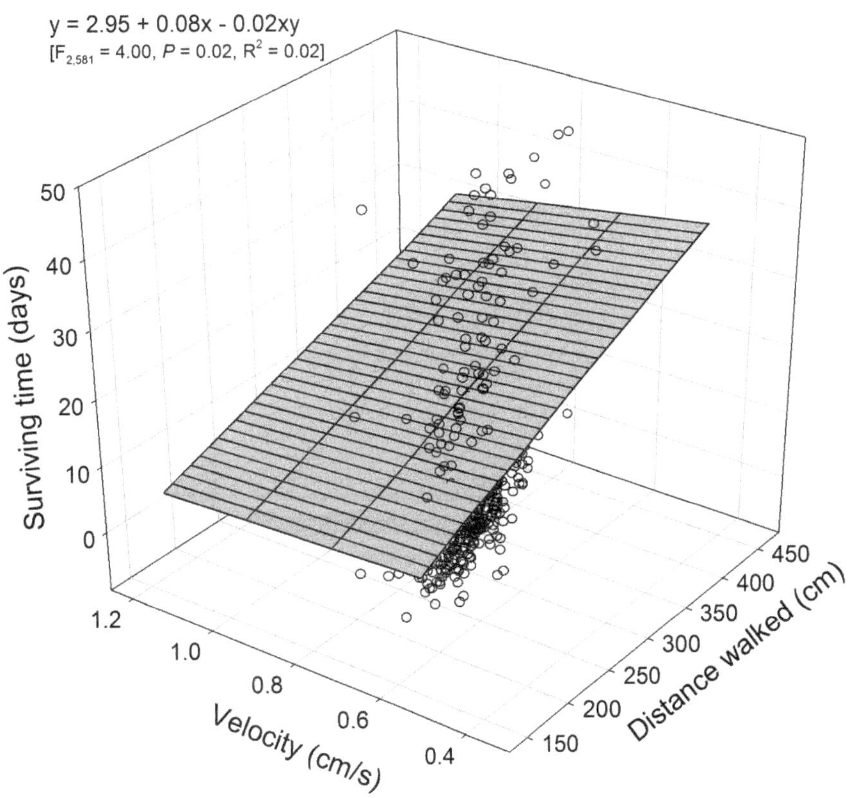

Figure 6. Effect of distance walked and walking velocity on survival time to deltamethrin exposure of adults from 15 populations of the maize weevil *Sitophilus zeamais*. The symbols indicate the observed data.

maize kernel infested with either a conspecific (for intraspecific interaction) or heterospecific (lesser grain borer *Rhyzopertha dominica*; for interspecific interaction) adult was placed at the center of the arena, and the weevil under investigation was placed at the edge of the arena. The weevil movement was recorded for 15 min, as well as the time (min) taken for the weevil to reach the infested kernel.

Insecticide Survival Bioassays

The pyrethroid insecticide deltamethrin (K-Obiol 25EC, Bayer Crop Science, São Paulo, Brazil) was used at its registered label rate for maize weevil control in Brazil (0.5 ppm) [61]. One milliliter of insecticide emulsion was sprayed on 500 g of maize kernels placed in a rotary stainless steel container for homogenizing the grain during the application and until the grain was dry (one hour later). An artist's air brush (Sagyma SW440A, Yamar Brasil, São Paulo, Brazil) coupled with an air pump (Prismatec 131A Tipo 2VC, Itu, SP, Brazil) was used for insecticide spraying, which was performed at a pressure of 0.7 kgf/cm^2.

Transparent glass vials (25 mL) containing 15 g of maize kernels (sprayed with the deltamethrin formulation) were used as experimental units and received one insect each. Parallel bioassays were performed with insects of similar age (one week old) from the same sex, and populations (although not subjected to the behavioral bioassays previously described) exposed to maize kernels that were sprayed with only water acted as controls to detect any problems of high natural mortality compromising the insecticide bioassay, which was not the case. The cap area of the upper portion of the vials had its inner walls coated with Teflon PTFE, and the vial opening was covered with organza tissue tightly attached with a rubber band to prevent insect escape and allow gas exchange. Insect survival was assessed hourly during the 1st 12 h and then at four hour intervals until 30 days after spraying. After 30 days, insect survival was recorded daily until each insect's eventual death.

Statistical Analyses

The results of the behavioral bioassays and their repetition (i.e., retesting) for each individual insect were subjected to correlation analysis to assess their consistency (PROC CORR in SAS (SAS Institute, Cary, NC, USA)) [62]. The behavioral traits assessed were subject to two sets of analyses, one focusing on the population (or inter-population) variation and the other focusing on individual (or within population) variation. Multivariate analysis of variance (complemented by univariate analysis of variance) and canonical variate analysis (CVA) were performed using population and sex as independent variables and the behavioral traits as dependent (response) variables, focusing on the population level variation using the CANDISC procedure in SAS [62]. For individual variation, the behavioral traits of individuals from both sexes and different populations were subjected to principal R-factor analysis using the FACTOR procedure in SAS with orthogonal (Varimax) rotation to reduce potential bias towards the first general factor [62]. Sampling adequacy was estimated using Kaiser's measure for the purpose, which should significantly exceed 0.50 if the correlation matrix generated were suitable [62]. Factor analysis was preferred instead of the usual principal component analysis because the former aims to explain the variation in the measured variables by constructing latent ones (i.e., factors), enabling comparisons and generalizations across bioassays instead of solely focusing on data reduction [11,63]. The relative contribution of populations and individuals within populations for the observed behavioral data variation was calculated using the modified analysis of variance proposed by Excoffier et al. [64].

The results of the survival bioassays were subjected to survival analysis using the non-parametric LIFETEST procedure in SAS [62], in which survival curves are obtained using Kaplan-Meyer estimators, allowing estimates of median survival time (LT_{50}) for each sex and each population. These estimates were subsequently used as dependent (response) variables in multiple regression analysis using either canonical variates or behavioral data responses as independent variables with the GLM procedure in SAS and both 'stepwise' and 'backward' selection statements. Initially, only the original variables were used to construct the model, and subsequently, their improvement was attempted considering the interactions between the main original variables [62]. This model-building approach was used for exploring the population variation data, but for the individual variation data, the survival time (days) was modeled instead of the LT_{50}, and the respective individual data, either from the main factors or behavioral data, were used for constructing multiple regression models with the GLM procedure in SAS and the 'stepwise' and 'backward' selection statements [62]. The assumptions of normality and homoscedasticity were evaluated before data analysis (UNIVARIATE procedure in SAS) [62], and wing beat and distance flown required transformation (log x+1) for the intended analyses.

Results

Repeatability and Multidimensional Behavioral Constructs

The results of testing and retesting the behavioral bioassays provided significant results when correlated with one another ($n = 630$, $p<0.001$), exhibiting correlation coefficients (r) ranging from 0.28 to 0.57 (Table 1). The multivariate analysis of variance performed indicated significant overall effects of population, sex and interaction of sex-population among the behavioral traits analyzed (Wilks' lambda <0.94, F >3.34, $p<0.001$) (Table 1). Subsequent univariate analyses of variance performed for each behavioral trait assessed indicated that all behavioral traits were significantly affected by population, sex and/or their interaction ($F_{29,600} = 1.93$, $p<0.003$), except latency for interacting with a conspecific ($F_{29,600} = 1.47$, $p = 0.06$) (Tables 1 and 2). These results indicate great overall variation among populations and sex, which is also true for each individual behavioral trait assessed, except latency of conspecific interaction.

The multidimensional behavioral constructs representing the population-level 'personality' of each sex from each population were obtained with a CVA of the behavioral traits assessed. The CVA ordination generated six significant axes ($p<0.05$), with the first two explaining 64.75% of the observed variance (Fig. 1). The variable with greater canonical loadings accounting for most of the divergence among sex and populations was the distance walked (1st canonical axis), followed by the duration of the death-feigning behavior and length of time to right the body (2nd canonical axis), with an opposite contribution mainly from the walking velocity (Table 3). The ordination diagram derived from the CVA representing the prevailing 'personality' of each sex from each population emphasizes the distinction of males from the population of Jacarezinho (by the 1st axis), mainly because of their extensive walking activity, and females from Jacuí (by the 2nd axis), which exhibited delayed body righting and recovery from death-feigning (Fig. 2; Table 1).

The multidimensional behavioral constructs representing individual-level weevil 'personality' were obtained with the R-factor analysis of the behavioral traits assessed. Such analysis was deemed suitable because the Kaiser's measure of sampling adequacy was

0.68, significantly exceeding the value of 0.5. Among the factors generated, the first two explained over 97% of the observed variance (Fig. 1). The behavioral traits explaining most of the divergence among individual weevils were the distance walked and walking velocity, with an opposing contribution from resting time (1st factor; Table 3). The 2nd factor exhibited a low Eigenvalue (<1.0) and was retained only aiming the graphical representation (Table 3). The ordination diagram that was obtained with the R-factor analysis representing the diversity of the individual weevil 'personalities' is exhibited in Fig. 3. The relative contributions of populations and of individuals within populations for the observed behavioral data variation were estimated to be 7.43 and 92.57%, respectively.

Insecticide Survival

The survival analysis of the data from weevils exposed to dried deltamethrin residues on maize kernels indicated significant differences among sexes and populations (Log-rank test, $\chi^2 = 223.67$, df = 29, $P<0.001$), allowing the estimation of the respective median lethal time (TL_{50}'s) using Kaplan-Meyer estimators (Fig. 4). These population estimates were obtained to allow subsequent testing through multiple regression of the potential role of population-based weevil 'personality' on the insecticidal efficacy of deltamethrin; this test was not performed for individual weevil 'personality', in which the survival time on sprayed kernels was used.

Weevil 'Personality' and Survival to Deltamethrin Exposure

The model-building procedures used for multiple regression analysis using canonical axes or factors as independent variables to estimate susceptibility to deltamethrin provided neither significant results in the case of factors ($p<0.05$), nor results more robust than when the behavioral traits were directly used in the analysis, which did occur for the canonical axes. Therefore, the behavioral traits were directly used in the multiple regression analysis instead of the multidimensional behavioral constructs obtained with CVA and R-factor analysis. Following this procedure, the distance walked was the only significant predictor of LT_{50} among weevil populations of both sexes, providing robust estimations ($R^2 = 0.35$; Fig. 5). When individual behavioral traits were considered to estimate deltamethrin susceptibility, the distance walked was again the main significant estimator of deltamethrin susceptibility, but with a significant contribution of its interaction with walking velocity (Fig. 6). However, the regression model obtained provided only poor estimations of survival ($R^2 = 0.02$).

Discussion

The 'personality' concept implies stability (or repeatability) in how the individual behaves, which can be estimated through test-retest correlations. Weevils exhibit significant correlations in behavioral traits subjected to test-retest, indicating that these traits are stable, but their reliability ranged from low to moderate. Low but significant reliability levels ($r < 0.25$) are frequent among animal behavioral traits. In our case, some of the behavioral traits measured for the weevils, such as death-feigning and walking-related traits, reached moderate levels ($r \approx 0.50$).

Weevils also exhibited multidimensional behavioral constructs both at the population and individual levels, which are basically diagrammatic representations of 'personality' as an individual's integrated set of behavioral tendencies inferred from multiple empirical measures. Therefore, as observed since the 1970s for vertebrates and the 1990s for invertebrates [65,66], weevils do exhibit 'personality' as defined here, joining the limited group of insects in which 'personality' and 'personality' variation have been detected [28–32]. Previous population-based studies with maize weevils, particularly in interactions with insecticides, previously indicated that some populations are noticeably and consistently more active than others [46,48,49,57,67]. The present study expands this recognition in maize weevils to the individual level and to multiple behavioral traits encompassing the three behavioral categories (activity, boldness/shyness and exploration/avoidance) explored within the five categories previously proposed (which also includes sociability and aggressiveness) [5]. The personality constructs obtained for weevils suggest that they may also exhibit suites of behavioral correlations across multiple contexts (i.e., behavioral syndromes), particularly between traits within the activity and boldness categories, which deserves future attention.

'Personality' is an individual attribute, and here, we observed that the bulk of the variation in the behavioral traits measured were because of individual variation within populations, with only a minor contribution from variation between populations. These results were reflected in the weevil personality constructs generated based on individual variation, which were simpler and more robust than those generated based on population variation, which required more ordination axes with less explicative power than the former. In fact, even the population-based diagram generated with CVA seems to be a rather crude approximation of the individual-based diagram generated with R-factor analysis. However, weevil individual 'personality' is roughly translated into weevil population 'personality', laying credence to previous population-based studies of weevil behavioral variation [46,48,49,57,67,68]. Such a finding also provides support for the potential use of the 'personality' concept in comparative studies of populations and even species [3–7,9,10].

The adaptive value of 'personality' is an emerging subject of attention. Among pest insects, the ability to withstand insecticidal applications has both theoretical and practical adaptive importance. Insect behavior is a determinant of insecticide exposure and activity and is also recognized as an insecticide resistance mechanism [38–49]. A parallel has been thoroughly explored in psychology, in which child personality has been associated not only with the risk of poisoning but also with injury risk [12,15,34–37]. By contrast, the potential association between 'animal personality' and chemical contamination remains largely unexplored despite its importance, especially for pest insects.

The potential association between maize weevil 'personality' as a determinant of its survival to the insecticide deltamethrin was tested in our study. The weevil personality constructs obtained were not associated with survival to deltamethrin exposure, which was better explained by individual behavioral traits, mainly the walking activity, when considering both populations and individuals. Therefore, simple behavioral traits, rather than complex 'personality' constructs, were more efficient determinants of survival to deltamethrin exposure, even when physiologically resistant insects were considered although the relationship was weak when individual variation was considered [52–55,68].

We initially expected that increased activity, mainly the walking activity, would be more likely to reduce survival to deltamethrin exposure based on previous studies of the maize weevil [46,48,57,68]. We presented this hypothesis because increased walking activity is expected to increase insecticide exposure and thus favor insecticidal activity. By contrast, our results show that survival to deltamethrin exposure was favored by higher walking activity, which was consistent for both population and individual weevil measurements. However, the general assumption of increased walking activity favoring insecticide exposure does not consider the possibility of escape to untreated surfaces or the possibility of insecticide-mediated behavioral avoidance. Both contingencies occurred in the present study. Only the maize kernels, rather than the vials, received deltamethrin application, allowing the insects to crawl on the vial's unsprayed inner walls. In addition, deltamethrin is known to frequently induce some behavioral avoidance in populations of the maize weevil [46,68,69], which also stimulates their escape to untreated areas, lowering insecticide exposure and explaining the observed association between an increase in the distance walked and an increase in survival to deltamethrin exposure. Furthermore, more active individuals able to cover greater distances are likely of better quality, healthier and thus potentially able to withstand the insecticide effects for longer.

Our experimental set-up was a simplification of field conditions, which may have led to an oversimplified outcome, minimizing the potential role of complex weevil 'personality' constructs in survival to deltamethrin exposure. Such confined set-up may also have restrained individual behavioral expression leading to a closer relationship between individual and population results than would take place under unrestrained conditions. Therefore, a more realistic field (or storage) setting for the insecticide survival test may highlight the manifestation of other behavioral traits that could also influence the survival of weevils.

In summary, we report on stable weevil 'personality' constructs and highlight the importance of the individual variation of behavioral traits in determining weevil 'personality'. Nonetheless, the individual weevil 'personality' can be roughly translated into weevil population 'personality'. These relatively complex constructs were not as efficient at determining survival to deltamethrin exposure as the walking activity alone, which plays a significant role in extending survival, and therefore fitness, under insecticide-contaminated conditions.

Acknowledgments

Technical assistance provided by N. Guedes and E. Cordeiro was greatly appreciated, as were the comments and suggestions provided by C. Cruz and L. Gontijo. The comments and suggestions provided by the editor and two anonymous reviewers were also greatly appreciated and acknowledged here.

Author Contributions

Conceived and designed the experiments: JAM TMCDL RNCG. Performed the experiments: JAM DGC. Analyzed the data: JAM RNCG. Contributed reagents/materials/analysis tools: TMCDL RNCG. Wrote the paper: JAM RNCG.

References

1. Gosling SD (2001) From mice to men: what can we learn about personality from animal research? Psychological Bulletin. 127: 45–86.
2. Sih A, Bell AM, Johnson JC, Ziembra RE (2004) Behavioral syndromes: an integrative overview. The Quarterly Review of Biology 79: 241–277.
3. Sih A, Bell A, Johnson JC (2004) Behavioral syndromes: an ecological and evolutionary overview. Trends in Ecology and Evolution 19: 372–378.
4. Dingemanse NJ, Réale D (2005) Natural selection and animal personality. Behaviour 142: 1165–1190.
5. Réale D, Reader SM, Sol D, McDougall PT, Dingemanse NJ (2007) Integrating animal temperament within ecology and evolution. Biological Reviews 82: 291–318.
6. Sih A, Bell AM (2008) Insights of behavioral ecology from behavioral syndromes. Advances in the Study of Behaviour 38: 227–281.
7. Chapple DG, Simmonds SM, Wong BBM (2011) Can behavioral and personality traits influence the success of unintentional species introductions? Trends in Ecology and Evolution 27: 57–64.
8. Webster MM, Ward AJW (2011) Personality and social context. Biological Reviews 86: 759–773.
9. Sih A, Cote J, Evans M, Fogarty S, Pruitt J (2012) Ecological implications of behavioral syndromes. Ecology Letters 15: 278–289.
10. Wolf M, Weissing FJ (2012) Animal personalities: consequences for ecology and evolution. Trends in Ecology and Evolution 27: 452–461.
11. Uher J (2011) Individual behavioral phenotypes: an integrative meta-theoretical framework. Why "behavioral syndromes" are not analogs of "personality". Developmental Psychobiology 53: 521–548.
12. Sibert JR, Newcombe RG (1977) Accidental ingestion of poisons and child personality. Postgraduate Medical Journal 53: 254–256.
13. Caspi A, Harrington H, Moffitt TE, Begg D, Dickson N, et al (1997) Personality differences predict health-risk behaviours in young adulthood: evidence from a longitudinal study. Journal of Personality and Social Psychology 73: 1052–1063.
14. Ellis BJ, Jackson JJ, Boyce WT (2006) The stress response system: universality and adaptive individual variation. Developmental Review 26: 175–212.
15. Garzon DL, Huang H, Todd RD (2008) Do attention deficit/hyperactivity disorder and oppositional defiant disorder influence preschool unintentional injury risk? Archives of Psychiatric Nursing 22: 288–296.
16. Boissy A, Bouissou MF (1995) Assessment of individual differences in behavioral reactions of heifers exposed to various fear-eliciting situations. Applied Animal Behavioral Science 46: 17–31.
17. Le Neindre P, Boivin X, Boissy A (1996) Handling of extensively kept animals. Applied Animal Behaviour Science 49: 73–81.
18. Carlstead K, Mellen J, Kleiman DG (1999) Black rhinoceros (*Diceros bicornis*) in US zoos: I. Inidividual behavior profiles and their relationship to breeding success. Zoo Biology 18: 17–34.
19. Trut LN (1999) Early canid comestication: the farm-fox experiment. American Scientist 87: 160–169.
20. Malmkvist J, Hansen SW (2001) The welfare of farmed mink (*Mustela vison*) in relation to behavioural selection: a review. Animal Welfare 10: 41–52.
21. Boissy A, Fisher AD, Bouix J, Hinch GN, Le Neindre P (2005) Genetics of fear in ruminant livestock. Livestock Production Science 93: 23–32.
22. McDougall PT, Réale D, Sol D, Reader SM (2006) Wildlife conservation and animal temperament: an evolutionary overview. Animal Conservation 9: 39–48.
23. Siebert C (2006) The animal self. The New York Times, January 22: 1–7. Available: http://www.nytimes.com/2006/01/22/magazine/22animal.html?pagewanted=all&_r=0. Accessed 2013 Jan 10.
24. Ogden LE (2012) Do animals have personality? BioScience 62: 533–537.
25. Johnson JC, Sih A (2007) Fear, food, sex and parental care: a syndrome of boldness in the fishing spider *Dolomedes triton*. Animal Behaviour 74: 1131–1138.
26. Pruit JN, Riechart SE, Jones TC (2008) Behavioural syndromes and their fitness consequences in a socially polymorphic spider *Anelosimus studiosus*. Animal Behaviour 76: 871–879.
27. Pruit JN, Riechart SE, Iturralde G, Vega M, Fitspatrick BM, et al (2010) Population differences in behaviour are explained by shared within-population trait correlations. Journal of Evolutionary Biology 23: 748–756.
28. Barth MB, Kellner K, Heinze J (2010) The police are not the army: context-dependent aggressiveness in a clonal ant. Biology Letters 6: 329–332.
29. Wray MK, Seeley TD (2011) Consistent personality differences in house-hunting behavior but not decision speed in swarms of honey bees (*Apis mellifera*). Behavioral Ecology and Sociobiology 65: 2061–2070.
30. Schuett W, Dall SRX, Baeumer J, Kloesener MH, Nakagawa S, et al (2011) "Personality" variation in a clonal insect: the pea aphid, *Acyrthosiphon pisum*. Developmental Psychobiology 53: 631–640.
31. Nakayama S, Sasaki K, Matsumura K, Lewis Z, Miyatake T (2012) Dopaminergic system as the mechanism underlying personality in a beetle. Journal of Insect Physiology 58: 750–755.
32. Pinter-Wollman N (2012) Personality in social insects: how does worker personality determine colony personality? Current Zoology 58: 580–588.
33. Pruitt JN, Riechert SE (2012) The ecological consequences of temperament in spiders. Current Zoology 58: 589–596.
34. Margolis R (1971) A psychological study of childhood poisoning. A five-year follow-up. Pediatrics 47: 439–444.
35. Wazana A (1997) Are there injury-prone children? A critical review of the literature. Canadian Journal of Psychiatry 42: 602–610.
36. Bijttebier P, Vertommen H, Florentie (2003) Risk-taking behavior as a mediator of the relationship between childreǹs temperament and injury liability. Psychology and Health 18: 645–653.
37. Katrivanou A, Lekka N, Beratis S (2004) Psychopathology and behavioural trends of children with accidental poisoning. Journal of Psychosomatic Research 57: 95–101.
38. Anonymous (1958) Exophily in anophelines and malaria control. WHO Chronicle 12: 81–82.
39. Akiyama J (1973) Exophily in *Anopheles gambiae* species B in Sudan. Transactions of the Royal Society of Tropical Medicine and Hygiene 67: 440.
40. Lockwood JA, Sparks TC, Story RN (1984) Evolution of insect resistance to insecticides: a reevaluation of the roles ofphysiology and behavior. Bulletin of the Entomological Society of America 30: 41–51.

41. Lockwood JA, Byford RL, Story RN, Sparks TC, Quisenberry SS (1985) Behavioral resistance to the pyrethroids in the horn fly, *Haematobia irritans* (Diptera: Muscidae). Environmental Entomology 14: 873–880.
42. Haynes KF (1988) Sublethal effects of neurotoxic insecticides on insect behavior. Annual Review of Entomology 33: 149–168.
43. Suiter KA, Gould F (1994) Physiological and behavioral avoidance responses to residues of four pesticides by six spider mite populations. Entomologia Experimentalis et Applicata 71: 1–14.
44. Wang C, Scharf ME, Bennett GW (2004) Behavioral and physiological resistance of the German cockroach to gel baits (Blattodea: Blatellidae). Journal of Economic Entomology 97: 2067–2072.
45. Muenworm V, Akaratanakul P, Bangs MJ, Parbaripai A, Chareonviriyaphap T (2006) Insecticide-induced behavioral responses in two populations of *Anopheles maculatus* and *Anopheles sawadwongporni*, malaria vectors in Thailand. Journal of the American Mosquito Control Association 22: 689–698.
46. Guedes NMP, Guedes NMP, Ferreira GH, Silva LB (2009) Flight take-off and walking behavior of insecticide-susceptible and -resistant strains of *Sitophilus zeamais* exposed to deltamethrin. Bulletin of Entomological Research 99: 393–400.
47. Cordeiro EMG, Corrêa AS, Venzon M, Guedes RNC (2010) Insecticidal survival and behavioral avoidance in the lacewings *Chrysoperla externa* and *Ceraeochrysa cubana*. Chemosphere 81: 1352–1357.
48. Braga LS, Corrêa AS, Pereira EJG, Guedes RNC (2011) Face or flee? Fenitrothion resistance and behavioral response in populations of the maize weevil, *Sitophilus zeamais*. Journal of Stored Products Research 47: 161–167.
49. Pimentel MAG, Faroni LRA, Corrêa AS, Guedes RNC (2012) Phosphine-induced walking response of the lesser grain borer (*Rhyzopertha dominica*). Pest Management Science 68: 1368–1373.
50. Jallow MFA, Hoy CW (2005) Phenotypic variationin adult behavioral response and offspring fitness in *Plutella xylostella* (Lepidoptera: Plutellidae) in response to permethrin. Journal of Economic Entomology 98: 2195–2202.
51. Martini X, Kincy N, Nansen C (2012) Quantitative impact assessment of spray coverage and pest behavior on contact pesticide performance. Pest Management Science 68: 1471–1477.
52. Guedes RNC, Lima JOG, Santos JP, Cruz CD (1995) Resistance to DDT and pyrethroids in Brazilian populations of *Sitophilus zeamais* Motsch. (Coleoptera: Curculionidae). Journal of Stored Products Research 31: 145–150.
53. Fragoso DB, Guedes RNC, Rezende ST (2003) Gluthatione *S*-transferase detoxification as a potential pyrethroid resistance mechanism in the maize weevil, *Sitophilus zeamais*. Entomologia Experimentalis et Apllicata 109: 21–29.
54. Fragoso DB, Guedes RNC, Oliveira MGA (2007) Partial characterization of glutathione *S*-transferases in pyrethroid-resistant and -susceptible populations of the maize weevil, *Sitophilus zeamais*. Journal of Stored Products Research 43: 167–170.
55. Araújo RA, Williamson MS, Bass C, Fields LM, Duce IR (2011) Pyrethroid resistance in *Sitophilus zeamais* is associated with a mutation (T929I) in the voltage-gated sodium channel. Insect Molecular Biology 20: 437–445.
56. Tolpo NC, Morrison EO (1965) Sex determination by snout characteristics of *Sitophilus zeamais* Motschulsky. The Texas journal of Science 7: 122–124.
57. Pereira CJ, Pereira EJG, Cordeiro EMG, Della Lucia TMC, Tótola MR, et al (2009) Organosphosphate resistance in the maize weevil *Sitophhilus zeamais*: magnitude and behavior. Crop Protection 28: 168–173.
58. Ohno T, Miyatake T (2007) Drop or fly? Negative genetic correlation between death-feigning intensity and flying ability as alternative anti-predator strategies. Proceedings of the Royal Society B 274: 555–560.
59. Riley JR, Dowham MCA, Cooter RJ (1997) Comparison of the performance of *Cicadulina* leafhoppers of flight mills with that to be expected in free flight. Entomologia Experimentalis et Applicata 83: 317–322.
60. Miyatake T, Tabuchi K, Sasaki K, Okada K, Katayama K, et al (2008) Pleiotropic antipredator strategies, fleeing and feigning death, correlated with dopamine levels in *Tribolium vastaneum*. Animal Behaviour 75: 113–121.
61. Ministério da Agricultura, Pecuária e Abastecimento (2012) *AGROFIT: Sistema de Agrotóxicos Fitossanitários*. Brasília, Brazil: MAPA/CGAF/DFIA/DAS. Available: http://extranet.agricultura.gov.br/agrofit_cons/principal_agrofit_cons. Accessed 2013 Jan 15.
62. SAS Institute (2008) SAS User's Guide: Statistics. SAS Institute, Cary, NC, USA.
63. Capitano JP (1999) Personality dimensions in adult male rhesus mocaques: prediction of behaviors across time and situation. American Journal of Primatology 47: 299–320.
64. Excoffier L, Smouse PE, Quattro JM (1992) Analysis of molecular variance inferred from metric distances among DNA haplotypes: application to human mitochondrial DNA restriction data. Genetics 131: 479–491.
65. Huntingford FA (1976) The relationship between anti-predaotr behavior and aggression among conspecifics in the three-spined stickleback, *Gasterosteus aculeatus*. Animal Behaviour 24: 245–260.
66. Mather JA, Anderson RC (1993) Personalities of octopuses (*Octopus rubescens*). Journal of Comparative Psychology 107: 336–340.
67. Sousa AH, Faroni LRA, Silva GN, Guedes RNC (2012) Ozone toxicity and walking response of populations of *Sitophilus zeamais*. Journal of Economic Entomology 105: 2187–2195.
68. Corrêa AS, Pereira EJG, Cordeiro EMG, Braga LS, Guedes RNC (2011) Insecticide resistance, mixture potentiation and fitness in populations of the maize weevil (*Sitophilus zeamais*). Crop Protection 30: 1655–1666.
69. Guedes NMP, Guedes RNC, Silva LB, Cordeiro EMG (2009) Deltamethrin-induced feeding plasticity in pyrthroid-susceptible and -resistant strains of the maize weevil, *Sitophilus zeamais*. Journal of Applied Entomology 133: 524–532.

9

Pesticide Exposure as a Risk Factor for Myelodysplastic Syndromes: A Meta-Analysis Based on 1,942 Cases and 5,359 Controls

Jie Jin[1,2¶], Mengxia Yu[1,2¶], Chao Hu[1,2¶], Li Ye[1,2], Lili Xie[1,2], Jin Jin[1,2], Feifei Chen[1,2], Hongyan Tong[1,2,3*]

1 Department of Hematology, the First Affiliated Hospital of Zhejiang University, Hangzhou, People's Republic of China, **2** Institute of Hematology, Zhejiang University School of Medicine, Hangzhou, People's Republic of China, **3** Myelodysplastic syndromes diagnosis and therapy center, Zhejiang University School of Medicine, Hangzhou, People's Republic of China

Abstract

Objective: Pesticide exposure has been linked to increased risk of cancer at several sites, but its association with risk of myelodysplastic syndromes (MDS) is still unclear. A meta-analysis of studies published through April, 2014 was performed to investigate the association of pesticide exposure with the risk of MDS.

Methods: Studies were identified by searching the Web of Science, Cochrane Library and PubMed databases. Summary odds ratios (ORs) with corresponding 95% confidence intervals (CIs) were calculated using random- or fixed-effect models.

Results: This meta-analysis included 11 case-control studies, all of which demonstrated a correlation between pesticide exposure and a statistically significant increased risk of MDS (OR = 1.95, 95% CI 1.23–3.09). In subgroup analyses, patients with pesticide exposure had increased risk of developing MDS if they were living in the Europe or Asia and had refractory anemia (RA) or RA with ringed sideroblasts (RARS). Moreover, in the analysis by specific pesticides, increased risk was associated with exposure to insecticides (OR = 1.71, 95% CI 1.22–2.40) but not exposure to herbicides or fungicides.

Conclusion: This meta-analysis supports the hypothesis that exposure to pesticides increases the risk of developing MDS. Further prospective cohort studies are warranted to verify the association and guide clinical practice in MDS prevention.

Editor: Shawn Hayley, Carleton University, Canada

Funding: This study was supported by grants from the Key Innovation Team Foundation of Zhejiang Province (2011R50015), major program fund of the Science Technology Department of Zhejiang Province (2013c03043-2), Zhejiang Province Fund for Distinguished Young Scholars (LR12H08001), National Public Health Grand Research Foundation (201202017) and the National Natural Science Foundation of China (No.30870914, No.81270582). The funders had no role in study design, data collection and analysis, decision to publish, or preparation of the manuscript.

Competing Interests: The authors have declared that no competing interests exist.

* Email: zjuhongyantong@163.com

¶ These authors are joint first authors on this work.

Introduction

Myelodysplastic syndromes (MDS) are a heterogeneous group of stem cell malignancies, characterized by ineffective hematopoiesis, and peripheral blood cytopenias. With disease progression, the risk of transformation into acute myeloid leukemia (AML) increased [1,2]. Despite development of new therapeutic methods in recent years, treatment of MDS is still limited and MDS remains incurable except in the case of the younger patients with good performance status, allogeneic stem cell transplantation eligibility, and adequate donor access [3]. As the whole population ages, MDS will become one of the most common myeloid malignancies. The societal impact and burden of the disease, measured in terms of the number of people affected yearly with a new diagnosis or who are living with the disease, is enormous and will continue to increase in the future. Therefore, a better comprehension of the etiology and further investigation of risk factors may significantly improve MDS prevention measures and reduce MDS incidence.

Since 1950, pesticide use has risen over 50% and pesticide toxicity has increased ten-fold [4]. Pesticide exposure is thought to increase cancer risk by promoting oxidative stress, chromosomal aberrations, cell signaling disturbances or gene mutations [5,6,7].

Over the past few decades, some epidemiological studies have analyzed the association between pesticide exposure and risk of MDS, but the findings are controversial. Five studies showed a positive association between incidence of MDS and pesticide exposure [8,9,10,11,12], and six studies illustrated no association [13,14,15,16,17,18]. Hence, the present meta-analysis was undertaken to further examine the potential involvement of pesticide exposure in MDS etiology.

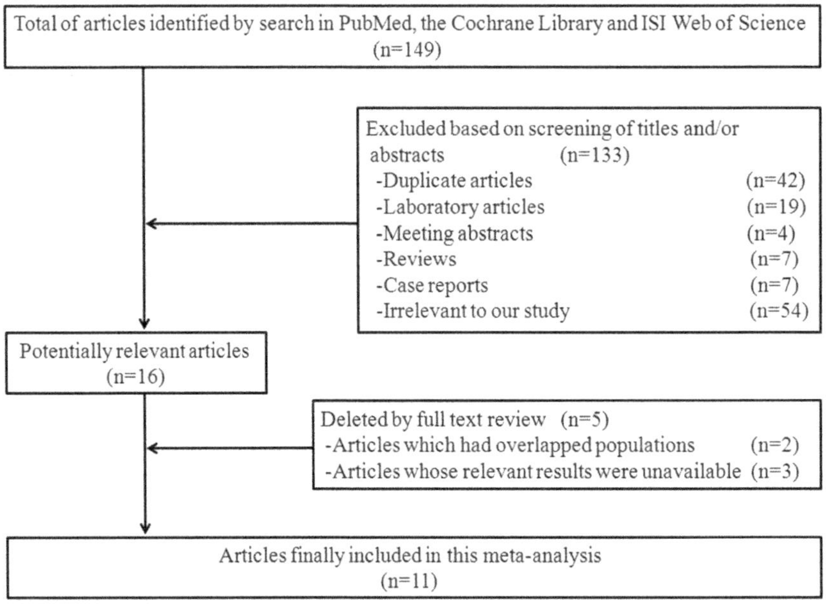

Figure 1. Process of study selection.

Materials and Methods

Literature research

A systematic literature search of the Web of Science, Cochrane Library and PubMed was executed by two independent reviewers (Chao Hu and Mengxia Yu). The following search strategy was used: (myelodysplastic syndrome OR MDS OR myelodysplastic OR myelodysplasia OR preleukemia) AND (pesticides OR herbicides OR fungicides OR insecticides). All relevant titles or abstracts were screened (see Study selection) to determine the suitability of each publication, and full-text articles were retrieved. We also checked the references from retrieved articles for additional studies not identified by database search.

Study selection

Studies included in this meta-analysis had to meet all the following criteria: (a) one of the exposures of interest was pesticide exposure; (b) one of the outcomes of interest was incidence of MDS; (c) a cohort design or case-control design; (d) providing the risk and corresponding 95% confidence intervals (CIs) or data to calculate these; (e) written in the English language. If there were multiple publications from the same study or overlapping study populations, the most recent and detailed study was eligible for inclusion in the meta-analysis.

Data extraction

Data were collected independently by two reviewers using a predefined data collection form. The following data were extracted from each study and included in the final analysis: the study name (together with the first author's name and year of publication), country of origin, gender, age, study design, source of patients, number of cases/controls, risk factor assessment, matching covariates, and adjusted covariates. We contacted the corresponding authors of the primary studies to acquire missing or insufficient data (when necessary), used group consensus and consulted a third reviewer to resolve discrepancies, and assigned scores of <7 and ≥ 7 for low- and high-quality studies, respectively, on the nine-score Newcastle-Ottawa Scale (NOS) [19,20].

Statistical analysis

To determine whether to use the fixed- or random-effects model, we measured statistical heterogeneity [21]. A fixed-effects model was used to calculate a pooled odds ratio (OR) with 95% CI when there was no heterogeneity. Otherwise, we calculated pooled ORs and confidence intervals assuming a random-effects model. The homogeneity of ORs across individual studies was quantified by the Q statistic and the I^2 score. $P>0.05$ for the Q-test was considered as a lack of heterogeneity among the studies. The I^2 values of 25%, 50%, and 75% represented mild, moderate, and severe heterogeneity, respectively [22]. Potential publication bias was assessed by using Begg's funnel plots (rank correlation method where an asymmetrical plot suggested possible publication bias) [23] and Egger's bias test (linear regression method where $P<0.05$ indicated the presence of statistically significant publication bias) [24]. Sensitivity analysis was conducted, in which the meta-analysis estimates were calculated by sequential omission of every study in turn, so as to reflect the influence of the data from individual studies on the pooled ORs and evaluate the stability of the results. Cumulative meta-analysis was also conducted by sorting the studies based on publication time. Subset analyses were performed by source of patients, disease subtype, geographic region, study quality, and type of pesticide. All of the statistical analyses were performed with STATA 11.0 (Stata Corporation, College Station, TX) using two-sided P-values, where $P<0.05$ was considered statistically significant.

Results

Literature search and study characteristics

The results of our literature search strategy and study selection process were detailed in Figure 1. We identified 11 case-control studies on the association of pesticide exposure with risk of MDS published between 1990 and 2011 [8,9,10,11,12,13,14,15,16,17,18]. A total of 1,942 MDS patients and 5,359 controls were included in the present meta-analysis. Among the 5,359 controls, 3853 persons were hospitalized patients without conditions related to hematological diseases and the remaining 1506 were recruited from healthy

Table 1. Main characteristics of studies evaluating the association between pesticides exposure and MDS.

Study	Country	Gender	Age	Study Design	Source of patients	Number of cases	Number of controls	Risk factor Assessment	Study Quality	Matching and Adjustments
Kokouva (2011)[13]	Greece	M/F	27–73	Case-control	Hospital-based	78	455	Questionnaire	5	Gender, age, smoking, family history
Lv (2011)[14]	China	M/F	20–88	Case-control	Hospital-based	403	806	Face-to-face Interview	6	Age, sex, anti-tb drugs, D860, traditional Chinese medicine, alcohol intake, benzene, gasoline, glues, hair dye, education, new building
Pekmezovic (2006)[8]	Serbia Montenegro	M/F	18–85	Case-control	Hospital-based	80	160	Interview	6	Age, sex
Strom (2005)[9]	United States	M/F	24–89	Case-control	Hospital-based	354	452	Mailed questionnaire	7	Age, sex, ethnicity, education, family history of hematopoietic cancer, alcohol intake, benzene, solvent, gasoline
Nisse (2001)[10]	France	M/F	NR	Case-control	Population-based	204	204	Interview	8	Agricultural workers, textile operators, health professionals, living next to an industrial plant, commercial and technical sales representatives, machine operators, oil use, smoking
Rigolin (1998)[11]	Italy	M/F	17–85	Case-control	Hospital-based	178	178	Interview and questionnaire	5	Age, sex
West (1995)[15]	UK	M/F	≥15	Case-control	Hospital-based	400	400	Interview and questionnaire	6	Age, sex, area of residence and hospital, year of diagnosis
Mele (1994)[16]	Italy	M/F	≥15	Case-control	Hospital-based	111	1161	Interview	6	Age, sex, education, and residence outside study town
Ciccone (1993)[12]	Italy	M/F	15–74	Case-control	Hospital-based and population-based	19	246	Interview	5	Sex, area of residence, age
Brown (1990)[17]	United States	M	≥30	Case-control	Population-based	63	1245	Interview	6	Vital status, age, state, tobacco use, family history of lymphopoietic cancer, high-risk occupations and high-risk exposure
Goldberg (1990)[18]	United States	NR	28–88	Case-control	Hospital-based	52	52	Interview	6	Age and sex

M: male; F: female; NR: not reported; tb: tuberculosis.

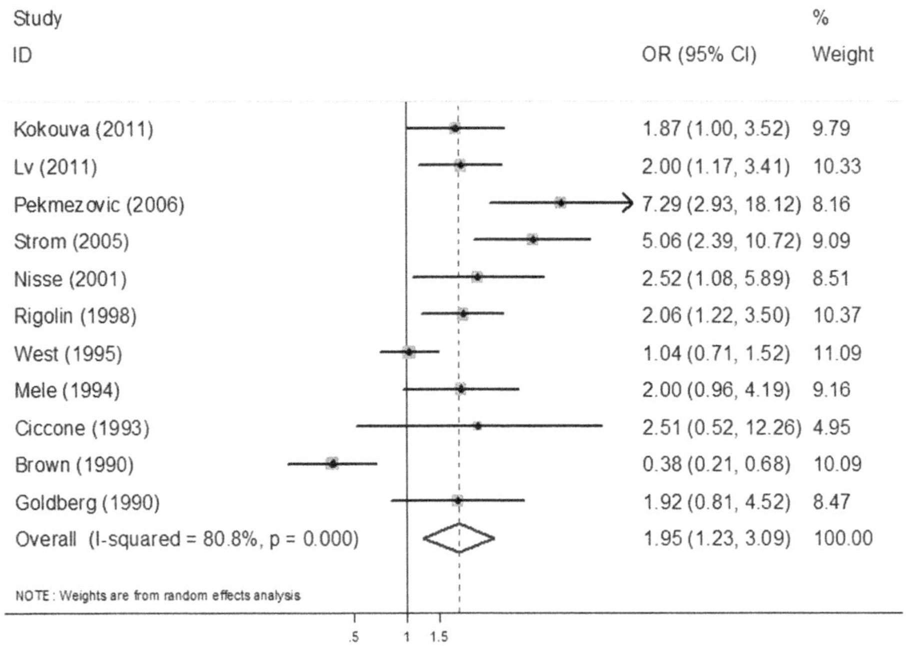

Figure 2. A forest plot illustrating risk estimates from included studies on the relationship between pesticide exposure and MDS risk.

people. A total of 1456 participants have exposed to pesticide, of whom 323 suffered from MDS. Among the chosen studies, seven were conducted in Europe, three in United States, and one in Asia. Pesticide exposure was ascertained by interview or questionnaire or both. The study quality was graded by the Newcastle-Ottawa Quality Assessment Scale, ranged from 5 to 8 (with a mean of 6). The main characteristics of the included articles were listed in Table 1.

Risk estimation

Our analysis demonstrated a significant adverse association between pesticide exposure (exposed vs. non-exposed status) and incidence of MDS (OR = 1.95, 95%CI 1.23–3.09) (Figure 2). Due to a statistically significant heterogeneity across studies ($I^2 = 80.8\%$, $P<0.001$), the summary OR were estimated using the DerSimonian and Laird random effects model [25]. A Galbraith plot identified four studies as major sources of heterogeneity (Figure 3A). After excluding these four studies [8,9,15,17], there was no study heterogeneity existed ($P = 0.999$, $I^2 = 0.0\%$) and the overall association became stronger (OR = 2.04, 95% CI 1.57–2.66).

Stratified analysis

Next, we pooled the OR estimates by patient source (population-based or hospital-based), MDS subtypes (refractory anemia (RA) and RA with ringed sideroblasts (RARS) or RA with excess blasts (RAEB) and RAEB in transformation (RAEBt)), geographic region (United States, Europe, or Asia), study quality (low or high), and type of pesticide (insecticide, herbicide or fungicide) (Table 2). When separated by patient source, the ORs (95% CI) were 2.26 (1.49–3.42) for hospital-based studies and 0.95 (0.15–6.06) for population-based studies. When stratified by MDS subtype, the associations were more positive for the RA/RARS sbutype (OR = 1.63, 95%CI 1.06–2.51) than the RAEB/RAEBt subtype (OR = 1.49, 95%CI 0.78–2.84). In the subset analyses stratified by geographic region, a statistically significant adverse effect of pesticide exposure on MDS was observed in Europe (OR = 2.13, 95%CI 1.35–3.36) and Asia (OR = 2.00, 95%CI 1.17–3.41), but not in United States (OR = 1.52, 95%CI 0.30–7.73). Furthermore, when stratified by study quality, the relationship was more significant in high quality studies (OR = 2.19, 95%CI 1.40–3.42) than in low quality studies (OR = 1.90, 95%CI 1.09–3.33). In addition, when analyzed by type of pesticide, the ORs (95% CI) for insecticides, herbicides, and fungicides were 1.71 (1.22–2.40), 1.16 (0.55–2.43) and 0.70 (0.20–3.20), respectively.

Sensitivity analysis

We also carried out sensitivity analysis by sequentially excluding one study at a time to detect the influence of a single study on the overall estimate. The results displayed that no study disproportionately affected the summary risk estimates in this meta-analysis (Figure 3B). The eleven study-specific ORs ranged from a low of 1.72 (95%CI 1.11–2.68) to 2.27 (95%CI 1.58–3.27) via the omission of the study by Pekmezovic et al. [8] and the study by Brown et al. [17], respectively.

Cumulative meta-analysis

Cumulative meta-analysis of the relationship between pesticide exposure and risk of MDS was also implemented by sorting the studies based on publication time. Figure 4 showed the results from the cumulative meta-analysis of this connection in chronologic order. The 95% CIs became increasingly narrower with each addition of more data, suggesting the precision of each estimate was progressively increasing with the addition of more cases.

Publication bias

As reflected by the funnel plot (Begg's test, $P = 0.350$) and the Egger's test ($P = 0.113$), there was no publication bias being discovered. The data witnessed our result was statistically robust.

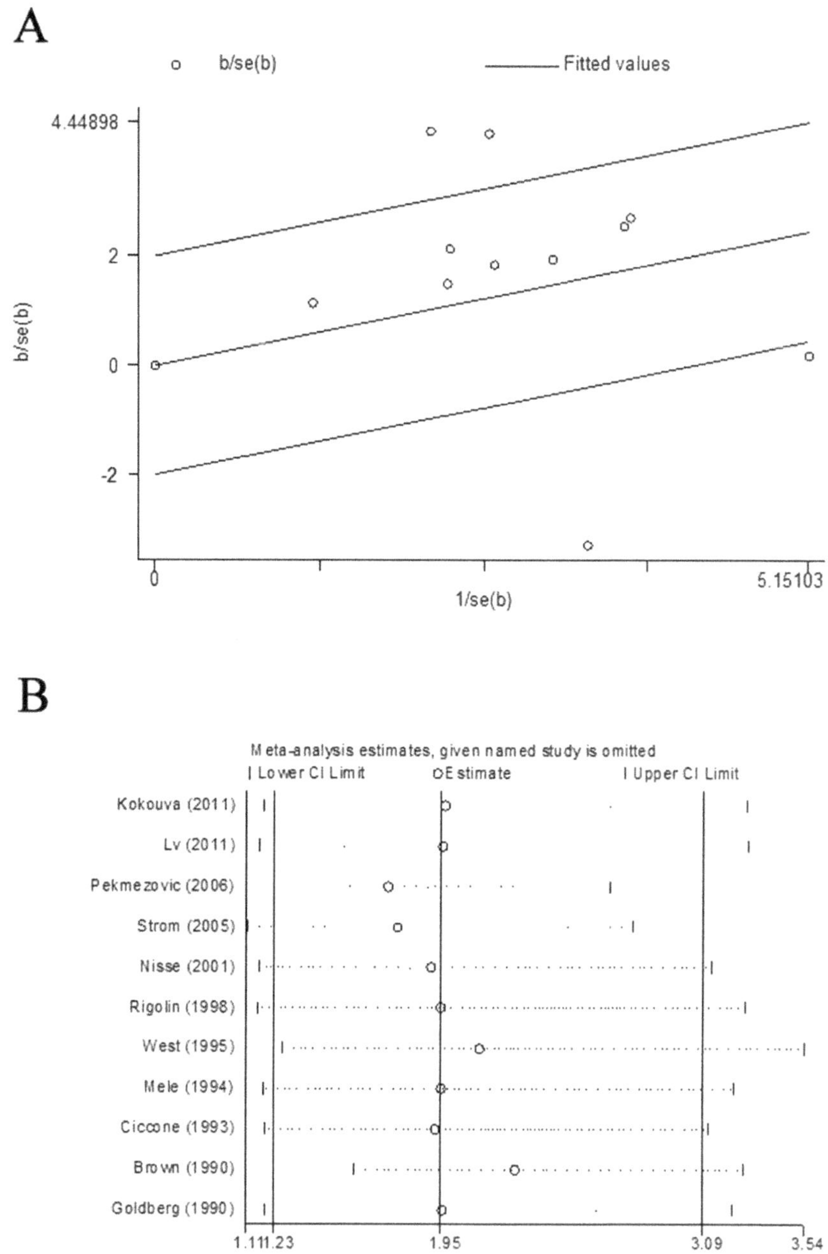

Figure 3. Evaluating the heterogeneity and the stability of the results. (A) Galbraith plot evaluating the heterogeneity; (B) Sensitivity analyses by sequential omission of individual studies in our study.

Discussion

The first myelodysplastic syndromes (MDS) case series was reported about 40 years ago [26]. Thus, the recognition of MDS is approximately 100 years behind the recognition of other hematologic malignancies. Similarly, the level of epidemiologic knowledge of MDS is far below that of other cancers. Therefore, further investigation of risk factors in MDS patients is needed to improve MDS prevention.

Pesticides are widely applied in agriculture all over the world. Three million cases of acute severe pesticide poisoning and over 200,000 deaths are reported annually [4]. Pesticides are thus considered a risk factor for some cancers. Moreover, two previous meta-analyses have been performed in hematological malignancies, which indicated that pesticide exposure could increase risk of non-Hodgkin lymphoma, leukemia and multiple myeloma [27,28]. However, the result about MDS and pesticide exposure was limited. Recent epidemiological studies have examined the potential association between pesticide exposure and the risk of MDS, but none of the results has been conclusive. We attempted to clarify this possible relationship through a meta-analysis of eleven case-control studies.

To the best of our knowledge, this is the first meta-analysis assessing the relationship between pesticide exposure and MDS. Several interesting points raised by our analysis are worth discussing. Firstly, our research demonstrated a significantly positive correlation between pesticide exposure and MDS, which indicated pesticide exposure was associated with a 95% increased

Table 2. Stratified pooled odds ratios of the relationship between pesticide exposure and risk of MDS.

Variables	Number of studies	Pooled OR (95%CI)	Q-test for heterogeneity *P* value (I^2 score)	Egger's test P value	Begg's test P value
Total	11 (8, 9, 10, 11, 12, 13, 14, 15, 16, 17, 18)	1.95 (1.23–3.09)	<0.001 (80.8%)	0.350	0.113
Source of patients					
Population based	2 (10, 17)	0.95 (0.15–6.06)	<0.001 (92.3%)	-	1.000
Hospital based	8 (8, 9, 11, 13, 14, 15, 16, 18)	2.26 (1.49–3.42)	0.001 (71.7%)	0.098	0.266
Disease subtype					
RA/RARS	3 (9, 11, 18)	1.63 (1.06–2.51)	0.258 (25.6%)	0.413	0.734
RAEB/RAEBt	4 (9, 11, 14, 16)	1.49 (0.78–2.84)	0.005 (70.4%)	0.734	0.452
Geographic region					
Europe	7 (8, 10, 11, 12, 13, 15, 16)	2.13 (1.35–3.36)	0.006 (66.8%)	0.133	0.057
Asia	1 (14)	2.00 (1.17–3.41)	–	–	–
United States	3 (9, 17, 18)	1.52 (0.30–7.73)	<0.001 (93.4%)	0.407	1.000
Study quality					
High	2 (10, 11)	2.19 (1.40–3.42)	0.698 (0.0%)	-	1.000
Low	9 (8, 9, 12, 13, 14, 15, 16, 17, 18)	1.90 (1.09–3.33)	<0.001 (83.9%)	0.155	0.348
Type of pesticides					
Insecticides	9 (10, 11, 12, 13, 14, 15, 16, 17, 18)	1.71 (1.22–2.40)	0.009 (60.8%)	0.147	0.348
Herbicides	4 (14, 15, 16, 17)	1.16 (0.55–2.43)	0.056 (60.3%)	0.203	0.089
Fungicides	1 (17)	0.70 (0.20–3.20)	–	–	–

RA: refractory anemia; RARS: RA with ringed sideroblasts; RAEB: RA with excess blasts (RAEB); RAEBt: RAEB in transformation.

risk of MDS. Sensitivity analysis and cumulative analysis confirmed the robustness of our outcomes. In addition, subgroup analyses showed a stronger effect of pesticide exposure on RA/RARS than on RAEB/RAEBt (i.e., exposed MDS patients had 63% increased risk of RA/RARS and 49% increased risk of RAEB/RAEBt, respectively). Our study also illustrated that exposure to insecticides can the increase risk of MDS by 71%, while exposure to herbicides (OR = 1.16, 95%CI 0.55–2.43) and fungicides (OR = 0.70, 95%CI 0.20–3.20), respectively, add no risk. Our subset analysis according to geographical region noted

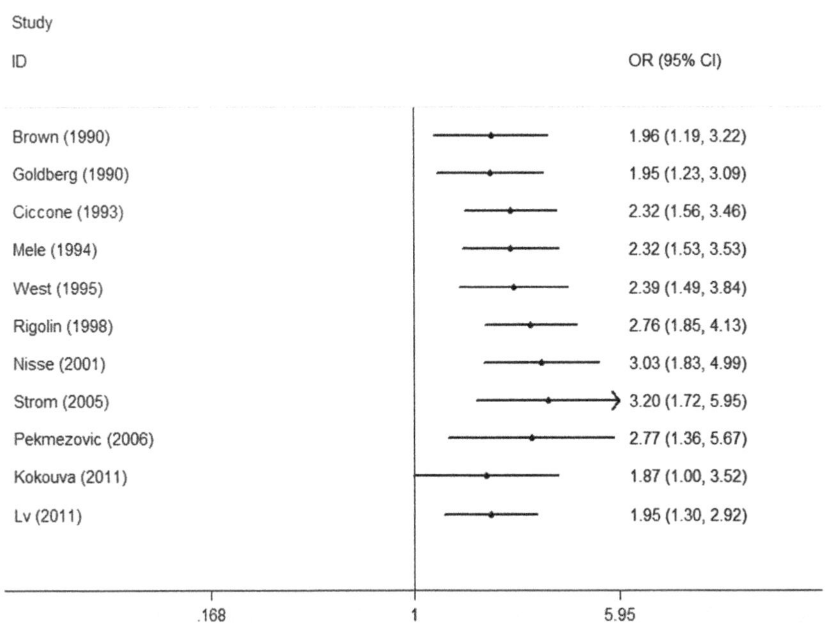

Figure 4. Forest plots showing the result of the cumulative meta-analysis.

higher risk of MDS in Europe (113%) than in Asia (100%) and the United States (52%).

The biological mechanism underlying the linkage of pesticide exposure to the pathogenesis of MDS remains largely unknown. However, several mechanisms are conceivable. Exposure to pesticides might cause overexpression of reactive oxygen species (ROS) sufficient to overwhelm antioxidant defense mechanisms and thereby lead to extensive DNA damage, protein damage, and hematopoietic irregularities [29]. On the other hand, pesticides might bind to and displace endogenous ligands of steroid nuclear receptors, including estrogen and androgen receptors, thus aberrantly activating receptor function and inducing changes in gene expression networks [30]. Recent in vitro mechanistic studies offer novel insight. For example, Boros and Williams reported that exposure of leukemic cell lines (K562) to increasing doses of an organophosphate pesticide (isofenphos) resulted in dose-dependent leukemic cell proliferation [31]. In addition, some previous studies demonstrating that pesticide exposure could induce chromosomal defects [32,33], might also suggest that pesticides could increase the risk of developing MDS. Further research is warranted to elucidate the likely biological mechanisms.

As a meta-analysis of previously published observational studies, our research has some limitations that influence interpretation of the results. First, although the present results seemed to suggest the absence of publication bias, our meta-analysis was vulnerable to publication bias, because only studies published in English were included. Limited resources prevented us from including articles published in other languages and databases. Second, no prospective studies of the association between pesticide exposure and MDS risk were available, and all included studies had a retrospective case-control design. Thus, owing to the limitations of case-control design, the possibility of undetected bias could not be excluded. Third, it is known farmers in China use large amounts of pesticides and this is an ideal population to study their effect on health. However, only one study from China was included in this meta-analysis. Fourth, too few original studies have separated biocides into insecticides, herbicides or fungicides to justify concluding the potential existence of a relationship between exposure to one or several categories of biocides and MDS. Significantly increased risks were observed, with an apparently higher increase when the exposure was to insecticides.

In summary, our findings support that pesticide exposure is associated with the increased risk of MDS, and this association varies widely across disease subtype, geographic region and specific biocide category. Larger and more rigorous analytical studies will be warranted to generate more robust conclusion to guide clinical practice in MDS prevention in the future.

Author Contributions

Conceived and designed the experiments: Jie Jin HT. Performed the experiments: CH MY. Analyzed the data: LY LX Jin Jin FC. Contributed reagents/materials/analysis tools: CH. Contributed to the writing of the manuscript: CH MY.

References

1. Kasner MT, Luger SM (2009) Update on the therapy for myelodysplastic syndrome. Am J Hematol 84: 177–186.
2. Newman K, Maness-Harris L, El-Hemaidi I, Akhtari M (2012) Revisiting use of growth factors in myelodysplastic syndromes. Asian Pac J Cancer Prev 13: 1081–1091.
3. Warlick ED, Smith BD (2007) Myelodysplastic syndromes: review of pathophysiology and current novel treatment approaches. Curr Cancer Drug Targets 7: 541–558.
4. Malek AM, Barchowsky A, Bowser R, Youk A, Talbott EO (2012) Pesticide exposure as a risk factor for amyotrophic lateral sclerosis: a meta-analysis of epidemiological studies: pesticide exposure as a risk factor for ALS. Environ Res 117: 112–119.
5. Infante-Rivard C, Weichenthal S (2007) Pesticides and childhood cancer: an update of Zahm and Ward's 1998 review. J Toxicol Environ Health B Crit Rev 10: 81–99.
6. Lafiura KM, Bielawski DM, Posecion NC, Jr., Ostrea EM, Jr., Matherly LH, et al. (2007) Association between prenatal pesticide exposures and the generation of leukemia-associated T(8;21). Pediatr Blood Cancer 49: 624–628.
7. Agopian J, Navarro JM, Gac AC, Lecluse Y, Briand M, et al. (2009) Agricultural pesticide exposure and the molecular connection to lymphomagenesis. J Exp Med 206: 1473–1483.
8. Pekmezovic T, Suvajdzic Vukovic N, Kisic D, Grgurevic A, Bogdanovic A, et al. (2006) A case-control study of myelodysplastic syndromes in Belgrade (Serbia Montenegro). Ann Hematol 85: 514–519.
9. Strom SS, Gu Y, Gruschkus SK, Pierce SA, Estey EH (2005) Risk factors of myelodysplastic syndromes: a case-control study. Leukemia 19: 1912–1918.
10. Nisse C, Haguenoer JM, Grandbastien B, Preudhomme C, Fontaine B, et al. (2001) Occupational and environmental risk factors of the myelodysplastic syndromes in the North of France. Br J Haematol 112: 927–935.
11. Rigolin GM, Cuneo A, Roberti MG, Bardi A, Bigoni R, et al. (1998) Exposure to myelotoxic agents and myelodysplasia: case-control study and correlation with clinicobiological findings. Br J Haematol 103: 189–197.
12. Ciccone G, Mirabelli D, Levis A, Gavarotti P, Rege-Cambrin G, et al. (1993) Myeloid leukemias and myelodysplastic syndromes: chemical exposure, histologic subtype and cytogenetics in a case-control study. Cancer Genet Cytogenet 68: 135–139.
13. Kokouva M, Bitsolas N, Hadjigeorgiou GM, Rachiotis G, Papadoulis N, et al. (2011) Pesticide exposure and lymphohaematopoietic cancers: a case-control study in an agricultural region (Larissa, Thessaly, Greece). BMC Public Health 11: 5.
14. Lv L, Lin G, Gao X, Wu C, Dai J, et al. (2011) Case-control study of risk factors of myelodysplastic syndromes according to World Health Organization classification in a Chinese population. Am J Hematol 86: 163–169.
15. West RR, Stafford DA, Farrow A, Jacobs A (1995) Occupational and environmental exposures and myelodysplasia: a case-control study. Leuk Res 19: 127–139.
16. Mele A, Szklo M, Visani G, Stazi MA, Castelli G, et al. (1994) Hair dye use and other risk factors for leukemia and pre-leukemia: a case-control study. Italian Leukemia Study Group. Am J Epidemiol 139: 609–619.
17. Brown LM, Blair A, Gibson R, Everett GD, Cantor KP, et al. (1990) Pesticide exposures and other agricultural risk factors for leukemia among men in Iowa and Minnesota. Cancer Res 50: 6585–6591.
18. Goldberg H, Lusk E, Moore J, Nowell PC, Besa EC (1990) Survey of exposure to genotoxic agents in primary myelodysplastic syndrome: correlation with chromosome patterns and data on patients without hematological disease. Cancer Res 50: 6876–6881.
19. Xu X, Cheng Y, Li S, Zhu Y, Zheng X, et al. (2014) Dietary carrot consumption and the risk of prostate cancer. Eur J Nutr.
20. Wells G, Shea B, O'Connell D (2009) Ottawa Hospital Research Institute: The Newcastle-Ottawa Scale (NOS) for assessing the quality of nonrandomized studies in meta-analyses. Available: http://www.ohri.ca/programs/clinical epidemiology/oxford.asp.
21. Hu ZH, Lin YW, Xu X, Chen H, Mao YQ, et al. (2013) Genetic polymorphisms of glutathione S-transferase M1 and prostate cancer risk in Asians: a meta-analysis of 18 studies. Asian Pac J Cancer Prev 14: 393–398.
22. Castillo JJ, Dalia S, Shum H (2011) Meta-analysis of the association between cigarette smoking and incidence of Hodgkin's Lymphoma. J Clin Oncol 29: 3900–3906.
23. Begg CB, Mazumdar M (1994) Operating characteristics of a rank correlation test for publication bias. Biometrics 50: 1088–1101.
24. Egger M, Davey Smith G, Schneider M, Minder C (1997) Bias in meta-analysis detected by a simple, graphical test. BMJ 315: 629–634.
25. DerSimonian R, Laird N (1986) Meta-analysis in clinical trials. Control Clin Trials 7: 177–188.
26. Saarni MI, Linman JW (1973) Preleukemia. The hematologic syndrome preceding acute leukemia. Am J Med 55: 38–48.
27. Van Maele-Fabry G, Duhayon S, Lison D (2007) A systematic review of myeloid leukemias and occupational pesticide exposure. Cancer Causes Control 18: 457–478.
28. Merhi M, Raynal H, Cahuzac E, Vinson F, Cravedi JP, et al. (2007) Occupational exposure to pesticides and risk of hematopoietic cancers: meta-analysis of case-control studies. Cancer Causes Control 18: 1209–1226.

29. Alavanja MC, Ross MK, Bonner MR (2013) Increased cancer burden among pesticide applicators and others due to pesticide exposure. CA Cancer J Clin 63: 120–142.
30. Schug TT, Janesick A, Blumberg B, Heindel JJ (2011) Endocrine disrupting chemicals and disease susceptibility. J Steroid Biochem Mol Biol 127: 204–215.
31. Boros LG, Williams RD (2001) Isofenphos induced metabolic changes in K562 myeloid blast cells. Leuk Res 25: 883–890.
32. Smith MT, McHale CM, Wiemels JL, Zhang L, Wiencke JK, et al. (2005) Molecular biomarkers for the study of childhood leukemia. Toxicol Appl Pharmacol 206: 237–245.
33. Chiu BC, Blair A (2009) Pesticides, chromosomal aberrations, and non-Hodgkin's lymphoma. J Agromedicine 14: 250–255.

10

Effects of Atrazine, Metolachlor, Carbaryl and Chlorothalonil on Benthic Microbes and Their Nutrient Dynamics

Daniel Elias*, **Melody J. Bernot**

Ball State University, Department of Biology, Muncie, Indiana, United States of America

Abstract

Atrazine, metolachlor, carbaryl, and chlorothalonil are detected in streams throughout the U.S. at concentrations that may have adverse effects on benthic microbes. Sediment samples were exposed to these pesticides to quantify responses of ammonium, nitrate, and phosphate uptake by the benthic microbial community. Control uptake rates of sediments had net remineralization of nitrate (-1.58 NO_3 µg gdm^{-1} h^{-1}), and net assimilation of phosphate (1.34 PO_4 µg gdm^{-1} h^{-1}) and ammonium (0.03 NH_4 µg gdm^{-1} h^{-1}). Metolachlor decreased ammonium and phosphate uptake. Chlorothalonil decreased nitrate remineralization and phosphate uptake. Nitrate, ammonium, and phosphate uptake rates are more pronounced in the presence of these pesticides due to microbial adaptations to toxicants. Our interpretation of pesticide availability based on their water/solid affinities supports no effects for atrazine and carbaryl, decreasing nitrate remineralization, and phosphate assimilation in response to chlorothalonil. Further, decreased ammonium and phosphate uptake in response to metolachlor is likely due to affinity. Because atrazine target autotrophs, and carbaryl synaptic activity, effects on benthic microbes were not hypothesized, consistent with results. Metolachlor and chlorothalonil (non-specific modes of action) had significant effects on sediment microbial nutrient dynamics. Thus, pesticides with a higher affinity to sediments and/or broad modes of action are likely to affect sediment microbes' nutrient dynamics than pesticides dissolved in water or specific modes of action. Predicted nutrient uptake rates were calculated at mean and peak concentrations of metolachlor and chlorothalonil in freshwaters using polynomial equations generated in this experiment. We concluded that in natural ecosystems, peak chlorothalonil and metolachlor concentrations could affect phosphate and ammonium by decreasing net assimilation, and nitrate uptake rates by decreasing remineralization, relative to mean concentrations of metolachlor and chlorothalonil. Our regression equations can complement models of nitrogen and phosphorus availability in streams to predict potential changes in nutrient dynamics in response to pesticides in freshwaters.

Editor: Christopher Joseph Salice, Texas Tech University, United States of America

Funding: Funding provided by Indiana Water Resources Research Consortium subaward from the United States Department of Interior, United States Geological Survey as part of the federal Water Resources Research Act of 1984 (Program 104B)"(MJB, DE), http://www.iwrrc.org/. The funders had no role in study design, data collection and analysis, decision to publish, or preparation of the manuscript.

Competing Interests: The authors have declared that no competing interests exist.

* Email: delias@bsu.edu

Introduction

Agricultural activities, such as crop protection via pesticides, are increasing in response to global human population growth (projected to reach 9 billion by 2050; [1]). In the last decade, U.S. pesticide sales have increased ~10% with 80% of these pesticides used for agricultural activities [1]. The continued growth of the human population, coupled with the need for more efficient agricultural practices, will undoubtedly yield future increases in the occurrence of pesticides in freshwaters. Further, despite decades of research on agricultural pesticides, recent calls to action have highlighted the need to fill critical knowledge gaps in our understanding of how pesticides may adversely affect aquatic ecosystems [2], [3].

Once in the aquatic ecosystem, pesticides may have adverse effects on organisms ranging from direct toxicity to indirect effects such as changes in growth or behavior [4]. At higher concentrations, such as those following spring runoff, direct mortality results across diverse organisms including tadpoles [4], bluegill [5], and benthic organisms (e.g., amphipods and chironomids, [6]). However, at lower concentrations, sub-lethal effects can result in altered respiration rates [7], organismal growth [4] and fecundity [8]. In streams, benthic microbes are an important component of aquatic ecosystems and are integral in nutrient and energy dynamics [9]. For example, nitrate, ammonium and phosphorus are cycled by benthic microbes through assimilation and remineralization [10]. These processes are influenced by N concentration in freshwaters [11], [12]. At saturated conditions, such as those in agricultural streams with high input of N from fertilizer runoff, more N is available, due to microbial uptake saturation (i.e. biota have reach their N demand; [13]) or an increase in heterotrophic mineralization [11]. Further, at increasing N concentrations, PO_4 often becomes a secondary limiting nutrient [12].

In the Midwestern U.S., two herbicides (atrazine and metolachlor), one insecticide (carbaryl) and one fungicide (chlorothalonil) have both high usage rates and prevalence in receiving waters [14–17]. Atrazine is a triazine herbicide used predominantly in corn production for control of broadleaf and grassy weeds [18] with a half-life in water at pH seven of 86 days [19]. Metolachlor is a chloroacetanilide herbicide that inhibits mitosis and cell division [19]. Metolachlor is stable in water at pH seven [19]. Atrazine and metolachlor were detected in U.S. freshwaters at peak concentrations of 201 μg/L and 77.6 μg/L (Table 1). Carbaryl is a carbamate family insecticide that inhibits the enzyme acetylcholinesterase [19], with a half-life in water of 12 days at pH seven [19]. Chlorothalonil is a fungicide used in U.S. agriculture; it is stable in water at pH seven [19]. Its mode of action is by binding to glutathione and negatively affecting cellular respiration [7]. Carbaryl and chlorothalonil were detected in U.S. freshwaters at peak concentrations of 4.8 μg/L and 0.3 μg/L (Table 1).

These pesticides are detected in freshwater ecosystems at concentrations that adversely affect biota and human health (Table 2). However, research has focused primarily on the impacts of agricultural pesticides on fish and invertebrates; little is known about how exposure to pesticides may directly influence benthic nutrient dynamics and overall ecosystem function [20]. For example, ecotoxicology studies addressing the effects of pesticides focus primarily on non-benthic vertebrates (e.g., bluegill, [5]), and benthic invertebrates (e.g., amphipods and chironomids, [6]) with few studies conducted on sediment microbial dynamics [21]. Benthic microbial communities influence nutrient cycling (e.g. uptake, remineralization) by affecting fluxes as consumers or sources. Thus, benthic microbial communities are an important component of the freshwater ecosystem [22]. These nutrient dynamics are affected by the presence of pesticides [23]. Specifically, pesticides can reduce microbial activity that contributes to nutrient cycling (e.g. *Volvox* spp., *Botryococcus* spp., *Synedra* spp.) [24], and change species composition by favoring microbes with enhanced pesticide degradation capacities. Also, pesticides can become nutrient sources by providing carbon, nitrogen or phosphorus to some microorganisms [25], and alter nitrogen and/or phosphorus cycles [26]. Thus, there is a need to understand the direct effect of pesticides on sediment nutrient dynamics, and how these changes can affect whole-ecosystem pools and fluxes of nutrients [27].

We measured the effects of atrazine, metolachlor, carbaryl and chlorothalonil on benthic microbial nutrient dynamics by quantifying net assimilation and remineralization rates of ammonium, nitrate and phosphate in laboratory mesocosms. We hypothesized that pesticides with a broad mode of action and higher affinity to organic matter such as chlorothalonil (disruption of cellular respiration, log K_{ow}: 2.9) and metolachlor (inhibitor of mitosis and cellular division, log K_{ow}: 3.4) would decrease microbial nitrate and phosphate uptake rates. In contrast, pesticides with more specific modes of action and higher affinity to the aqueous phase (atrazine: blocks photosynthesis, log K_{ow}: 2.7 and carbaryl: inhibitor of synaptic activity, log K_{ow}: 2.4) were predicted to have no effect on nutrient cycling. Further, natural resources managers and stakeholders would be able to make general predictions of agricultural pesticides effects on the microbial community based on the mode of action and water/solid affinities of these pesticides.

Materials and Methods

Experimental mesocosms

Sediment and water collection was conducted in May 2012 at Ball State University field station property of Jakes Creek - Cooper farm/Skinner field (40.234493, −85.45235) and approval for these experiments was received following the appropriate procedures. This sample collection did not involve endangered or protected species. Jakes Creek is a 3rd order agriculturally-influenced stream with adjacent row crops (i.e. corn and soybean) in Muncie, Indiana within the Upper White River Watershed (UWRW). During the sampling time, Jakes Creek water temperature was 17°C, pH 7.92, depth 5 cm (at sampling location), and discharge was 37 L/s. -While stream samples were not collected for pesticides analysis at the time of this experiment; stream water and sediment samples were collected one week prior to this study at the same site that shows atrazine and metolachlor concentrations were below detection limits. Filtered water samples (ten 1000 mL and one 200 mL) were collected from the stream thalweg using a 60 mm syringe and subsequently filtered (Whatman© glass fiber filter; 0.7 μm nominal pore size) into acid-washed Nalgene© bottles. The 200 mL Nalgene plastic bottle was used to determine initial concentrations of nitrate, ammonium, and phosphate. A composite sediment sample (~2000 cm^3) was randomly collected from the top 5 cm of the stream benthos and placed into three Nalgene plastic bottles. Sediment samples were transported on ice and subsequently combined and homogenized using a USGS no. 5 sieve in the laboratory. Homogenized sediment (20 cm^3) and 60 ml filtered stream water were placed into each of 160 laboratory mesocosms (Fisherbrand sterile urine cup, 120 ml).

Stock solutions were prepared for atrazine (Atrazine 4L, 42.2% purity, Loveland, CO, US), metolachlor (Me-too-lachlor II, 84.4% purity, Drexel Chemical Company, TN, US), carbaryl (Sevin XLR Plus, 44.1% purity, Bayer, NC, US), and chlorothalonil (Bravo, 54% purity, Syngenta, NC, US) to achieve final stock concentrations of 10,000 μg/L for atrazine and metolachlor, 5,000 μg/L for carbaryl, and 8,000 μg/L for chlorothalonil. Aliquots from each stock solution were added to mesocosms to

Table 1. Detection frequency and concentrations of atrazine, metolachlor, carbaryl, and chlorothalonil in U.S. freshwaters.

Compound	Detection frequency (%)	Mean concentration (μg/L)	Maximum concentration (μg/L)	References
Atrazine (Herbicide)	78.1	2.4	201	[15–17]
Metolachlor (Herbicide)	71.1	1.2	77.6	[15–17]
Carbaryl (Insecticide)	18.1	0.013	4.78	[15–17]
Chlorothalonil (Fungicide)	0.033	<0.07	0.29 (290*)	[15–17]

Detection frequency and concentrations of atrazine, metolachlor, carbaryl, and chlorothalonil in U.S. freshwaters. Detection frequency was estimated throughout the U.S. across 50 basins (33 agricultural, 10 urban and 7 mixed); mean and maximum concentrations correspond to 83 agricultural streams. Annual mean detection frequencies for each compound at each site provide the proportion of water samples that have detectable levels of pesticides for a year period. *Chlorothalonil was detected at concentrations of 290 μg/L in run-off near golf courses.

Table 2. Toxicity and Octanol-water partition coefficient of atrazine, metolachlor, carbaryl and chlorothalonil to daphnids, green algae, and humans.

Compound	Daphnids	Green algae	Humans	Octanol-Water partition coefficient
	(mg/L)	(mg/L)	(mg/kg/bw/d)	
	NOEC	NOEC	ADI	Log K_{ow}
Atrazine	0.25	0.1	0.02	2.7
Metolachlor	0.7	57.1*	0.1	3.4
Carbaryl	0.25	-	0.0075	2.4
Chlorothalonil	0.009	0.033	0.015	2.9

Toxicity of atrazine, metolachlor, carbaryl and chlorothalonil in mg/L to daphnids and green algae, and mg/kg of body weight (bw) per day (d) to humans. No observed effect concentrations (NOEC) for daphnids and green algae were calculated by chronic tests of 21 days and 96 hours, respectively. Acceptable daily intake (ADI).* Half maximal effective concentration (EC50) of metolachlor on growth after 72 hours (19).

reach ten target treatment concentrations for each pesticide with four replicates for each treatment.

Mean and peak atrazine (2 µg/L, 201 µg/L) and metolachlor (1 µg/L, 78 µg/L) are detected at concentrations ~2–3 orders of magnitude higher than carbaryl (0.01 µg/L, 5 µg/L) or chlorothalonil (0.07 µg/L, 0.3 µg/L) throughout the U.S. (14–17). Thus, the treatment concentrations used were selected to include these environmentally relevant concentrations and appropriately represent pesticide occurrence in streams. Treatment solutions used in this study ranged from 0 µg/L (control) to the maximum concentrations detected in U.S. freshwaters for atrazine (200 µg/L), metolachlor (80 µg/L), carbaryl (4 µg/L) and chlorothalonil (0.5 µg/L) (Table 2).

Water from each mesocosm was removed after 24 h using a 10 mL syringe, subsequently filtered as above and placed into vials (two analytical replicates ~5 ml) for analysis of nitrate and phosphate via ion chromatograph (DIONEX, ICS-3000). The colorimetric phenol-hypochlorite technique [28], [29] was used to quantify ammonium concentrations. Initial concentrations (background) of nitrate, ammonium, and phosphate were also analyzed following the analytical methods above. Sediment dry mass in each mesocosm was quantified using an analytical balance (OHAUS, Adventurer SL AS64).

Data analysis

Nutrient uptake rates were calculated for phosphate, ammonium and nitrate as changes in concentration over time (24 h) per g of dry mass (sediment) in response to treatments as (22):

$$Nutrient\,uptake\,rate = \frac{(C_f - C_i)\cdot V}{T\cdot gdm}$$

Where: Cf = Final concentration (mg/L); Ci = Initial concentration (mg/L); V = Volume (L) in the jar$T$ = time (h); gdm = g dry mass (g). Negative nutrient uptake rates indicated net remineralization of nutrients and positive nutrient uptake rates indicated net assimilation of nutrients (Table 3). Uptake rates were divided by the average of the control treatment for each pesticide (N = 16) to assess the effects of different treatments for a particular pesticides. Thus, in this study, a response ratio relative to controls was used as the response variable. Further, data were log-transformed to meet normality assumptions for statistical analyses. SigmaPlot© 12.0 software was used for linear and nonlinear regression analyses of response to pesticide concentration. The Akaike Information criterion (AIC) was used to select the best fit model among the different polynomial candidate models. Further, to develop predictive models of microbial response to pesticides in agricultural waters mean and peak concentrations of metolachlor and chlorothalonil throughout U.S. freshwaters (Table 1) were used to calculate nitrogen and phosphorus uptake rates.

Results

Nitrate, phosphate, and ammonium uptake rates varied less than one order of magnitude across pesticide treatments (Table 3). Phosphate uptake rates were three orders of magnitude greater than ammonium uptake rates across treatments, though both phosphate (mean = 180.56 µg gdm^{-1} h^{-1}) and ammonium (mean = 0.34 µg gdm^{-1} h^{-1}) uptake rates were net assimilative (Table 3). In contrast, we observed net remineralization of nitrate (mean = −32.22 µg gdm^{-1} h^{-1}) across pesticides (Table 3). Overall, these nutrient dynamics are expected in nitrogen-saturated agricultural ecosystems (i.e., microbial nitrate assimilation is saturated and PO_4 becomes a limiting nutrient).

Nitrate dynamics

Control nitrate uptake rate was ~20x higher on average than nitrate uptake influenced by pesticides (i.e., increasing remineralization in presence of pesticides). Nitrate uptake rates in response to atrazine, metolachlor, and carbaryl treatments exposure ranged from net assimilation (consumption/removal) to remineralization (source/addition) in the water column of our mesocosms (Table 3). In contrast, increasing concentrations of chlorothalonil yielded increasing nitrate remineralization ($p = 0.02$, $r^2 = 0.83$, Fig. 1A). No other pesticides had significant effects on nitrate uptake rates ($p > 0.05$).

Phosphate dynamics

Phosphate uptake rates in response to pesticides exposure was ~100x higher than control phosphate uptake rates. Thus, phosphate assimilation increased in the presence of atrazine, metolachlor, carbaryl, and chlorothalonil relative to the control (Table 3). Further, phosphate uptake was negatively related to increasing chlorothalonil concentration ($p < 0.001$, $r^2 = 0.87$, Fig. 1B). Similarly, the metolachlor effect on phosphate uptake rate ($p = 0.005$, $r^2 = 0.91$, Fig. 1C) followed a cubic relationship with increasing phosphate uptake at lower concentrations (0 to 10 µg/L), and decreasing rates at higher concentrations (10 to 80 µg/L). No other pesticides resulted in significant effects on phosphate uptake rates ($p > 0.05$).

Table 3. Uptake rates for nitrate, phosphate, and ammonium in response to pesticides exposure.

Pesticide	Nitrate (μg gdm^{-1} h^{-1})	Phosphate (μg gdm^{-1} h^{-1})	Ammonium (μg gdm^{-1} h^{-1})
Atrazine	−25.88 (−45.2-4.30)	126.93 (83.1-169)	0.36 (0.23-0.45)
Metolachlor	−21.34 (−40.9-2.10)	169.54 (120-214)	0.39 (0.03-0.53)
Carbaryl	−25.13 (−56.6-8.80)	192.11 (37.9-260)	0.31 (0.13-0.53)
Chlorothalonil	−56.54 (−83.1-−3.7))	233.67 (39.8-601)	0.28 (0.13-0.47)
Mean uptake rate	**−32.22**	**180.56**	**0.34**
Control	−1.58 (−1.62-−1.49)	1.34 (7.03E-05-4.4)	0.03 (0.01-0.04)

Uptake rates for nitrate, phosphate, and ammonium (μg gdm^{-1} h^{-1}) across pesticide and control (no pesticide) treatments. Range noted in parentheses. Mean uptake rate was calculated for each nutrient across pesticides.

Ammonium dynamics

Ammonium uptake rates in response to pesticides exposure was ~10x higher than control ammonium uptake rate. Thus ammonium assimilation increased in the presence of atrazine, metolachlor, carbaryl, and chlorothalonil relative to the control (Table 3). The ammonium uptake rate varied in response to metolachlor treatments ($p = 0.023$, $r^2 = 0.83$, Fig. 1D), increasing with lower concentrations of metolachlor (0 to 10 μg/L), followed by a decline at higher concentrations (10 to 80 μg/L). No other pesticides had significant effects on ammonium uptake rates ($p > 0.05$).

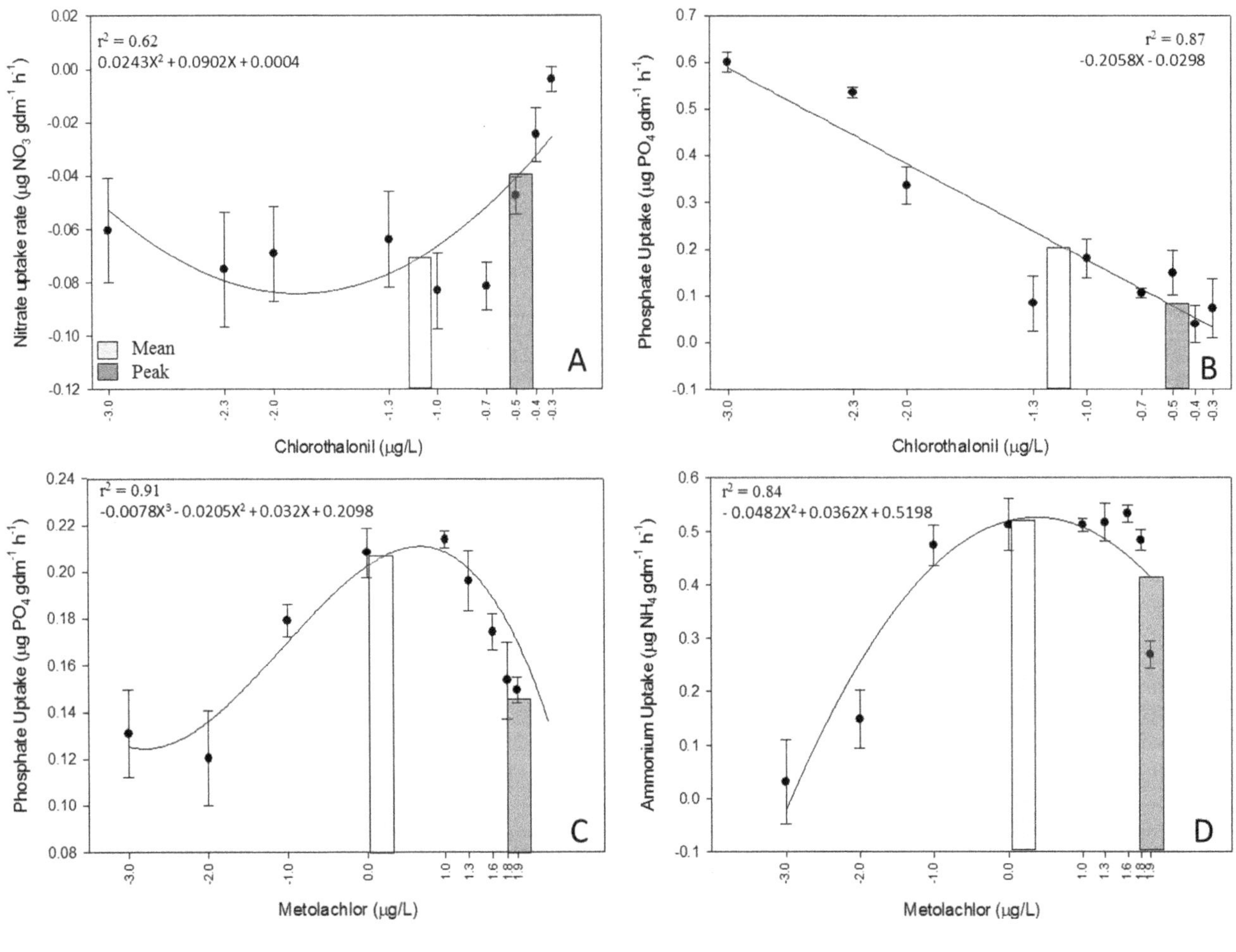

Figure 1. Nutrient uptake rates (mean +/− SE) response to pesticide concentrations after 24 h incubation (4 replicates, 10 treatments, N = 40). X-axis is in a Log10 scale. A: Nitrate uptake in response to chlorothalonil concentrations. B: Phosphate uptake rate response to chlorothalonil concentrations. C: Phosphate uptake rate response to metolachlor concentrations. D: Ammonium uptake rate response to metolachlor concentrations. Columns represent predicted uptake rates for each nutrient calculated at mean and peak concentrations of metolachlor and chlorothalonil measure in U.S. freshwaters (Table 2).

Predicting nutrient response to pesticides

Changes in stream ecosystem nitrogen and phosphorus uptake rates were modeled across metolachlor and chlorothalonil concentrations detected throughout the U.S. (Fig. 1). These changes were generated with polynomial regressions from this experiment ($p<0.05$).

The predicted ammonium uptake (0.52 NH_4 μg gdm^{-1} h^{-1}) in response to mean metolachlor concentrations was ~20% higher than ammonium uptake (0.42 NH_4 μg gdm^{-1} h^{-1}) at peak metolachlor concentrations. At mean concentrations of metolachlor, the predicted phosphate uptake (212.2 PO_4 μg gdm^{-1} h^{-1}) was ~40% lower than phosphate uptake (372.41 PO_4 μg gdm^{-1} h^{-1}) at peak metolachlor concentrations. At mean and peak chlorothalonil concentrations, there is net remineralization of nitrate (−71.99 NO_3 μg gdm^{-1} h^{-1} and −41.87 NO_3 μg gdm^{-1} h^{-1}, respectively). Further, there was net assimilation of phosphate in response to mean and peak concentrations of chlorothalonil (207.88 and 77.81 PO_4 μg gdm^{-1} h^{-1}, respectively).

Overall, there was decreased ammonium assimilation and remineralization of nitrate in the presence of peak concentrations of metolachlor and chlorothalonil (Fig. 1). Phosphate assimilation increased at peak concentrations of metolachlor and decreased at peak concentrations of chlorothalonil (Fig. 1). Peak chlorothalonil is predicted to have the greatest effect on phosphate and nitrate uptake rates, by decreasing net assimilation and remineralization, respectively, over 50% relative to mean concentrations of these pesticides.

Discussion

Research focuses primarily on the impacts of agricultural pesticides on fish and invertebrates [5], [6]; however, little is known about effects on microbial communities [20]. Our ecotoxicological research showed that benthic microbes' nutrient uptake rate response is likely a function of pesticide chemical characteristics, and how these changes can affect nutrient dynamics, due to nutrient availability.

Variation in nutrient uptake rates is likely a result of differences in baseline nutrient concentrations and the biotic community. For example, control ammonium uptake rates in this study (0.026 μg NH_4 gdm^{-1} h^{-1}) were ~3x lower than control ammonium uptake rates in Bunch and Bernot [30]. These results suggest dissimilar nutrient demand due to unique biotic communities and nutrient conditions across sampling sites and temporal variability.

In our study, mean ammonium uptake rate was ~1.5x higher than ammonium uptake under enriched nitrate conditions and ~2.5x lower under enriched ammonium conditions reported in Bunch and Bernot [30]. Thus, ammonium uptake is higher in the presence of pesticides than under enriched nitrate conditions and lower under enriched ammonium conditions, due to microbial adaptations to nutrient availability and metabolic responses in presence of pesticides. Our nitrate uptake results were similar to yields with net remineralization under enriched ammonium conditions reported in Bunch and Bernot [30], though remineralization was 10 times lower than rates measured in response to pesticides in our study.

The effects of metolachlor and chlorothalonil on nutrient uptake rates suggest a unique biotic community at this site, represented by mostly heterotrophic benthic microbes. Our results suggest a toxic effect of metolachlor and chlorothalonil on the benthic microbial community that is reflected by the increased nitrate remineralization, and reduced assimilation of ammonium and phosphate. This increase in remineralization rates could be an outcome of cellular lysis or a stress mechanism [31]. Further, a biotic community characterized by autotrophs [22] had ammonium and nitrate uptake rates three orders of magnitude higher than benthic uptake in our study; possibly, under these study conditions, primary producers have a higher assimilation rate than heterotrophic benthic microbes in response to the available forms of nitrogen [22]. Further, control phosphate uptake rates in our study were ~5x lower than rates previously measured with phosphorus enrichment (0.1–2 mg/L; 34). Thus, phosphate is likely a limiting nutrient in our system. However, in presence of pesticides, phosphate uptake was two orders of magnitude higher than the rates measured under phosphorus limiting conditions [32] possibly, in addition to available phosphorus in the water column, microbes were potentially degrading pesticides as a source of phosphorus [33].

Nutrient dynamics in the presence of pesticides are dependent on the physicochemical characteristics (e.g. sorption kinetics, modes of action) of each pesticide. Sorption kinetics of pesticides and their corresponding index (Octanol-water partition coefficient, K_{ow}) determine the affinity of organic contaminants to either the water column or sediments [34]. Atrazine and carbaryl have a higher affinity to the aqueous phase (Log K_{ow}: 2.7 and 2.4, respectively), relative to metolachlor and chlorothalonil (Log K_{ow}: 3.4 and 2.9, respectively). Thus, atrazine and carbaryl are likely more prevalent in the water column and less available to the sediment microbial communities, with minimal effect on benthic microbial activity. In contrast, metolachlor and chlorothalonil have a higher affinity to solids and higher prevalence in sediment, potentially affecting benthic microbial activity. Our interpretations of pesticide availability based on their water/solid affinities supports our results of no effects for atrazine and carbaryl (Log $K_{ow}<2.7$, $p>0.05$) within the tested range, decreasing nitrate remineralization and decreasing phosphate uptake in response to chlorothalonil, and decreasing ammonium and phosphate uptake in response to metolachlor (Log $K_{ow}>2.9$, $p<0.05$).

Nutrient dynamics are also affected by the pesticide mode of action. Atrazine and carbaryl have specific modes of action; atrazine blocks photosynthesis, and carbaryl inhibits the activity of acetylcholinesterase, an enzyme of insects, fish, mammals [35]. Thus, the specificity of atrazine and carbaryl, and a sediment microbial community dominated by heterotrophs with no synaptic activity [36] may explain the lack of significant effects of these pesticides on benthic microbes within the tested range (Table 3). In contrast, metolachlor and chlorothalonil are broad spectrum pesticides [19]. In our study, ammonium and phosphate uptake rates decreased with increasing metolachlor concentrations. These decreasing ammonium and phosphate uptake rates could be due to metolachlor inhibition of mitosis and cell division. Similarly, phosphate and nitrate uptake rates were affected by chlorothalonil; possibly this fungicide affects benthic microbes by disrupting cellular respiration. Thus, pesticides with non-specific modes of action (e.g. metolachlor, chlorothalonil) are more likely to have a significant effect on nutrient dynamics of sediment microbes, consistent with our study results.

Metolachlor and chlorothalonil not only affect nutrient dynamics of the sediment microbial community; they can also affect other processes. In our study, at increasing concentrations of chlorothalonil, there is a decrease in phosphate uptake, thus there is more phosphate available for organismal consumption. In agricultural streams, where phosphorus is the limiting nutrient [37], increasing availability of phosphate can lead to algal blooms, eutrophication, hypoxia, loss of biodiversity (e.g. fish kills), and loss of aesthetic value of these habitats [38]. In contrast, in the presence of peak concentrations of metolachlor, there is an increase in phosphate assimilation, which could mitigate the excess phosphate in streams [39]. Similarly, there is an increase of ammonium availability due to inhibited uptake rates in response to metolachlor, which in turn could potentially increase biological activity of pesticide resistant microbes [21], [25], [26], [33]. Nitrate remineralization also decreases in response to chlorothalonil, reducing nitrate availability, and potentially further mitigating excess nitrogen in these habitats [40].

Mesocosms have a critical role in understanding the mechanisms driving ecological processes [42], [43] and provide a bridge between smaller, better controlled experiments and the larger freshwater ecosystems [44]. For example, [43] revealed that the effects of nutrients on primary producers are similar in artificial habitats across five orders of magnitude in size. Also, [44] mentioned that mesocosms help disentangle direct from indirect effects over scale. Thus, our conclusions try to bridge what we observed in the laboratory level and what could potentially occur at an agricultural influenced stream level.

Pesticide occurrence and concentrations in streams are dictated primarily by land-use. Streams receiving run-off from agricultural landscapes frequently have the highest concentrations of pesticides, compared to forested, mixed-use and urban lands [41]. These high concentrations occur as pulses that are coupled by the seasonality of agricultural practices [16]. Peak concentrations of metolachlor and chlorothalonil are detected from April to August [16]. Thus, the effects of these pesticides on nutrient dynamics are highest during this critical time. At peak metolachlor and chlorothalonil concentrations there is a decrease of predicted ammonium assimilation and nitrate remineralization. Further, at peak concentrations of metolachlor there is increased phosphate assimilation and at peak concentrations of chlorothalonil there is decreased phosphate assimilation (Fig. 1).

Our findings demonstrate individual effects of these pesticides on sediment nutrient dynamics that are likely driven by a pesticides' mode of action and water/sediment affinities. More studies are required to understand the net effect on ecosystems and to address their synergistic or antagonistic effects as mixtures. The regression equations we generated can complement models of nitrogen and phosphorus availability in streams to predict the potential changes in nutrient dynamics in response to increasing presence of pesticides in lotic ecosystems.

Acknowledgments

We thank James Justice, Mark Pyron and two anonymous reviewers for helpful comments.

Author Contributions

Conceived and designed the experiments: DE MJB. Performed the experiments: DE. Analyzed the data: DE. Contributed reagents/materials/analysis tools: DE MJB. Wrote the paper: DE MJB.

References

1. Enserink M, Hines PJ, Vignieri SN, Wigginton NS, Yeston JS (2013) Smarter pest control. The pesticide paradox. Introduction. Science 341: 728–729.
2. Köhler H-R, Triebskorn R (2013) Wildlife ecotoxicology of pesticides: can we track effects to the population level and beyond? Science 341: 759–765.
3. Murray KE, Thomas SM, Bodour AA (2010) Prioritizing research for trace pollutants and emerging contaminants in the freshwater environment. Environmental Pollution 158: 3462–3471.
4. Relyea RA (2006) The effects of pesticides, pH, and predatory stress on amphibians under mesocosm conditions. Ecotoxicology 15: 503–511.
5. Munn MD, Gilliom RJ, Moran PW, Nowell LH (2001) Pesticide toxicity index for freshwater aquatic organisms. Water Resources Investigations Report United States Geological Survey: 66.
6. Liess M, Ohe PC (2005) Analyzing effects of pesticides on invertebrate communities in streams. Environmental Toxicology and Chemistry 24: 954–965.
7. McMahon TA, Halstead NT, Johnson S, Raffel TR, Romansic JM, et al. (2012) Fungicide-induced declines of freshwater biodiversity modify ecosystem functions and services. Ecology Letters 15: 714–722.
8. Kashian DR, Dodson SI (2002) Effects of common-use pesticides on developmental and reproductive processes in Daphnia. Toxicology and Industrial Health 18: 225–235.
9. Covich AP, Palmer MA, Crowl TA (1999) The role of benthic invertebrate species in freshwater ecosystems: zoobenthic species influence energy flows and nutrient cycling. BioScience 49: 119–127.
10. Dodds WK, Lopez AJ, Bowden WB, Gregory S, Grimm NB, et al. (2002) N uptake as a function of concentration in streams. Journal of the North American Benthological Society 21: 206–220.
11. Bernot MJ, Dodds WK (2005) Nitrogen Retention, Removal, and Saturation in Lotic Ecosystems. Ecosystems 8: 442–453.
12. Earl SR, Valett HM, Webster JR (2006) Nitrogen saturation in stream ecosystems. Ecology 87: 3140–3151.
13. Davis JC, Wayne MG (1999) Nitrogen and phosphorus uptake in two Idaho (USA) headwater wilderness streams. Oecologia 119: 247–255.
14. United States Environmental Protection Agency (1987) Chlorothalonil: health advisory. U.S. Environmental Protection Agency, Washington, D.C.
15. Kolpin DW, Barbash JE, Gilliom RJ (1998) Occurrence of pesticides in shallow groundwater of the United States: Initial results from the National Water-Quality Assessment Program. Environ. Sci. Technol. 32: 558–566.
16. Larson SJ, Gilliom RJ, Capel PD (1999) Pesticides in Streams of the United States—Initial Results from the National Water-Quality Assessment Program. U.S. Department of the Interior, US Geological Survey.
17. Gilliom RJ, Barbash JE, Crawford CG, Hamilton PA, Martin JD, et al. (2006) Pesticides in the nation's streams and ground water, 1992–2001. U.S. Geological Survey Circular 1291: 172.
18. Kruger EL, Coats JR, Zhu BE (1996) Relative mobilities of atrazine, five atrazine degradates, metolachlor, and simazine in soils of Iowa. Environmental Toxicology and Chemistry 15: 691–695.
19. University of Hertfordshire (2013). The Pesticide Properties DataBase (PPDB) developed by the Agriculture & Environment Research Unit (AERU), University of Hertfordshire, 2006–2013.
20. Beketov MA, Kefford BJ, Schäfer RB, Liess M (2013) Pesticides reduce regional biodiversity of stream invertebrates. Proceedings of the National Academy of Sciences.
21. Satsuma K (2006) Characterisation of new strains of atrazine-degrading Nocardioides sp. isolated from Japanese riverbed sediment using naturally derived river ecosystem. Pest Management Science 62: 340–349.
22. Kemp MJ, Dodds WK (2002) The influence of ammonium, nitrate, and dissolved oxygen concentrations on uptake, nitrification, and denitrification rates associated with prairie stream substrata. Limnology and Oceanography 47: 1380–1393.
23. Gramlich J, Davis D (1967) Effect of atrazine on nitrogen metabolism of resistant species. Weeds: 157–160.
24. Neumann M, Dudgeon D (2002) The impact of agricultural runoff on stream benthos in Hong Kong, China. Water Research 36: 3103–3109.
25. Tappin AD, Loughnane JP, McCarthy AJ, Fitzsimons MF (2012) Removal of atrazine from river waters by indigenous microorganisms. Environmental Chemistry Letters 10: 89–96.
26. Doddamani HP, Ninnekar HZ (2001) Biodegradation of carbaryl by a Micrococcus species. Current Microbiology 43: 69–73.
27. Brown JH, Gillooly JF, Allen AP, Savage VM, West GB (2004) Toward a metabolic theory of ecology. Ecology 85: 1771–1789.
28. American Public Health Association AWWAWEF (1995) Standard methods for the examination of water and wastewater. Washington, D.C.: American Public Health Association.
29. Aminot A, Kirkwood DS, Kerouel R (1997) Determination of ammonia in seawater by the indophenol-blue method: Evaluation of the ICES NUTS I/C 5 questionnaire. Marine Chemistry 56: 59–75.
30. Bunch ND, Bernot MJ (2012) Nitrate and ammonium uptake by natural stream sediment microbial communities in response to nutrient enrichment. Research in Microbiology 163: 137–141.
31. Bronk D, Ward B (1999) Gross and net nitrogen uptake and DON release in the euphotic zone of Monterey Bay, California. Limnology and Oceanography 44: 573–585.
32. Klotz R (1985) Factors controlling phosphorus limitation in stream sediments. Limnology and Oceanography 30: 543–553.
33. Cook AM, Daughton CG, Alexander M (1978) Phosphorus-containing pesticide breakdown products: quantitative utilization as phosphorus sources by bacteria. Applied and Environmental Microbiology 36: 668–672.
34. Wauchope RD, Yeh S, Linders JBHJ, Kloskowski R, Tanaka K, et al. (2002) Pesticide soil sorption parameters: theory, measurement, uses, limitations and reliability. Pest Management Science 58: 419–445.
35. Durieux ED, Farver TB, Fitzgerald PS, Eder KJ, Ostrach DJ (2011) Natural factors to consider when using acetylcholinesterase activity as neurotoxicity biomarker in Young-Of-Year striped bass (Morone saxatilis). Fish Physiology and Biochemistry 37: 21–29.

36. Pester M, Friedrich MW, Schink B, Brune A (2004) pmoA-based analysis of methanotrophs in a littoral lake sediment reveals a diverse and stable community in a dynamic environment. Applied and Environmental Microbiology 70: 3138–3142.
37. Correll D (1999) Phosphorus: a rate limiting nutrient in surface waters. Poultry Science 78: 674–682.
38. Conley DJ, Paerl HW, Howarth RW, Boesch DF, Seitzinger SP, et al. (2009) Controlling eutrophication: nitrogen and phosphorus. Science 323: 1014–1015.
39. Gächter R, Meyer J (1993) The role of microorganisms in mobilization and fixation of phosphorus in sediments. In: Boers PCM, Cappenberg TE, van Raaphorst W, editors. Proceedings of the Third International Workshop on Phosphorus in Sediments: Springer Netherlands. pp. 103–121.
40. Zogg GP, Zak DR, Pregitzer KS, Burton AJ (2000) Microbial immobilization and the retention of anthropogenic nitrate in a northern hardwood forest. Ecology 81: 1858–1866.
41. Rinella FA, Janet ML (1998) Seasonal and spatial variability of nutrients and pesticides in streams of the Willamette Basin, Oregon, 1993–95: U.S. Department of the Interior, U.S. Geological Survey.
42. Frost T, Ulanowicz R, Blumenshine S, Allen T, Taub F, et al. (2001) Scaling issues in experimental ecology: freshwater ecosystems. Scaling relations in experimental ecology Columbia University Press, New York: 253–280.
43. Spivak AC, Vanni MJ, Mette EM (2011) Moving on up: can results from simple aquatic mesocosm experiments be applied across broad spatial scales? Freshwater Biology 56: 279–291.
44. Stewart RIA, Dossena M, Bohan DA, Jeppesen E, Kordas RL, et al. (2013) Mesocosm Experiments as a Tool for Ecological Climate-Change Research. Advances in Ecological Research: Global Change in Multispecies Systems, Pt 3 48: 71–181.

11

Honeybee Colony Disorder in Crop Areas: The Role of Pesticides and Viruses

Noa Simon-Delso[1,2]*, Gilles San Martin[3], Etienne Bruneau[1], Laure-Anne Minsart[1], Coralie Mouret[1], Louis Hautier[3]

1 Beekeeping Research and Information Centre, Louvain la Neuve, Belgium, **2** Environmental Sciences, Copernicus Institute, Utrecht University, Utrecht, The Netherlands, **3** Plant Protection and Ecotoxicology Unit, Life Sciences Department, Walloon Agricultural Research Centre, Gembloux, Belgium

Abstract

As in many other locations in the world, honeybee colony losses and disorders have increased in Belgium. Some of the symptoms observed rest unspecific and their causes remain unknown. The present study aims to determine the role of both pesticide exposure and virus load on the appraisal of unexplained honeybee colony disorders in field conditions. From July 2011 to May 2012, 330 colonies were monitored. Honeybees, wax, beebread and honey samples were collected. Morbidity and mortality information provided by beekeepers, colony clinical visits and availability of analytical matrix were used to form 2 groups: healthy colonies and colonies with disorders (n = 29, n = 25, respectively). Disorders included: (1) dead colonies or colonies in which part of the colony appeared dead, or had disappeared; (2) weak colonies; (3) queen loss; (4) problems linked to brood and not related to any known disease. Five common viruses and 99 pesticides (41 fungicides, 39 insecticides and synergist, 14 herbicides, 5 acaricides and metabolites) were quantified in the samples.The main symptoms observed in the group with disorders are linked to brood and queens. The viruses most frequently found are Black Queen Cell Virus, Sac Brood Virus, Deformed Wing Virus. No significant difference in virus load was observed between the two groups. Three acaricides, 5 insecticides and 13 fungicides were detected in the analysed samples. A significant correlation was found between the presence of fungicide residues and honeybee colony disorders. A significant positive link could also be established between the observation of disorder and the abundance of crop surface around the beehive. According to our results, the role of fungicides as a potential stressor for honeybee colonies should be further studied, either by their direct and/or indirect impacts on bees and bee colonies.

Editor: Fabio S. Nascimento, Universidade de São Paulo, Faculdade de Filosofia Ciências e Letras de Ribeirão Preto, Brazil

Funding: The Walloon Region was the only funder of this study. The funders had no role in study design, data collection and analysis, or preparation of the manuscript.

Competing Interests: The authors have declared that no competing interests exist.

* Email: simon@cari.be

Introduction

The evolution of pollinator populations has been the subject of an increasing number of studies, most of them showing worrying negative trends [1–4]. Furthermore, beekeepers have long been notifying enhanced bee winter losses and disorders [5–8]. Bees contribute to ecosystem services and their decline thus threatens pollination of both wild and cultured plants, endangering biodiversity, food and fibre production [9,10]. This decline may also have an impact on the production of other goods with pharmacological uses (e.g. honey, propolis) [11,12], and scientific and technological inspiration (e.g. development of visual guided flight robotics) [13]. Bees are also part of our culture (e.g. culinary, hobby occupation, etc.), contributing to the dynamism of rural and urban areas [14] and providing a source of inspiration and well-being for many [15].

Belgium shares the trends observed worldwide both in terms of wild and reared pollinators [8,16], with enhanced winter mortality observed from 1999 [8,17,18]. However, apart from colony mortality, Walloon beekeepers have reported a number of imprecise symptoms: colony weakness, mainly in spring; fast renewal of young queens; rapid loss of individuals in the colony, mainly foragers, or slow loss of individuals in the colony. In many cases brood and food remains in the colony. Sometime a small cluster of bees with the queen survives [17]. Some beekeepers described unspecific brood abnormalities not characteristic to any known disease (Baudoin and Lequeux, pers. com.). Other studies describe similar mortality trends [8,19,20] as well as unspecific symptoms: increased colony mortality and/or weakening [21–23], queen failure [21,23–25] or honey yield reduction [21].

Multifarious factors have been proposed to provide a cause to such a phenomenon. Climate change is proposed as one of them, together with a decrease of genetic variability of the bee colonies, electromagnetic radiation, pathogens and parasites or the impact of intensive agricultural systems (nutritional lack, GM plants or pesticides) [26]. Previous studies developed in the region discard some of these factors as causes of bee losses, specifically *Nosema* spp and American Foulbrood [27]. However, pesticide residues and certain viruses were detected in bee colonies [28,27]. These elements were the most relevant to us.

Countless studies have shown lethal and sub-lethal effects of pesticides on bees [29–31]. Insecticides are often the most studied for obvious reasons. However other substances (fungicides, herbicides) deserve analysis for their specific toxicity, individually or in synergy with other substances [32,33] or pathogens [34–36], or their extensive exposure given their large scale and/or repeated use. Indeed, residue analyses show that these types of substance are found in the hives, even though the crops in which they were applied would suggest no bee exposure [37–40].

Beside pesticides, bee viruses are often mentioned as a putative cause of decline of the colonies, or at least to reside among the presumed multi-factorial causes [41]. In Europe, at least 12 viruses infecting bees have been compiled [21]. Many honeybee viruses commonly occur in seemingly healthy populations that continue to run well. Viruses may remain latent and confined within certain cells or tissues with no active replication and no disruption of cellular function. Likewise, they may be replicating at low level in permissive cells but in non-vital sites or in honeybee life stages that do not exhibit any symptoms or obvious pathology [21,42,43]. Nevertheless, two viruses, ABPV and DWV, are able of inducing serious disorders to honeybees and have been shown to cause -in association with *V. destructor* - winter losses in Germany [23].

In this paper, we first studied the relationship of in-hive viral prevalence as well as the presence of pesticide residues on honeybee colonies' health. As a next step, we focused on the relationship between the environment surrounding the monitored apiaries and the health condition of their colonies.

Materials and Methods

Field work – colony follow up

A group of voluntary beekeepers were requested to participate in the study. All of them shared the following criteria: (1) have a minimum of 5 production colonies per apiary in June 2011; (2) regularly follow up the health status and development of their colonies; and (3) monitor the varroa infestation level and carry out officially recommended varroa treatments (treatment in July-August with veterinary medicaments based on thymol and winter treatment with veterinary medicaments based on oxalic acid). A total of 330 colonies distributed among 66 apiaries (5 colonies/apiary) in Walloon and Brussels regions (Belgium) were followed. In those apiaries composed of 5 or more colonies, the 5 well-developed colonies at the beginning of the study were selected.

Colonies received three visits along the study, the first one from mid-July to mid-August 2011, the second one from mid-September to mid-October 2011 and the third one from March to April 2012. State official beekeeping technicians specialised in bee health were trained in the framework of the project in order to minimise as far as possible the variability of the results due to handling and observations. For each of the visits, beekeepers were requested to fulfil a form asking information about the health history of the apiary and of the followed-up colonies, as well as about their beekeeping practices. Honeybee colony disorder symptoms were reported in the questionnaire. These include the following symptoms for which no causal agent could obviously be identified:

(1) dead and disappeared colonies : (1a) death of part or the whole colony, where dead bees can be found close to the hive or inside it. Beekeepers also describe the (1b) disappearance of part or the whole colony, leaving behind food reserves and brood, a phenomenon similar to the one described in North-American apiaries (Colony Collapse Disorder (CCD) [24];
(2) weak colonies : weakening of the colony, showing in occasions a slow development in spring under optimal conditions (e.g. optimal weather, low varroa pressure, etc) with as consequence the loss of the spring production, but in occasions showing abnormally small colonies or colonies with low activity;
(3) queen loss : replacement of young queens sometimes leading to queen-less colonies or interruption of the egg-laying activity of the queen [17];
(4) problems linked to brood and not related to any known disease.

The form also included requests about other symptoms typically linked to known diseases (e.g. diarrhoea, mummified larvae, varroa presence, presence of deformed wing bees, etc). Each of the questions requested further characterisation of the symptoms (e.g. population size, population dynamics, bees behaviour, etc.).

In addition, a thorough clinical inspection was carried out for each colony in the mentioned period. Special attention was given to the strength of the colony in terms of bee numbers, brood surface and reserves content, the presence of the queen and the varroa infestation level. Finally, two different samples were collected before and after the winter: (1) in-hive bees (minimum of 100 bees); (2) a section of the frame containing beebread and honey (about 100 cm^2). Samples were collected in hermetic plastic bags, cooled after collection and stored at $-20°C$.

Case choice – hierarchical sample clustering

Information available from bee colonies comprised field observations, beekeeper answers to the questionnaire and results from the analyses of the samples of beekeeping matrices (honey, beebread, wax and bees). Colonies for which this information was available were considered for our analyses. Colonies with well identified problems (heavy varroa infestation, lack of food or drone-laying queens) were discarded. After that, two groups were made: a group with disorders and another with healthy colonies. In order to constitute these groups and to limit variations due to potential different bee management, a hierarchical classification was made. The criteria used were the amount of food stores before the winter, the year of colony creation, the subspecies and the age of the queen. This classification was made by using Ward aggregation method. This method allows to build group with the lower variation within the group and the higher variation between groups [44].

Virus analyses

The viruses under investigation were the Black Queen Cell Virus (BQCV), the Chronic Bee Paralysis Virus (CBPV), the Acute Bee Paralysis Virus (ABPV), the Deformed Wings Virus (DWV), and the Sac Brood Virus (SBV). Viral analyses were conducted on the honeybee workers collected before the winter, during the first and second periods (July–August and September–October). The samples were analysed with a quantitative RT-PCR by the National Bee Unit laboratory, Food and Environment Research Agency (Sand Hutton, York, United-Kingdom).

Total Nucleic acid (TNA) was extracted from 60 bees homogenised for 12 minutes with 20 ml GITC Lysis Buffer (5 M Guandine Thiocynate, 0.05 M Tris base, 0.02 M EDTA, pH 8.0) in a 30 ml bottle containing 3, 7/16″ ball bearings. GITC Lysis buffer also contained 17.3 mM SDS buffer (173 mM Sodium dodecyl sulphate (SDS) in 100 ml MGW). The SDS buffer is added prior to use in warmed GITC Lysis Buffer. The homogenate was then incubated at 65°C for 40 minutes and then spun at 6189 g for 5 minutes. Polypropylene 96 -deep well plates (DT850301 Elkay Laboratory Products Ltd) were prepared as follows (1 well/extract); plate A: 800 µl extract, and 100 µl MagneSil^{TM} beads (MD1441, Promega); plate B: 1 ml of GITC wash buffer (5 M Guandine Thiocynate, 0.05 M Tris base); plates C, D, E: 1 ml 70%v/v ethanol (E/00665DF/17, Fisher Scientific); plate F: 1 ml 5 M Betaine (B2629, Sigma), plate G, 300 ul 1×TE

Buffer (stock 100X TE (EC-862 National Diagnostics). Plates were loaded onto the Kingfisher Flex and processed as follows: Plate A – Bind 5 mins (fast dual mix), Plate B – Wash 3 mins (fast dual mix), Plate C, D, E – Wash 2 mins (fast dual mix), collect beads at 1 min intervals, Plate F – Wash 20 secs (medium), without releasing beads, Plate G – Mix 1 min then incubate at 65°C for 5 mins with mixing. All steps of the process are looped through twice. TNA was collected from plate G of each reaction and stored at −80°C prior to use in real-time PCR.

Reactions were set up in 96 well reaction plates using Absolute Blue QPCR ROX mix (AB-4139, Thermo Scientific) following the protocols supplied. 0.1 mM of 0.1 M Dithiothreitol (165680250, Arcos Organics) and 0.33 unit of MMLV (EPO441: Fermentas) were added to each reaction. The primers (Table 1) were all used at 400 nM and probes at 200 nM final concentration. Total nucleic acid (5 µl) was added to each reaction, giving a final reaction volume of 20 µl. Plates were cycled for 48°C/30 min, 95°C/10 min and 40 cycles of 60°C/1 min, 95°C/15 sec within the 7900 HT Sequence Detection System (Applied Biosystems), using real time data collection. The results were recorded as the cycle threshold (Ct) or cycle number after which a significant accumulation of florescence over the baseline was observed; an average (of duplicate wells) Ct value below 40 was regarded as a positive result with a threshold ΔRn setting of 0.2. Given the absence of internal standard, we assumed that the extraction method had the same efficiency towards bee DNA/RNA and viral RNA.

Virus primers and probes used for all pathogens tested have been previously described in Chantawannakul et al., 2006 [45]. Additional APBV primers are described in Martin et al., 2012. [46].

Pesticide analyses

Samples collected before the winter were sent to Eurofins Chemiphar NV, Brugge, Belgium and analysed by SOFIA GmbH Chemisches Labor für Softwareentwicklung und Intelligente Analytik, Berlin, Germany. Two multi-residues methods (SF146 and SF150) were used searching for 99, 93 and 96 pesticides residues in wax (54 samples), beebread (108 samples) and honey (107 samples – one sample did not contain enough matrix) respectively (LOQ in Table 2).

For SF146 method, 10 ml of water was added to 10 g of samples. Methanol was added and the preparation was mixed. The mixture was filtered and centrifuged. Next, for analysis by gas chromatography coupled with mass spectrometry (GC-MS), the filtrate was mixed with sodium chloride and ethyl acetate solution 1:1 (v/v). All was dried with sodium sulfate and filtrated. This filtrate was then concentrated and analysed by GC-MS. For analysis with liquid chromatography coupled with tandem mass spectrometry (LC-MS/MS), 5 ml of sample were transferred in ChemElut column and eluted with dichloromethane. After concentration, 1 ml of water-methanol 1:1 (v/v) was added and this solution was analysed by LC-MS/MS. For the wax, some modifications were made. For GC-MS, sodium chloride was replaced by hexane to dissolve wax and the mixture was frozen until a precipitate appears. For LC-MS/MS, 0.2 g of sample were extracted with 30 ml of mixture of naphtha:acetonitrile 1:2 (v/v). 10 ml of acetonitrile phase was isolated and concentrated before the analysis by LC-MS/MS.

For SF150 method, 10 ml of water was added to 10 g of samples as well as methanol and hydrochloric acid. All was mixed, filtrated and centrifuged. A solution of sodium chloride (20%) was added and 5 ml of sample were transferred in ChemElut column and eluted with dichloromethane. After concentration, 1 ml of water-methanol 1:1 (v/v) was added and this solution was analysed by LC-MS/MS.

Descriptive analyses

A descriptive analysis of virus prevalence and pesticide residues was first carried out. The relationship between disorder and these stressing factors was then assessed. For each virus type, a linear model (two way ANOVA) was used, with the cycle threshold value as dependent variable and the visit (first or second), group (with disorder or healthy) and their interaction as explanatory variables. For the pesticides residues, we used a similar model with the total number of residues as dependent variable and type (fungicide or insecticide/acaricide), group (with disorder or healthy) and their interaction as explanatory variable. Based on these models, we constructed a contrast matrix to test explicit post-hoc hypotheses (e.g. compare virus load between groups within each visit...). As several dependent variables had a strongly asymmetric distribution (i.e. non-normal), we computed all p-values by permutation ($n = 1000$). We applied a Bonferoni correction on the post-hoc tests p-values to take into account the multiplicity of the tests.

Relationship between bee colony disorders and potential stressors or surrounding environment of the apiary

We used two separate generalized linear mixed models with a binomial distribution and logit link function. In both models, the "group" (with disorder or healthy) was used as dependent binary variable and the apiary was used as random effect (grouping

Table 1. List of primers used for virus analyses.

Target	Primer Name	Sequence (5′-3′)
BQCV	BQCV 9195F	GGT GCG GGA GAT GAT ATG GA
	BQCV 8265R	GCC GTC TGA GAT GCA TGA ATA C
	BQCV 8217T	TTT CCA TCT TTA TCG GTA CGC CGC C
SBV	SBV 311F	AAG TTG GAG GCG CGy AAT TG
	SBV 380R	CAA ATG TCT TCT TAC dAG AGG yAA GGA TTG
	SBV 331T (MGB)	CGG AGT GGA AAG AT
CPV	CPV 304F	TCT GGC TCT GTC TTC GCA AA
	CPV 371R	GAT ACC GTC GTC ACC CTC ATG
	CPV 325T	TGC CCA CCA ATA GTT GGC AGT CTG C

Table 2. Active ingredients or metabolite included in the multi-residue analyses by beekeeping matrix (A = Acaricide; F = Fungicide; H = Herbicide; I = Insecticide; S = Synergist).

Active ingredient or metabolite*	Class	LOQ (mg/kg)		
		Wax	Beebread	Honey
2,4-D	H	0.1	–	0.01
Abamectin	I	0.1	0.1	0.005
Acetamiprid	I	0.1	0.1	0.005
Aldicarb	I	0.1	0.1	0.005
Alpha-cypermethrin	I	0.1	0.1	0.005
Amitraz	A	0.1	0.1	0.01
Azoxystrobin	F	0.1	0.1	0.005
Bentazone	H	0.2	–	0.02
Benthiavalicarb	F	0.1	0.1	0.005
Beta-cyfluthrin	I	0.1	0.1	0.01
Bifenthrin	I	0.1	0.1	0.005
Boscalid	F	0.1	0.1	0.005
Captan	F	0.1	0.1	0.01
Carbaryl	I	0.1	0.1	0.005
Chlorpyriphos	I	0.1	0.1	0.005
Chlorpyriphos-methyl	I	0.1	0.1	0.005
Chlorothalonil	F	0.1	0.1	0.005
Clothianidin	I	0.1	0.1	0.01
Coumaphos	A	0.1	0.1	0.05
Coumaphos oxon *	A	0.1	0.1	–
Coumaphos phenolic*	A	0.1	0.1	–
Cyazofamid	F	0.2	0.2	0.02
Cyfluthrin	I	0.1	0.1	0.01
Cymoxanil	F	0.1	0.1	0.01
Cypermethrin	I	0.1	0.1	0.005
Cyproconazole	F	0.1	0.1	0.005
Cyprodinil	F	0.1	0.1	0.01
DDT	I	0.1	0.1	0.005
Deltamethrine	I	0.1	0.1	0.005
Dichlorprop-P	H	0.1	–	0.01
Difenoconazole	F	0.1	0.1	0.005
Diflubenzuron	I	0.2	0.2	0.02
Dimethenamid-P	H	0.1	0.1	0.005
Dimethoate	I	0.1	0.1	0.005
Dimethomorph	F	0.1	0.1	0.01
Dimoxystrobin	F	0.1	0.1	0.005
Epoxiconazole	F	0.1	0.1	0.005
Esfenvalerate	I	0.1	0.1	0.005
Ethofumesate	H	0.5	0.5	0.05
Famoxadone	F	0.1	0.1	0.01
Fenhexamid	F	0.1	0.1	0.005
Fenoxycarb	F	0.1	0.1	0.005
Fenpropidin	F	0.1	0.1	0.01
Fenpropimorph	F	0.1	0.1	0.005
Fipronil	I	0.1	0.1	0.01
Flonicamid	I	2	2	0.2
Fluazinam	F	0.1	0.1	0.01
Flufenacet	H	0.1	0.1	0.01

Table 2. Cont.

Active ingredient or metabolite*	Class	LOQ (mg/kg)		
		Wax	Beebread	Honey
Fluopicolide	F	0.1	0.1	0.01
Flusilazole	F	0.1	0.1	0.005
Heptenophos	I	0.1	0.1	0.01
Imidacloprid	I	0.1	0.1	0.005
Indoxacarb	I	0.1	0.1	0.005
Iprodione	F	0.1	0.1	0.005
Kresoxim-methyl	I	0.1	0.1	0.005
Lambda-cyhalothrin	I	0.1	0.1	0.005
Lindane	I	0.1	0.1	0.005
Linuron	H	0.1	0.1	0.005
MCPA	H	0.1	–	0.01
Mecoprop-P	H	0.1	–	0.01
Metalaxyl-M	F	0.1	0.1	0.005
Metamitron	H	0.1	0.1	0.005
Metconazole	F	0.2	0.2	0.02
Methiocarb	I	0.1	0.1	0.005
Methoxyfenozide	F	0.1	0.1	0.005
Metribuzin	H	0.1	0.1	0.005
Penconazole	F	0.1	0.1	0.005
Pendimethalin	H	0.1	0.1	0.005
Permethrin	I	0.2	0.2	0.02
Picoxystrobin	F	0.1	0.1	0.005
Piperonyl butoxide	S	0.1	0.1	0.01
Pirimicarb	I	0.1	0.1	0.005
Prochloraz	F	0.1	0.1	0.005
Propamocarb	F	0.1	0.1	0.005
Propiconazole	F	0.1	0.1	0.01
Propyzamide	F	0.1	0.1	0.01
Pymetrozine	I	0.1	0.1	0.005
Pyraclostrobin	F	0.1	0.1	0.005
Pyrethrin	I	0.2	0.2	0.02
Pyrimethanil	F	0.1	0.1	0.005
Rotenone	I	0.1	0.1	0.01
Spinosad	I	0.1	0.1	–
Spirodiclofen	I	0.1	0.1	0.01
Spirotetramat	F	0.1	0.1	0.01
Spiroxamine	F	0.1	0.1	0.005
Sulcotrione	H	0.1	–	0.01
Tau-fluvalinate	IA	0.1	0.1	0.01
Tebuconazole	F	0.1	0.1	0.005
Tebufenozide	I	0.1	0.1	0.005
Tefluthrine	I	0.1	0.1	0.005
Terbuthylazine	H	0.1	0.1	0.01
Tetraconazole	F	0.1	0.1	0.01
Thiabendazole	F	0.1	0.1	0.005
Thiacloprid	I	0.1	0.1	0.005
Thiamethoxam	I	0.1	0.1	0.005
Thiophanate-methyl	F	0.1	0.1	0.005
Trifloxystrobin	F	0.1	0.1	0.01

Table 2. Cont.

Active ingredient or metabolite*	Class	LOQ (mg/kg)		
		Wax	Beebread	Honey
Zeta-cypermethrin	I	0.1	0.1	0.005
Zoxamide	F	0.1	0.1	0.01

factor) to take into account the non-independence between colonies from the same apiary and allow the intercept to vary between apiaries [47].

With the first model, we explored the relationship between the probability of disorder in a colony and potential stress factors, i.e. pesticides or viruses total load into the colonies. We used as explanatory variables the total number of fungicides residues, the total number of insecticides or acaricides (pooled together) residues, the total number of viruses detected for both visits and all first level interactions between these three explanatory variables. The insecticides and acaricides were pooled together because the most frequently found active ingredient, namely tau-fluvalinate, is authorized in Belgium for both purposes.

With the second model, we investigated the relationship between the probability of disorder in a colony and the structure of the agricultural landscape around the beehive. The surfaces of all different kinds of agricultural soil occupancy in a circle with a radius of 1500 m around each apiary were calculated according to the information provided by the farmers to the Walloon administration (Land Parcel Identification System). The different soil occupancy categories were pooled into two categories defined according to the potential frequency of plant protection product use: (1) grasslands (low pesticides use), (2) crops *sensu lato* (potentially higher pesticides use), including major crops (mainly: cereals, potatoes, beet, oilseed rape, maize, flax, …) but also surfaces dedicated to fruit or vegetables production, fodder production (mainly legumes) and to horticulture. The surfaces of these two land use options, grasslands and crops (*sensu lato*), were used as explanatory variables in the model. The 1500 m radius was chosen accordingly to the mean pollen and nectar collecting distance for the honeybee [48,49]. As the agricultural soil occupancy data was available only for Wallonia, we eliminated from the analysis the apiaries whose buffer had less than 70% of their surface within the administrative boundaries of Wallonia (n = 2).

We used Likelihood Ratio (LR) Tests to evaluate the significance of the explanatory variables [47,50]. We respected the marginality rules, i.e. all main effects were tested after removing from the model the interactions in which they are involved [51].

All analyses were performed with R (R Core Team 2013) and the mixed models were fitted with the package lme4 [52].

Results

Description of honeybee colony disorders

We gathered all sources of information, i.e. questionnaires, analytical matrices and information from clinical visits, for 173 colonies. After the clustering, 54 colonies coming from 21 apiaries (Figure 1) were considered for the study: 25 presenting bee disorders and 29 not presenting them, defined as healthy group. Their data and samples were collected by 8 beekeeping technicians.

In the group presenting disorders, 6 colonies were dead or unviable -with only a handful of bees remaining with the queen- at the spring visit (Table 3). Pre-wintering weakness, winter worker losses, late-season re-queening were reported in these cases. One colony was dead due to queen failure without re-queening in spring. Furthermore, a number of colonies were weak in spring. They were characterised by a low number of individuals and a slow development. In total, nine colonies lost their queens during the study, five of them leading to queenless colonies. These were considered as "disorder colonies" in the framework of the study. Finally, brood abnormalities not linked to known disease were observed in 10 colonies, most of them both before and after the winter.

Symptoms of other diseases or bee parasite observation remained low within the selected colonies. Wax moths were observed in two cases, one in each group. *Varroa destructor* was present in all colonies of the study. All of them were treated according to official veterinary advice. Diarrhoea was observed in one of the colonies showing brood abnormalities. Anatomical modifications (i.e. deformed or undeveloped wings, small abdomens) were observed in 6 colonies, half of them being healthy and half of them showing disorders. No symptoms of any of the foulbroods were observed.

Virus content

One bees sample from the "disorder" group could not be analysed due to a lack of sufficient matrix. The three most abundantly found viruses were the Deformed Wings Virus (DWV), the Black Queen Cell Virus (BQCV) and the Sac Brood Virus (SBV) (Figure 2). The Acute Bee Paralysis Virus (ABPV) and the Chronic Bee Paralysis Virus (CBPV) were found only in a very

Figure 1. Spatial distribution of selected apiaries.

Table 3. Symptoms observed in the group with disorders.

Symptoms of disorder	Frequency
Mortality	2*
Weakening	3
Queen failure	5
Brood problems	9
Mortality+Weakening	2
Mortality+Weakening+Queen failure	2
Mortality+Queen failure	1
Queen failure+Brood problems	1
Total	25

* One of these colonies not considered in the model due to a lack of viral results.

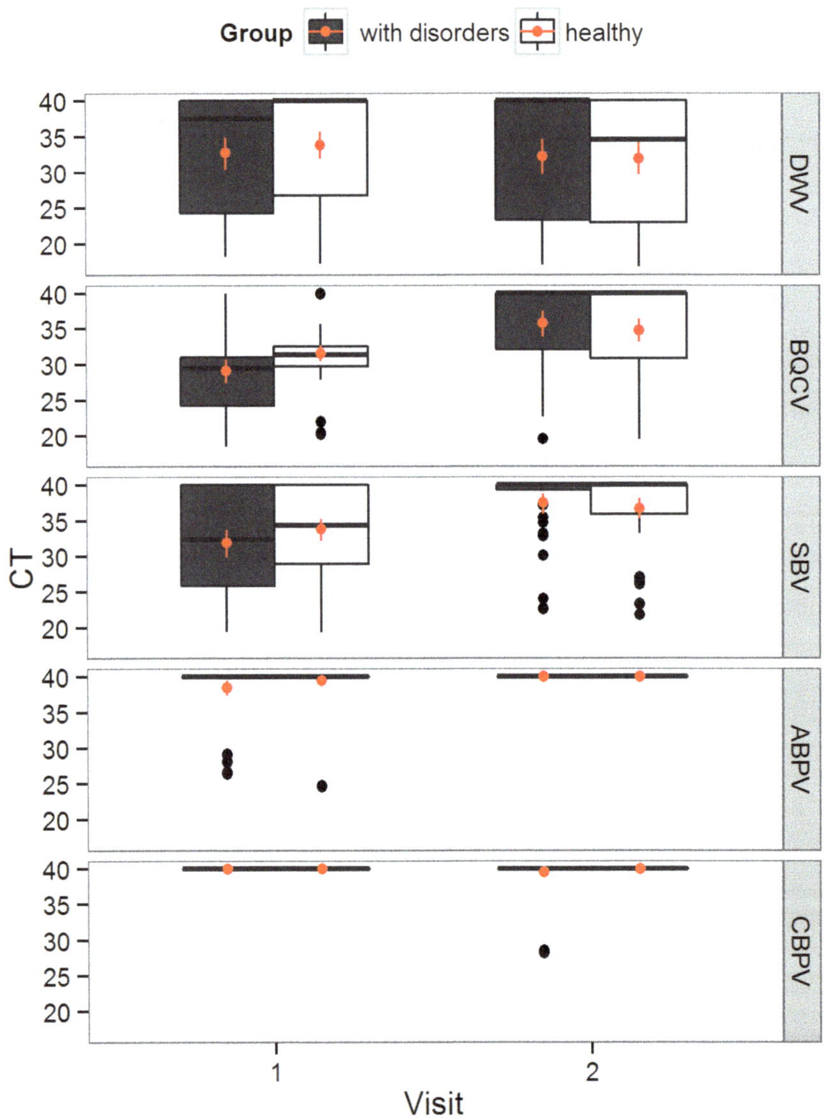

Figure 2. Virus content according to the Cycle Threshold (CT) for the groups "with disorder" and "healthy". Boxplot of Cycle Threshold for the first and second visits (visit 1 - mid-July to mid-august and visit 2 - mid-September to mid-October) and the group with disorders (grey, n = 24 colonies) and the healthy one (white, n = 29 colonies). Deformed Wings Virus (DWV), Black Queen Cell Virus (BQCV), SacBrood Virus (SBV), Acute Bee Paralysis Virus (ABPV), Chronic Bee Paralysis Virus (CBPV). In red, mean with confidence interval estimated by boostrap method. Note: CT values below 40 were regarded as positive results and the lowest CT values correspond to the higher virus contents.

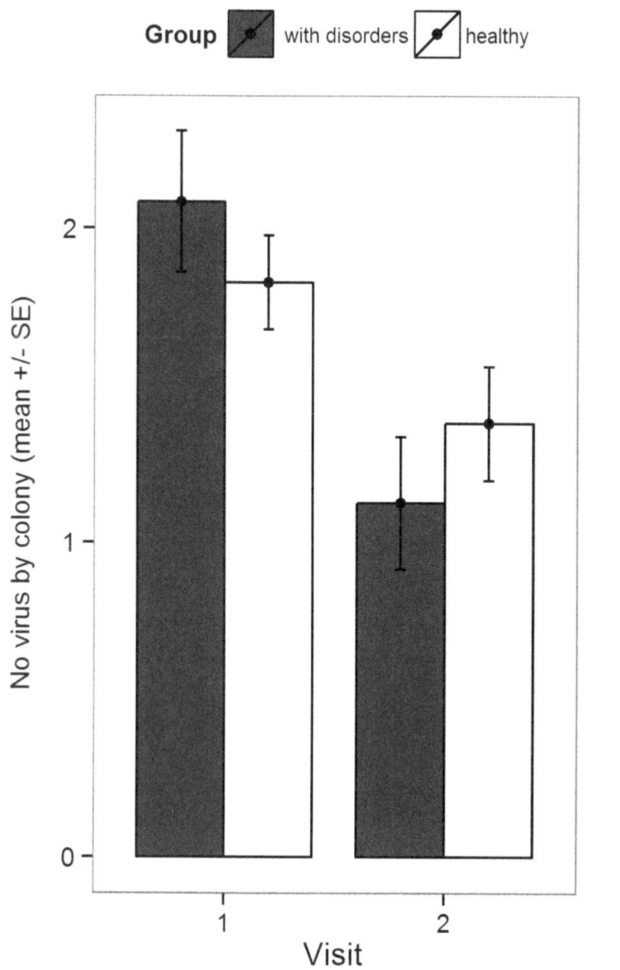

Figure 3. Average number of different viruses per colony. Data shown for the first and second visits (visit 1 - mid-July to mid-august and visit 2 - mid-September to mid-October) and the group with disorders (grey, n = 24 colonies) and the healthy one (white, n = 29 colonies). Whiskers show the standard error (SE).

limited number of samples and did not allowed particular statistical analyses on these two viruses.

There was no significant difference in viral content between the group with disorders or the healthy one, and independently of the visit for any of the three most abundant viruses (2-way ANOVA tested by permutations, DWV: p = 0.731, BQCV: p = 0.373, SBV: p = 0.54). We observed a decrease of virus content from summer to fall 2011, independently of the group, for BQCV (p<0.001) and SBV (p<0.001) but not for DWV (p = 0.271). For BQCV only, we observed a significant (p = 0.036) group×visit interaction indicating that the decrease of the virus abundance from visit 1 to visit 2 was higher in the group "with disorder" than in the group "healthy". Overall, the number of virus changed between visit 1 and visit 2 and a significant decrease was observed in the group with disorders (p<0.001) (Figure 3).

Pesticide analysis

172 agrochemical residues of 23 different active ingredients were detected in 94 out of 269 samples. Wax was the most contaminated matrix: 109 residues of 15 different active ingredients; while 39 and 24 residues of 10 and 8 substances were detected in beebread and honey, respectively (Figures 4 & 5).

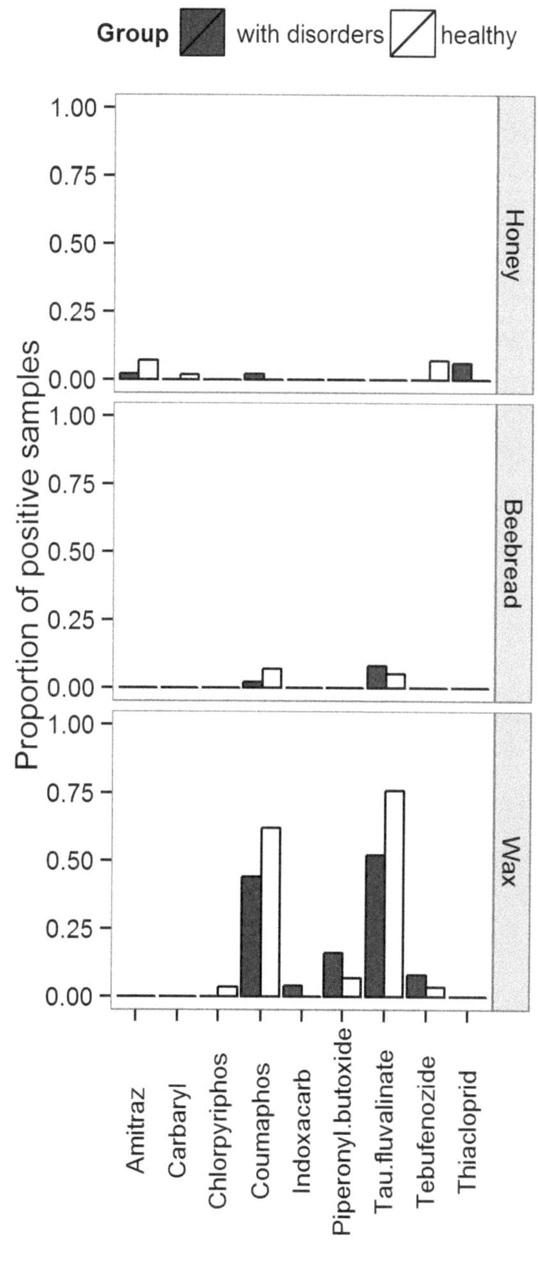

Figure 4. Proportion of samples containing residues of acaricides/insecticides in the different beekeeping matrices (honey, beebread and wax). Data shown for the group with disorders (grey, n = 25 colonies) and the healthy one (white, n = 29 colonies).

Residue levels contained in wax and beebread were higher (0.21–3.1 mg/kg) than those in honey (0.001–0.058 mg/kg).

For the subsequent statistical analysis, the results obtained for insecticide and acaricides residues were pooled together because the most frequently found active ingredient, tau-fluvalinate (n = 46), can be used in Belgium as an acaricide against varroa mite and as an insecticide to control *Meligethes aeneus* in rape. The second most frequently found residue was the coumaphos (n = 35), followed by two fungicides, boscalid (n = 33), iprodione (n = 13) (Table 4). Some residues of neonicotinoid insecticides, synergist and herbicide were also detected: thiacloprid (n = 3), piperonyl butoxide (n = 6), terbuthylazine (n = 1). The highest

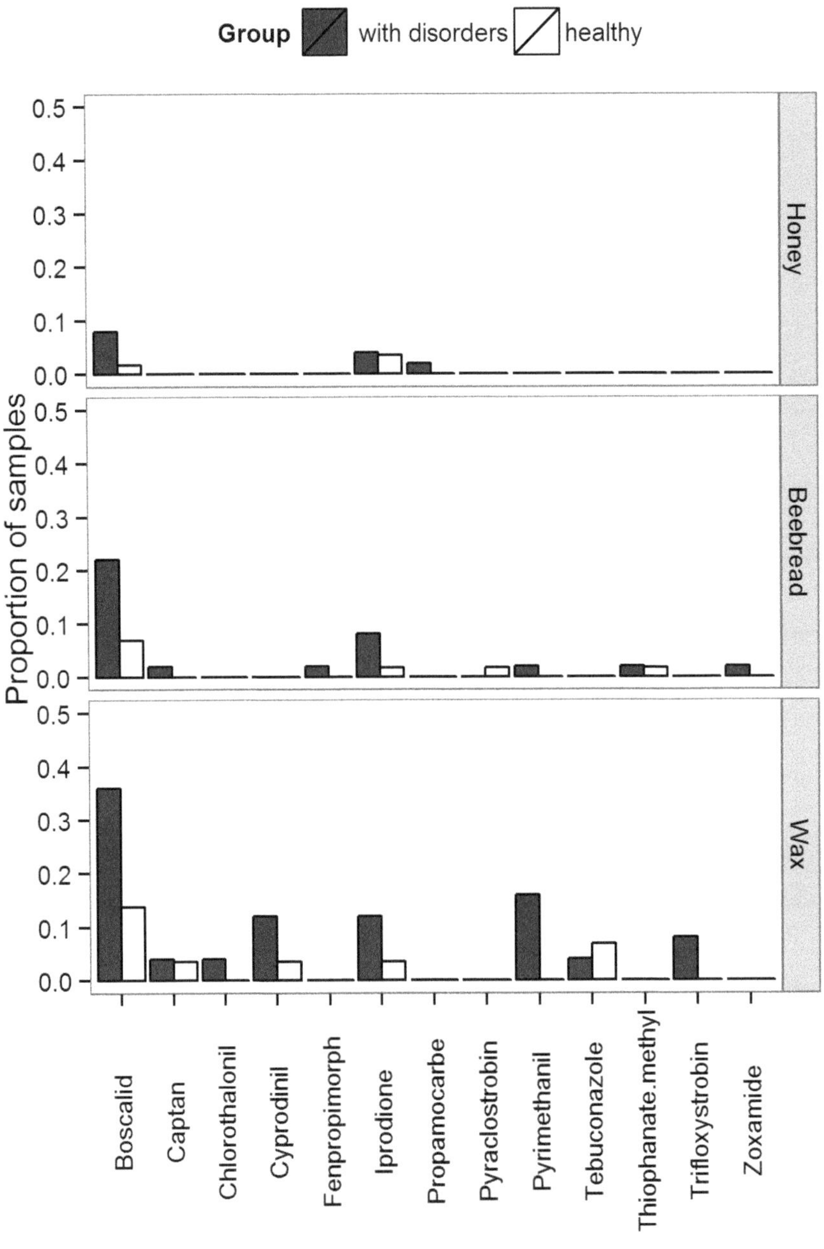

Figure 5. Proportion of samples containing residues of fungicides in the different beekeeping matrices (honey, beebread and wax). Data shown for the group with disorders (grey, n = 25 colonies) and the healthy one (white, n = 29 colonies).

residue level concerned captan with 3.1 and 1.9 mg/kg in the wax and bee bread of the same colony, respectively. Despite of being the most frequently found in matrices, tau-fluvalinate and coumaphos residues never exceeded 0.71 and 0.58 mg/kg, respectively. Boscalid, the most commonly found fungicide, ranged from 0.005 to 1.3 mg/kg.

There was a significant difference ($p = 0.01$) in terms of number of fungicide substances found between the "disorder" group (mean = 2.0) and the "healthy" one (mean = 0.7) as presented in Figure 6. The total number of insecticides/acaricides residues was slightly higher in the disorder group (mean = 2.1) than in the healthy group (mean = 1.64) but this difference is not statistically significant ($p = 0.79$). The most frequent fungicides in the group with disorders were boscalid and iprodione, detected in the three investigated matrices (Figure 5). However, for single active substances, no significant difference of number of residues was observed between the groups "with disorder" and "healthy".

Probability of disorders in relation to potential stressors

There is clearly a significant positive relationship between the probability of a colony showing disorders and the total number of fungicides (Figure 7, LR = 7.128, df = 1, $p = 0.008$). The estimated probability for a colony to be in the "with disorder" group is 0.26 without fungicides residues, 0.60 with 2 fungicides residues and 0.88 with 4 fungicides residues when the insecticides-acaricides residues number and virus load are fixed to their observed mean value (Figure 7). None of the other variables (total number of viruses, total number of insecticides-acaricides) and none of the first level interactions seem to have any explanatory power on the probability of disorders in a colony (Table 5). No direct link could

Table 4. Residues of active ingredients found in wax, beebread and honey samples from colonies with disorders (Group D, n = 25 colonies) and healthy ones (Group H, n = 29 colonies).

Active ingredient	Group	Wax							Bee bread							Honey						
		LOQ (mg/kg)	Number of samples (n = 54)			Residues amount (mg/kg)			LOQ (mg/kg)	Number of samples (n = 108)			Residues amount (mg/kg)			LOQ (mg/kg)	Number of samples (n = 107)			Residues amount (mg/kg)		
			<LOD	Detected	Quantified	Range	Mean	s.d.		<LOD	Detected	Quantified	Range	Mean	s.d.		<LOD	Detected	Quantified	Range	Mean	s.d.
Amitraz	D	0.1	–	–	–				0.1	–	–	–				0.01	49	–	1	0.022	0.02	
	H		–	–	–					–	–	–					53	–	4	0.01–0.028	0.02	0.01
Boscalid	D	0.1	16	8	1	0.29	0.29	–	0.1	39	7	4	0.4–1.3	0.68	0.42	0.005	46	–	4	0.005–0.026	0.01	0.01
	H		25	4	–	–	–	–		54	4	–	–	–	–		56	–	1	0.058	0.06	–
Captan	D	0.1	24	–	1	3.1	3.1	–	0.1	48	–	1	1.9	1.90	–	0.01	–	–	–			
	H		28	1	–	–	–	–		58	–	–	–	–	–		–	–	–			
Carbaryl	D	0.1	–	–	–				0.1	–	–	–				0.005	50	–				
	H		–	–	–					–	–	–					56	–	1	0.02	0.02	–
Chlorpyriphos	D	0.1	25	–	–				0.1	–	–	–				0.005	–	–	–			
	H		28	1	–					–	–	–					–	–	–			
Chlorothalonil	D	0.1	24	1	–				0.1	–	–	–				0.005	–	–	–			
	H		29	–	–					–	–	–					–	–	–			
Coumaphos	D	0.1	14	9	2	0.32–0.34	0.33	0.01	0.1	48	1	–				0.05	49	–	1	0.012	0.01	–
	H		11	10	8	0.23–0.58	0.37	0.01		54	4	–					57	–	–			
Cyprodinil	D	0.1	22	3	–				0.1	–	–	–				0.01	–	–	–			
	H		28	1	–					–	–	–					–	–	–			
Fenpropimorph	D	0.1	–	–	–				0.1	49	1	–				0.005	–	–	–			
	H		–	–	–					58	–	–					–	–	–			
Indoxacarb	D	0.1	24	1	–				0.1	–	–	–				0.005	–	–	–			
	H		29	–	–					–	–	–					–	–	–			
Iprodione	D	0.1	22	1	2	0.24–1.5	0.87	0.89	0.1	45	–	4	0.34–1.5	0.90	0.48	0.005	48	–	2	0.017–0.022	0.02	0.00
	H		28	1	–	–	–	–		57	1	–	–	–	–		55	–	2	0.022–0.04	0.03	0.01
Piperonyl butoxide	D	0.1	21	4	–				0.1	–	–	–				0.01	–	–	–			
	H		27	2	–					–	–	–					–	–	–			
Propamocarbe	D	0.1	–	–	–				0.1	–	–	–				0.005	49	–	1	0.008	0.01	0.00

Table 4. Cont.

Active ingredient	Group	Wax							Bee bread							Honey						
		LOQ (mg/kg)	Number of samples (n = 54)			Residues amount (mg/kg)			LOQ (mg/kg)	Number of samples (n = 108)			Residues amount (mg/kg)			LOQ (mg/kg)	Number of samples (n = 107)			Residues amount (mg/kg)		
			<LOD	Detected	Quantified	Range	Mean	s.d.		<LOD	Detected	Quantified	Range	Mean	s.d.		<LOD	Detected	Quantified	Range	Mean	s.d.
	H		–	–	–					–	–	–					57	–	–	–	–	–
Pyraclostrobin	D	0.1	–	–	–				0.1	50	–	–				0.005	–	–	–			
	H		–	–	–					57	1	–					–	–	–			
Pyrimethanil	D	0.1	21	4	–				0.1	49	1	–				0.005	–	–	–			
	H		29	–	–					58	–	–					–	–	–			
Tau-fluvalinate	D	0.1	12	9	4	0.29–0.46	0.4	0.08	0.1	45	4	–				0.01	50	–	–	–	–	–
	H		7	13	9	0.21–0.71	0.5	0.08		55	3	–					53	–	4	0.011–0.02	0.02	0.00
Tebuconazole	D	0.1	24	1	–				0.1	–	–	–				0.005	–	–	–			
	H		27	2	–					–	–	–					–	–	–			
Tebufenozide	D	0.1	23	2	–				0.1	–	–	–				0.005	–	–	–			
	H		28	1	–					–	–	–					–	–	–			
Terbuthylazine	D	0.1	24	1	–				0.1	–	–	–				0.01	–	–	–			
	H		29	–	–					–	–	–					–	–	–			
Thiacloprid	D	0.1	–	–	–				0.1	–	–	–				0.005	47	–	3	0.009–0.013	0.01	0.00
	H		–	–	–					–	–	–					57	–	–	–	–	–
Thiophanate-methyl	D	0.1	–	–	–				0.1	49	1	–				0.005	–	–	–			
	H		–	–	–					57	–	1	0.38	0.38	–		–	–	–			
Trifloxystrobin	D	0.1	23	2	–				0.1	–	–	–	–	–	–	0.01	–	–	–			
	H		29	–	–					–	–	–					–	–	–			
Zoxamide	D	0.1	–	–	–				0.1	49	1	–				0.01	–	–	–			
	H		–	–	–					58	–	–					–	–	–			

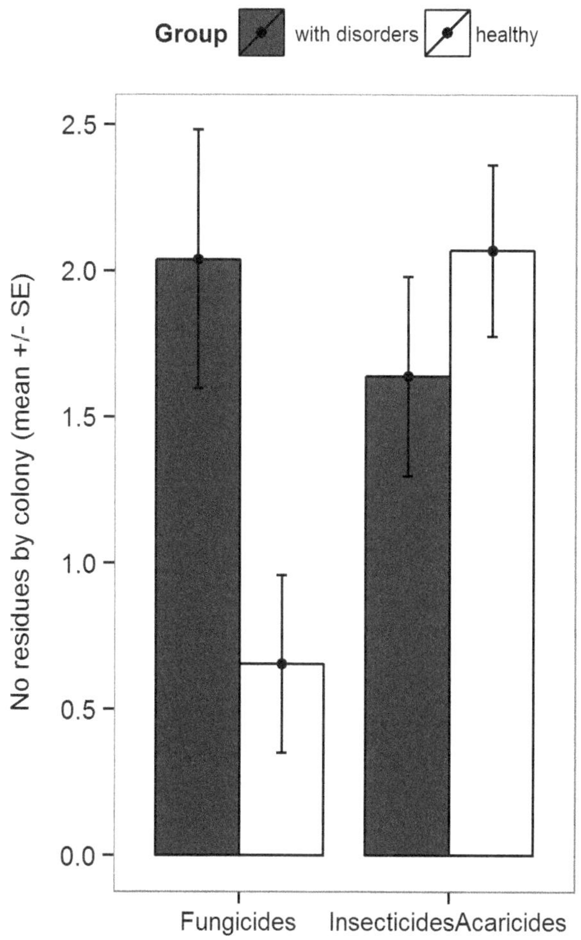

Figure 6. Average number of residues per colony. Data shown for the group with disorders (grey, n = 25 colonies) and the healthy ones (white, n = 29 colonies). Whiskers show the standard error (SE).

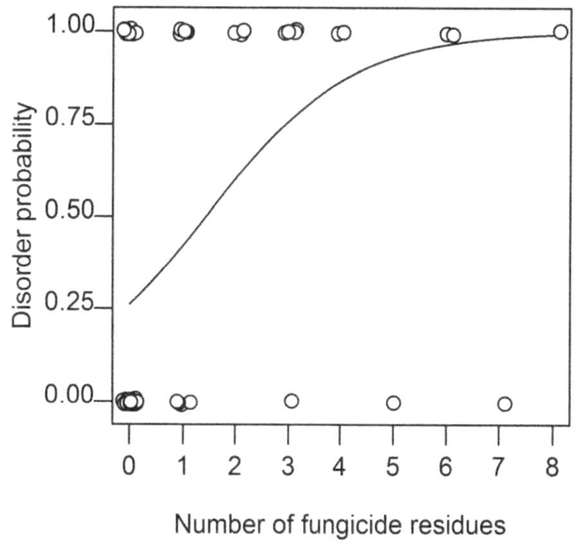

Figure 7. Probability of honeybee colony disorders depending on the number of fungicide residues detected. Model based on averaged coefficients and median value both for the number of insecticides-acaricides residues and total number of virus detected for both visits (n = 53 colonies).

be established between bee colony disorders and the amount (in µg/kg) of residues found in the matrices.

Probability of disorders and crop/grassland surface

Our data clearly show a significant increase in disorder probability with the increase of crop surfaces (*sensu lato*, i.e. including fruit, vegetables, fodder production and horticulture) in the surrounding of the apiary (Figure 8, LR = 8.052, df = 1, p = 0.0045). The predicted probability of disorders is close to 0.1 for a crop surface of 0 ha in the radius of 1500 m and increases up to 0.8 for a surface of 500 ha of crops when fixing the grasslands surface to its observed mean.

On the contrary, the probability of disorders strongly decreases when the grassland surfaces increases (Figure 8, LR = 14.527, df = 1, p = 0.0001) after controlling for the crop surface. The predicted probability of disorders is close to 1 for a grassland surface of 0 ha in a radius of 1500 m and drops to ~0.1 for a surface of 150 ha of grasslands when fixing the crops surface to its observed mean.

Very similar results were found with a 3000 m buffer around the apiary (results not shown).

Discussion

The bee disorders observed after the winter in the Walloon region happened despite of normal climatic conditions. The autumn of 2011 was dry (140.4 mm rainfall water from October to December, average being 219.9 mm), sunny and warm for Belgian conditions (12.4°C on average, which is normally 10.9°C for this period), followed by average winter and spring 2012 in terms of temperature, rainfall, with the exception of a cold wave lasting twelve days in February [53]. These conditions would not induce, *a priori*, a risk for honeybee colonies. Furthermore, based on studies carried out on the palynological diversity of the pollens collected by these colonies before the winter (Table S1), the hypothesis of nutritional lack is unlikely. All samples analysed contained at least 8 different botanic sources of pollen, including rich protein content as oilseed rape, ivy and *Phacelia*.

Viruses infections has often been mentioned as a source of stress for honeybee colonies [7]. However no significant difference in quantity or occurrence was observed between the healthy and the "with disorder" groups. DWV is one of the most commonly detected virus in *A. mellifera*. The prevalence of this virus is even more important at colonies infested by *Varroa destructor*, a well-established vector of this virus [7]. In accordance with the present study, further studies run on Belgian apiaries show DWV, BQCV and SBV as the most frequently found viruses [54]. Unlike DWV, we observed that the amount of SBV and BQCV has dropped significantly between the first and second visits for each group. In the case of BQCV, this is of no surprise as the cycle of incidence of this virus has been shown to increase in late winter, with a peak in May or June followed by a rapid decline [55]. The observed decrease of SBV has also been reported in other studies [42] in which the authors suggested that bees could develop a molecular defensive mechanism to reduce virus multiplication, or that the change in bee susceptibility to the virus could result from environmental conditions such as the quality of pollen.

We cannot prove any causal relationship between any of these viruses analysed and bee disorders, nor does the interaction

Table 5. Analysis of deviance table for generalized linear binomial models describing the relationships between the colony disorder probability and three variables: the total number of (1) fungicide residues (totfungicides), (2) insecticide-acaricide residues (totinsaca) and (3) virus detected for both visits (totvirus).

	LR	df	p(>Chisq)
totfungicides	7.13	1	0.008
totinsaca	0.005	1	0.943
totvirus	0.136	1	0.712
totfungicides:totvirus	2.222	1	0.136
totinsaca:totvirus	0.975	1	0.323
totfungicides:totinsaca	0.901	1	0.342

LR = likelihood ratio.

between these two factors. Cox-Foster et al., 2007 [41] found no clear correlation between a variety of pathogens, including *Nosema* spp., DWV, CBPV, ABPV, BQCV, *Mellisococcus pluton* and *Paenibacillus larvae* ssp and CCD. No specific spore counts were carried out in our study. However, no signs of nosemosis or foulbrood could be linked to colonies presenting bee disorders. It is noteworthy that a monitoring run at Belgian level found *Nosema* spp spores in one out of four colonies [56]. Nevertheless, Cox-Foster et al., 2007 [41], show a positive correlation between IAPV and CCD, which *a priori* would not be relevant in our conditions given the low prevalence of IAPV in this country [27,54]. A recent publication [36] shows a positive correlation between the presence of fungicides in pollen loads and the presence of spores of *Nosema ceranae*. We do not exclude the potential involvement of *Nosema* spp. in the case of bee disorders. However, in the framework of our study, *N. ceranae* seems either to play a role in the development of this weakening, while remaining asymptomatic, or not to play a decisive role in it.

When considering pesticide residues, no direct link could be established between bee colony disorders and the amount (in µg/kg) of residues found in the matrices. Neither could we identify any specific molecule as cause of bee disorders. Nevertheless, the study of the residue load of pesticides, specifically fungicides, opens new avenues for a better understanding of honeybee colony disorders.

Even if insecticides/acaricides were the most numerous residues detected in hives mainly in wax, no significant difference in the number of accumulated residues was observed between colonies with disorders and the healthy ones. Indeed the two most abundant active ingredients, tau-fluvalinate and coumaphos, came most probably from varroa control measures even if tau-fluvalinate is used as an insecticide (Mavrik 2F) against *Meligethes aeneus* in oilseed rape. These active ingredients seem to be a frequent outcome of residue analyses studies [57,58,39,38,40,59]. Synergistic effects have been shown between acaricides and other molecules [33,60]. Nevertheless, our modelling did not suggest any synergistic effects in the appearance of bee disorders occurring between residues of fungicides and insecticides-acaricides. Residues of synergist, piperonyl butoxide, were also detected that indicates a prior exposure to synthetic or natural pyrethroids even though residues were not found in the analysed matrix, probably due to a fast degradation of these active ingredients [61]. A non-authorised active ingredient was also detected: carbaryl, forbidden in Belgium since 2007, indicating an illegal agricultural or gardening use.

The only neonicotinoid found, thiacloprid, was detected in honey during the sampling period of July-August in two colonies. However, the limits of detection achieved in our study do not allow to determine the presence of neonicotinoids -with the exception of acetamiprid- or fipronil at levels in the range of the acute toxicity of these substances (30–40 µg/Kg). Residues of these substances could be present at lower levels and thus an exposure to these substances cannot be excluded.

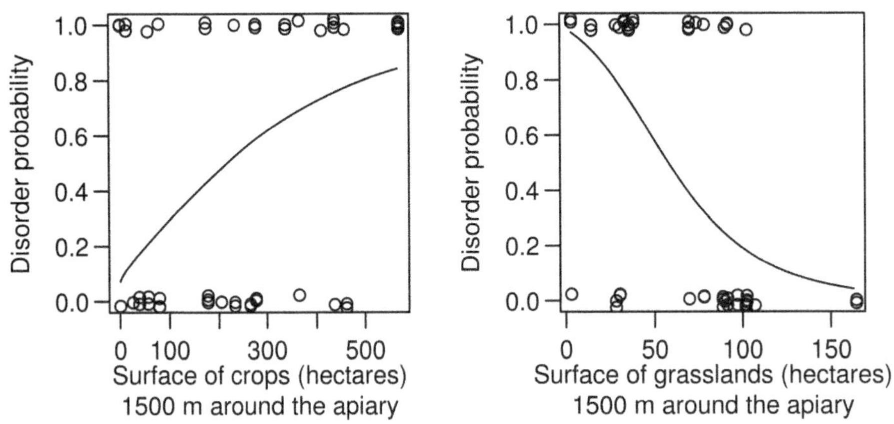

Figure 8. Probability of honeybee colony disorders depending on the apiary's environment. Consideration of crop surface *vs.* grassland surface in a radius of 1500 m around the apiary. Crops include fruit, vegetables, fodder production and horticulture. For each graph, the value of the variable not displayed is fixed to its observed mean (n = 53 colonies).

According to our results, the number of fungicide residues seems to plays a role in the appearance of honeybee colony disorder.

Significantly, more fungicides residues were detected in colonies with disorders than in the healthy ones. Mainly four active ingredients, frequently used in plant protection, were detected: boscalid, iprodione, pyrimethanil and cyprodinil. Fungicides are often considered safe for honeybees based on their acute toxicity. However some direct toxic effects on bees either by the mother compounds or their metabolites have been reported in the past [62]. For example, boscalid shows low acute toxicity to bees [63], despite the fact that beekeepers in the USA have already reported losses and adverse effects on bee brood development related to the foliar application of this systemic active ingredient [64]. Incidents were often linked to the use of a co-formulation boscalid – pyraclostrobin. However, both molecules appeared above detection levels in only two colonies of our study and could not explain the general trend. Interestingly, one metabolite of boscalid, the 2-chloronicotinic acid, is similar to the 6-chloronicotinic acid, a metabolite of imidacloprid. The latter has proved to be lethal to bees at low concentrations (0.1 μg/L) following chronic exposure [65]. Further research should be carried out in order to clarify the mechanism of bee toxicity of boscalid. Indeed, boscalid has proved toxic to other aquatic invertebrates, reducing daphnid fecundity and *Chironomid* emergence [64]. In addition, synergism with other active ingredients like insecticides are possible and increase the toxicity for honeybee [66,32,67]. Two other fungicides were also detected at very high levels in wax and beebread: iprodione and captan. The former is known for its synergistic effects in collaboration with insecticidal compounds [33]. The latter has been shown to induce effects on growth and development of larval honeybees [68,69]. Chronic and larval toxicity studies would be interesting at this stage in order to evaluate possible direct toxic effects on bee individuals and their behaviour. Indeed, a recent study showed increased larval mortality following chronic exposure to tau-fluvalinate, coumaphos, chlorpyriphos and chlorotalonyl or some of their combinations [70]. All these substances were found in our study. Increased mortality rates in the fall may compromise the size and age structure of the wintering cluster, which could lead to winter losses.

The indirect effects of fungicides on bees or on bee colonies are relatively little-known to date.

Fungicides may have an impact on the colony by modifying the existing microflora present in the food stores or in the bee intestinal tract [71]. Studies have already shown the possible modification of microbial composition both at beebread level [62,72] and at intestinal level [73]. This modification in the composition of microbiota may lead to dysbiosis [74]. The impact that such an unbalance in the bee gut microflora may have on bee health has already been considered. The link between the unspecific symptoms observed in our study and a possible microbial alteration could be subject of further research.

In parallel, the potential impact of microbial modification on digestibility and availability of nutrients should be a target for further research. Indeed, the content of essential amino acids might be altered when beebread is contaminated with fungicides (DeGrandi-Hoffman, 2013, pers com.). Given the importance of nutrition, especially pollen, in the good development of the colony [75] alterations in composition or lack of essential nutrients would put the homeostasis of the colony at stake. Some studies have already shown the impact of nutritional lack on bee development and health [76]. Provided that pollen is the unique source of amino acids for honeybees, royal jelly production could also be affected [77,78] with unexpected potential consequences for its main consumers, larvae and the queen. A poor nutrition of the queen, could have as a result an impact in its activity. Likewise, a poor nutrition of the larvae has been shown to impact their development [76]. As a result, the presence of fungicides on beebread and honey may have both a direct effect on their health, but also an indirect one on the colony development.

Fungicides are widely used in agriculture and are broadly present in bee matrices, sometimes at high concentrations at levels of mg/kg [79,80,38,59]. Boscalid, cyprodinil, iprodione are used to control a broad range of fungal pathogens including *Botrytis* spp., *Alternaria* spp. and *Sclerotinia* spp. on a wide range of crops including fruits, vegetables and ornamentals. Pyrimethanil is more specifically used to control grey mould on fruits, vegetables and ornamentals, and leaf scab on apple trees [81]. These active ingredients were already reported in bee matrices in Europe and in the USA [23,39,62]. Their frequent presence in bee matrices might be an indication of chemical plant protection intensity in all agricultural landscapes. Fungicides could also be markers of exposure to other active ingredients with higher toxicity to bees. Mixes of products like fungicides – insecticides are often applied to reduce the number of spray applications. As a result, other pesticides often used in combination with the fungicides found or applied in the same crops could have been at the origin of the effects observed in this study. However, the sensibility of the residue analyses used in our study might explain the lack of detection of such components. Further intensive monitoring and a thorough record of the agricultural practices on pesticide application in tank mixes would help clarifying this alternative explanation.

Factors different from fungicides are most likely involved in the development of bee disorders. According to our model, the disorder probability was not absent when fungicides residues were not detected in presence of insecticide residues and virus. Non identified pathogens, chemicals or factors of different nature could be operating as silent stressors. Ravoet and colleagues (2013) [82] reported the presence of new pathogens in Belgium that were not taken into consideration in our analyses (i.e. *Crithidia mellificae* or the Lake Sinai Virus (LSV)). Other stressing factors could be also linked to the intensive agriculture. In fact, we observed an increase of colony disorders in the area with high density of crops in comparison with areas with grassland. Some studies and beekeeper claims go in line with our outcome regarding the concentration of bee problems in areas with intensive agriculture [83–86]. Furthermore, the same negative trends on pollinators and biodiversity in areas with intensive agriculture have already been described as the result of as increased pesticide use, decreased landscape heterogeneity, loss and fragmentation of natural habitat [87,88].

In conclusion, the five virus studied (ABPV, CBPV, QBCV, SBV, DWV) do not seem determinant in the appearance of bee disorders in our study. These disorders seem clearly linked to the environment of the apiaries and were observed mainly in agricultural crop areas. We observed also that the number of fungicide residues appears as the main potential stress factor linked to bee disorder. However other stressing factors could be acting or interacting at the same time: insecticides exposure, a lack of amino acids and oligo-elements, etc. Our results open new avenues for future research in order to better understand the side effects of fungicides on the bee colony and questions the sustainability of intensive agriculture model and its impact on bees. Specific toxicological studies on both adult bees and larvae would be recommendable in order to better characterise the toxicity of fungicides.

Acknowledgments

We would like to salute and to thank the beekeepers and beekeeping technicians for their dedication and active contribution to our study. Likewise, we thank Gloria DiGrandi-Hoffman, Gerard Arnold, Szaniszlo Szöke and Martin Dermine for their critical and constructive review of the article. We would like also to thank the Direction Générale Opérationnelle Agriculture, Ressources Naturelles et Environnement (DGO3), Département des Aides (D4), Direction des Surfaces (D42), Service 42/3 - LPIS (Land Parcel Identification System) - Service Public de Wallonie for the spatial data. We would also like to thank the Molecular Technology Unit at The Food and Environment Research Agency, Sand Hutton, York, YO41 1LZ for Bee Pathogen Screening Services. Finally, we would like to thank the reviewers for their helpful comments.

Author Contributions

Conceived and designed the experiments: NS LH EB. Performed the experiments: NS LH CM. Analyzed the data: GSM LH. Wrote the paper: LH NS LAM GSM.

References

1. Biesmeijer JC, Roberts SPM, Reemer M, Ohlemüller R, Edwards M, et al. (2006) Parallel declines in pollinators and insect-pollinated plants in Britain and the Netherlands. Science 313: 351–354.
2. Potts SG, Biesmeijer JC, Kremen C, Neumann P, Schweiger O, et al. (2010) Global pollinator declines: trends, impacts and drivers. Trends Ecol Evol 25: 345–353.
3. Goulson D, Lye GC, Darvill B (2008) Decline and conservation of bumble bees. Annu Rev Entomol 53: 191–208.
4. Carvalheiro LG, Kunin WE, Keil P, Aguirre-Gutiérrez J, Ellis WN, et al. (2013) Species richness declines and biotic homogenisation have slowed down for NW-European pollinators and plants. Ecol Lett 16: 870–878. doi:10.1111/ele.12121.
5. Kluser S, Peduzzi P (2007) Global pollinator decline: a literature review. UNEP/GRID- Europe. © UNEP 2007. Switzerland.
6. Maxim L, van der Sluijs JP (2010) Expert explanations of honeybee losses in areas of extensive agriculture in France: Gaucho ® compared with other supposed causal factors. Environ Res Lett 5: 014006. doi:10.1088/1748-9326/5/1/014006.
7. vanEngelsdorp D, Meixner MD (2010) A historical review of managed honey bee populations in Europe and the United States and the factors that may affect them. J Invertebr Pathol 103: S80–S95.
8. Van der Zee R, Pisa L, Andonov S, Brodschneider R, Charriere J-D, et al. (2012) Managed honey bee colony losses in Canada, China, Europe, Israel and Turkey, for the winters of 2008–9 and 1009–10. J Apic Res Bee World 51: 100–114.
9. Klein A-M, Müller C, Hoehn P, Kremen C (2009) Understanding the role of species richness for crop pollination services. Biodiversity, ecosystem function and human wellbeing. New York. 195–208.
10. Lautenbach S, Seppelt R, Liebscher J, Dormann CF (2012) Spatial and temporal trends of global pollination benefit. PloS One 7: e35954.
11. Banskota AH, Tezuka Y, Kadota S (2001) Recent progress in pharmacological research of propolis. Phytother Res 15: 561–571. doi:10.1002/ptr.1029.
12. Jull AB, Rodgers A, Walker N (2008) Honey as a topical treatment for wounds. Cochrane Database of Systematic Reviews. Chichester, UK: John Wiley & Sons, Ltd. 1–43.
13. Srinivasan MV (2011) Honeybees as a Model for the Study of Visually Guided Flight, Navigation, and Biologically Inspired Robotics. Physiol Rev 91: 413–460. doi:10.1152/physrev.00005.2010.
14. Bradber N (1990) Beekeeping in rural development. International Bee Research Association. 15 pp.
15. UNEP (2005) Ecosystems and Human Well-Being: Our Human Planet: Summary for Decision Makers. Island Press. Washington: Island Press.
16. Rasmont P, Pauly A, Terzo M, Patiny S, Michez D, et al. (2005) The survey of wild bees (Hymenoptera, Apoidea) in Belgium and France. Food Agric Organ Rome: 18.
17. Lefevbre M, Bruneau E (2004) État des lieux du phénomène de dépérissement des ruchers en Région wallonne. Louvain la Neuve.
18. Nguyen BK, Mignon J, Laget D, de Graaf D, Jacobs F, et al. (2010) Honey bee colony losses in Belgium during the 2008–9 winter. J Apic Res 49: 337–339. doi:10.3896/IBRA.1.49.4.07.
19. Neumann P, Carreck NL (2010) Honey bee colony losses. J Apic Res 49: 1–6.
20. Vanengelsdorp D, Hayes J, Underwood RM, Caron D, Pettis J (2011) A survey of managed honey bee colony losses in the USA, fall 2009 to winter 2010. J Apic Res 50: 1–10.
21. Aubert M, Faucon J-P, Chauzat M-P (2008) Enquête prospective multi-factorielle: influence des agents microbiens et parasitaires, et des résidus de pesticides sur le devenir de colonies d'abeilles domestiques en conditions naturelles. AFSSA.
22. Higes M, Martín-Hernández R, Garrido-Bailón E, González-Porto AV, García-Palencia P, et al. (2009) Honeybee colony collapse due to *Nosema ceranae* in professional apiaries. Environ Microbiol Rep 1: 110–113. doi:10.1111/j.1758-2229.2009.00014.x.
23. Genersch E, von der Ohe W, Kaatz H, Schroeder A, Otten C, et al. (2010) The German bee monitoring project: a long term study to understand periodically high winter losses of honey bee colonies. Apidologie 41: 332–352.
24. vanEngelsdorp D, Speybroeck N, Evans JD, Nguyen BK, Mullin C, et al. (2010) Weighing Risk Factors Associated With Bee Colony Collapse Disorder by Classification and Regression Tree Analysis. J Econ Entomol 103: 1517–1523. doi:10.1603/EC09429.
25. Pettis JS (2013) The Role Of Pesticides In Queen Health & Sperm Viability. Apimondia 2014, Kiev.
26. Farooqui T (2013) A potential link among biogenic amines-based pesticides, learning and memory, and colony collapse disorder: A unique hypothesis. Neurochem Int 62: 122–136. doi:10.1016/j.neuint.2012.09.020.
27. Nguyen BK, Ribiere M, Vanengelsdorp D, Snoeck C, Saegerman C, et al. (2011) Effects of honey bee virus prevalence, Varroa destructor load and queen condition on honey bee colony survival over the winter in Belgium. J Apic Res 50: 195–202.
28. Nguyen BK, Saegerman C, Pirard C, Mignon J, Widart J, et al. (2009) Does imidacloprid seed-treated maize have an impact on honey bee mortality? J Econ Entomol 102: 616–623.
29. Desneux N, Decourtye A, Delpuech J-M (2007) The Sublethal Effects of Pesticides on Beneficial Arthropods. Annu Rev Entomol 52: 81–106. doi:10.1146/annurev.ento.52.110405.091440.
30. Belzunces LP, Tchamitchian S, Brunet J-L (2012) Neural effects of insecticides in the honey bee. Apidologie 43: 348–370. doi:10.1007/s13592-012-0134-0.
31. Van der Sluijs JP, Simon-Delso N, Goulson D, Maxim L, Bonmatin J-M, et al. (2013) Neonicotinoids, bee disorders and the sustainability of pollinator services. Curr Opin Environ Sustain 5: 293–305. doi:10.1016/j.cosust.2013.05.007.
32. Pilling ED, Jepson PC (1993) Synergism between EBI fungicides and a pyrethroid insecticide in the honeybee (Apis mellifera). Pestic Sci 39: 293–297.
33. Fischer R, Wachendorff-Neumann U (2003) Active ingredient combinations with insecticidal, fungicidal and acaricidal properties.
34. Alaux C, Brunet J-L, Dussaubat C, Mondet F, Tchamitchan S, et al. (2010) Interactions between *Nosema* microspores and a neonicotinoid weaken honeybees (*Apis mellifera*). Environ Microbiol 12: 774–782. doi:10.1111/j.1462-2920.2009.02123.x.
35. Vidau C, González-Polo RA, Niso-Santano M, Gómez-Sánchez R, Bravo-San Pedro JM, et al. (2011) Fipronil is a powerful uncoupler of oxidative phosphorylation that triggers apoptosis in human neuronal cell line SHSY5Y. NeuroToxicology 32: 935–943. doi:10.1016/j.neuro.2011.04.006.
36. Pettis JS, Lichtenberg EM, Andree M, Stitzinger J, Rose R (2013) Crop pollination exposes honey bees to pesticides which alters their susceptibility to the gut pathogen Nosema ceranae. PloS One 8: e70182.
37. Bogdanov S (2005) Contaminants of bee products. Apidologie 37: 1–18. doi:10.1051/apido:2005043.
38. Johnson RM, Ellis MD, Mullin CA, Frazier M (2010) Pesticides and honey bee toxicity – USA. Apidologie 41: 312–331. doi:10.1051/apido/2010018.
39. Mullin CA, Frazier M, Frazier JL, Ashcraft S, Simonds R, et al. (2010) High Levels of Miticides and Agrochemicals in North American Apiaries: Implications for Honey Bee Health. PLoS ONE 5: e9754. doi:10.1371/journal.pone.0009754.
40. Chauzat M-P, Martel A-C, Cougoule N, Porta P, Lachaize J, et al. (2011) An assessment of honeybee colony matrices, Apis mellifera (Hymenoptera: Apidae) to monitor pesticide presence in continental France. Environ Toxicol Chem 30: 103–111. doi:10.1002/etc.361.
41. Cox-Foster DL, Conlan S, Holmes EC, Palacios G, Evans JD, et al. (2007) A metagenomic survey of microbes in honey bee colony collapse disorder. Science 318: 283–287.
42. Tentcheva D, Gauthier L, Zappulla N, Dainat B, Cousserans F, et al. (2004) Prevalence and seasonal variations of six bee viruses in Apis mellifera L. and Varroa destructor mite populations in France. Appl Environ Microbiol 70: 7185–7191.
43. Gauthier L, Tentcheva D, Tournaire M, Dainat B, Cousserans F, et al. (2007) Viral load estimation in asymptomatic honey bee colonies using the quantitative RT-PCR technique. Apidologie 38: 426–435.
44. Zuur AF, Ieno EN, Smith GM (2007) The Analysis of Ecological Data. Springer-Verlag.
45. Chantawannakul P, Ward L, Boonham N, Brown M (2006) A Scientific Note on the Detection of Honeybee Viruses Using real-time PCR (TaqMan®) in Varroa

Mites Collected from a Thai Honeybee (Apis mellifera) Apiary. Journal of Invertebrate Pathology: 91: 69–73.
46. Martin SJ, Highfield AC, Brettell L, Villalobos EM, Budge GE, et al. (2012) Global honey bee viral landscape altered by a parasitic mite. Science 336: 1304–1306.
47. Zuur AF, Ieno EN, Elphick CS (2010) A protocol for data exploration to avoid common statistical problems. Methods Ecol Evol 1: 3–14.
48. Beekman M, Ratnieks FLW (2000) Long-range foraging by the honey-bee, Apis mellifera L. Funct Ecol 14: 490–496.
49. Steffan-dewenter I, Kuhn A (2003) Honeybee foraging in differentially structured landscapes. Proc R Soc Lond B Biol Sci 270: 569–575.
50. Pinheiro JC, Bates DM (2000) Mixed-Effects Models in S and S-PLUS. Springer. 560 p.
51. Fox J (2002) An R and S-Plus companion to applied regression. Sage Publications, Inc. Available: http://books.google.be/books?hl=en&lr=&id=xWS8kgRjGcAC&oi=fnd&pg=PR9&dq=Fox+2002+An+R+and+S+Plus+&ots=o5qCjIMtOb&sig=lvi2jQFv4iDIBrv0QtaWiRIlRBM. Accessed 2012 Sep 18.
52. Bates D (2013) Linear mixed model implementation in lme4. Ms Univ Wis. Available: http://debianmirror.wwi.dk/cran/web/packages/lme4/vignettes/Implementation.pdf. Accessed 2014 Feb 3.
53. IRM IR de M (2013) Bilan climatologique annuel. Available: http://www.meteo.be/meteo/view/fr/1317239-Bilan+climatologique+annuel.html. Accessed 2013 Dec 15.
54. De Smet L, Ravoet J, de Miranda JR, Wenseleers T, Mueller MY, et al. (2012) BeeDoctor, a versatile MLPA-based diagnostic tool for screening bee viruses. PloS One 7: e47953.
55. Bailey L, Ball BV, Perry JN (1981) The prevalence of viruses of honey bees in Britain. Ann Appl Biol 97: 109–118.
56. Verhoeven B (2014) Projet pilote de surveillance de la mortalité des abeilles 2012–2013 Premiers résultats. AFSCA. Journé apicole 26 janvier 2014, Namur.
57. Chauzat M-P, Faucon J-P, Martel A-C, Lachaize J, Cougoule N, et al. (2006) A survey of pesticide residues in pollen loads collected by honey bees in France. J Econ Entomol 99: 253–262.
58. Chauzat M-P, Faucon J-P (2007) Pesticide residues in beeswax samples collected from honey bee colonies (Apis mellifera L.) in France. Pest Manag Sci 63: 1100–1106. doi:10.1002/ps.1451.
59. Wiest L, Buleté A, Giroud B, Fratta C, Amic S, et al. (2011) Multi-residue analysis of 80 environmental contaminants in honeys, honeybees and pollens by one extraction procedure followed by liquid and gas chromatography coupled with mass spectrometric detection. J Chromatogr A 1218: 5743–5756. doi:10.1016/j.chroma.2011.06.079.
60. Johnson RM, Pollock HS, Berenbaum MR (2009) Synergistic interactions between in-hive miticides in Apis mellifera. J Econ Entomol 102: 474–479.
61. Yu SJ (2008) The toxicology and biochemistry of insecticides. CRC N Y NY USA. Available: http://www.lavoisier.fr/livre/notice.asp?ouvrage=1795057. Accessed 2014 Jan 31.
62. DeGrandi-Hoffman G, Chen Y, Simonds R (2013) The Effects of Pesticides on Queen Rearing and Virus Titers in Honey Bees (Apis mellifera L.). Insects 4: 71–89. doi:10.3390/insects4010071.
63. Pesticide Properties Database (PPDB) (2014) Boscalid (Ref: BAS 510F). Available: http://sitem.herts.ac.uk/aeru/ppdb/en/86.htm. Accessed 2014 Jan 3.
64. Aubee C, Lieu D (2010) Environmental Fate and Ecological Risk Assessment for Boscalid New Use on Rapeseed, Including Canola (Seed Treatment). U.S. EPA - Washington.
65. Suchail S, Guez D, Belzunces LP (2001) Discrepancy between acute and chronic toxicity induced by imidacloprid and its metabolites in Apis mellifera. Environ Toxicol Chem SETAC 20: 2482–2486.
66. Belzunces LP, Colin ME (1992) Les phenoménes d'affaiblissement de ruchers.
67. Meled M, Thrasyvoulou A, Belzunces LP (1998) Seasonal variations in susceptibility of Apis mellifera to the synergistic action of prochloraz and deltamethrin. Environ Toxicol Chem 17: 2517–2520.
68. Thompson HM (2003) Behavioural effects of pesticides in bees–their potential for use in risk assessment. Ecotoxicology 12: 317–330.
69. Mussen EC, Lopez JE, Peng CY (2004) Effects of selected fungicides on growth and development of larval honey bees, Apis mellifera L.(Hymenoptera: Apidae). Environ Entomol 33: 1151–1154.
70. Zhu W, Schmehl DR, Mullin CA, Frazier JL (2014) Four Common Pesticides, Their Mixtures and a Formulation Solvent in the Hive Environment Have High Oral Toxicity to Honey Bee Larvae. PloS One 9: e77547.
71. Batra LR, Batra SWT, Bohart GE (1973) The mycoflora of domesticated and wild bees (Apoidea). Mycopathol Mycol Appl 49: 13–44.
72. Yoder JA, Jajack AJ, Rosselot AE, Smith TJ, Yerke MC, et al. (2013) Fungicide Contamination Reduces Beneficial Fungi in Bee Bread Based on an Area-Wide Field Study in Honey Bee, *Apis mellifera*, Colonies. J Toxicol Environ Health A 76: 587–600. doi:10.1080/15287394.2013.798846.
73. Anderson KE, Sheehan TH, Eckholm BJ, Mott BM, DeGrandi-Hoffman G (2011) An emerging paradigm of colony health: microbial balance of the honey bee and hive (Apis mellifera). Insectes Sociaux 58: 431–444. doi:10.1007/s00040-011-0194-6.
74. Sartor RB (2008) Therapeutic correction of bacterial dysbiosis discovered by molecular techniques. Proc Natl Acad Sci 105: 16413–16414.
75. Di Pasquale G, Salignon M, Le Conte Y, Belzunces LP, Decourtye A, et al. (2013) Influence of Pollen Nutrition on Honey Bee Health: Do Pollen Quality and Diversity Matter? PLoS ONE 8: e72016. doi:10.1371/journal.pone.0072016.
76. Brodschneider R, Crailsheim K (2010) Nutrition and health in honey bees. Apidologie 41: 278–294.
77. Standifer LN (1967) A comparison of the protein quality of pollens for growth-stimulation of the hypopharyngeal glands and longevity of honey bees, Apis mellifera L. (Hymenoptera: Apidae). Insectes Sociaux 14: 415–425. doi:10.1007/BF02223687.
78. Liming W, Jinhui Z, Xiaofeng X, Yi L, Jing Z (2009) Fast determination of 26 amino acids and their content changes in royal jelly during storage using ultra-performance liquid chromatography. J Food Compos Anal 22: 242–249.
79. Kubik M, Nowacki J, Pidek A, Warakomska Z, Michalczuk L, et al. (1999) Pesticide residues in bee products collected from cherry trees protected during blooming period with contact and systemic fungicides. Apidologie 30: 12. doi:10.1051/apido:19990607.
80. Kubik M, Nowacki J, Pidek A, Warakomska Z, Michalczuk L, et al. (2000) Residues of captan (contact) and difenoconazole (systemic) fungicides in bee products from an apple orchard. Apidologie 31: 531–542.
81. Pesticide Properties Database (PPDB) (2014) Pyrimethanil (Ref: SN 100309). Available: http://sitem.herts.ac.uk/aeru/ppdb/en/573.htm. Accessed 2014 Jan 3.
82. Ravoet J, Maharramov J, Meeus I, De Smet L, Wenseleers T, et al. (2013) Comprehensive bee pathogen screening in Belgium reveals Crithidia mellificae as a new contributory factor to winter mortality. PloS One 8: e72443.
83. Oomen PA, Belzunces L, Pélissier C, Lewis G (2001) Honey bee poisoning incidents over the last ten years, as reported by bee keepers in The Netherlands. Les Colloques 98: 129–136.
84. Brittain CA, Vighi M, Bommarco R, Settele J, Potts SG (2010) Impacts of a pesticide on pollinator species richness at different spatial scales. Basic Appl Ecol 11: 106–115.
85. CRA-API (2011) Effects of coated maize seed on honey bees. Report based on results obtained from the first year of activity of the APENET project. Italy.
86. Maxim L, van der Sluijs J (2013) Seed-dressing systemic insecticides and honeybees. Late Lessons from Early Warnings: Science, Precaution, Innovation. Copenhagen. 401–438. Available: http://www.beekeeping.com/articles/us/late_lessons_from_early_warnings_2.pdf. Accessed 2014 Feb 10.
87. Holzschuh A, Steffan-Dewenter I, Tscharntke T (2008) Agricultural landscapes with organic crops support higher pollinator diversity. Oikos 117: 354–361.
88. Kleijn D, Kohler F, Báldi A, Batáry P, Concepcion ED, et al. (2009) On the relationship between farmland biodiversity and land-use intensity in Europe. Proc R Soc B Biol Sci 276: 903–909.

12

Islands and Stepping-Stones: Comparative Population Structure of *Anopheles gambiae sensu stricto* and *Anopheles arabiensis* in Tanzania and Implications for the Spread of Insecticide Resistance

Deodatus Maliti[1,2]*, Hilary Ranson[3], Stephen Magesa[5], William Kisinza[6], Juma Mcha[4], Khamis Haji[4], Gerald Killeen[1,3], David Weetman[3]

1 Ifakara Health Institute, Environmental Health and Ecological Sciences Thematic Group, Ifakara, Morogoro, United Republic of Tanzania, **2** University of Glasgow, Institute of Biodiversity Animal Health and Comparative Medicine, Glasgow, Lancashire, United Kingdom, **3** Department of Vector Biology, Liverpool School of Tropical Medicine, Merseyside, Liverpool, United Kingdom, **4** Zanzibar Malaria Elimination Programme, Unguja, Zanzibar, United Republic of Tanzania, **5** RTI International, Global Health Division, Dar es Salaam, United Republic of Tanzania, **6** National Institute for Medical Research, Amani Research Center, Muheza, Tanga, United Republic of Tanzania

Abstract

Population genetic structures of the two major malaria vectors *Anopheles gambiae* s.s. and *An. arabiensis*, differ markedly across Sub-Saharan Africa, which could reflect differences in historical demographies or in contemporary gene flow. Elucidation of the degree and cause of population structure is important for predicting the spread of genetic traits such as insecticide resistance genes or artificially engineered genes. Here the population genetics of *An. gambiae* s.s. and *An. arabiensis* in the central, eastern and island regions of Tanzania were compared. Microsatellite markers were screened in 33 collections of female *An. gambiae* s.l., originating from 22 geographical locations, four of which were sampled in two or three years between 2008 and 2010. *An. gambiae* were sampled from six sites, *An. arabiensis* from 14 sites, and both species from two sites, with an additional colonised insectary sample of each species. Frequencies of the knock-down resistance (*kdr*) alleles 1014S and 1014F were also determined. *An. gambiae* exhibited relatively high genetic differentiation (average pairwise $F_{ST}=0.131$), significant even between nearby samples, but without clear geographical patterning. In contrast, *An. arabiensis* exhibited limited differentiation (average $F_{ST}=0.015$), but strong isolation-by-distance (Mantel test $r=0.46$, $p=0.0008$). Most time-series samples of *An. arabiensis* were homogeneous, suggesting general temporal stability of the genetic structure. *An. gambiae* populations from Dar es Salaam and Bagamoyo were found to have high frequencies of *kdr* 1014S (around 70%), with almost 50% homozygote but was at much lower frequency on Unguja Island, with no. *An. gambiae* population genetic differentiation was consistent with an island model of genetic structuring with highly restricted gene flow, contrary to *An. arabiensis* which was consistent with a stepping-stone model of extensive, but geographically-restricted gene flow.

Editor: Igor V. Sharakhov, Virginia Tech, United States of America

Funding: The authors thank the Wellcome Trust for funding this work through the Wellcome Trust Masters of Science Fellowship Programme (award number 08703/Z/08/Z). The authors thank the Bill and Melinda Gates Foundation through the Malaria Transmission Consortium (award number 45114) for supporting the initial processes of this work. The funders had no role in study design, data collection and analysis, decision to publish, or preparation of the manuscript.

Competing Interests: The authors have declared that no competing interests exist.

* Email: dmaliti@ihi.or.tz

Introduction

In sub-Saharan Africa, a dramatic increase in household ownership of long-lasting insecticidal nets (LLINs), is considered one of the major factors contributing to the fall in malaria cases over the last decade [1]. Sustainability of LLINs as a frontline control strategy against malaria is threatened by growing *Anopheles* resistance to pyrethroids [2], the only class of insecticides licensed for LLIN treatment. Improved understanding of the mechanisms responsible for insecticide resistance in *Anopheles* malaria vectors, and development of reliable diagnostics (such as those available for *kdr* knockdown resistance mutations [3]) are considered important goals to prolong the efficacy of pyrethroids for mosquito control [4]. A less well-studied aspect of vector control concerns how resistance spreads, though such information is important to permit optimised targeting of interventions using complementary insecticides and insecticide combinations. Genetic data can aid predictions of the spread of resistance alleles via inference of vector population structure, which can be compared to the spatial or temporal distribution of diagnostic markers for specific resistance mechanisms where these are available [5]. In addition, vector population genetic data could potentially give insight into connectivity of disease transmission dynamics [6], and is also an essential prerequisite for rational planning of vector control strategies that focus on release of sterilised or genetically manipulated mosquitoes [7].

Until recently, *An. gambiae* s.s. was considered the principal malaria vector across most parts of Sub-Saharan Africa, but several areas of East Africa have shown a recent frequency shift towards *An. arabiensis*. Though a causal link remains to be demonstrated, this is coincident with scaling-up of LLIN distribution [8–10]. Potentially paradoxically, *An. arabiensis* is typically less resistant to pyrethroids than *An. gambiae* s.s. when sympatric populations are compared, and usually shows lower frequencies (often complete absence) of known target site resistance mechanisms, such as the *kdr* 1014 mutations [11–14]. However, *An. arabiensis* is considered more adaptable in blood-feeding behaviour in being more zoophagic, exophagic and exophilic [15–20]. *An. arabiensis* also exhibits greater resilience to arid conditions [21] than *An. gambiae* s.s. (i.e. S-molecular form; [22]), and appears to avoid dramatic changes in effective population size (N_e) across wet and dry periods, even in extremely seasonal environments [23]. As a consequence, *An. arabiensis* might be predicted to exhibit more widespread homogeneity in population structure than *An. gambiae s.s.*

In West Africa, the marked population structure in *An. gambiae s.s.* tends to be associated with divergence between molecular and chromosomal forms [24–26], although the level of differentiation detected can depend upon the type of markers studied (e.g. microsatellites *vs* single nucleotide polymorphisms, SNPs) and/or their genomic location [27–29]. The inhospitable environments of the Rift Valley Complex have been identified as a major barrier partitioning East African *An. gambiae s.s.* populations [30,31] but appear to have relatively little impact on gene flow in *An. arabiensis* [32,33], perhaps reflecting greater ecological tolerance. In contrast, geographical isolation on islands can cause substantial differentiation and reduced diversity, relative to mainland populations, in both species [34–36]. Ecological zonation has also occasionally been associated with strong population structure ($F_{ST}>0.1$; where F_{ST} is a widely applied metric measuring within *vs* among population diversity) in country-wide surveys of *An. gambiae* s.s. (M and S forms) in Ghana [37] and *An. arabiensis* in Nigeria [38]. Moreover, the Nigerian study was one of few in either species to detect significant isolation by distance, although in this case patterns of genetic diversity suggest that historical range expansion [39], may have played a greater role than contemporary geographical restriction of gene flow [38]. At small spatial scales (e.g. <200 km; [40]), and in the absence of variation in molecular or chromosomal forms in West African *An. gambiae* s.s., differentiation is usually low in both species [32,40–42], often falling below the limits of detection of the microsatellite marker panels applied. Therefore studies to date partially support a hypothesis of weaker population structure in *An. arabiensis* than *An. gambiae* s.s., a difference which might be compounded by relative range expansion and contraction in the latter. In contrast, a recent study of comparative population genetic structure in the Kilombero valley of Tanzania reported little differentiation among three *An. gambiae* s.s. samples but strong population structure in *An. arabiensis*, with $F_{ST}>0.1$ at a scale of <100 km, and even suggestion of differentiation within sympatric samples [43].

While *An. gambiae* s.s. populations have recently declined dramatically in many parts of East Africa [8–10,44,45], this is not consistently the case throughout the region and *An. arabiensis* has proven remarkably persistent despite high coverage of LLINs and, in some areas Indoor Residual Spraying (IRS). While physiological resistance to pyrethroids is now clearly emerging in both species [46], *An. arabiensis* also appears to exhibit a number of behaviours that render it relatively unresponsive to LLINs and IRS, such behaviours include early exit, outdoor resting, outdoor feeding and feeding upon animals [44,47,48]. These front line strategies therefore need to be supplemented with complementary vector control measures that improve on the levels of control achieved outdoors [49,50] and others which target mosquitoes outdoors and/or at source [51–53]. Clarification of vector population connectivity within each species in countries like Tanzania, where 73% of the human population live in high malaria transmission areas [4], can aid targeting of interventions and planning of management strategies to combat the spread of insecticide resistance.

Here we present a comparative study of population genetic structure in *An. arabiensis* and *An. gambiae* s.s. across Tanzania to: (1) investigate spatial and temporal population structure; (2) identify possible barriers to gene flow; and (3) determine inter-relationships of (1) and (2) with the frequency of insecticide resistance-associated *kdr* alleles. We report that most *An. gambiae* s.s. samples were differentiated, in some cases strongly, but without clear geographical patterning, consistent with an island model of genetic structure. By contrast, *An. arabiensis* exhibited weak differentiation with strong isolation-by-distance, concordant with a stepping-stone model of extensive, but geographically-restricted gene flow.

Materials and Methods

Ethics statement

All mosquitoes were either collected through routine physiological surveillance activities of the National Institute for Medical Research (NIMR) of Tanzania and the Zanzibar National Malaria Elimination Programme (ZAMEP), or through research protocols implemented by the Ifakara Health Institute (IHI) that were approved by both the IHI internal institutional review board (Reference IHI/IRB/A.50) and the Medical Research Coordination Committee at NIMR (Reference NIMR/HQ/R.8a/Vol. IX/801). Informed written consents were obtained from the household owners for permission to perform sampling in and around households.

Sample sites and species identification

The study was conducted in three regions of Tanzania (Fig. 1): the south-central area, which includes the highly malaria-endemic Kilombero Valley; the Indian Ocean coast, including three districts of urban Dar es Salaam, and Bagamoyo 60 km to the north; and the Zanzibar islands of Unguja and Pemba. A total of 33 collections, comprising of nine *An. gambiae* s.s. samples and 24 *An. arabiensis* samples were included in the study. Of the 16 collections of *An. arabiensis* from the Kilombero Valley, nine formed a temporal series from the villages of Idete, Namawala and Lupiro sampled between 2008 and 2010. In addition, we included samples from IHI insectary colonies of *An. gambiae* s.s. (colonised in 1996) and *An. arabiensis* (colonised in 2008) as entirely isolated out-groups. A total of 1429 *An. gambiae s.l.* mosquitoes (identified using morphological keys [54]) were collected between 2008 and 2010 using human landing catches (HLC), Centre for Diseases Control (CDC) light traps, Ifakara Tent Traps (ITT), window exit traps, and resting catches inside households using mouth aspirators and back-pack aspirators. All samples were stored dry over silica gel. DNA was extracted from whole *An. gambiae* s.l. using the Livak method [55] and re-suspended in 100 µl of water. Species identity as *An. gambiae* s.s. or *An. arabiensis* was diagnosed using a standard allele-specific PCR method [56] with visualisation of amplicons on a 2% agarose gel.

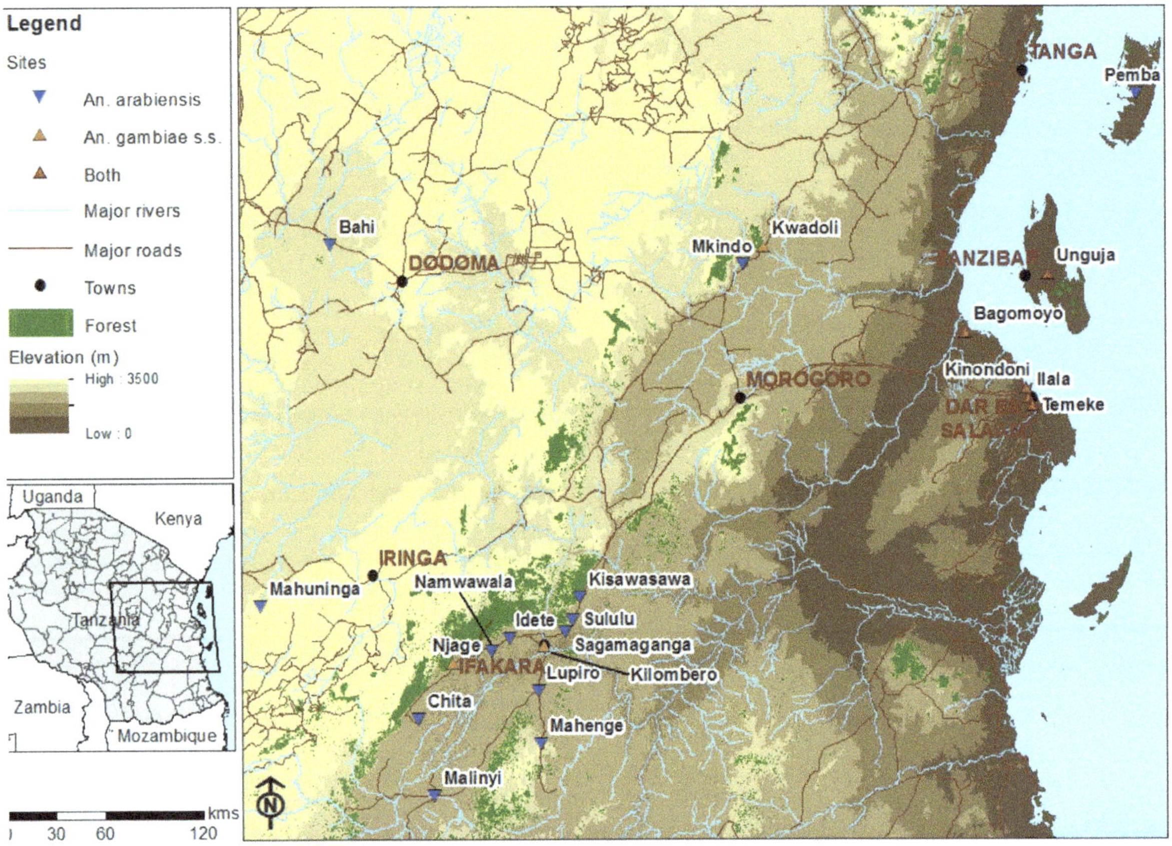

Figure 1. Map of central-eastern Tanzania showing sampling sites.

Microsatellite and *kdr* genotyping

Twelve microsatellite loci spanning all three chromosomes were genotyped: AGXH678 and AGXH7 from the X chromosome, AG2H79, AG2H786, AG2H799 and 2R_Si_5 from chromosome 2, and AG3H812, AG3H119, AG3H577, AG3H811, AG3H765 and 33C1 from chromosome 3. Primers for loci beginning with the prefix AG were developed by [57], for 33C1 by [58], and for 2R_Si_5 by DW (primers given in [59]). Each locus was amplified in a 15 μl reaction containing 1.5 μl 10X PCR buffer with 1.5 mM $MgCl_2$, 0.3 μl 10 mM dNTPs, 0.2 μl 10 mM cy5- or cy5.5-labelled forward primer, 0.15 μl of 10 mM reverse primer, 0.2 μl of Taq, 11.15 μl of PCR-quality water and 1.5 μl of DNA extract. PCR cycles included initial denaturation at 95°C for 5 minutes, followed by 35 cycles of denaturation at 94°C for 30 seconds, annealing at 55°C or 58°C for 1 minute (depending on the optimal annealing temperature of the primers), extension at 72°C for 1 minute and a final extension step of 72°C for 10 minutes. Three pairs of primers with the same annealing temperature, different base pair sizes and different fluorescent labels were amplified in each reaction. PCR products were run on a Beckman-Coulter CEQ 8000 capillary sequencer and sizes scored automatically by comparison with the Beckman-Coulter DNA size standard 400 with all alleles checked manually. *Kdr* L1014F and L1014S genotyping was performed on a randomly-selected sample of 20 individuals from each collection site using TaqMan qPCR [3]. PCR reactions were carried out in 20 μl each containing 10 μl of SensiMix (Bioline), 900 nM of primer, 900 nM of probe, 8.5 μls of PCR quality water and 1 μl of DNA. Samples were run on a Stratagene 3005 (Agilent Technologies) with cycling conditions: 10 minutes at 95°C followed by 40 cycles of 95°C of 10 seconds, and 60°C for 45 seconds.

Statistical analysis

Microchecker 2.2.3 [60] was used to identify possible scoring errors. Deviation from neutrality of loci was examined using LOSITAN, which uses an F_{ST} outlier approach to detect loci showing extreme variation given their level of polymorphism [61]. Linkage disequilibrium among loci was tested using the exact tests in GENEPOP 4.0 [62], with default settings. Hardy-Weinberg (H-W) equilibrium was tested using FSTAT 1.2 via permutation tests based on the positive or negative magnitude of F_{IS}. Genetic diversity was measured by expected heterozygosity (H_e) and allelic richness (R_s) computed by FSTAT 1.2 [63], the latter based on a minimum number of genotypes in any population of nine for *An. gambiae* and 14 for *An. arabiensis*. FSTAT was also used to generate pairwise F_{ST} values between sample sites and to test for population differentiation using the G-test genotypic permutation procedure. Following pooling of temporal samples, isolation-by-distance was examined by comparison of matrices for linearized F_{ST} ($F_{ST}/1\text{-}F_{ST}$) and the natural logarithm (ln) of geographical distance using a Mantel test with 10 000 permutations implemented by the Poptools add-in for Excel [64]. Insectary samples were excluded from this test. PHYLIP 3.68 [65] was used to produce a neighbour-joining tree from F_{ST} values, again with pooled temporal samples, which was visualised using FIGTREE 1.3 [66]. The Bonferroni procedure was applied throughout to correct for multiple testing. Bayesian clustering analysis of data was performed using two models implemented by BAPS [67]. The

Table 1. F_{IS} and Lositan values for *An. gambiae s.s.*(A) and *An. arabiensis* (B).

(A) *Anophelesgambiae*

Lositan P	F_{IS}	Unguja (2010)	Ilala (2008)	Kinondoni (2008)	Temeke (2008)	Bagamoyo (2008)	Kwadoli (2009)	Kilombero (2009)	Njage (2009)	Insectary (2011)
		F_{ST}								
0.722	AGH765	0.13	0.18	0.22	0.15	0.29	0.07	0.01	0.21	−0.02
0.494	AG3H811	0.08	0.11	0.06	0.18	0.14	0.14	0.11	0.21	0.07
0.952	AG2H79	−0.09	−0.01	−0.02	−0.11	−0.17	−0.19	**−0.34**	−0.22	0.11
0.154	AGXH7	0.42	**0.73**	**0.70**	**0.52**	0.20	**0.52**	0.25	0.12	−0.31
0.702	AG3H812	0.57	0.13	0.16	0.03	0.18	0.20	0.12	0.11	0.25
0.750	AG3H119	−0.08	0.22	0.25	0.04	**0.41**	0.14	0.16	0.11	−0.01
0.790	AG3H577	**0.59**	0.06	0.02	0.02	−0.07	−0.01	0.11	0.08	−0.21
0.999	33C1	−0.02	−0.09	−0.14	0.23	−0.13	0.03	−0.15	−0.01	−0.02
0.384	2R_Si_5	0.14	0.19	0.24	0.07	0.10	−0.05	0.10	−0.05	−0.09
0.054	AG2H786	−0.31	0.09	0.01	0.03	0.19	0.12	0.20	−0.02	0.11
0.952	AGXH678	−0.02	0.08	0.04	−0.01	−0.03	−0.11	0.09	−0.10	**0.80**

(B) *Anopheles arabiensis*

Lositan P	F_{IS}	Pemba (2010)	Unguja (2008)	Unguja (2010)	Dar (2008)	Bagamoyo (2008)	Idete (2008)	Idete (2009)	Idete (2010)	Namawala (2008)	Namawala (2009)	Namawala (2010)	Lupiro (2008)	Lupiro (2009)	Lupiro (2010)	Sululu (2010)	Sawasawa (2010)	Chita (2010)	Malinyi (2010)	Mahenge (2011)	Sagamanganga (2010)	Mkindo (2009)	Mahuninga (2011)	Bahi (2011)	Insectary (2011)
												F_{ST}													
1.000	AGH765	−0.02	−0.09	1.00	0.21	−0.01	−0.05	−0.01	−0.13	−0.30	−0.08	**−0.32**	−0.08	−0.15	0.15	−0.07	−0.01	−0.02	−0.02	−0.04	0.11	0.29	−0.05	0.49	0.45
0.144	AG3H811	**0.42**	**0.65**	0.40	0.08	0.42	**−0.55**	**0.41**	**0.49**	0.21	0.21	0.30	**0.25**	0.21	0.25	0.05	**0.36**	**0.41**	**0.30**	0.28	**0.38**	0.29	**0.42**	**0.37**	**0.40**
0.411	AG2H79	0.11	0.01	0.47	−0.09	−0.24	−0.12	0.20	−0.14	−0.16	0.15	−0.23	0.15	0.19	−0.19	0.21	0.20	0.29	0.23	0.07	0.06	0.03	−0.07	−0.08	0.06
0.043	AGXH7	0.00	0.34	0.20	−0.23	−0.34	−0.11	0.11	−0.14	0.13	−0.03	−0.27	0.16	0.08	−0.17	−0.03	−0.07	−0.07	−0.20	−0.14	−0.06	0.02	0.11	0.13	−0.08
0.001	AG3H812	−0.09	0.15	0.24	−0.02	−0.15	−0.16	−0.01	−0.14	0.01	−0.13	−0.05	0.12	0.16	0.01	−0.23	0.00	0.05	0.09	0.01	0.00	0.01	−0.01	0.06	0.10
0.000	AG3H119	0.05	0.10	0.27	0.00	0.31	0.06	−0.03	−0.14	−0.05	−0.02	−0.05	0.17	0.02	0.22	0.14	−0.04	−0.09	0.01	0.18	−0.07	0.27	0.13	−0.06	0.11
0.001	AG3H577	0.08	0.07	0.42	0.02	0.00	0.21	−0.18	−0.14	−0.14	0.03	−0.04	0.06	0.03	−0.24	−0.09	−0.14	−0.03	−0.07	−0.02	0.18	0.12	0.01	0.02	0.06
0.993	33C1	−0.14	−0.12	−0.10	−.08	−0.15	0.07	−0.01	−0.14	0.07	−0.11	−0.17	0.03	−0.08	0.00	0.02	0.08	0.06	0.10	0.04	−.04	0.13	0.01	0.01	0.31

Table 1. Cont.

(B) *Anopheles arabiensis*																									
Lositan P	F_{IS}	Pemba (2010)	Unguja (2008)	Unguja (2010)	Dar (2008)	Bagamoyo (2008)	Idete (2008)	Idete (2009)	Idete (2010)	Namawala (2008)	Namawala (2009)	Namawala (2010)	Lupiro (2008)	Lupiro (2009)	Lupiro (2010)	Sululu (2010)	Sawasawa (2010)	Chita (2010)	Malinyi (2010)	Mahenge (2011)	Sagamaganga (2010)	Mkindo (2009)	Mahuninga (2011)	Bahi (2011)	Insectary (2011)
												F_{ST}													
0.080	2R_Si_5	0.02	0.14	−0.13	0.09	−0.10	−0.11	−0.11	−0.14	0.05	0.05	0.15	0.05	0.25	0.03	−0.05	−0.11	0.12	0.07	0.18	0.07	−0.03	0.17	0.02	0.08
0.995	AG2H786	0.02	−0.02	−0.07	−0.19	−0.16	**−0.37**	**−0.55**	−0.14	**−0.46**	−0.20	**−0.63**	−0.10	**−0.67**	0.54	0.04	−0.03	0.15	0.07	0.34	0.08	−0.08	0.07	0.19	0.13
1.000	AGXH678	0.07	0.05	0.10	−0.07	−0.22	0.10	**0.43**	−0.14	0.38	**−0.65**	−0.12	**−0.41**	0.14	−0.12	−0.04	**−0.61**	**−0.59**	−0.38	**−0.40**	**−0.49**	−0.37	0.01	0.23	0.06

FIS values in bold are significantly greater or larger than expected (following Bonferroni correction).
Lositan P values in red indicate loci showing patterns of differentiation exceeding neutral expectations; these loci were excluded.

first, used for both species, applies a spatially-conditioned non-admixture model to determine clusters of sample sites; an individual-based spatially-conditioned model was also applied to *An. arabiensis*. Both clustering models incorporate geographical location as a prior to aid cluster determination and the individual model works best when individual coordinates are used. Since these were not available we produced them by randomly generated coordinates from within a 10 km radius from the location of each site to assign to individuals based on observations of maximal anopheline flight distances [68]. For the population-level clustering analyses temporal samples from single sites were pooled, however, this does not apply to the individual analysis, which does not use population information as a prior. For all sample collections we attempted estimation of variance effective population sizes via the linkage disequilibrium method implemented in the software LDNE [69], and utilising only alleles with frequency greater than 5%, and tested for mutation-drift equilibrium using the Wilcoxon test in Bottleneck [70], with default settings for each mutation model. Principal component analysis (in SPSS 20) was used to generate a single axis summarising geographical position from latitude and longitude data.

Results

Data quality control

Of the microsatellite loci genotyped, only AGH799 proved to be impossible to score reliably and was excluded prior to any analysis. Microchecker highlighted multiple instances of scoring errors, primarily as null alleles, and scoring was checked wherever potential problems were highlighted. Nevertheless, 8 out of 99 tests for H–W disequilibrium in *An. gambiae* s.s. were significant following correction for multiple testing. All but one indicated a deficit of heterozygotes and each was in a different sample, negating the likelihood of within-population structure as an explanation. However, four were significant for locus AGXH7, suggesting the presence of null alleles (Table 1A). Owing to the moderate number of loci available, AGXH7 was retained in the analysis uncorrected but its impact was monitored subsequently. Lositan [61] indicated that locus 33c1 gave a signal of excessive differentiation (Table 1A), and it was removed from subsequent analyses. None of the tests for linkage disequilibrium in *An. gambiae* s.s. were significant following multiple-testing correction, so the loci included were considered to be segregating independently. Data for *An. arabiensis* samples proved more problematic, with 27 tests for H–W disequilibrium significant following correction for multiple testing (Table 1B). Of those indicating a deficit of heterozygotes, all but one involved locus AG3H811, overwhelmingly suggesting null alleles rather than within-population structure as an explanation. Again, AG3H811 was retained in the analysis uncorrected, but its impact was monitored subsequently. Of the 14 significant tests for heterozygote excess, 12 involved loci AG2H78 and AGXH67. These loci, in addition to AG3H765 and 33c1 were identified by Lositan as exhibiting deviations from neutral expectations, thereby unduly influencing estimated differentiation and consequently were excluded. Only one of over 500 tests for linkage disequilibrium was significant following correction for multiple testing, suggesting overall that included loci were segregating independently. Following these quality control procedures, the final dataset was reduced to ten microsatellites for *An. gambiae* s.s. and seven for *An. arabiensis* (Data S1).

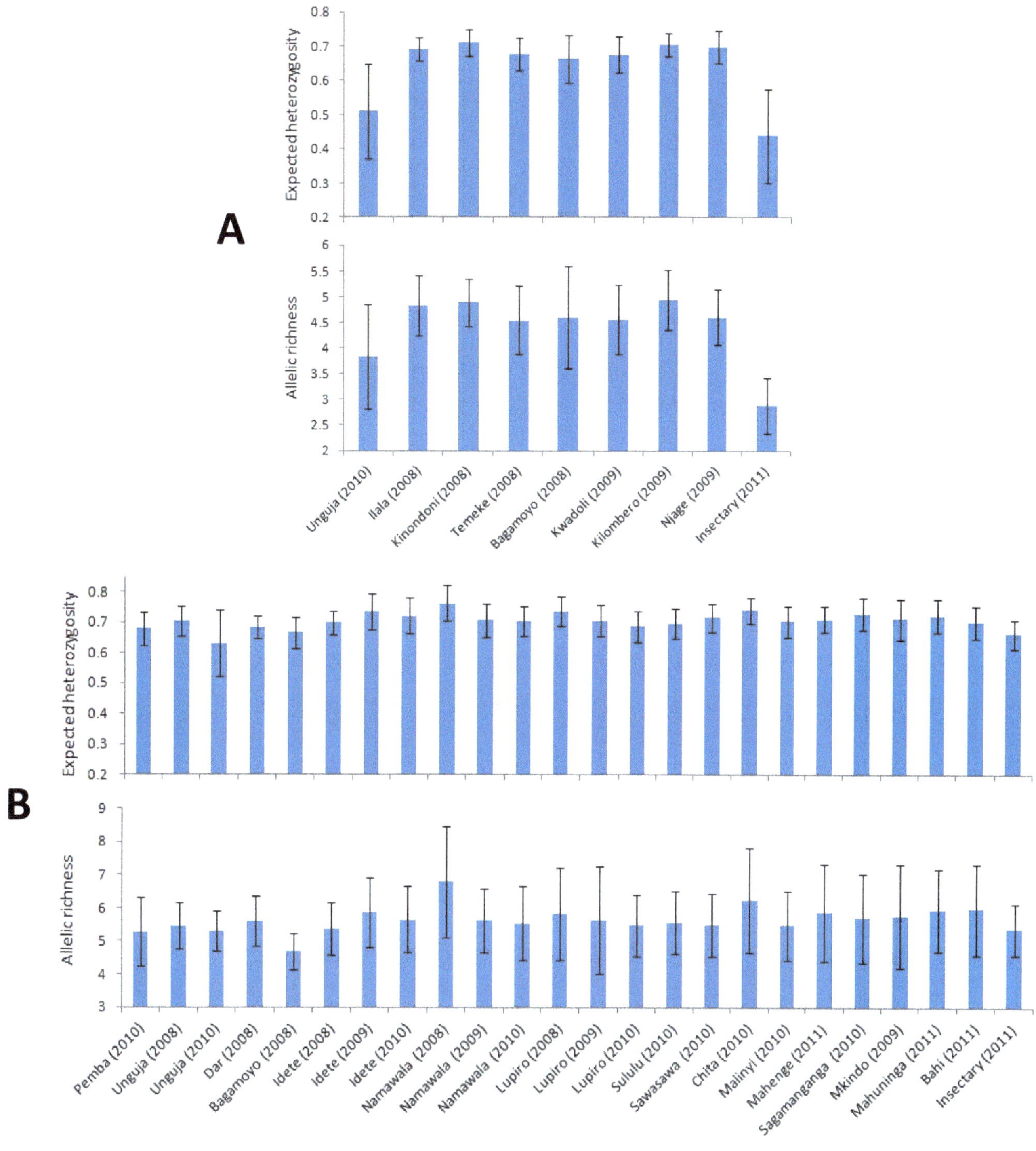

Figure 2. Genetic diversity in (A) *An. gambiae s.s.* and (B) *An. arabiensis*. Each plot shows the mean expected heterozygosity or allelic richness across loci with 95% confidence intervals.

Genetic diversity and differentiation

Whether measured by expected heterozygosity (H_e) or allelic richness (R_s), diversity was generally lower in *An. gambiae* s.s. samples from the island of Unguja and, as expected, the Ifakara insectary (Fig. 2A). In contrast, genetic diversity varied little across *An. arabiensis* samples (Fig. 2B), with even the Ifakara insectary sample exhibiting levels of H_e and R_s comparable to wild populations. *An. gambiae* s.s. exhibited generally moderate population differentiation but most pairwise tests of differentiation were significant (Table 2A). This was highlighted both by F_{ST} levels and by BAPS group-level cluster analysis (which tends to detect higher-level structure) which partitioned Unguja, the insectary sample, and also Bagamoyo as each being distinct from the other *An. gambiae* s.s. samples (Fig. 3). AGXH7, for which null allele(s) were suspected, did not show especially high or low differentiation and its exclusion had no effect on BAPS results. When all data were included, there was no relationship between genetic differentiation and geographic distance (Mantel test r =

Table 2. (A) *Anopheles gambiae* pairwise F_{ST} values.

Sample	N	Latitude	Longitude	PC1	1	2	3	4	5	6	7	8	9
Unguja (2010)	31	−6.171	39.314	0.912		*	*	*	*	*	*	*	*
Ilala (2008)	48	−.816	39.266	0.473	0.319		NS	NS	*	*	*	*	*
Kinondoni (2008)	39	−6.792	39.255	0.484	0.306	0.000		*	*	*	*	*	*
Temeke (2008)	49	−6.841	39.287	0.465	0.322	0.001	0.007		*	*	*	*	*
Bagamoyo (2008)	46	−6.453	38.897	0.556	0.327	0.017	0.017	0.026		*	*	*	*
Kwadoli (2009)	46	−6.053	37.634	0.293	0.328	0.010	0.014	0.026	0.024		*	*	*
Kilombero (2009)[k]	45	−8.140	36.681	−1.460	0.314	0.010	0.017	0.021	0.032	0.023		NS	*
Njage (2009)[k]	43	−8.230	36.190	−1.722	0.324	0.013	0.028	0.026	0.040	0.031	0.002		*
Insectary (2011)[k]	51	n/a	n/a	n/a	0.485	0.250	0.268	0.273	0.281	0.247	0.229	0.223	

(B) ***Anopheles arabiensis***

Sample	N	Latitude	Longitude	PC1	1	2	3	4	5	6	7	8	9	10	11	12	13	14	15	16	17	18	19	20	21	22	23	24
Pemba (2010)	46	−5.179	39.763	2.380		NS	*	*	*	*	*	*	*	*	*	*	*	*	*	*	*	*	*	*	*	*	NS	*
Unguja (2008)	51	−6.171	39.314	1.678	0.016		NS	NS	*	*	NS	NS	NS	NS	NS	*	NS	NS	*	*	NS	*	*	*	NS	NS	NS	*
Unguja (2010)	42	−6.171	39.314	1.678	0.039	0.014		NS	NS	*	*	NS	*	NS	NS	*	NS	NS	NS	*	NS	NS	NS	*	NS	*	NS	*
Dar (2008)	31	−6.841	39.287	1.320	0.049	0.024	0.034		*	*	*	NS	*	*	*	*	*	NS	NS	*	*	*	*	*	NS	*	*	*
Bagamoyo (2008)	24	−6.453	38.897	1.356	0.026	0.031	0.032	0.038		*	NS	NS	*	NS	NS	NS	NS	NS	NS	NS	NS	NS	NS	NS	NS	NS	NS	*
Idete (2008)[k]	37	−8.099	36.484	−0.514	0.060	0.051	0.083	0.054	0.020		*	*	*	*	*	*	*	*	NS	*	*	*	*	*	*	*	*	*
Idete (2009)[k]	50	−8.099	36.484	−0.514	0.039	0.017	0.045	0.034	0.014	0.020		NS	NS	NS	*	NS	NS	NS	NS	NS	NS	NS	NS	NS	NS	*	*	*
Idete (2010)[k]	47	−8.099	36.484	−0.514	0.020	0.006	0.034	0.016	0.013	0.018	0.004		NS	NS	NS	NS	NS	NS	NS	NS	NS	NS	NS	NS	NS	NS	NS	*
Namawala (2008)[k]	40	−8.157	36.578	−0.504	0.034	0.021	0.052	0.039	0.027	0.030	0.010	0.009		NS	NS	NS	NS	*	NS	*	NS	*	*	NS	NS	*	NS	*
Namawala (2009)[k]	38	−8.157	36.578	−0.504	0.038	0.017	0.050	0.038	0.022	0.024	0.004	0.000	0.007		NS	NS	NS	NS	NS	NS	NS	NS	NS	NS	NS	NS	NS	*
Namawala (2010)[k]	48	−8.157	36.578	−0.504	0.020	0.019	0.040	0.028	0.013	0.030	0.016	0.004	0.016	0.013		*	NS	NS	NS	*	NS	NS	*	*	NS	NS	NS	*
Lupiro (2008)[k]	51	−8.385	36.678	−0.580	0.040	0.018	0.044	0.032	0.027	0.021	0.002	0.003	0.010	0.004	0.012		NS	NS	NS	NS	NS	NS	NS	NS	NS	NS	*	*
Lupiro (2009)[k]	43	−8.385	36.678	−0.580	0.030	0.021	0.046	0.036	0.014	0.020	0.007	−0.002	0.005	0.001	0.009	0.003		NS	NS	NS	NS	NS	NS	NS	NS	NS	NS	*
Lupiro (2010)[k]	42	−8.385	36.678	−0.580	0.030	0.026	0.029	0.021	0.004	0.027	0.014	0.003	0.021	0.007	0.004	0.014	0.005		NS	NS	NS	NS	NS	NS	NS	NS	NS	*
Sululu (2010)[k]	34	−8.003	36.837	−0.315	0.033	0.025	0.028	0.015	0.005	0.021	0.019	0.004	0.025	0.016	0.003	0.014	0.012	−0.007		NS	NS	NS	NS	NS	NS	NS	NS	*
Sawasawa (2010)[k]	51	−7.883	36.867	−0.241	0.036	0.016	0.027	0.014	0.009	0.023	0.003	0.001	0.011	0.010	0.015	0.005	0.005	0.005	0.007		NS	NS	NS	NS	NS	*	*	*
Chita (2010)[k]	45	−8.292	35.596	−0.989	0.035	0.022	0.042	0.025	0.016	0.020	0.004	0.001	0.002	0.002	0.002	−0.002	0.000	0.001	0.005	0.003		NS	NS	NS	NS	NS	*	*
Malinyi (2010)[k]	48	−8.943	36.139	−1.096	0.035	0.025	0.035	0.026	0.014	0.023	0.011	0.003	0.013	0.008	0.004	0.003	0.006	−0.002	0.000	0.006	−0.003		NS	NS	NS	*	*	*
Mahenge (2011)[k]	49	−8.897	36.717	−0.827	0.047	0.030	0.036	0.027	−0.001	0.019	0.013	0.007	0.020	0.015	0.020	0.018	0.010	0.004	0.004	0.003	0.011	0.014		NS	NS	*	*	*
Sagamanganga (2010)[k]	47	−8.067	36.800	−0.364	0.041	0.019	0.040	0.022	0.015	0.022	0.007	0.006	0.008	0.014	0.019	0.006	0.011	0.013	0.014	0.000	0.004	0.007	0.006		NS	*	*	*
Mkindo (2009)	38	−6.254	37.549	0.889	0.040	0.024	0.040	0.023	0.001	0.025	0.008	0.008	0.014	0.011	0.012	0.014	0.011	0.003	0.006	0.005	0.005	0.008	−0.001	−0.002		*	*	*

Table 2. Cont.

(B) *Anopheles arabiensis*																											
Mahuninga (2011)	48	−7.928	35.652	−0.778	0.007	0.007	0.040	0.028	0.015	0.028	0.015	−0.002	0.017	0.012	0.004	0.016	0.005	0.010	0.010	0.015	0.011	0.014	0.021	0.020	0.017	NS	*
Bahi (2011)	46	−5.950	35.312	0.100	0.003	0.017	0.043	0.035	0.017	0.042	0.024	0.007	0.019	0.018	0.003	0.027	0.010	0.008	0.012	0.023	0.015	0.017	0.028	0.028	0.021	−0.005	*
Insectary (2011)[k]	46	n/a	n/a	n/a	0.120	0.083	0.111	0.059	0.088	0.084	0.066	0.072	0.085	0.087	0.066	0.078	0.099	0.073	0.059	0.073	0.071	0.064	0.083	0.073	0.069	0.078	0.088

Boxed F_{ST} and significance values are from temporal populations. The principal component (PC) reflects location explaining 87% and 84% of variation in latitude and longitude for *An. gambiae s.s.* (A) and *An. arabiensis* (B) collection sites respectively.
Significant and non-significant pairwise F_{ST} values are represented by an asterisk (*) and NS respectively. [k]Samples from the Kilombero Valley.

−0.05, p = 0.38) in *An. gambiae* s.s (Fig. 4A). Following exclusion of the sample from Unguja, for which all pairwise F_{ST} values were much higher than others (Table 2A), a highly significant relationship between genetic differentiation and distance was detectable for *An. gambiae* s.s. (Fig. 4B). In *An. arabiensis*, the majority of pairwise comparisons were not significant, and most F_{ST} values were low to moderate (Table 2B), with the exception of comparisons involving the insectary sample, which was the only one to partition separately in cluster analysis. There was no significant differentiation among the time-series samples of *An. arabiensis* from Unguja, Lupiro or Namawala, nor between Idete samples from 2009 and 2010, though the Idete sample from 2008 was significantly different from 2009 and 2010 (Table 2B).

The relationship between genetic differentiation and distance was also highly significant in *An. arabiensis* (Fig. 4C), and in contrast to *An. gambiae* s.s., the greater number and continuity of samples permitted closer inspection of the relationship. At the smallest spatial scale of distance comparisons (up to 100 km) there was no relationship between distance and differentiation, but in each broader scale category thereafter, the slope of the isolation by distance (IBD) relationship was quite consistent (Fig. 5). Therefore, apart from at a fine scale, distance was a reasonable predictor of genetic differentiation, concordant with gene flow limited by distance. Owing to this clear IBD relationship and near-complete lack of clustering using the BAPS group-level analysis, we also performed individual-level clustering analysis for *An. arabiensis* data (again using spatial information as a prior) to determine whether locations appeared especially differentiated. The resultant solution was dominated by one major cluster containing almost 83% of all individual samples, 10 very small clusters, which we pooled together to aid interpretation, and two similar clusters, which when pooled yielded a similar overall size to the 10 minor clusters. Though not conclusive, samples from the island of Pemba were less represented in the dominant cluster (Fig. 6), consistent with slightly greater differentiation than observed among the rest of the dataset.

Evidence of population stability and *kdr* distributions

Owing to the fragmented nature of the *An. gambiae* s.s. distribution and known declines in frequency, we examined evidence for population instability via bottleneck tests. *An. arabiensis* populations which have not obviously declined in the same manner as *An. gambiae s.s.* have, were also tested, though consequently with a contrasting expectation. In both species, tests proved inconclusive, with results entirely dependent on the mutation model applied in simulations (Table S1A, B). We also attempted estimation of effective population size, *Ne*. All samples of *An. gambiae* s.s., with the exception of the insectary sample, exhibited an upper confidence interval of infinity, highlighting poor reliability of the estimates. Nevertheless it is interesting to note that, after the insectary, the Unguja and Bagamoyo samples were also the next most differentiated and exhibited the next lowest point *Ne* estimates, suggesting a possible role for isolation and genetic drift (Table S1C).

Kdr 1014F was absent in all samples genotyped in this study. *Kdr* 1014S was almost entirely absent from *An. arabiensis*, with just a single heterozygote detected in Dar es Salaam (from 693 genotyped individuals), and a second heterozygote in one of the two *An. gambiae* s.s. x *An. arabiensis* hybrids found in the Dar es Salaam collections. In *An. gambiae* s.s., *kdr* 1014S was found at highly variable frequencies among sites, exceeding 70% in the three samples from Dar es Salaam and also Bagomoyo, but only 16% on the nearby island of Unguja, where only *kdr* heterozygotes were present. *Kdr* was absent from three additional sample sites,

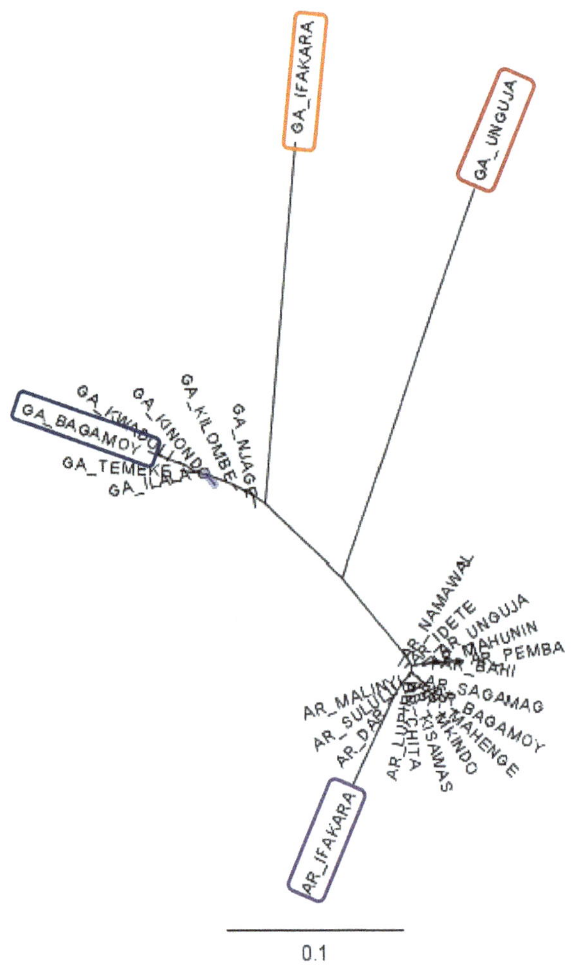

Figure 3. Neighbour-joining tree based on linearised F_{ST}. Sample names with the prefix GA are *An. gambiae s.s.*, and those with the prefix AR are *An. arabiensis*. Samples labelled IFAKARA are from insectary colonies. Within each species, samples identified as distinct clusters by BAPS are circled; others fall within a single cluster in each species.

Njage, Kwadoli and Kilombero (Fig. 7). In all samples, *kdr* was in Hardy-Weinberg equilibrium. Although the relationship between *kdr* frequency and geographical location (measured using a principal component) was not significant (Spearman's $\rho = 0.46$, $p = 0.30$), too few sample sites were available to confidently reject a hypothesis of distance-restricted *kdr* distribution. However, the extremely strong differentiation of Unguja from all mainland *An. gambiae* s.s. samples is more consistent with positive selection driving *kdr* frequencies following introduction of the allele by very occasional migration, rather than recurrent gene flow from the mainland.

Discussion

Physical barriers, distance and environmental adaptation have all been implicated previously as causal factors reducing gene flow in *An. gambiae* s.l., but few studies have compared their roles in *An. gambiae* s.s. and *An. arabiensis*. Comparison of the population structure of each species in Tanzania was hindered by the relative rarity of *An. gambiae* s.s. in many areas. Indeed the more fragmented nature of the *An. gambiae s.s.* sampling scheme than in *An. arabiensis* in our study is a direct consequence of this. In spite of this limitation, both similarities and differences between the species were apparent. *An. gambiae* s.s. sample sites further apart were generally more differentiated than those nearby, with the notable exception of Unguja. This island population bore the hallmarks of isolation, specifically extremely high differentiation and much reduced genetic diversity. Furthermore, the absence of any clear signal of a population bottleneck suggests this may reflect a history of limited gene flow from the mainland. This extreme isolation of Unguja actually masked a strong correlation between distance and differentiation in *An. gambiae* s.s., though we suggest caution in interpretation of the underlying causation. IBD is expected when gene flow is limited by dispersal distance, leading to a stepping-stone population model whereby neighbouring locations are much more likely to exchange migrants [71]. IBD can also be indicative of migration-drift equilibrium [72], a state at which the link between differentiation and gene flow (c.f. differentiation and population history) becomes much closer [73], making patterns of differentiation easier to interpret for practical applications. However, with relatively limited and discontinuous sampling, and most inter-sample distances far exceeding the plausible dispersal range of an organism [74], such interpretation is problematic. This is the case for *An. gambiae* s.s. in our study, and thus we cannot conclude that distance is the causal factor in creating differentiation, or in limiting gene flow. Local factors influencing immigration or demography may also be important. In this context the relatively high differentiation of the coastal Bagamoyo sample, located approximately 60 km from those in Dar es Salaam, and the low but significant differentiation of the Temeke from Kinondoni and Ilala samples (7 km apart within Dar es Salaam) may be of note. Other East African studies (conducted in Kenya) have also reported relatively high genetic differentiation in *An. gambiae* s.s. populations sampled at small spatial scales, e.g. 50 km apart [75,76].

An. gambiae s.s. has been systematically experiencing a remarkably rapid decline in the Kilombero Valley, with recent research showing this species comprises less than 1% of the *An. gambiae* complex in some villages where it used to be the dominant species [77]. Thus it might be expected that *An. gambiae* s.s. population in the valley would be experiencing a more isolated or patchy existence. However, we did not detect significant differentiation between samples within the valley. Though based on only a single pair of sites, this mirrors findings from another recent study in the area [43], which reported very low differentiation (average $F_{ST} = 0.006$).

In contrast to *An. gambiae* s.s., genetic diversity was invariant among *An. arabiensis* samples, most pairwise comparisons were not significant, and, with the exception of the insectary sample, cluster analysis failed to detect any significant partitions in the data. Even application of the potentially more sensitive individual-based spatial clustering method was inconclusive, though there was some suggestion of at least minor separation of the Pemba island sample. Does such weak structure reflect extensive gene flow, or even near-panmixia? The larger and more continuous sample set for *An. arabiensis* helps to answer this question. The relationship between genetic differentiation and distance was similarly strong for *An. gambiae* s.s. and *An. arabiensis*, but for the latter it was possible to examine the relationship at different spatial scales. This revealed a consistent IBD pattern for all but the finest scale of comparisons, consistent with migration-drift equilibrium [72], and in the absence of patterns of variation in genetic diversity, alternative explanations related to population spread and/or colonisation time are not supported [39,78,79]. Despite this support for a broad equilibrium scenario, which permits

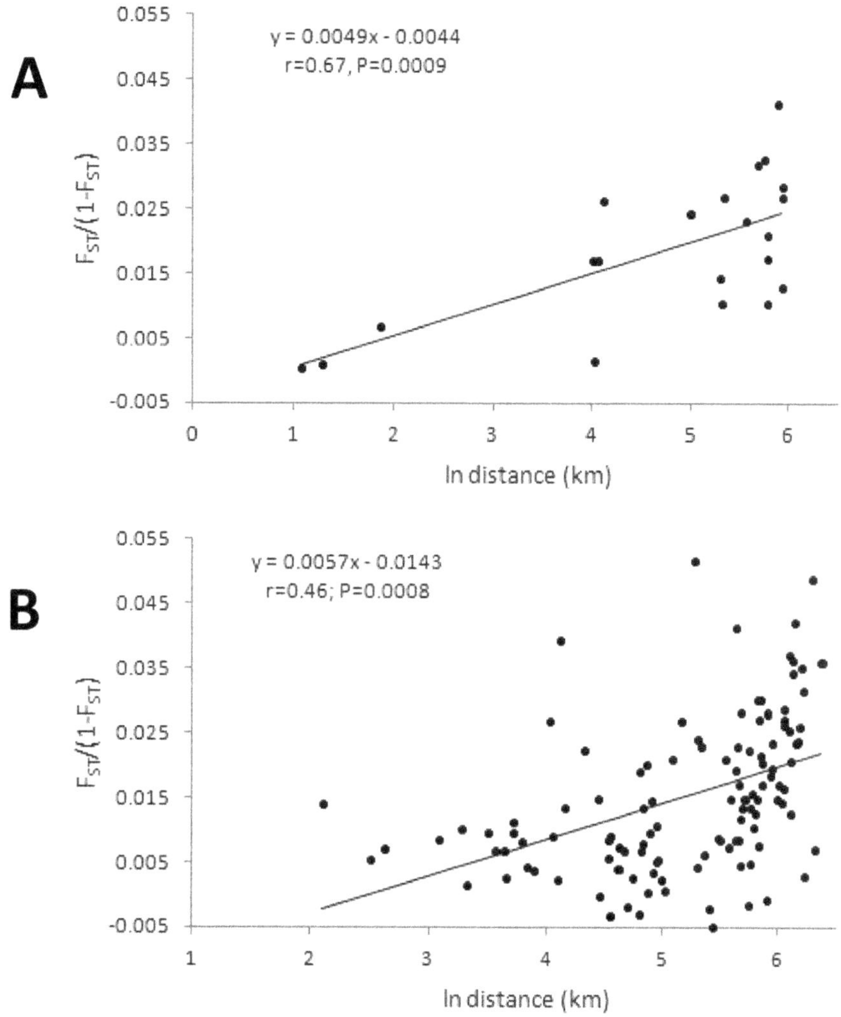

Figure 4. Isolation by distance (IBD) plots in *An. gambiae s.s.* (A, B) and *An. arabiensis* (C). Each plot shows linearised pairwise F_{ST} against the natural logarithm of pairwise distance between sample sites. In (A) all *An. gambiae* sites are included, and the massive differentiation of Unguja (pairwise point in dashed circle) from all other sites obscures the IBD relationship visible in (B) once pairwise comparisons involving Unguja were excluded. In all plots insectary samples are excluded and in (C) temporal samples from the same location were pooled for the analysis. P-values are from Mantel tests.

inference of gene flow from differentiation, direct conversion of F_{ST} to number of effective migrants per generation (*Nm*), though still commonplace in literature, is probably unwise for pairwise comparisons because high sampling error in F_{ST} [80] is compounded when converting to *Nm* [73]. However, as a rough indicator, average F_{ST} values for the distance classes translate to *Nm* estimates of between approximately 9 and 25, though literal interpretation as numbers of migrating individuals is unwise. In spite of widespread lack of significant differentiation, it is important to appreciate that *An. arabiensis* in Tanzania and the associated major islands are not panmictic, but rather conform to a stepping-stone model of semi-continuous population structure, with considerable gene flow limited primarily by distance. Such continuity suggests that vector control applied at a local scale could often be hampered by persistent re-colonisation, and potentially to a much greater extent than for Tanzanian *An. gambiae*. Again, in contrast to *An. gambiae* s.s. differentiation of Unguja was entirely unexceptional, and that of the more isolated island, Pemba, only slightly elevated above the majority of inter-site comparisons. This observation could reflect an impact of the sea as a partial barrier to gene flow, or more simply might arise from the location of Pemba at the end of a chain of sample sites; thus with more limited potential for immigration as a result of distance alone. Differentiation between the Zanzibar islands and mainland populations of *An. arabiensis* is dramatically less than that found in a previous study of differentiation among Madagascar, Reunion and Mauritius [34], perhaps reflecting the much closer proximity to the mainland of the Zanzibar islands.

Genetic differentiation among populations in the Kilombero Valley was generally very limited. Ng'habi et al [43] reported a very much higher level of genetic differentiation among *An. arabiensis* sample sites within the Kilombero Valley ($F_{ST}=0.066$) which was attributed to the presence of a separate genetic cluster of *An. arabiensis*, which were in some sites but common in others, and highly divergent ($F_{ST}>0.1$). It is possible a novel cryptic subgroup was discovered in the valley which gave rise to these results [43], but the cause of multiple deviations from Hardy-Weinberg equilibrium and extremely unusual patterns of linkage disequilibrium among markers does not appear to have been fully explored. Therefore, null alleles and scoring errors should first be

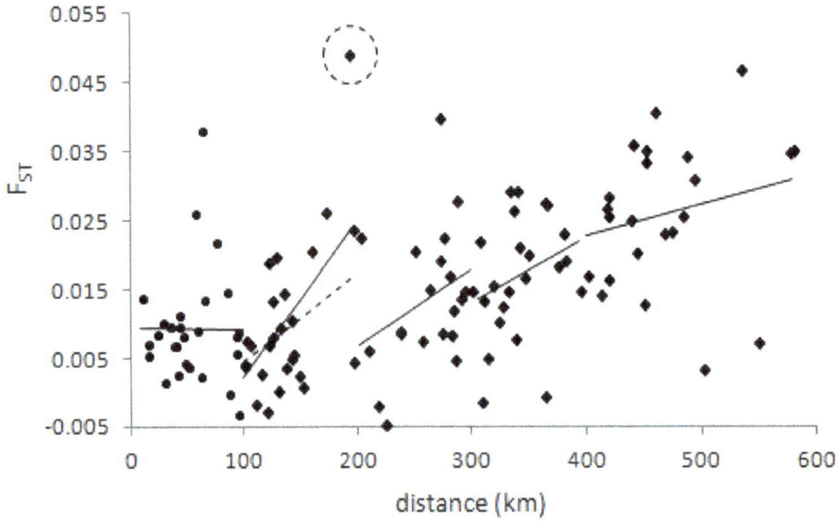

Figure 5. Variation in IBD slope with geographic scale in *An. arabiensis*. The isolation by distance slope is calculated separately for each 100 km class of pairwise distances (or 200 km class for 400–600 km, owing to fewer points). For the 101–200 km class the dashed line shows the slope if the outlying point (enclosed in dashed circle) is excluded.

discounted, especially (a) given the scoring problems we observed for some of the *An. gambiae* s.s. microsatellites when used to amplify *An. arabiensis*, and (b) the sensitivity of clustering to small numbers of highly differentiated markers [29].

There was slight, but significant genetic differentiation [Table 2B] among sequences of samples from *An. arabiensis* populations collected between 2008 and 2010 from Idete. Lupiro and Namawala had temporal populations collected within similar periods but did not show significant genetic differentiation to each other across time. Temporal sequences of samples from populations of *An. arabiensis* collected in 2008 and 2010 in Unguja Island were not significantly different from each other. Observation of absence of genetic differentiation among temporal populations of *An. arabiensis* have been reported by [23,42,81], although a study on temporal population structure of *An. gambiae* s.s. revealed significant genetic differentiation between successive monthly collections [75]. Lower levels of differentiation among *An. arabiensis* in our study may be explained by the larger effective population size evidenced in *An. arabiensis*, which maintains stable genotype frequencies across wet and dry seasons [23].

LLINs are widely used in most parts of Tanzania and especially in the Kilombero Valley. Behavioural differences in the two species, which render *An. gambiae* s.s. more amenable to control by LLINs, may have contributed to the decrease in *An. gambiae* s.s. numbers relative to its sister species *An. arabiensis* [8–10] as described above. Persistence of *An. gambiae* s.s. in more genetically isolated populations than found in *An. arabiensis* is consistent with the findings of the present study.

The pyrethroid and DDT-linked resistance mutation *kdr* 1014S [12,82,83] was found at extremely low frequency in *An. arabiensis*. This mirrors results from other studies in Tanzania [84,85]. However, *kdr* 1014F was absent from all samples we tested, yet was recently found in samples of *An. arabiensis* from Dar es Salaam collected only three years later, (but not in *An. gambiae* s.s.) [85] at frequencies approaching those in *An. arabiensis* from Ethiopia and Sudan [86,87]. *An. gambiae* s.s. populations from Dar es Salaam and Bagamoyo were found to have high

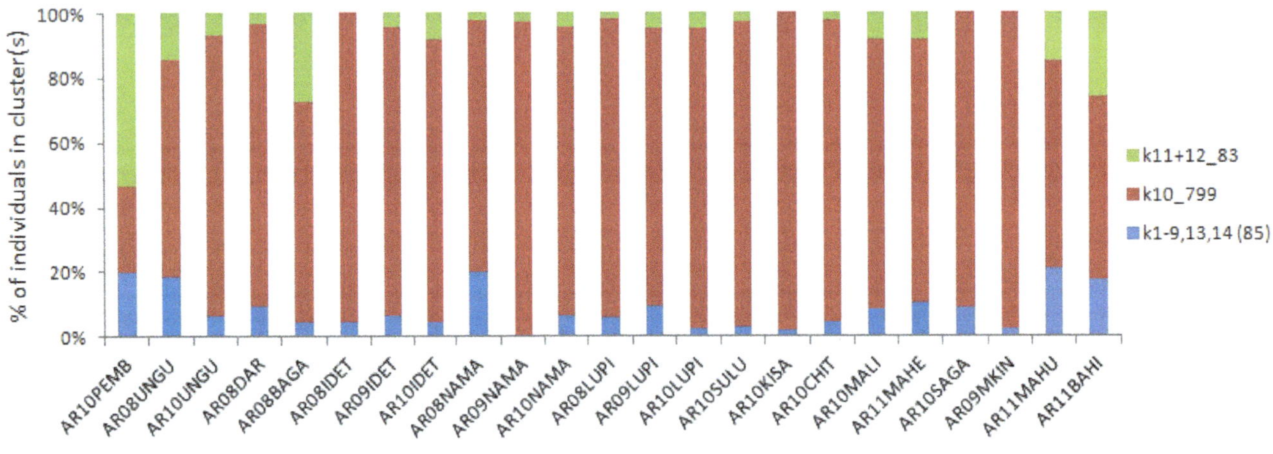

Figure 6. Individual-based BAPS spatially-conditioned clustering. Cluster identification numbers (k*n*) are shown, with number of individuals after the underscore.

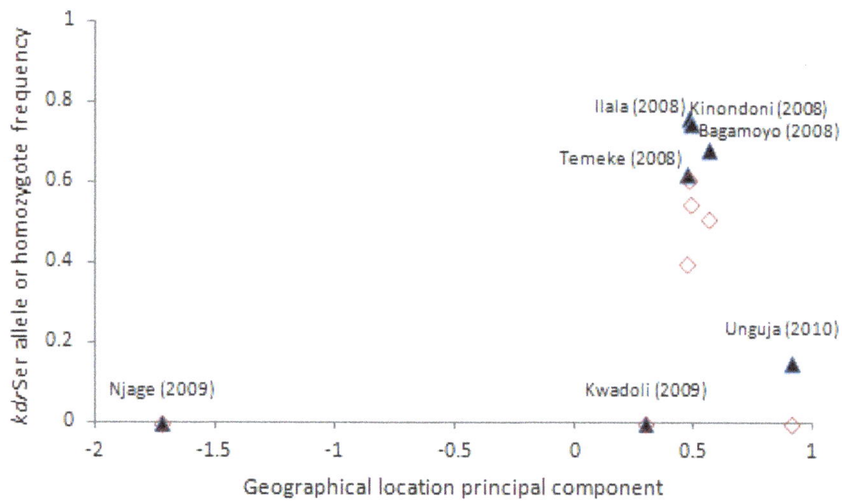

Figure 7. *Kdr* L1014S in *An. gambiae s.s.* Serine allele (filled triangles) and serine homozygote (open diamonds) frequencies are plotted against a principal component reflecting location (explaining 87% of variation in latitude and longitude).

frequencies of *kdr* 1014S (around 70%), with approaching 50% homozygotes. In Unguja Island 1014S was at much lower frequency, with no homozygotes. The other populations examined including one in the Kilombero Valley were wild type for *kdr*. The finding of 1014S at high frequency in this study is consistent with other recent reports in East Africa [14,88,89].

Absence of *kdr* from the Kwadoli and Njage (Kilombero valley) samples of *An. gambiae* s.s. highlights that, despite extensive use of LLINs for nearly 10 years, *kdr* has either failed to migrate into these populations or the populations have been subjected to reduced selection pressure compared to those on the coast. The link between amount of gene flow and transfer of insecticide resistance mutations in *Anopheles* is not well understood, though it seems that any quantitative connection may be weak. In West Africa, *kdr* 1014F has introgressed from S to M forms of *An. gambiae* (now termed *An. gambiae* s.s. and *An. coluzzii*), between which hybridisation is rare, and has subsequently risen to high frequency [90,91]. Similarly in this study, differentiation of Unguja from Bagomoyo and Dar es Salaam is high enough to suggest a major barrier to recent gene flow, yet *kdr*, which we assume came from the mainland, is present at appreciable frequency. Preventing the spread of resistance alleles which are under sufficient selection pressure to almost ensure establishment in a population, which appears to be true for both of the *kdr* 1014 mutations, presents a difficult challenge.

Supporting Information

Table S1 Supplementary table showing results for bottleneck and effective population size tests. BOTTLENECK (A, B) was run with three contrasting mutations with default settings for the TPM. A significant heterozygote excess ($P < 0.05$, relative to equilibrium model expectations) provides evidence of a bottleneck, whereas a heterozygote deficit suggests population expansion. Effective population size estimates were calculated by LDNA using a minimum permissible allele frequency of 0.05. Confidence limits calculated by two methods (parametric and jackknifing are shown).

Data S1 Genotype and collection sites for all samples.

Data S2 Allele count summaries for each collection site and locus (note that not all loci were included in final analysis - see main text).

Acknowledgments

We thank Donall Cross of Ifakara Health Institute and Aberystwyth University for helping to produce a map of the study sites.

Author Contributions

Conceived and designed the experiments: DM GK HR. Performed the experiments: DM. Analyzed the data: DW. Contributed reagents/materials/analysis tools: HR DW JM KH SM WK. Contributed to the writing of the manuscript: DM DW. Reviewed and approved the final manuscript: DM DW HR GK JM KH SM WK.

References

1. WHO (2012) Global plan for insecticide resistance management in malaria vectors (GPIRM).
2. Ranson H, N'Guessan R, Lines J, Moiroux N, Nkuni Z, et al. (2011) Pyrethroid resistance in African anopheline mosquitoes: what are the implications for malaria control? Trends Parasitol 27: 91–98.
3. Bass C, Nikou D, Donnelly MJ, Williamson MS, Ranson H, et al. (2007) Detection of knockdown resistance (kdr) mutations in Anopheles gambiae: a comparison of two new high-throughput assays with existing methods. Malar J 6: 111.
4. WHO (2012) WHO Malaria Report.
5. Cartaxo MF, Ayres CF, Weetman D (2011) Loss of genetic diversity in Culex quinquefasciatus targeted by a lymphatic filariasis vector control program in Recife, Brazil. Trans R Soc Trop Med Hyg 105: 491–499.
6. Donnelly MJ, Pinto J, Girod R, Besansky NJ, Lehmann T (2004) Revisiting the role of introgression vs shared ancestral polymorphisms as key processes shaping genetic diversity in the recently separated sibling species of the Anopheles gambiae complex. Heredity (Edinb) 92: 61–68.
7. Harris AF, Nimmo D, McKemey AR, Kelly N, Scaife S, et al. (2011) Field performance of engineered male mosquitoes. Nat Biotechnol 29: 1034–1037.
8. Bayoh MN, Mathias DK, Odiere MR, Mutuku FM, Kamau L, et al. (2010) Anopheles gambiae: historical population decline associated with regional

distribution of insecticide-treated bed nets in western Nyanza Province, Kenya. Malar J 9: 62.
9. Russell TL, Lwetoijera DW, Maliti D, Chipwaza B, Kihonda J, et al. (2010) Impact of promoting longer-lasting insecticide treatment of bed nets upon malaria transmission in a rural Tanzanian setting with pre-existing high coverage of untreated nets. Malar J 9: 187.
10. Derua YA, Alifrangis M, Hosea KM, Meyrowitsch DW, Magesa SM, et al. (2012) Change in composition of the Anopheles gambiae complex and its possible implications for the transmission of malaria and lymphatic filariasis in north-eastern Tanzania. Malar J 11: 188.
11. Ndjemai HN, Patchoke S, Atangana J, Etang J, Simard F, et al. (2009) The distribution of insecticide resistance in Anopheles gambiae s.l. populations from Cameroon: an update. Trans R Soc Trop Med Hyg 103: 1127–1138.
12. Ramphul U, Boase T, Bass C, Okedi LM, Donnelly MJ, et al. (2009) Insecticide resistance and its association with target-site mutations in natural populations of Anopheles gambiae from eastern Uganda. Trans R Soc Trop Med Hyg 103: 1121–1126.
13. Badolo A, Traore A, Jones CM, Sanou A, Flood L, et al. (2012) Three years of insecticide resistance monitoring in Anopheles gambiae in Burkina Faso: resistance on the rise? Malar J 11: 232.
14. Mawejje HD, Wilding CS, Rippon EJ, Hughes A, Weetman D, et al. (2013) Insecticide resistance monitoring of field-collected Anopheles gambiae s.l. populations from Jinja, eastern Uganda, identifies high levels of pyrethroid resistance. Med Vet Entomol 27: 276–283.
15. Lunde TM, Balkew M, Korecha D, Gebre-Michael T, Massebo F, et al. (2013) A dynamic model of some malaria-transmitting anopheline mosquitoes of the Afrotropical region. II. Validation of species distribution and seasonal variations. Malar J 12: 78.
16. Dia I, Ba H, Mohamed SA, Diallo D, Lo B, et al. (2009) Distribution, host preference and infection rates of malaria vectors in Mauritania. Parasit Vectors 2: 61.
17. Mahande A, Mosha F, Mahande J, Kweka E (2007) Feeding and resting behaviour of malaria vector, Anopheles arabiensis with reference to zooprophylaxis. Malar J 6: 100.
18. van den Broek IV, den Otter CJ (1999) Olfactory sensitivities of mosquitoes with different host preferences (Anopheles gambiae s.s., An. arabiensis, An. quadriannulatus, An. m. atroparvus) to synthetic host odours. J Insect Physiol 45: 1001–1010.
19. Duchemin JB, Tsy JM, Rabarison P, Roux J, Coluzzi M, et al. (2001) Zoophily of Anopheles arabiensis and An. gambiae in Madagascar demonstrated by odour-baited entry traps. Med Vet Entomol 15: 50–57.
20. Faye O, Konate L, Mouchet J, Fontenille D, Sy N, et al. (1997) Indoor resting by outdoor biting females of Anopheles gambiae complex (Diptera:Culicidae) in the Sahel of northern Senegal. J Med Entomol 34: 285–289.
21. Lindsay SW, Parson L, Thomas CJ (1998) Mapping the ranges and relative abundance of the two principal African malaria vectors, Anopheles gambiae sensu stricto and An. arabiensis, using climate data. Proc Biol Sci 265: 847–854.
22. Coetzee M, Koekemoer LL (2013) Molecular systematics and insecticide resistance in the major African malaria vector Anopheles funestus. Annu Rev Entomol 58: 393–412.
23. Simard F, Lehmann T, Lemasson JJ, Diatta M, Fontenille D (2000) Persistence of Anopheles arabiensis during the severe dry season conditions in Senegal: an indirect approach using microsatellite loci. Insect Mol Biol 9: 467–479.
24. Wondji C, Frederic S, Petrarca V, Etang J, Santolamazza F, et al. (2005) Species and populations of the Anopheles gambiae complex in Cameroon with special emphasis on chromosomal and molecular forms of Anopheles gambiae s.s. J Med Entomol 42: 998–1005.
25. Costantini C, Ayala D, Guelbeogo WM, Pombi M, Some CY, et al. (2009) Living at the edge: biogeographic patterns of habitat segregation conform to speciation by niche expansion in Anopheles gambiae. BMC Ecol 9: 16.
26. Simard F, Ayala D, Kamdem GC, Pombi M, Etouna J, et al. (2009) Ecological niche partitioning between Anopheles gambiae molecular forms in Cameroon: the ecological side of speciation. BMC Ecol 9: 17.
27. Temu EA, Yan G (2005) Microsatellite and mitochondrial genetic differentiation of Anopheles arabiensis (Diptera: Culicidae) from western Kenya, the Great Rift Valley, and coastal Kenya. Am J Trop Med Hyg 73: 726–733.
28. Tripet F, Dolo G, Lanzaro GC (2005) Multilevel analyses of genetic differentiation in Anopheles gambiae s.s. reveal patterns of gene flow important for malaria-fighting mosquito projects. Genetics 169: 313–324.
29. Weetman D, Wilding CS, Steen K, Pinto J, Donnelly MJ (2012) Gene flow-dependent genomic divergence between Anopheles gambiae M and S forms. Mol Biol Evol 29: 279–291.
30. Lehmann T, Hawley WA, Grebert H, Danga M, Atieli F, et al. (1999) The Rift Valley complex as a barrier to gene flow for Anopheles gambiae in Kenya. J Hered 90: 613–621.
31. Lehmann T, Licht M, Elissa N, Maega BT, Chimumbwa JM, et al. (2003) Population Structure of Anopheles gambiae in Africa. J Hered 94: 133–147.
32. Kamau L, Mukabana WR, Hawley WA, Lehmann T, Irungu LW, et al. (1999) Analysis of genetic variability in Anopheles arabiensis and Anopheles gambiae using microsatellite loci. Insect Mol Biol 8: 287–297.
33. Nyanjom SR, Chen H, Gebre-Michael T, Bekele E, Shililu J, et al. (2003) Population genetic structure of Anopheles arabiensis mosquitoes in Ethiopia and Eritrea. J Hered 94: 457–463.
34. Simard F, Fontenille D, Lehmann T, Girod R, Brutus L, et al. (1999) High amounts of genetic differentiation between populations of the malaria vector Anopheles arabiensis from West Africa and eastern outer islands. Am J Trop Med Hyg 60: 1000–1009.
35. Kayondo JK, Mukwaya LG, Stump A, Michel AP, Coulibaly MB, et al. (2005) Genetic structure of Anopheles gambiae populations on islands in northwestern Lake Victoria, Uganda. Malar J 4: 59.
36. Moreno M, Salgueiro P, Vicente JL, Cano J, Berzosa PJ, et al. (2007) Genetic population structure of Anopheles gambiae in Equatorial Guinea. Malar J 6: 137.
37. Yawson AE, Weetman D, Wilson MD, Donnelly MJ (2007) Ecological zones rather than molecular forms predict genetic differentiation in the malaria vector Anopheles gambiae s.s. in Ghana. Genetics 175: 751–761.
38. Onyabe DY, Conn JE (2001) Population genetic structure of the malaria mosquito Anopheles arabiensis across Nigeria suggests range expansion. Mol Ecol 10: 2577–2591.
39. Ramachandran S, Deshpande O, Roseman CC, Rosenberg NA, Feldman MW, et al. (2005) Support from the relationship of genetic and geographic distance in human populations for a serial founder effect originating in Africa. Proc Natl Acad Sci U S A 102: 15942–15947.
40. Donnelly MJ, Townson H (2000) Evidence for extensive genetic differentiation among populations of the malaria vector Anopheles arabiensis in Eastern Africa. Insect Mol Biol 9: 357–367.
41. Donnelly MJ, Cuamba N, Charlwood JD, Collins FH, Townson H (1999) Population structure in the malaria vector, Anopheles arabiensis patton, in East Africa. Heredity (Edinb) 83 (Pt 4): 408–417.
42. Czeher C, Labbo R, Vieville G, Arzika I, Bogreau H, et al. (2010) Population genetic structure of Anopheles gambiae and Anopheles arabiensis in Niger. J Med Entomol 47: 355–366.
43. Ng'habi KR, Knols BG, Lee Y, Ferguson HM, Lanzaro GC (2011) Population genetic structure of Anopheles arabiensis and Anopheles gambiae in a malaria endemic region of southern Tanzania. Malar J 10: 289.
44. Mutuku FM, King CH, Mungai P, Mbogo C, Mwangangi J, et al. (2011) Impact of insecticide-treated bed nets on malaria transmission indices on the south coast of Kenya. Malar J 10: 356.
45. Mwangangi JM, Mbogo CM, Orindi BO, Muturi EJ, Midega JT, et al. (2013) Shifts in malaria vector species composition and transmission dynamics along the Kenyan coast over the past 20 years. Malar J 12: 13.
46. Kabula B, Tungu P, Malima R, Rowland M, Minja J, et al. (2013) Distribution and spread of pyrethroid and DDT resistance among the Anopheles gambiae complex in Tanzania. Med Vet Entomol.
47. Russell TL, Govella NJ, Azizi S, Drakeley CJ, Kachur SP, et al. (2011) Increased proportions of outdoor feeding among residual malaria vector populations following increased use of insecticide-treated nets in rural Tanzania. Malar J 10: 80.
48. Reddy MR, Overgaard HJ, Abaga S, Reddy VP, Caccone A, et al. (2011) Outdoor host seeking behaviour of Anopheles gambiae mosquitoes following initiation of malaria vector control on Bioko Island, Equatorial Guinea. Malar J 10: 184.
49. Okumu FO, Mbeyela E, Lingamba G, Moore J, Ntamatungiro AJ, et al. (2013) Comparative field evaluation of combinations of long-lasting insecticide treated nets and indoor residual spraying, relative to either method alone, for malaria prevention in an area where the main vector is Anopheles arabiensis. Parasit Vectors 6: 46.
50. Okumu FO, Kiware SS, Moore SJ, Killeen GF (2013) Mathematical evaluation of community level impact of combining bed nets and indoor residual spraying upon malaria transmission in areas where the main vectors are Anopheles arabiensis mosquitoes. Parasit Vectors 6: 17.
51. Govella NJ, Chaki PP, Killeen GF (2013) Entomological surveillance of behavioural resilience and resistance in residual malaria vector populations. Malar J 12: 124.
52. Russell TL, Lwetoijera DW, Knols BG, Takken W, Killeen GF, et al. (2013) Geographic coincidence of increased malaria transmission hazard and vulnerability occurring at the periphery of two Tanzanian villages. Malar J 12: 24.
53. Durnez L, Mao S, Denis L, Roelants P, Sochantha T, et al. (2013) Outdoor malaria transmission in forested villages of Cambodia. Malar J 12: 329.
54. Gillies MT, Coetzee M (1987) A Supplement of the Anophelinae of Africa south of the Sahara (Afrotropical region). South African Institute for Medical Research 55.
55. Collins FH, Mendez MA, Rasmussen MO, Mehaffey PC, Besansky NJ, et al. (1987) A ribosomal RNA gene probe differentiates member species of the Anopheles gambiae complex. Am J Trop Med Hyg 37: 37–41.
56. Scott JA, Brogdon WG, Collins FH (1993) Identification of single specimens of the Anopheles gambiae complex by the polymerase chain reaction. Am J Trop Med Hyg 49: 520–529.
57. Zheng L, Benedict MQ, Cornel AJ, Collins FH, Kafatos FC (1996) An integrated genetic map of the African human malaria vector mosquito, Anopheles gambiae. Genetics 143: 941–952.
58. Romans P, Seeley DC, Kew Y, Gwadz RW (1991) Use of a restriction fragment length polymorphism (RFLP) as a genetic marker in crosses of Anopheles gambiae (Diptera: Culicidae): independent assortment of a diphenol oxidase RFLP and an esterase locus. J Med Entomol 28: 147–151.

59. Witzig C, Wondji CS, Strode C, Djouaka R, Ranson H (2013) Identifying permethrin resistance loci in malaria vectors by genetic mapping. Parasitology 140: 1468–1477.
60. Oosterhou C, Hutchison W, Wills B, Shipley P (2004) PROGRAM NOTE MICRO-CHECKER : software for identifying and correcting genotyping errors in microsatellite data Molecular Ecology Notes;Blackwell Publishing Ltd 4: 535–538.
61. Antao T, Lopes A, Lopes RJ, Beja-Pereira A, Luikart G (2008) LOSITAN: a workbench to detect molecular adaptation based on a Fst-outlier method. BMC Bioinformatics 9: 323.
62. Rousset F (2008) genepop'007: a complete re-implementation of the genepop software for Windows and Linux. Mol Ecol Resour 8: 103–106.
63. Goudet J, Perrin N, Waser P (2002) Tests for sex-biased dispersal using bi-parentally inherited genetic markers. Mol Ecol 11: 1103–1114.
64. Hood GM (2008) PopTools Version 2.5.9. Available:http://www.cse.csiro.au/poptools.
65. Felsenstein J (2008) PHYLIP (Phylogeny Inference Package) version 3.68. Distributed by the author. Department of Genome Sciences, University of Washington, Seattle.
66. Rambaut A (2008) FigTree 1.3.1. Available:http://tree.bio.ed.ac.uk/software/figtree/.
67. Corander J, Marttinen P, Siren J, Tang J (2008) Enhanced Bayesian modelling in BAPS software for learning genetic structures of populations. BMC Bioinformatics 9: 539.
68. Pinto J, Yawson E, Vicente JL, Gomes B, Santolamazza F, et al. Geographic population structure of the African malaria vector Anopheles gambiae suggests a role for the forest-savannah biome transition as a barrier to gene flow. evolutionary Applications 6: 910–924.
69. Waples RS (2008) ldne: a program for estimating effective population size from data on linkage disequilibrium. Mol Ecol Resour 8: 753–756.
70. Cornuet JM, Luikart G (1996) Description and power analysis of two tests for detecting recent population bottlenecks from allele frequency data. Genetics 144: 2001–2014.
71. Slatkin M (1993) Isolation by distance in equilibrium and non-equilibriun populations. Evolution 47: 264–279.
72. Hutchison WD, Templeton AR (1999) Correlation of pairwise genetic and geographical distance measures: infering the relative influences of gene flow and the distribution of genetic variability. Evolution 53.
73. Whitlock MC, McCauley DE (1999) Indirect measures of gene flow and migration: FST not equal to 1/(4Nm +1). Heredity (Edinb) 82 (Pt 2): 117–125.
74. Balding D, Bishop M, Cannings CE (2001) Handbook of statistical genetics. John Wiley&Sons.
75. Midega JT, Muturi EJ, Baliraine FN, Mbogo CM, Githure J, et al. (2010) Population structure of Anopheles gambiae along the Kenyan coast. Acta Trop 114: 103–108.
76. Kamau L, Lehmann T, Hawley WA, Orago AS, Collins FH (1998) Microgeographic genetic differentiation of Anopheles gambiae mosquitoes from Asembo Bay, western Kenya: a comparison with Kilifi in coastal Kenya. Am J Trop Med Hyg 58: 64–69.
77. Matowo NS, Moore J, Mapua S, Madumla EP, Moshi IR, et al. (2013) Using a new odour-baited device to explore options for luring and killing outdoor-biting malaria vectors: a report on design and field evaluation of the Mosquito Landing Box. Parasit Vectors 6: 137.
78. Hanfling B, Weetman D (2006) Concordant genetic estimators of migration reveal anthropogenically enhanced source-sink population structure in the river sculpin, Cottus gobio. Genetics 173: 1487–1501.
79. Herborg LM, Weetman D, van Oosterhout C, Hanfling B (2007) Genetic population structure and contemporary dispersal patterns of a recent European invader, the Chinese mitten crab, Eriocheir sinensis. Mol Ecol 16: 231–242.
80. Waples RS (1998) Separating the wheat from the chaff: patterns of genetic differentiation in high gene flow species. Journal of Heredity 89: 438–450.
81. Kent RJ, Mharakurwa S, Norris DE (2007) Spatial and temporal genetic structure of Anopheles arabiensis in Southern Zambia over consecutive wet and drought years. Am J Trop Med Hyg 77: 316–323.
82. Ranson H, Jensen B, Wang X, Prapanthadara L, Hemingway J, et al. (2000) Genetic mapping of two loci affecting DDT resistance in the malaria vector Anopheles gambiae. Insect Mol Biol 9: 499–507.
83. Reimer L, Fondjo E, Patchoke S, Diallo B, Lee Y, et al. (2008) Relationship between kdr mutation and resistance to pyrethroid and DDT insecticides in natural populations of Anopheles gambiae. J Med Entomol 45: 260–266.
84. Jones CM, Haji KA, Khatib BO, Bagi J, Mcha J, et al. (2013) The dynamics of pyrethroid resistance in Anopheles arabiensis from Zanzibar and an assessment of the underlying genetic basis. Parasit Vectors 6: 343.
85. Kabula B, Kisinza W, Tungu P, Ndege C, Batengana B, et al. (2014) Co-occurrence and distribution of East (L1014S) and West (L1014F) African knock-down resistance in Anopheles gambiae sensu lato population of Tanzania. Trop Med Int Health 19: 331–341.
86. Yewhalaw D, Bortel WV, Denis L, Coosemans M, Duchateau L, et al. (2010) First evidence of high knockdown resistance frequency in Anopheles arabiensis (Diptera: Culicidae) from Ethiopia. Am J Trop Med Hyg 83: 122–125.
87. Abuelmaali SA, Elaagip AH, Basheer MA, Frah EA, Ahmed FT, et al. (2013) Impacts of agricultural practices on insecticide resistance in the malaria vector Anopheles arabiensis in Khartoum State, Sudan. PLoS One 8: e80549.
88. Kawada H, Futami K, Komagata O, Kasai S, Tomita T, et al. (2011) Distribution of a knockdown resistance mutation (L1014S) in Anopheles gambiae s.s. and Anopheles arabiensis in western and southern Kenya. PLoS One 6: e24323.
89. Mathias DK, Ochomo E, Atieli F, Ombok M, Bayoh MN, et al. (2011) Spatial and temporal variation in the kdr allele L1014S in Anopheles gambiae s.s. and phenotypic variability in susceptibility to insecticides in Western Kenya. Malar J 10: 10.
90. Weetman D, Wilding CS, Steen K, Morgan JC, Simard F, et al. (2010) Association mapping of insecticide resistance in wild Anopheles gambiae populations: major variants identified in a low-linkage disequilbrium genome. PLoS One 5: e13140.
91. Lynd A, Weetman D, Barbosa S, Egyir Yawson A, Mitchell S, et al. (2010) Field, genetic, and modeling approaches show strong positive selection acting upon an insecticide resistance mutation in Anopheles gambiae s.s. Mol Biol Evol 27: 1117–1125.

13

Non-Hodgkin Lymphoma Risk and Insecticide, Fungicide and Fumigant Use in the Agricultural Health Study

Michael C. R. Alavanja[1]*, Jonathan N. Hofmann[1], Charles F. Lynch[2], Cynthia J. Hines[3], Kathryn H. Barry[1], Joseph Barker[4], Dennis W. Buckman[4], Kent Thomas[5], Dale P. Sandler[6], Jane A. Hoppin[6], Stella Koutros[1], Gabriella Andreotti[1], Jay H. Lubin[1], Aaron Blair[1], Laura E. Beane Freeman[1]

1 Division of Cancer Epidemiology and Genetics, National Cancer Institute, Rockville, Maryland, United States of America, **2** College of Public Health, University of Iowa, Iowa City, Iowa, United States of America, **3** National Institute for Occupational Safety and Health, Cincinnati, Ohio, United States of America, **4** IMS, Inc, Calverton, Maryland, United States of America, **5** National Exposure Research Laboratory, U.S. Environmental Protection Agency, Research Triangle Park, North Carolina, United States of America, **6** Epidemiology Branch, National Institute for Environmental Health Sciences, Research Triangle Park, North Carolina, United States of America

Abstract

Farming and pesticide use have previously been linked to non-Hodgkin lymphoma (NHL), chronic lymphocytic leukemia (CLL) and multiple myeloma (MM). We evaluated agricultural use of specific insecticides, fungicides, and fumigants and risk of NHL and NHL-subtypes (including CLL and MM) in a U.S.-based prospective cohort of farmers and commercial pesticide applicators. A total of 523 cases occurred among 54,306 pesticide applicators from enrollment (1993–97) through December 31, 2011 in Iowa, and December 31, 2010 in North Carolina. Information on pesticide use, other agricultural exposures and other factors was obtained from questionnaires at enrollment and at follow-up approximately five years later (1999–2005). Information from questionnaires, monitoring, and the literature were used to create lifetime-days and intensity-weighted lifetime days of pesticide use, taking into account exposure-modifying factors. Poisson and polytomous models were used to calculate relative risks (RR) and 95% confidence intervals (CI) to evaluate associations between 26 pesticides and NHL and five NHL-subtypes, while adjusting for potential confounding factors. For total NHL, statistically significant positive exposure-response trends were seen with lindane and DDT. Terbufos was associated with total NHL in ever/never comparisons only. In subtype analyses, terbufos and DDT were associated with small cell lymphoma/chronic lymphocytic leukemia/marginal cell lymphoma, lindane and diazinon with follicular lymphoma, and permethrin with MM. However, tests of homogeneity did not show significant differences in exposure-response among NHL-subtypes for any pesticide. Because 26 pesticides were evaluated for their association with NHL and its subtypes, some chance finding could have occurred. Our results showed pesticides from different chemical and functional classes were associated with an excess risk of NHL and NHL subtypes, but not all members of any single class of pesticides were associated with an elevated risk of NHL or NHL subtypes. These findings are among the first to suggest links between DDT, lindane, permethrin, diazinon and terbufos with NHL subtypes.

Editor: Suminori Akiba, Kagoshima University Graduate School of Medical and Dental Sciences, Japan

Funding: This work was supported by the Intramural Research Program of the NIH, National Cancer Institute, Division of Cancer Epidemiology and Genetics (Z01CP010119) and The National Institutes of Environmental Health Sciences (Z01ES049030). The funders had no role in study design, data collection and analysis, decision to publish, or preparation of the manuscript. IMS, Inc, provided support in the form of salaries for authors Joseph Barker and Denis W Buckman, but did not have any additional role in the study design, data collection and analysis, decision to publish, or preparation of the manuscript. The specific roles of these authors are articulated in the 'author contributions' section.

Competing Interests: The authors have the following interests. Joseph Barker and Dennis W. Buckman are employed by IMS, Inc. There are no patents, products in development or marketed products to declare.

* Email: alavanjm@mail.nih.gov

Introduction

Since the 1970s, epidemiologic studies of non-Hodgkin lymphoma (NHL) and multiple myeloma (MM) have shown increased risk among farmers and associations with the type of farming practiced [1–6]. While farmers are exposed to many agents that may be carcinogenic [7]; there has been a particular focus on pesticides. Studies from around the world have suggested increased risk of NHL or MM [8,9] and other NHL subtypes [10] in relation to the use of specific pesticides in different functional classes (i.e., insecticides, fungicides, fumigants and herbicides). A meta-analysis of 13 case-control studies published between 1993–2005 observed an overall significant meta-odds ratio (OR) between occupational exposure to pesticides and NHL (OR = 1.35; 95% CI: 1.2–1.5) [11]. This risk was greater among individuals with more than 10 years of exposure (OR = 1.65; 95% CI: 1.08–1.95) [11], but the meta-analysis lacked details about the use of specific pesticides and other risk factors [11]. Although the International Agency for Research on Cancer (IARC) has classified "Occupational exposures in spraying and application of non-arsenical insecticides" as "probably carcinogenic to humans", the human evidence for the 17 individual pesticides evaluated in this monograph was determined to be inadequate for nine and there were no epidemiological studies for eight pesticides [12]. Since then, more studies have focused on cancer risk from specific pesticides, although the information is still relatively limited for many cancer-pesticide combinations [8,9].

To help fill the current information gap we evaluated the relationships between the use of specific insecticides, fungicides and fumigants and NHL in the Agricultural Health Study (AHS), a prospective cohort of licensed private (i.e., mostly farmer) and commercial pesticide applicators. Because the etiology of NHL and its B and T cell subtypes may differ by cell type[13], we also evaluated risk by subtype while controlling for potential confounding factors suggested from the literature [13], and the AHS data.

Novelty and Impact

These findings on occupationally exposed pesticide applicators with high quality exposure information are among the first to suggest links between DDT, lindane, permethrin, diazinon and terbufos and specific NHL subtypes in a prospective cohort study.

Materials and Methods

Study Population

The AHS is a prospective cohort study of 52,394 licensed private pesticide applicators (mostly farmers) in Iowa and North Carolina and 4,916 licensed commercial applicators in Iowa (individuals paid to apply pesticides to farms, homes, lawns, etc.), and 32,346 spouses of private applicators. Only applicators are included in this analysis. The cohort has been previously described in detail [14,15] and study questionnaires are available on the AHS website (www.aghealth.nih.gov). Briefly, individuals seeking licenses to apply restricted use pesticides were enrolled in the study from December 1993 through December 1997 (82% of the target population enrolled). At enrollment, subjects did not sign a written informed consent form. However, the cover letter of the questionnaire booklet informed subjects of the voluntary nature of participation, the ability to not answer any question, and it provided an assurance of confidentiality (including a Privacy Act Notification statement). The letter also included a written summary of the purpose of research, time involved, benefits of research, and a contact for questions about the research. The cover letter to the take-home questionnaire included all of the above and also informed the participant that they had the right to withdraw at any time. Finally, subjects were specifically informed that their contact information (including Social Security Number) would be used to search health and vital records in the future. The participants provided consent by completing and returning the questionnaire booklet. These documents and procedures were approved in 1993 by all relevant institutional review boards (i.e., National Cancer Institute Special Studies Institutional Review Board, Westat Institutional Review Board, and the University of Iowa Institutional Review Board-01).

Excluded from this analysis were study participants who had a history of any cancer at the time of enrollment (n = 1094), individuals who sought pesticide registration in Iowa or North Carolina but did not live in these states at the time of registration (n = 341) and were thus outside the catchment area of these cancer registries and individuals that were missing information on potential confounders (i.e., race or total herbicides application days [n = 1,569]). This resulted in an analysis sample of 54,306. We obtained cancer incidence information by regular linkage to the population-based cancer registry files in Iowa and North Carolina. In addition, we linked cohort members to state mortality registries of Iowa and North Carolina and the nation-wide National Death Index to determine vital status, and to the nation-wide address records of the Internal Revenue Service, state-wide motor vehicle registration files, and pesticide license registries of state agricultural departments to determine residence in Iowa or North Carolina. The current analysis included all incident primary NHL, as well as CLL and MM (which are now classified as NHL) [13] (*n* = 523) diagnosed from enrollment (1993–1997) through December 31, 2010 in North Carolina and from enrollment (1993–1997) through December 31, 2011 in Iowa, the last date of complete cancer incidence reports in each state. We ended follow-up and person-year accumulation at the date of diagnosis of any cancer, death, movement out of state, or December 31, 2010 in North Carolina and December 31, 2011 in Iowa, whichever was earlier.

Tumor Characteristics

Information on tumor characteristics was obtained from state cancer registries. We followed the definition of NHL and six subtypes of NHL used by the Surveillance Epidemiology and End Results (SEER) coding scheme [16] which was based on the Pathology Working Group of the International Lymphoma Epidemiology Consortium (ICD-O-3 InterLymph modification) classification (Table S1 in File S1, [17], i.e., 1. Small B-cell lymphocytic lymphomas (SLL)/chronic B-cell lymphocytic lymphomas (CLL)/mantle-cell lymphomas (MCL); 2. Diffuse large B-cell lymphomas; 3. Follicular lymphomas; 4. 'Other B-cell lymphomas' consisting of a diverse set of B-cell lymphomas; 5. Multiple myeloma; and 6. T-cell NHL and undefined cell type). There were too few T-cell NHL cases available for analysis [n = 19] so this cell type was not included in the subtype analysis). The ICD-O-3 original definition (used in many earlier studies of pesticides and cancer) of NHL [18] was also evaluated in relation to pesticide exposure to allow a clearer comparison of our results with previous studies.

Exposure Assessment

Initial information on lifetime use of 50 specific pesticides (Table S2 in File S1), including 22 insecticides, 6 fungicides and 4 fumigants was obtained from two self-administered questionnaires [14,15] completed during cohort enrollment (Phase 1). All 57,310 applicators completed the first enrollment questionnaire, which inquired about ever/never use of 50 pesticides, as well as duration (years) and frequency (average days/year) of use for a subset of 22 pesticides including 9 insecticides, 2 fungicides and 1 fumigant. In addition, 25,291 (44%) of the applicators returned the second (take-home) questionnaire, which inquired about duration and frequency of use for the remaining 28 pesticides, including 13 insecticides, 4 fungicides and 3 fumigants.

A follow-up questionnaire, which ascertained pesticide use since enrollment, was administered approximately 5 years after enrollment (1999–2005, Phase 2) and completed by 36,342 (63%) of the original participants. The full text of the questionnaires is available at www.aghealth.nih.gov. For participants who did not complete the Phase 2 questionnaire (20,968 applicators, 37%), a data-driven multiple imputation procedure which used logistic regression and stratified sampling [19] was employed to impute use of specific pesticides in Phase 2. Information on pesticide use from Phase 1, Phase 2 and imputation for Phase 2 was used to construct three cumulative exposure metrics: (i) lifetime days of pesticide use (i.e., the product of years of use of a specific pesticide and the number of days used per year); (ii) intensity-weighted lifetime days of use (i.e., the product of lifetime days of use and a measure of exposure intensity) and (iii) ever/never use data for each pesticide. Intensity was derived from an exposure-algorithm, which was based on exposure measurements from the literature and individual information on pesticide use and practices (e.g., whether or not they mixed pesticides, application method, whether or not they repaired equipment and use of personal protective equipment) obtained from questionnaires completed by study participants [20].

Statistical Analyses

We divided follow-up time into 2-year intervals to accumulate person-time and update time-varying factors, such as attained age and pesticide use. We fit Poisson models to estimate rate ratios (RRs) and 95% confidence intervals (95% CI) to evaluate the effects of pesticide use on rates of overall NHL and the five NHL subtypes.

We evaluated pesticides with 15 or more exposed cases of total NHL, thereby excluding aluminum phosphide, carbon tetrachloride/carbon disulfide, ethylene dibromide, trichlorfon, and ziram leaving 26 insecticides, fungicides and fumigants for analysis (permethrin for animal use and crop use were combined into one category, all insecticides, fungicides and fumigants are listed in Table S2 in File S1). For each pesticide, we evaluated ever vs. never exposure, as well as tertiles of exposure which were created based on the distribution of all NHL exposed cases and compared to those unexposed. In the NHL subtype analysis and in circumstances where multiple pesticides were included in the model we categorized exposure for each pesticide into unexposed (i.e., never users) and two exposed groups (i.e., low and high) separated at the median exposure level. The number of exposed cases included in the ever/never analysis and in the trend analysis can differ because of the lack of information necessary to construct quantitative exposure metrics for some individuals.

Several lifestyle and demographic factors associated with NHL in the AHS cohort or previously suggested as possible confounders in the NHL literature[13] were evaluated as potential confounders in this analysis. These included: age at enrollment, gender, race, state, license type, education, autoimmune diseases, family history of lymphoma in first-degree relatives, body mass index, height, cigarette smoking history, alcohol consumption per week and several occupational exposures[1–13] including number of livestock, cattle, poultry, whether they raised poultry, hogs or sheep, whether they provided veterinary services to their animals, number of acres planted, welding, diesel engine use, number of years lived on the farm, total days of any pesticide use, and total days of herbicide use. However, since most of these variables did not change the risk estimates for specific pesticides, we present results adjusted for age, race, state and total days of herbicide use, which impacted risk estimates by more than 10% for some subtypes. We also performed analyses adjusting for specific insecticides, fungicides and fumigants shown to be associated with NHL or a specific NHL subtype in the current analysis. Tests for trend used the median value of each exposure category. All tests were two-sided and conducted at $\alpha = 0.05$ level. Analysis by NHL subtype was limited to insecticides, fungicides, and fumigants with 6 or more exposed cases.

We also fit polytomous logit models, where the dependent variable was a five-level variable (i.e., five NHL subtypes) and a baseline level (i.e., no NHL) to estimate exposure-response odds ratios (ORs) and 95% confidence intervals (CIs) for each subtypes of NHL. We then used polytomous logit models to estimate exposure-response trend while adjusting for age, state, race and total days of herbicide use, as in the Poisson models, and tested homogeneity among the 5 NHL subtypes.

Poisson models were fit using the GENMOD procedure and polytomous logit models were fit using the LOGISTIC procedure of the SAS 9.2 statistical software package (SAS Institute, Cary, NC). Summary estimates of NHL and NHL subtype risks for both Poisson models and polytomous logit models incorporated imputed data and were calculated along with standard error estimates, confidence intervals, and p-values, using multiple imputation methods implemented in the MIANALYZE procedure of SAS 9.2.

We also evaluated the impact of the additional pesticide exposure information imputed for Phase 2 on risk estimates. We compared risk estimates for those who completed both the phase 1 enrollment and take-home questionnaires and the phase 2 questionnaires (n = 17,545) with risk estimates obtained from the combined completed questionnaire data plus the imputed phase 2 data (n = 54,306). We also explored the effect of lagging exposure data 5 years because recent exposures may not have had time to have an impact on cancer development. For comparison to previous studies, we also assessed the exposure-response association for NHL using the original ICD-O-3 definition of NHL [18] and the new definition [16] in Table S3 in File S1. Unless otherwise specified, reported results show un-lagged exposure information from both Phase 1 and Phase 2 including Phase 2 imputed data for lifetime exposure-days and intensity-weighted lifetime days of use and NHL defined by the InterLymph modification of ICD-O-3 [17]. Data were obtained from AHS data release versions P1REL201005.00 (for Phase 1) and P2REL201007.00 (for Phase 2).

Results

The 54,306 applicators in this analysis contributed 803,140 person-years of follow-up from enrollment through December 31, 2010 in North Carolina and December 31, 2011 in Iowa (Table 1). During this period, there were 523 incident cases of NHL, including 148 SLL/CLL/MCL, 117 diffuse large B-cell lymphomas, 67 follicular lymphomas, 53 'other B-cell lymphomas' (consisting of a diverse set of B-cell lymphomas) and 97 cases of MM. Another 41 cases consisting of T-cell lymphomas (n = 19) and non-Hodgkin lymphoma of unknown lineage (n = 22) were excluded from cell type-specific analyses because of small numbers of cases with identified cell types. Between enrollment and the end of follow-up, 6,195 individuals were diagnosed with an incident cancer other than NHL, 4,619 died without a record of cancer in the registry data, and 1,248 cohort members left the state and could not be followed-up for cancer. Person-years of follow-up accumulated for all of these study participants after enrollment until they were censored for the incident cancer, death or moving out of the state (data not shown). The risk of NHL increased significantly and monotonically with age in the AHS cohort in this analysis (p = 0.001) and age-adjusted risks were significant for state and NHL overall and race for multiple myeloma (data not shown). Total days of herbicide use had a small but significant effect on the risk of some NHL subtypes, but not on NHL overall. No other demographic or occupational factors showed evidence of confounding so they were not included in the final models.

In Table 2 we present ever/never results for 26 insecticides, fungicides and fumigants by total NHL and by NHL subtype adjusted for age, race, state and herbicide use (total life-time days). Terbufos was the only pesticide associated with an increased risk of total NHL in the ever/never use analysis (RR = 1.2 [1.0–1.5]), although the trend for increasing use and risk of total NHL was not significant (p trend = 0.43) (Table 3). In contrast, there were a few chemicals that were not associated with ever/never use, but

Table 1. Baseline characteristics of AHS study participants in the NHL incidence analysis[1,2].

Variables	All NHL cases (%)	Cohort Person-years.
Age at Enrollment		
<45	84 (16.1)	426,288
45–49	51 (9.8)	101,018
50–54	75 (14.3)	84,998
55–59	90 17.2)	74,440
60–64	78 (14.9)	56,978
65–69	79 (15.1)	35,071
≥70	66 (12.6)	24,347
Race		
White	509 (97.3)	787,799
Black	14 (2.7)	15,341
State		
IA	332 (63.5)	537,252
NC	191 (36.5)	265,888
Lifetime Total Herbicide Exposure Days		
0–146 days	170 (32.5)	251,401
147–543 days	169 (32.3)	273,107
544–2453 days	184 (35.2)	278,632

[1]During the period from enrollment (1993–1997) to December 31, 2010 in NC and December 31, 2011 in Iowa.
[2]Individuals with missing ever/never exposure information or missing confounding variable information were not included in the table.

did show evidence of an exposure-response association. Lindane was the only pesticide that showed a statistically significant increasing trend in risk for NHL with both exposure metrics, for lifetime-days of lindane use the RR were = 1.0 (ref), 1.2 (0.7–1.9), 1.0 (0.6–1.7), 2.5 (1.4–4.4); p trend = 0.004 and intensity-weighted lifetime-days of use the: RR were: = 1.0 (ref), 1.3 (0.8–2.2), 1.1 (0.7–1.8), 1.8 (1.0–3.2); p trend = 0.04. DDT showed a significant trend for NHL risk with life-time days of use RR = 1.0 (ref), 1.3 (0.9–1.8), 1.1 (0.7–1.7), 1.7 (1.1–2.6); p trend = 0.02, while the intensity weighted lifetime days of use of DDT was of borderline significance: RR = 1.0 (ref), 1.2 (0.8–1.8), 1.1 (0.8–1.7), 1.6 (1.0–2.3); p trend = 0.06. The number of lifetime days of use of lindane and DDT was weakly correlated (coefficient of determination = 0.04), and the pattern of NHL risk showed little change when both were included in the model. The results for lindane adjusted for DDT were, RR = 1.0 (ref), 1.2 (0.7–2.0), 1.0 (0.5–1.8), 1.6 (0.9–3.3); p trend = 0.07 and the results for DDT adjusted for lindane were, RR = 1.0 (ref), 1.3 (0.9–2.0), 0.9 (0.6–1.6), 1.6 (0.9–2.6); p trend = 0.08).

We also evaluated pesticides by NHL sub-type. In the ever/never analyses (Table 2), permethrin was significantly associated with multiple myeloma, RR = 2.2 (1.4–3.5) and also demonstrated an exposure-response trend (RR = 1.0 (ref), 1.4 (0.8–2.7), 3.1 (1.5–6.2); p trend = 0.002) (Table 4). Similarly, there was an elevated risk of SLL/CLL/MCL with terbufos in ever/never analyses RR = 1.4 (0.97–2.0) and an exposure response trend (RR = 1.0 (ref), 1.3 (0.8–2.0), 1.6 (1.0–2.5); p trend = 0.05). For follicular lymphoma, lindane showed an elevated but non-significant association for ever use, RR = 1.7 (0.96–3.2) and a significant exposure-response association (RR = 1.0 (ref), 4.9 (1.9–12.6), 3.6 (1.4–9.5); p trend = 0.04). There were also two chemicals with evidence of exposure-response that were not associated with specific subtypes in the ever/never analyses: DDT (Dichlorodiphenyltrichloroethane) with SLL/CLL/MCL (RR = 1.0 (ref), 1.0 (0.5–1.8), 2.6 (1.3–4.8; p trend = 0.04); and diazinon with follicular lymphoma (RR = 1.0 (ref), 2.2 (0.9–5.4), 3.8 (1.2–11.4); p trend = 0.02) (Table 4).

The pattern of increased CLL/SLL/MCL risk with increased use of DDT and terbufos remained after both insecticides were placed in our model concurrently. CLL/SLL/MCL risk increased with DDT use (RR = 1.0 (ref), 0.9 (0.5–4.7); 2.4 (1.1–4.7); p trend = 0.04), and a pattern of increased CLL/SLL/MCL risk was also observed with terbufos use (RR = 1.0 (ref), 1.1 (0.6–2.1), 1.7 (0.9–3.3) p trend = 0.07), although the trend was not significant for terbufos. Similarly, the pattern of increased follicular lymphoma risk with lindane use and diazinon use remained after both insecticides were placed in our model concurrently. Follicular lymphoma risk increased with diazinon use (RR = 1.0 (ref), 4.1 (1.5–11.1); 2.5 (0.9–7.2); p trend = 0.09), and a similarly, pattern of increased follicular lymphoma risk was observed with lindane use (RR = 1.0 (ref), 1.6 (0.6–4.1), 2.6 (0.8–8.3) p trend = 0.09), although neither remained statistically significant (Table 4).

Three chemicals showed elevated risks in ever/never analyses for certain subtypes, with no apparent pattern in exposure-response analyses: metalaxyl and chlordane with SLL/CLL/MCL, RR = 1.6 (1.0–2.5) and RR = 1.4 (0.97–2.0) respectively, and methyl bromide with diffuse large B-cell lymphoma RR = 1.9 (1.1–3.3). Although there was evidence of association by subtype, and polytomous logit models indicated homogeneity across subtypes for lindane (p = 0.54), DDT (p = 0.44) and any other pesticide evaluated in this study (e.g., permethrin (p = 0.10), diazinon (p = 0.09), terbufos (p = 0.63), (last column in Table 4).

There was no evidence of confounding of the total NHL associations with either lindane or DDT. We also calculated RR for those who completed both the phase 1 enrollment and take-home questionnaires and the phase 2 questionnaire (n = 17,545) and found no meaningful difference in the RR that also included imputed exposures, although there was an increase in precision of

Table 2. Pesticides exposure (ever/never) and adjusted Relative Risk of total NHL and NHL Subtype[1].

Insecticide

	Total NHL Cases[2]		SLL/CLL/MCL Cases[2]		Diffuse Large B-Cell Cases[2]		Follicular B-Cell Cases[2]		Other B-cell Cases[2]		Multiple Myeloma Cases[2]	
Pesticide (chemical-functional class)	**Ever/Never Exposed**	**RR[3,4] (95% CI)**	**Ever/Never Exposed**	**RR[3,4] (95% CI)**	**Ever/Never Exposed**	**RR[3,4] (95% CI)**	**Ever/Never Exposed**	**RR[3,4] (95% CI)**	**Ever/Never Exposed**	**RR[3,4] (95% CI)**	**Ever/Never Exposed**	**RR[3,4] (95% CI)**
Aldicarb	47/435	1	14/124	1.1	8/98	0.7	6/54	0.9	7/41	1.6	10/82	1.2
(carbamate-insecticide)		(0.7–1.4)		(0.6–1.8)		(0.4–1.5)		(0.3–2.2)		(0.7–3.5)		(0.6–2.2)
Carbofuran	147/317	1.1	48/86	1.2	26/78	0.8	18/39	1	13/31	0.8	31/56	1.3
(carbamate-insecticide)		(0.9–1.3)		(0.8–1.8)		(0.5–1.3)		(0.5–1.7)		(0.4–1.6)		(0.8–2.1)
Carbaryl	272/225	1	75/66	1	58/53	0.8	37/24	0.8	24/28	0.9	58/34	0.9
(carbamate-insecticide)		(0.8–1.2)		(0.7–1.5)		(0.5–1.3)		(0.5–1.3)		(0.5–1.6)		(0.6–1.4)
Chlorpyrifos	210/300	1	62/84	1	44/70	0.9	32/33	1.3	21/31	0.8	36/58	1
(organophosphate-insecticide)		(0.8–1.2)		(0.7–1.4)		(0.6–1.4)		(0.8–2.2)		(0.5–1.5)		(0.6–1.5)
Coumaphos	46/411	1.1	15/120	1.2	10/93	1	8/48	1.6	5/40	xxx	7/78	1
(organophos-phate-insecticide)		(0.8–1.5)		(0.7–2.1)		(0.5–2.1)		(0.8–3.5)				(0.1–2.1)
DDVP	55/407	1.1	13/124	0.8	10/93	1	8/48	1.3	6/39	1	12/73	1.7
(dimethyl phosphate-insecticide)		(0.8–1.5)		(0.5–1.5)		(0.5–1.9)		(0.6–2.7)		(0.4–2.4)		(0.9–3.2)
Diazinon	144/342	1	46/93	1.3	30/78	0.9	22/38	1.3	12/37	0.8	27/64	1
(organophosphorous-insecticide)		(0.8–1.3)		(0.9–1.9)		(0.6–1.4)		(0.7–2.3)		(0.4–1.6)		(0.6–1.6)
Fonofos	115/349	1.1	35/100	1.1	25/81	1.2	13/45	0.9	15/30	1.3	19/66	1.3
(organophosphorous-insecticide)		(0.9–1.4)		(0.7–1.6)		(0.7–1.9)		(0.5–1.7)		(0.7–2.5)		(0.8–2.3)
Malathion	332/163	0.9	99/43	1	72/37	0.9	46/14	1.3	30/21	0.6	61/32	0.9
(organophosphorous-insecticide)		(0.8–1.1)		(0.7–1.4)		(0.6–1.4)		(0.7–2.4)		(0.3–1.0)		(0.6–1.5)
Parathion (ethyl or methyl)	69/411	1.1	20/117	1	14/91	1	10/48	1.1	7/44	1.1	14/77	1
(organophosphorous insecticide		(0.8–1.4)		(0.7–1.4)		(0.6–1.4)		(0.8–1.5)		(0.7–1.5)		(0.8–1.5)
Permethrin (animal and crop applications)	112/363	1.1	32/106	1	18/81	0.7	18/81	1.1	9/14	0.8	20/72	**2.2**
(pyrethroid insecticide)		(0.8–1.3)		(0.6–1.5)		(0.4–1.2)		(0.6–2.0)		(0.4–1.6)		**(1.4–3.5)**
Phorate	160/325	1	53/87	1.1	31/76	0.9	20/40	0.9	19/31	0.9	26/64	1
(organophosphorous-insecticide)		(0.8–1.2)		(0.8–1.6)		(0.5–1.3)		(0.5–1.6)		(0.5–1.6)		(0.6–1.7)
Terbufos	201/267	**1.2**	64/72	1.4	42/63	1.1	31/26	1.2	26/19	1.8	32/59	1.2
(organophosphorous-insecticide)		**(1.0–1.5)**		(0.97–2.0)		(0.7–1.7)		(0.7–2.1)		(0.94–3.2)		(0.7–1.9)
Chlorinated Insecticides												
Aldrin	116/364	0.9	53/99	0.9	15/91	0.8	13/45	0.8	12/37	0.6	29/62	1.5
(chlorinated insecticide)		(0.7–1.1)		(0.6–1.4)		(0.4–1.6)		(0.4–1.6)		(0.3–1.3)		(0.9–2.5)
Chlordane	136/344	1	49/90	1.4	20/86	0.6	18/41	1.2	13/36	1	31/60	1.2
(chlorinated insecticide)		(0.8–1.3)		(0.99–2.1)		(0.4–1.0)		(0.7–2.1)		(0.7–2.0)		(0.8–1.9)
DDT	182/300	1	59/79	1.2	34/73	0.8	18/41	0.9	20/31	1.1	40/50	1.1

Table 2. Cont.

Insecticide												
	Total NHL Cases[2]		SLL/CLL/MCL Cases[2]		Diffuse Large B-Cell Cases[2]		Follicular B-Cell Cases[2]		Other B-cell Cases[2]		Multiple Myeloma Cases[2]	
Pesticide (chemical-functional class)	**Ever/Never Exposed**	**RR[3,4]**	**Ever/Never Exposed**	**RR[3,4]**	**Ever/Never Exposed**	**RR[3,4]**	**Ever/Never Exposed**	**RR[3,4]**	**Ever/Never Exposed**	**RR[3,4]**	**Ever/Never Exposed**	**RR[3,4]**
		(95% CI)		**(95% CI)**		**(95% CI)**		**(95% CI)**		**(95% CI)**		**(95% CI)**
(chlorinated insecticide)		(0.8–1.3)		(0.8–1.8)		(0.5–1.3)		(0.5–1.6)		(0.6–2.1)		(0.7–1.8)
Dieldrin	35/442	0.9	5/130	xxx	4/101	xxx	4/54	xxx	7/42	1	10/81	0.9
(chlorinated insecticide)		(0.6–1.2)								(0.7–2.0)		(0.5–1.4)
Heptachlor	90/384	1	33/104	1.1	10/95	1.1	9/48	1.1	13/36	0.9	17/72	1.1
(chlorinated insecticide)		(0.7–1.2)		(0.7–3.0)		(0.3–3.1)		(0.5–3.2)		(0.5–2.7))		(0.6–2.0)
Lindane	85/396	1	27/113	1.2	12/95	0.6	16/41	1.7	9/40	0.7	13/73	1.1
(chlorinated insecticide)		(0.8–1.2)		(0.6–1.5)		(0.3–1.1)		(0.96–3.2)		(0.4–1.2)		(0.5–2.0)
Toxaphene	79/397	1	21/116	0.9	14/90	0.8	9/47	1	10/40	1.1	19/73	1.1
(chlorinated insecticide)		(0.7–1.2)		(0.5–1.5)		(0.4–1.4)		(0.6–2.0)		(0.6–2.0)		(0.6–1.9)
Fungicides												
Benomyl	54/428	1.1	18/123	1.2	12/95	1.1	4/51	xxx	4/51	xxx	11/80	1.1
(carbamate fungicide)		(0.8–1.5)		(0.7–2.0)		(0.6–1.9)						(0.6–2.0)
Captan	60/406	1.1	18/118	1.1	12/91	0.9	5/51	xxx	6/39	1.1	12/76	1.2
(phthalimide fungicide)		(0.8–1.4)		(0.6–1.8)		(0.5–1.8)				(0.5–2.7)		(0.6–2.2)
Chloro-thalonil	35/474	0.8	9/135	0.9	6/107	0.5	5/60	xxx	2/50	xxx	11/84	1.2
(poly-chlorinated aromatic thalonitrile fungicide)		(0.5–1.2)		(0.4–1.9)		(0.2–1.3)						(0.6–2.3)
Maneb/	44/437	0.9	13/127	1.1	12/95	1.1	4/60	xxx	5/49	xxx	10/79	0.8
Mancozeb		(0.7–1.3)		(0.6–2.1)		(0.6–2.1)						(0.4–1.7)
(dithiocarbamate fungicide)												
Metalaxyl	108/381	1	34/106	**1.6**	27/82	1.1	10/48	0.7	10/40	0.9	21/71	0.8
(acylalanine fungicide)		(0.8–1.3)		**(1.0–2.5)**		(0.6–1.8)		(0.4–1.4)		(0.4–1.7)		(0.4–1.3)
Fumigant												
Methyl bromide	85/425	1.1	18/126	0.9	28/86	**1.9**	7/58	0.6	8/44	2.2	19/76	1
(methyl halide fumigant)		(0.9–1.5)		(0.5–1.7)		**(1.1–3.3)**		(0.2–1.4)		(0.9–5.7)		(0.6–1.8)

[1]During the period from enrollment (1993–1997) to December 31, 2010 in NC and December 31, 2011 in Iowa.
[2]Numbers of cases by NHL subtype do not sum to total number of NHL cases (n = 523) due to missing data.
[3]Adjusted RR: age (<45, 45–49, 50–54, 55–59, 60–64, 65–69, ≥70), State (NC vs. IA), Race (White vs. Black), AHS herbicides (tertiles of total herbicide use-days). Statistically significant RR and 95% confidence limits are bolded.
[4]RR was not calculated if the number of exposed cases in a pesticide-NHL subtype cell was <6 and the missing RR was marked with an XXX. Statistically significant RRs and 95% confidence limits are bolded.

Table 3. Pesticide exposure (lifetime-days & intensity weighted life-time days) and adjusted risks of total NHL incidence[1].

Insecticides						
Pesticide (chemical-functional class)	**NHL Cases[2]**	**Non-Cases[2]**	**RR[3,4] (95% CI) by Total Days of Exposure**	**NHL**	**Non-Cases**	**RR[3,4] (95% CI)**
[days of lifetime exposure for each category]				**Cases[2,]**		**Intensity-weighted days of exposure**
Aldicarb (carbamate-insecticide)						
None	238	21557	1.0 (ref)	238	21557	1.0 (ref)
Low [≤8.75]	7	633	1.1 (0.5–2.3)	6	383	1.3 (0.6–3.3))
Medium [>8.75–25.5]	5	522	0.9 (0.3–2.5)	6	853	0.9 (0.4–1.9)
High [>25.5–224.75]	5	1266	0.5 (0.2–1.3)	5	1183	0.5 (0.2–1.3)
			P trend = 0.23			P trend = 0.22
Carbofuran (carbamate-insecticide)						
None	317	36296	1.0 (ref)	317	36296	1.0 (ref)
Low [≤8.75]	63	4775	1.2 (0.9–1.6)	46	3695	1.2 (0.9–1.6)
Medium [>8.75–38.75]	32	3648	0.8 (0.6–1.2)	46	4590	1.0 (0.7–1.3)
High [>38.75–767.25]	44	4370	0.97 (0.7–1.4)	45	4477	1.0 (0.7–1.4)
			P trend = 0.69			P trend = 0.74
Carbaryl (carbamate-insecticide)						
None	128	12864	1.0 (ref)	128	12864	1.0 (ref)
Low [≤8.75]	54	4128	1.1 (0.7–1.6)	46	3962	1.0 (0.7–1.5)
Medium [8.75–56]	43	5096	0.9 (0.6–1.2)	45	4433	0.9 (0.7–1.5)
High [>56–737.5]	39	3281	1.0 (0.7–1.6)	44	4029	1.0 (0.6–1.5)
			P trend = 0.87			P trend = 0.94
Chlorpyrifos (organophosphate-insecticide)						
None	300	30393	1.0 (ref)	300	30393	1.0 (ref)
Low [≤8.75]	71	6493	1.1 (0.9–1.5)	61	6383	1.1 (0.8–1.4)
Medium [>8.75–44]	65	6892	1.1 (0.8–1.4)	60	7549	0.9 (0.7–1.2)
High [>44–767.25]	67	9380	0.8 (0.6–1.1)	60	7044	1.0 (0.7–1.3)
			P trend = 0.11			P trend = 0.85
Coumaphos (organophosphate-insecticide)						
None	411	44846	1.0 (ref)	411	44846	1.0 (ref)
Low [<8.75]	16	1510	1.0 (0.6–1.7)	15	1132	1.3 (0.8–2.1)
Medium [>8.75–38.75]	14	1076	1.2 (0.7–2.1)	14	1452	1.0 (0.6–1.6)
High [>38.75–1627.5]	13	1175	1.2 (0.7–2.0)	14	1170	1.2 (0.7–2.1)
			P for trend = 0.50			P trend = 0.48
DDVP (dimethyl phosphate-insecticide)						
None	407	44551	1.0 (ref)	407	44551	1.0 (ref)
Low [≤8.75]	19	1342	1.4 (0.9–2.1)	18	1281	1.4 (0.9–2.3)
Medium [>8.75–87.5]	17	1519	1.2 (0.7–1.9)	18	1633	1.1 (0.7–1.8)
High [>87.5–2677.5]	17	1893	0.9 (0.6–1.5)	17	1824	1.0 (0.6–1.6)
			P trend = 0.78			P trend = 0.83
Diazinon (organophosphorous-insecticide)						
None	187	17943	1.0 (ref)	187	17943	1.0 (ref)
Low [≤8.75]	28	2506	1.1 (0.7–1.6)	23	2047	1.1 (0.7–1.8)
Medium [>8.75–25]	19	1515	1.0 (0.6–1.8)	24	2246	0.9 (0.5–1.5)
High [>25–457.25]	23	1990	1.2 (0.7–1.9)	22	1708	1.3 (0.8–2.1)
			P trend = 0.52			P trend = 0.33

Table 3. Cont.

Insecticides						
Pesticide (chemical-functional class)	**NHL Cases[2]**	**Non-Cases[2]**	**RR[3,4] (95% CI) by Total Days of Exposure**	**NHL**	**Non-Cases**	**RR[3,4] (95% CI)**
[days of lifetime exposure for each category]				**Cases[2,]**		**Intensity-weighted days of exposure**
Fonofos (organophosphorous-insecticide)						
None	349	39570	1.0 (ref)	349	39570	1.0 (ref)
Low [≤20]	47	3812	1.3 (0.96–1.8)	37	2906	1.4 (0.97–1.9)
Medium [>20–50.75]	28	2819	1.1 (0.7–1.6)	38	3487	1.1 (0.8–1.6)
High [>50.75–369.75]	37	3385	1.1 (0.7–1.5)	36	3606	1.0 (0.7–1.4)
			P trend = 0.83			P trend = 0.87
Malathion (organophosphorous-insecticide)						
None	90	8368	1.0 (ref)	90	8368	1.0 (ref)
Low [≤8.75]	75	7284	0.97 (0.7–1.3)	60	5535	1.0 (0.7–1.4)
Medium [>8.75–38.75]	47	5779	0.7 (0.5–1.1)	59	6899	0.8 (0.6–1.1)
High [>38.75–737.5]	57	5037	0.9 (0.6–1.3)	59	5588	0.9 (0.6–1.2)
			P trend = 0.63			P trend = 0.46
Parathion (ethyl or methyl) (organophosphorous insecticide)						
None	228	21457	1.0 (ref)	228	21457	1.0 (ref)
Low [≤8.75]	9	693	1.0 (0.5–2.0)	7	612	0.9 (0.4–2.0)
Medium [>8.75–24.5]	6	351	1.4 (0.6–3.2)	8	462	1.4 (0.7–2.9)
High [>.24.5–1237.5]	6	652	0.8 (0.3–1.8)	6	621	0.8 (0.4–1.9)
			P trend = 0.64			P trend = 0.74
Permethrin (animal and crop applications)						
(pyrethroid insecticide)						
None	371	37496	1.0 (ref)	371	37496	1.0 (ref)
Low [≤8.75]	38	4315	1.1 (0.8–1.5)	33	4263	0.9 (0.6–1.3)
Medium [>8.75–50.75]	31	4611	0.8 (0.5–1.2)	33	4200	1.0 (0.7–1.4)
High [>50.75–1262.25]	33	4121	1.2 (0.8–1.7)	32	4553	1.0 (0.7–1.5)
			P trend = 0.54			P trend = 0.99
Phorate (organophosphorous-insecticide)						
None	171	16834	1.0 (ref)	171	16834	1.0 (ref)
Low [≤8.75]	27	2621	0.8 (0.5–1.2)	26	2320	0.9 (0.6–1.4)
Medium [8.75–24.5]	33	1819	1.4 (0.96–2.1)	27	1951	1.1 (0.7–1.7)
High [>24.5–224.75]	18	2246	0.6 (0.4–1.1)	25	2409	0.8 (0.5–1.3)
			P trend = 0.25			P trend = 0.44
Terbufos (organophosphorous-insecticide)						
None	267	31076	1.0 (ref)	267	31076	1.0 (ref)
Low [≤24.5]	82	8410	1.2 (0.9–1.5)	64	6895	1.1 (0.9–1.5)
Medium [>24.5–56]	54	3925	1.6 (1.2–2.1)	64	4642	1.6 (1.2–2.2)
High [>56–1627.5]	57	6080	1.1 (0.8–1.5)	63	6842	1.1 (0.8–1.5)
			P trend = 0.43			P trend = 0.44
Chlorinated Insecticides						
Aldrin (chlorinated insecticide)						
None	193	19743	1.0 (ref)	193	19743	1.0 (ref)
Low [≤8.75]	27	1613	0.9 (0.6–1.4)	20	1212	0.9 (0.6–1.4)
Medium [>8.75–24.5]	16	1002	0.8 (0.5–1.3)	20	1279	0.8 (0.5–1.3)

Table 3. Cont.

Insecticides						
Pesticide (chemical-functional class)	**NHL Cases[2]**	**Non-Cases[2]**	**RR[3,4] (95% CI) by Total Days of Exposure**	**NHL**	**Non-Cases**	**RR[3,4] (95% CI)**
[days of lifetime exposure for each category]				**Cases[2,]**		**Intensity-weighted days of exposure**
High [>24.5–457.25]	17	903	0.9 (0.5–1.5)	19	1026	0.9 (0.6–1.5)
			P trend = 0.58			P trend = 0.74
Chlordane (chlorinated insecticide)						
None	179	19115	1.0 (ref)	179	19115	1.0 (ref)
Low [≤8.75]	47	2687	1.3 (0.97–1.9)	23	1303	1.4 (0.9–2.2)
Medium[5]	0	0	xxx	24	1747	1.0 (0.6–1.5)
High [>8.75–1600]	23	1450	1.1 (0.7–1.7)	22	1085	1.4 (0.9–2.2)
			P trend = 0.43			P trend = 0.16
DDT (chlorinated insecticide)						
None	152	18543	1.0 (ref)	152	18543	1.0 (ref)
Low [≤8.75]	43	2121	1.3 (0.9–1.8)	33	1601	1.2 (0.8–1.8)
Medium [>8.75–56]	28	1598	1.1 (0.7–1.7)	32	1760	1.1 (0.8–1.7)
High [>56–1627.5]	27	953	1.7 (1.1–2.6)	32	1305	1.6 (1.0–2.3)
			P trend = 0.02			P trend = 0.06
Dieldrin (chlorinated insecticide)						
None	235	22510	1.0 (ref)	235	22510	1.0 (ref)
Low [≤8.75]	7	472	0.7 (0.3–1.5)	6	363	0.8 (0.4–1.8)
Medium [>8.75–24.5]	8	154	2.3 (1.1–4.7)	5	106	2.2 (0.9–5.3)
High [>24.5–224.75]	2	140	0.7 (0.2–2.9)	5	298	0.8 (0.3–2.0)
			P trend = 0.47			P trend = 0.84
Heptachlor (chlorinated insecticide)						
None	205	20844	1.0 (ref)	205	20844	1.0 (ref)
Low [≤8.75]	21	1261	1.0 (0.6–1.6)	15	1110	0.8 (0.5–1.4)
Medium [>8.75–24.5]	18	679	1.5 (0.9–2.4)	16	425	2.0 (1.2–3.4)
High [>24.5–457.25]	7	600	0.7 (0.3–1.4)	14	1001	0.8 (0.5–1.4)
			P trend = 0.82			P trend = 0.88
Lindane (chlorinated insecticide)						
None	205	20375	1.0 (ref)	205	20375	1.0 (ref)
Low [≤8.75]	18	1285	1.2 (0.7–1.9)	15	976	1.3 (0.8–2.2)
Medium [>8.75–56]	13	1103	1.0 (0.6–1.7)	16	1205	1.1 (0.7–1.8)
High [>56–457.25]	14	467	2.5 (1.4–4.4)	14	673	1.8 (1.0–3.2)
			P trend = 0.004			**P trend = 0.04**
Toxaphene (chlorinated insecticide)						
None	214	20911	1.0 (ref)	214	20911	1.0 (ref)
Low [≤8.75]	14	1198	0.8 (0.5–1.4)	11	630	1.3 (0.7–2.3)
Medium [>8.75–24.5]	13	564	1.5 (0.9–2.7)	12	931	0.9 (0.5–1.6)
High [>24.5–457.25]	6	686	0.6 (0.3–1.4)	10	886	0.8 (0.4–1.5)
			P trend = 0.50			P trend = 0.38
Fungicides						
Benomyl (carbamate fungicide)						
None	219	21425	1.0 (ref)	219	21425	1.0 (ref)
Low [≤12.25]	14	896	1.7 (0.9–2.9)	9	432	2.2 (1.1–4.3)
Medium [>12.25–24.5]	4	214	2.4 (0.9–6.6)	10	732	1.7 (0.9–3.2)

Table 3. Cont.

Insecticides						
Pesticide (chemical-functional class)	**NHL Cases[2]**	**Non-Cases[2]**	**RR[3,4] (95% CI) by Total Days of Exposure**	**NHL**	**Non-Cases**	**RR[3,4] (95% CI)**
[days of lifetime exposure for each category]				**Cases[2,]**		**Intensity-weighted days of exposure**
High [>24.5–457.25]	8	834	1.0 (0.5–2.1)	7	779	0.9 (0.4–2.0)
			P trend = 0.93			P trend = 0.75
Captan (phthalimide fungicide)						
None	407	43433	1.0 (ref)	407	43433	1.0 (ref)
Low [≤0.25]	15	2334	0.8 (0.5–1.4)	15	2108	0.9 (0.6–1.5)
Medium [>0.25–12.25]	16	1004	1.5 (0.8–2.6)	15	1171	1.2 (0.7–2.2)
High [>12.25–875]	14	1823	0.8 (0.5–1.5)	14	1805	0.8 (0.5–1.5)
			P trend = 0.69			P trend = 0.52
Chlorothalonil (polychlorinated aromatic thalonitrile fungicide)						
None	474	48442	1.0 (ref)	474	48442	1.0 (ref)
Low [≤12.25]	13	1509	0.9 (0.5–1.6)	10	1800	0.6 (0.3–1.2)
Medium [>12.25–64]	9	1492	0.8 (0.4–1.6)	11	1501	0.9 (0.5–1.7)
High [>64–395.25]	9	1678	0.6 (0.3–1.3)	9	1362	0.8 (0.4–1.6)
			P trend = 0.16			PP trend = 0.52
Maneb/Mancozeb (dithiocarbamate fungicide)						
None	228	21512	1.0 (ref)	228	21512	1.0 (ref)
Low [≤7]	8	400	1.9 (0.9–3.9)	8	486	1.6 (0.8–3.3)
Medium [>7–103.25]	9	990	0.9 (0.4–1.7)	9	680	1.3 (0.6–2.6)
High [>103.25–737.5]	7	454	1.4 (0.6–2.9)	7	677	0.9 (0.4–1.9)
			P trend = 0.49			P trend = 0.78
Metalaxyl (acylalanine fungicide)						
None	209	18833	1.0 (ref)	209	18833	1.0 (ref)
Low [≤6]	16	1439	1.0 (0.6–1.8)	15	1079	1.3 (0.8–2.2)
Medium [>6–28]	15	2182	0.7 (0.4–1.3)	15	2203	0.8 (0.4–1.3)
High [>28–224.75]	13	1566	1.1 (0.6–2.1)	14	1893	0.9 (0.5–1.6)
			P trend = 0.76			P trend = 0.63
Fumigant						
Methyl bromide (methyl halide fumigant)						
None	425	45265	1.0 (ref)	425	45265	1.0 (ref)
Low [≤8]	37	2060	2.0 (1.4–2.9)	26	1680	1.8 (1.2–2.7)
Medium [>8–28]	24	3011	0.9 (0.6–1.4)	25	2501	1.1 (0.7–1.8)
High [>28–387.5]	17	2768	0.6 (0.4–1.0)	25	3571	0.8 (0.5–1.2)
			P trend = 0.04			P trend = 0.10

[1]During the period from enrollment (1993–1997) to December 31, 2010 in NC and December 31, 2011 in Iowa.
[2]Numbers of cases in columns do not sum to total number of NHL cases (n = 523) due to missing data. In the enrollment questionnaire, lifetime-days & intensity weighted life-time days of pesticide use was obtained for the insecticides: carbofuran, chlorpyrifos, coumaphos, DDVP, fonofos, permethrin and terbufos; the fungicides: captan, chlothalonil and the fumigant: methyl bromide. In the take home questionnaire lifetime-days & intensity weighted life-time days of pesticide use were obtained for the insecticides: aldicarb, carbaryl, diazinon, malathion, parathion, and phorate, the chlorinated insecticides: aldrin, chlordane, DDT, dieldrin, heptachlor, lindane, and toxaphene, the fungicides: benomyl, maneb/mancozeb and metalaxyl, therefore, numbers of NHL cases can vary among pesticides listed in the table.
[3]Adjusted RR: age (<45, 45–49, 50–54, 55–59, 60–64, 65–69, ≥70), State (NC vs. IA), Race (White vs. Black), AHS herbicides (tertiles of total herbicide use-days). Statistically significant P trends are bolded.
[4]Permethrin for animal use and crop use were combined into one category.
[5]The distribution of life-time days of chlordane exposure was clumped into two exposed groups those who with, ≤8.75 life-time days of exposure and those with >8.75 life-time days of exposure.

Table 4. Pesticide exposure (Lifetime-Days of Exposure) and adjusted risks for NHL Subtypes.

Insecticides	SLL, CLL, MCL		Diffuse Large B-cell		Follicular B-cell		Other B-cell types		Multiple Myeloma		
	RR[3,4] (95% CI)	N[2]	RR[3,4] (95% CI)	N[2]	RR[3,4] (95% CI)	N[2]	RR[3,4] (95% CI)	N[2]	RR[3,4] (95% CI)	N[2]	NHL subtype Homo-geneity Test (p-value)
Carbaryl											
None	1.0 (ref)	42	1.0 (ref)	29	1.0 (ref)	11	1.0 (ref)	14	1.0 (ref)	22	
Low	1.1 (0.6–2.2)	19	0.8 (0.4–1.6)	17	1.6 (0.6–3.9)	10	1.8 (0.7–4.3)	10	0.7 (0.3–1.4)	14	
High	0.6 (0.3–1.3)	15	1.3 (0.6–2.8)	15	2.8 (1.0–7.4)	10	0.4 (0.1–1.5)	3	1.1 (0.7–1.8)	13	
	p trend = 0.16		p trend = 0.33		p trend = 0.06		p trend = 0.63		p trend = 0.98		0.19
Carbofuran											
None	1.0 (ref)	87	1.0 (ref)	78	1.0 (ref)	39	1.0 (ref)	33	1.0 (ref)	56	
Low	1.1 (0.7–1.8)	28	0.9 (0.5–1.7)	13	1.3 (0.7–2.4)	15	0.8 (0.4–1.8)	8	1.9 ((1.1–3.3)	16	
High	1.5 (0.9–2.5)	19	0.8 (0.5–1.3)	13	0.4 (0.1–1.4)	3	0.7 (0.2–2.0)	4	0.9 (0.4–1.6)	12	
	p trend = 0.16		p trend = 0.37		p trend = 0.31		p trend = 0.46		p trend = 0.57		0.52
Chlorpyrifos											
None	1.0 (ref)	84	1.0 (ref)	70	1.0 (ref)	33	1.0 (ref)	31	1 (ref)	58	
Low	1.2 (0.8–1.8)	31	0.9 (0.6–1.5)	22	1.6 (0.9–2.9)	20	1.2 (0.6–2.2)	14	1.0 (0.6–1.8)	17	
High	0.9 (0.6–1.3)	30	1.1 (0.6–1.7)	22	1.0 (0.5–2.1)	11	0.5 (0.2–1.3)	7	0.7 (0.4–1.3)	14	
	p trend = 0.45		p trend = 0.80		p trend = 0.94		p trend = 0.13		p trend = 0.27		0.90
Coumaphos											
None	1.0 (ref)	120	1.0 (ref)	92	1.0 (ref)	48	1.0 (ref)	40	1.0 (ref)	78	
Low	1.1 (0.5–2.2)	8	0.7 (0.3–1.9)	4	2.1 (0.7–5.8)	4	xxx-	4	0.7 (0.2–2.2)	3	
High	1.5 (0.6–3.4)	6	1.6 (0.6–4.5)	4	1.4 (0.5–4.0)	4	xxx-	1	1.2 (0.4–4.0)	3	
	p trend = 0.35		p trend = 0.42		p trend = 0.47		p trend = xxx		p trend = 0.84		0.63
Diazinon											
None	1.0 (ref)	53	1.0 (ref)	40	1.0 (ref)	15	1.0 (ref)	20	1.0 (ref)	41	
Low	1.4 (0.7–2.7)	14	1.5 (0.7–3.2)	9	2.2 (0.9–5.4)	8	xxx	3	0.4 (0.1–1.2)	4	
High	1.9 (0.98–3.6)	12	1.1 (0.5–2.4)	8	3.8 (1.2–11.4)	7	xxx	2	0.5 (0.2–1.7)	3	
	p trend = 0.06		p trend = 0.72		**p trend = 0.02**		p trend = xxx		p trend = 0.35		0.09
DDVP											
None	1.0 (ref)	124	1.0 (ref)	93	1.0 (ref)	48	1.0 (ref)	39	1.0 (ref)	73	
Low	0.8 (0.4–1.9)	6	1.1 (0.4–2.7)	5	1.5 (0.6–3.9)	5	1.1 (0.4–3.7)	3	2.7 (1.2–5.8)	7	
High	0.7 (0.3–1.7)	6	0.9 (0.4–2.3)	5	1.0 (0.3–3.4)	3	0.9 (0.3–3.1)	3	1.0 (0.3–2.7)	4	
	p trend = 0.49		p trend = 0.87		p trend = 0.90		p trend = 0.91		p trend = 0.81		0.96
Fonofos											
None	1.0 (ref)	100	1.0 (ref)	81	1.0 (ref)	45	1.0 (ref)	30	1.0 (ref)	66	
Low	1.2 (0.7–2.0)	20	1.2 (0.7–2.2)	13	1.5 (0.8–3.0)	11	1.4 (0.6–3.1)	8	1.2 (0.6–2.5)	9	
High	1.0 (0.6–1.8)	15	1.2 (0.6–2.3)	11	0.3 (0.1–1.2)	2	1.1 (0.4–2.7)	6	1.4 (0.7–3.0)	9	
	p trend = 0.96		p trend = 0.65		p trend = 0.19		p trend = 0.84		p trend = 0.33		0.35
Malathion											
None	1.0 (ref)	27	1.0 (ref)	20	1.0 (ref)	6	1.0 (ref)	11	1.0 (ref)	17	
Low	0.7 (0.4–1.3)	29	0.96 (0.5–1.8)	23	1.0 (0.4–2.9)	12	1.0 (0.5–2.4)	11	1.0 (0.5–2.1)	18	
High	1.0 (0.6–1.8)	22	1.0 (0.5–2.0)	20	1.6 (0.6–4.4)	11	0.3 (0.1–0.8)	6	1.0 (0.5–2.0)	17	
Ever/Never	1.0 (0.7–1.4)		0.9 (0.6–1.4)		1.3 (0.7–2.4)		0.6 (0.3–1.0)		0.9 (0.6–1.5)		
	p trend = 0.65		p trend = 0.88		p trend = 0.25		p trend = 0.17		p trend = 0.86		0.33
Permethrin											

Table 4. Cont.

Insecticides											
	SLL, CLL, MCL		**Diffuse Large B-cell**		**Follicular B-cell**		**Other B-cell types**		**Multiple Myeloma**		
	RR[3,4] (95% CI)	**N[2]**	**RR[3,4] (95% CI)**	**N[2]**	**RR[3,4] (95% CI)**	**N[2]**	**RR[3,4] (95% CI)**	**N[2]**	**RR[3,4] (95% CI)**	**N[2]**	**NHL subtype Homo-geneity Test (p-value)**
None	1.0 (ref)	108	1.0 (ref)	89	1.0 (ref)	41	1.0 (ref)	38	1.0 (ref)	64	
Low	1.1 (0.6–2.0)	15	0.6 (0.3–1.2)	8	1.3 (0.6–2.7)	8	0.9 (0.3–2.7)	5	1.4 (0.8–2.7)	13	
High	0.8 (0.5–1.5)	15	1.0 (0.5–2.1)	8	1.0 (0.5–2.4)	8	0.5 (0.2–1.7)	4	3.1 (1.5–6.2)	12	
	p trend = 0.53		p trend = 0.99		p trend = 0.88		p trend = 0.28		**p trend = 0.002**		0.10
Phorate											
None	1.0 (ref)	48	1.0 (ref)	37	1.0 (ref)	20	1.0 (ref)	16	1.0 (ref)	36	
Low	1.0 (0.6–1.9)	14	1.4 (0.7–2.7)	15	1.1 (0.4–3.0)	5	0.9 (0.3–2.2)	6	0.7 (0.3–1.8)	6	
High	0.8 (0.4–1.6)	11	0.7 (0.3–2.1)	4	0.8 (0.3–2.2)	5	1.1 (0.4–3.5)	4	0.8 (0.3–2.4)	4	
	p trend = 0.51		p trend = 0.80		p trend = 0.67		p trend = 0.91		p trend = 0.73		0.77
Terbufos											
None	1.0 (ref)	72	1.0 (ref)	63	1.0 (ref)	31	1.0 (ref)	19	1.0 (ref)	59	
Low	1.3 (0.8–2.0)	32	1.2 (0.8–1.9)	29	1.6 (0.9–3.1)	15	1.8 (0.9–3.6)	17	1.1 (0.6–1.9)	12	
High	1.6 (1.0–2.5)	31	1.0 (0.5–2.0)	12	0.8 (0.4–1.7)	10	1.6 (0.7–3.9)	8	1.3 (0.7–2.7)	5	
	p trend = 0.05		p trend = 0.90		p trend = 0.48		p trend = 0.29		p trend = 0.42		0.63
Chlorinated Insecticides											
Aldrin											
None	1.0 (ref)	53	1.0 (ref)	46	1.0 (ref)	22	1.0 (ref)	20	1.0 (ref)	34	
Low	1.0 (0.5–2.0)	11	xxx	2	1.2 (0.4–3.8)	4	0.4 (0.1–1.5)	3	2.1 (0.9–4.7)	8	
High	1.0 (0.5–2.0)	10	xxx	3	0.8 (0.3–2.5)	4	1.1 (0.3–3.9)	3	1.2 (0.5–3.2)	6	
	p trend = 0.70		p trend = xxx		p trend = 0.21		p trend = 0.67		p trend = 0.40		0.98
Chlordane											
None	1.0 (ref)	48	1.0 (ref)	42	1.0 (ref)	20	1.0 (ref)	21	1.0 (ref)	32	
Low	1.8 (1.0–3.1)	16	1.0 (0.5–2.2)	8	1.7 (0.7–4.3)	6	xxx	2	1.7 (0.9–3.3)	13	
High	1.5 (0.7–3.3)	8	1.4 (0.6–3.3)	7	1.3 (0.4–4.6)	3	xxx	2	0.7 (0.2–2.2)	3	
	p trend = 0.34		p trend = 0.69		p trend = 0.70		p trend = xxx		p trend = 0.57		0.85
DDT											
None	1.0 (ref)	42	1.0 (ref)	34	1.0 (ref)	17	1.0 (ref)	16	1.0 (ref)	28	
Low	1.0 (0.5–1.8)	16	1.6 (0.4–3.1)	2	3.3 (1.4–8.1)	9	0.4 (0.3–2.5))	5	1.2 (0.6–2.6)	10	
High	2.6 (1.3–4.8)	15	1.4 (0.6–3.5)	3	1.1 (0.3–3.6)	4	2.1 (0.7–6.5)	5	0.8 (0.4–1.8)	9	
	p trend = 0.04		P trend = 0.17		p trend = 0.80		p trend = 0.64		p trend = 0.37		0.44
Heptachlor											
None	1.0 (ref)	58	1.0 (ref)	47	1.0 (ref)	24	1.0 (ref)	21	1.0 (ref)	40	
Low	1.1 (0.5–2.3)	9	xxx	3	xxx	2	xxx	3	1.3 (0.4–3.8)	4	
High	1.4 (0.7–3.0)	9	xxx	1	xxx	1	xxx	2	1.2 (0.4–3.6)	4	
	p trend = 0.16		p trend = xxx		p trend = xxx		p trend = xxx		p trend = 0.91		0.68
Lindane											
None	1.0 (ref)	57	1.0 (ref)	49	1.0 (ref)	16	1.0 (ref)	21	1.0 (ref)	43	
Low	1.2 (0.6–2.5)	10	0.6 (0.2–1.7)	4	4.9 (1.9–12.6)	6	xxx	2	xxx	3	
High	2.6 (1.2–5.6)	9	2.0 (0.6–6.5)	3	3.6 (1.4–9.5)	6	xxx	1	xxx	2	
	p trend = 0.13		p trend = 0.96		**p trend = 0.04**		p trend = xxx		p trend = xxx		0.54
Toxaphene											
None	1.0 (ref)	68	1.0 (ref)	47	1 (ref)	23	1.0 (ref)	22	1.0 (ref)	40	

Table 4. Cont.

	SLL, CLL, MCL		Diffuse Large B-cell		Follicular B-cell		Other B-cell types		Multiple Myeloma		
Insecticides	**RR[3,4] (95% CI)**	**N[2]**	**RR[3,4] (95% CI)**	**N[2]**	**RR[3,4] (95% CI)**	**N[2]**	**RR[3,4] (95% CI)**	**N[2]**	**RR[3,4] (95% CI)**	**N[2]**	**NHL subtype Homo-geneity Test (p-value)**
Low	0.9 (0.4–2.3)	5	1.3 (0.5–3.3)	5	xxx	2	xxx	3	0.7 (0.2–2.0)	4	
High	0.4 (0.1–1.6)	2	0.9 (0.3–3.0)	3	xxx	2	xxx	2	0.7 (0.2–2.9)	2	
	p trend = 0.08		p trend = 0.77		p trend = xxx		p trend = xxx		p trend = 0.64		0.34
Fungicides											
Captan											
None	1.0 (ref)	118	1.0 (ref)	91	1.0 (ref)	52	1.0 (ref)	39	1.0 (ref)	76	
Low	0.9 (0.4–1.9)	7	1.1 (0.5–2.4)	7	xxx	2	xxx	3	1.4 (0.5–3.4)	5	
High	1.1 (0.5–2.6)	7	0.7 (0.1–3.1)	4	xxx	1	xxx	2	1.2 (0.5–2.9)	5	
	p trend = 0.78		p trend = 0.58		p trend = xxx		p trend = xxx		p trend = 0.75		0.92
Chlorothalonil											
None	1.0 (ref)	135	1.0 (ref)	107	1.0 (ref)	60	1.0 (ref)	50	1.0 (ref)	84	
Low	0.9 (0.4–2.3)	5	1.1 (0.4–3.1)	4	xxx	3	−xxx	1	1.1 (0.4–2.8)	5	
High	1.1 (0.4–3.3)	4	0.3 (0.1–1.2)	2	xxx	2	−xxx	1	0.7 (0.6–2.3)	3	
	p trend = 0.83		p trend = 0.09		p trend = xxx		p trend = xxx		p trend = 0.56		0.76
Metalaxyl											
None	1.0 (ref)	60	1.0 (ref)	45	1.0 (ref)	25	1.0 (ref)	23	1.0 (ref)	39	
Low	2.8 (1.4–5.8)	9	1.1 (0.4–2.6)	7	xxx	3	−xxx	2	0.4 (0.1–1.1)	4	
High	1.1 (0.4–2.8)	6	1.0 (0.4–2.7)	5	xxx	2	−xxx	1	1.1 (0.4–3.2)	4	
	p trend = 0.99		p trend = 0.97		p trend = xxx		p trend = xxx		p trend = 0.87		0.92
Maneb/ Mancozeb											
None	1.0 (ref)	69	1.0 (ref)	49	1.0 (ref)	25	1.0 (ref)	26	1.0 (ref)	41	
Low	2.1 (0.7–6.0)	4	4.0 (1.4–11.6)	4	xxx	2	−xxx	0	1.0 (0.4–2.5)	5	
High	1.2 (0.3–4.0)	3	0.9 (0.3–3.1)	3	−xxx	1	−xxx	0	2.2 (0.5–9.5)	2	
	p trend = 0.84		p trend = 0.74		p trend = xxx		p trend = xxx		p trend = 0.28		0.82
Fumigant											
Methyl Bromide											
None	1.0 (ref)	126	1.0 (ref)	86	1.0 (ref)	58	1.0 (ref)	44	1.0 (ref)	76	
Low	1.1 (0.5–2.2)	9	4.0 (2.2–7.4)	15	1.4 (0.5–4.2)	4	3.6 (1.3–9.8)	5	1.0 (0.5–2.1)	8	
High	0.8 (0.4–1.8)	8	1.0 (0.5–2.1)	11	0.3 (0.1–1.1)	3	1.3 (0.3–5.0)	3	0.8 (0.4–1.8)	8	
	p trend = 0.58		p trend = 0.67		p trend = 0.08		p trend = 0.56		p trend = 0.63		0.59

[1]During the period from enrollment (1993–1997) to December 31, 2010 in NC and December 31, 2011 in Iowa.
[2]Numbers of cases in columns do not sum to total number of NHL cases (n = 523) due to missing data. Ever/never use of all 26 pesticides (table 3) do not always match with exposure-response data in table 4 because of missing data to calculate lifetime-days of use.
[3]Adjusted for age (<45, 45–49, 50–54, 55–59, 60–64, 65–69, ≥70), State (NC vs. IA), Race (White vs. Black), AHS herbicides (in tertiles of total herbicide use-days). Significant RR and 95% confidence limits are bolded.
[4]RR was not calculated if the number of exposed cases for any NHL subtype was <6 and these cells are marked XXX. Four pesticides included in Table 2 (i.e., aldicarb, benomyl, dieldrin and parathion) were not included in Table 4 because no NHL subtype included≥6 cases of a specific cell types with lifetime-days of exposure.

risk estimates (i.e., narrower confidence intervals) when we included phase 2 imputed data (n = 54,306) (data not shown). Lagging exposures by five years did not meaningfully change the association between lindane or DDT and total NHL (data not shown). The significant exposure-response trends linking use of a particular pesticide to NHL and certain NHL subtypes did not always correspond to a significant excess risk among those who ever used the same pesticide. For chemicals for which the detailed information was only asked about in the take-home questionnaire, we evaluated potential differences between the ever/never analyses based on the enrolment questionnaire and data from the same sub-set of participants who completed the exposure-response in the take-home questionnaire and found no meaningful differences in the results. We also evaluated the impact of using an updated definition of NHL; when using the original ICD-O-3 definition of NHL[19], lifetime-days of lindane use remained significantly associated with NHL risk (RR = 1.0 (ref), 1.3 (0.7–2.6), 1.2 (0.6–2.8), 2.7 (1.3–5.4), p trend = 0.006). The trend between total NHL and lifetime-days of DDT, however, was less clear and not statistically significant (RR = 1.0 (ref) 1.3 (0.9–1.8), 1.1 (0.5–2.1), 1.4 (0.8–2.6), p trend = 0.32) [Table S3 in File S1]. Carbaryl and diazinon showed non-significant trends with the older definition of NHL, but not with the newer definition used here.

Discussion

A significant exposure–response trend for total NHL was observed with increasing lifetime-days of use for two organochlorine insecticides, lindane and DDT, although RRs from ever/never comparisons were not elevated. On the other hand, terbufos use showed a significant excess risk with total NHL in ever vs. never exposed analysis, but displayed no clear exposure-response trend. Several pesticides showed significant exposure-response trends with specific NHL subtypes however, when polytomous models were used to test the difference in parametric estimates of trend among the five NHL subtypes, there was no evidence of heterogeneity in the sub-types for specific chemicals. The subtype relationships that looked particularly interesting were DDT and terbufos with the SLL/CLL/MCL subtype, lindane and diazinon with the follicular subtype, and permethrin with MM. These pesticide-NHL links should be evaluated in future studies.

Lindane (gamma-hexachlorocyclohexane) is a chlorinated hydrocarbon insecticide. Production of lindane was terminated in the United States in 1976, but imported lindane was used to treat scabies and lice infestation and for agricultural seed treatment [21] until its registration was cancelled in 2009 [22], the same year production was banned worldwide [23]. In our study, 3,410 people reporting ever using lindane (6%) prior to enrollment, 433 reported use at the phase 2 questionnaire (1%), indicating that use had dropped substantially. Oral administration of lindane has increased the incidence of liver tumors in mice and less clearly, thyroid tumors in rats [24]. Lindane produces free radicals and oxidative stress (reactive oxygen species [ROS]) [25] and has been linked with chromosomal aberrations in human peripheral lymphocytes in vitro [26].

Lindane has been linked with NHL in previous epidemiologic studies. A significant association between lindane use and NHL was observed in a pooled analysis of three population-based case-control studies conducted in the Midwestern US, with stronger relative risks observed for greater duration and intensity of use [27]. NHL was also associated with lindane use in a Canadian case-control study [28]. Lindane was significantly associated with NHL risk in an earlier report from the AHS [29]. We are not aware of any previous study that assessed the association between a NHL subtype and lindane use. The exposure-response pattern with total NHL and the follicular lymphoma subtype indicates a need for further evaluation of lindane and NHL.

DDT is an organochlorine insecticide that was used with great success to control malaria and typhus during and after World War II [29] and was widely used for crop and livestock pest control in the United States from the mid-1940s to the 1960s [30]. Its registration for crop use was cancelled in the US in 1972 [30] and banned worldwide for agricultural use in 2009, but continues to be used for disease vector control in some parts of the world [23]. In our study, 12,471 participants (23%) reported ever using DDT prior to enrollment; 12%, 8.7% and 2.3% responding to the take-home questionnaire reported their first use occurred prior to the 1960s, during the 1960s, and during the 1970s, respectively. The National Toxicology Program classifies DDT as "reasonably anticipated to be a human carcinogen" [31] and IARC classifies DDT as a "possible human carcinogen (2B)" [12], both classifications were based on experimental studies in which excess liver tumors were observed in two rodent species. Epidemiology data on the carcinogenic risk of DDT is inconsistent. NHL was not associated with use of DDT in a pooled analysis of three case-control studies in the U.S. where information on exposure was obtained from farmers by questionnaire [32]. There also was no association between the use of DDT and NHL in our study when we used an earlier definition of NHL [18], suggesting some of the inconsistency may be due to disease definition. In the large Epilymph study, no meaningful links between DDT and the risk of NHL, or diffuse large B cell lymphoma were observed, and only limited support was found for a link to CLL [33], although a case-control study of farmers in Italy suggested increased risk of NHL and CLL with DDT exposure [34]. NHL was not associated with serum levels of DDT in a prospective cohort study from the U.S. [35], but NHL was associated with the DDT-metabolite p, p'-DDE, as well as chlordane and heptachlor-related compounds (oxychlordane, heptachlor epoxide) and dieldrin, in a study with exposure measured in human adipose tissue samples [36]. In a Danish cohort, a higher risk of NHL was associated with higher prediagnostic adipose levels of DDT, cis-nonachlor, and oxychlordane [37]. In a Canadian study, analytes from six insecticides/insecticide metabolites (beta-hexachlorocyclohexane, p, p'-dichloro-DDE, hexachlorobenzene (HCB), mirex, oxychlordane and transnonachlor) were linked with a significant increased risk with NHL [38]. However, in an analysis of plasma samples from a case-control study in France, Germany and Spain, the risk of NHL did not increase with plasma levels of hexachlorobenzene, beta-hexachlorobenzene or DDE [39]. In this analysis, NHL was significantly associated with reported use of DDT, but not with the other organochlorine insecticides studied (i.e., aldrin, chlordane, dieldrin, heptachlor, toxaphene). Our findings add further support for an association between DDT and total NHL and our results on SLL/CLL/MCL are novel and should be further explored.

Permethrin is a broad-spectrum synthetic pyrethroid pesticide widely used in agriculture and in home and garden use as an insecticide and acaricide, as an insect repellant, and as a treatment to eradicate parasites such as head lice or mites responsible for scabies [40]. This synthetic pyrethroid was first registered for use in the United States in 1979 [40]. The U.S. Environmental Protection Agency classified permethrin as "likely to be carcinogenic to humans" largely based on the observed increase incidence of benign lung tumors in female mice, liver tumors in rats and liver tumors in male and female mice [41]. Permethrin was not associated with NHL overall in our study, nor in pooled case-control studies of NHL from the U.S (the NHL definition in use at the time of the study did not include MM) [42]. In our analysis, however, the risk of MM increased significantly with lifetime-days of exposure to permethrin, as had been noted in an earlier analysis of AHS data [43]. We are unaware of other studies that have found this association.

Terbufos is an organophosphate insecticide and nematicide first registered in 1974 [44]. The EPA classifies terbufos as Group E, i.e., "Evidence of Non-Carcinogenicity for Humans" [44]. We found some evidence for an association between terbufos use and NHL, particularly for the SLL/CLL/MCL subtype. NHL was not associated with terbufos in the pooled case-control studies from the

U.S. [42] but there was a non-significant association between terbufos and small cell lymphocytic lymphoma [10].

Diazinon is an organophosphate insecticide registered for a variety of uses on plants and animals in agriculture [45]. It was commonly used in household insecticide products until the EPA phased out all residential product registrations for diazinon in December 2004 [45.46]. In an earlier evaluation of diazinon in the AHS, a significant exposure-response association was observed for leukemia risk with lifetime exposure-days [47]. While there was no link between diazinon and NHL overall in this analysis, there was a statistically significant exposure-response association between diazinon and the follicular lymphoma subtype and an association with the SLL/CLL/MCL subtype that was not statistically significant. Diazinon was previously associated with NHL in pooled case-control studies from the U.S. and particularly with SLL [10].

Several other insecticides, fungicides and fumigants cited in recent reviews of the pesticide-cancer literature suggested etiological associations with total NHL [8,9], these include: oxychlordane, trans-nonachlor, and cis-nonachlor which are metabolites of chlordane; and dieldrin and toxaphene among NHL cases with t(14,18) translocations. We did not find a significant association between chlordane and total NHL nor with any NHL subtype, but we did not have information about chlordane metabolites to make a more direct comparison. Similarly we did not observe a significant association between dieldrin nor toxaphene and total NHL nor with any NHL subtypes. Mirex (1,3-cyclopentadiene), an insecticide, and hexachlorobenzene, a fungicide, were also associated with NHL risk [8,9] but we did not examine these compounds in the AHS.

This study has a number of strengths. It is a large population of farmers and commercial pesticide applicators who can provide reliable information regarding their pesticide use history [48]. Information on pesticide use and application practices was obtained prior to onset of cancer. An algorithm that incorporated several exposure determinants which predicted urinary pesticide levels was used to develop an intensity-weighted exposure metric in our study [20]. Exposure was ascertained prior to diagnosis of disease, which should eliminate the possibility of case-response bias [14]. Because of the detailed information available on pesticide use, we were able to assess the impact for the use of multiple pesticides. For example, we evaluated total pesticide use-days, and specific pesticides found to be associated with NHL or its subtypes in the AHS. We found no meaningful change in the associations with DDT, lindane, permethrin, diazinon and terbufos from such adjustments. Information on many potential NHL risk factors was available and could be controlled in the analysis.

Most epidemiological investigations of NHL prior to 2007 [17] did not include CLL and MM as part of the definition. These two subtypes made up 37% (193/523) of the NHL cases in this analysis. This is a strength of our study in that the definition of NHL used here is based on the most recent classification system [16,17] and will be relevant for comparisons with future studies. On the other hand, the inclusion of MM and CLL in the recent definition of NHL makes comparisons of our findings with earlier literature challenging, because the NHL subtypes may have different etiologies. For example, DDT was not significantly associated with NHL using the older definition, but was significantly associated with the NHL using the most recent definition of NHL because of its association with the SLL/CLL/MCL subtype (Table S1 in File S1). On the other hand, carbaryl and diazinon were associated with the old definition of NHL (although non-significantly) but not with the new definition. Lindane, however, was associated with both definitions of NHL. Lindane was significantly associated with the follicular lymphoma subtype and this subtype was included in the older and newer definition of NHL. No other pesticides were significantly associated with NHL under the old definition (Table S3 in File S1).

Although this is a large prospective study, limitations should be acknowledged. A small number of cases exposed to some specific pesticides could lead to false positive or negative findings. We also had reduced statistical power to evaluate some pesticides for total days of use and intensity-weighted days of use because some participants did not complete the phase one take-home questionnaire and the tests of homogeneity between specific pesticides and specific NHL subtypes were underpowered. Some chance associations could occur because of multiple testing, i.e., a number of pesticides, several NHL subtypes, and more than one exposure metric. Despite the generally high quality of the information on pesticide use provided by AHS participants [48,50], misclassification of pesticide exposures can occur and can have a sizeable impact on estimates of relative risk, which in a prospective cohort design would tend to produce false negative results [49].

Conclusion

Our results showed pesticides from different chemical and functional classes were associated with an excess risk of NHL and NHL subtypes, but not all members of any single class of pesticides were associated with an elevated risk of NHL or NHL subtypes, nor were all chemicals of a class included on our questionnaire. Significant pesticide associations were between total NHL and reported use of lindane and DDT. Links between DDT and terbufos and SLL/CLL/MCL, lindane and diazinon and follicular lymphoma, and permethrin and MM, although based on relatively small numbers of exposed cases, deserve further evaluation. The epidemiologic literature on NHL and these pesticides is inconsistent and although the findings from this large, prospective cohort add important information, additional studies that focus on NHL and its subtypes and specific pesticides are needed. The findings from this large, prospective cohort add important new information regarding the involvement of pesticides in the development of NHL. It provides additional information regarding specific pesticides and NHL overall and some new leads regarding possible links with NHL subtypes that deserve evaluation in future studies.

Supporting Information

File S1 This file contains Table S1, Table S2, and Table S3. Table S1, Frequency of NHL in Agricultural Health Study applicators using New (Interlymph hierarchical classification of lymphoid neoplasms) and Older Definitions (ICD-O-3). Table S2, Pesticides included in the Agricultural Health Study questionnaires by Chemical/Functional Class. Table S3, Pesticide exposure (lifetime-days) and adjusted risks of total NHL incidence (Older definition [ICD-O-3]).

Acknowledgments

Disclaimer: The findings and conclusions in this report are those of the author(s) and do not necessarily represent the views of the National Institute for Occupational Safety and Health. The United States Environmental Protection Agency through its Office of Research and Development partially funded and collaborated in the research described here under Contracts 68-D99–011 and 68-D99–012, and through Interagency Agreement DW-75–93912801–0. It has been subjected to Agency review and approved for publication.

This work was supported by the Intramural Research Program of the NIH, National Cancer Institute, Division of Cancer Epidemiology and

Genetics (Z01CP010119) and the National Institutes of Environmental Health Science (Z01ES049030). The authors have no conflicts of interest in connection with this manuscript.

Ms. Marsha Dunn and Ms. Kate Torres, (employed by Westat, Inc. Rockville, Maryland) are gratefully acknowledged for study coordination. The ongoing participation of the Agricultural Health Study participants is indispensable and sincerely appreciated.

Author Contributions

Conceived and designed the experiments: MCA DPS AB. Performed the experiments: MCA CFL KT CJH. Analyzed the data: MCA JNH CFL CJH KHB JB DWB KT DPS JAH SK GA JHL AB LEB. Contributed reagents/materials/analysis tools: MCA JB DWB CFL. Wrote the paper: MCA LEBF JNH CFL CJH KT AB DWB JHL. Designed the software: JB DWB.

References

1. Milham S (1971) Leukemia and multiple myeloma in farmers. Am J Epidemiol 94: 507–510.
2. Cantor KP (1982) Farming and mortality from non-Hodgkin's lymphoma: a case-control study. Int J Cancer 29: 239–247.
3. Blair A, Malker H, Cantor KP, Burmeister L, Wiklund K (1985) Cancer among farmers: A review. Scand J Work Environ Health 11: 397–407.
4. Pearce NE, Smith AH, Fischer DO (1985) Malignant lymphoma and multiple myeloma linked with agricultural occupation in a New Zealand cancer registration-base-study. Am J Epidemiol 121: 235–237.
5. Baris D, Silverman DT, Brown LM, Swanson GM, Hayes RB, et al. (2004) Occupation, pesticide exposure and risk of multiple myeloma. Scand J Work Environ Health 30(3): 215–222.
6. Beane Freeman LE, DeRoos AJ, Koutros S, Blair A, Ward MH, et al. (2012) Poultry and livestock exposure and cancer risk among farmers in the agricultural health study. Cancer Causes Control 23: 663–670.
7. Cordes DH, Rea DF (1991) Farming: A Hazardous Occupation. In: Health Hazards of Farming. Occupational Medicine: State of the Art Reviews. Vol6(3). Hanley & Belfus, Inc., Philadelphia, PA.
8. Alavanja M, Bonner M (2012) Occupational pesticide exposure and cancer risk. A Review. J. Toxicol Environ Health B Critic Review. 1594: 238–263.
9. Alavanja MCR, Ross MK, Bonner MR (2013) Increased cancer burden among pesticide applicators and others due to pesticide exposure. CA, Cancer J Clin; 63(2): 120–142.
10. Waddell BL, Zahm SH, Baris D, Weisenburger DD, Holmes F, et al. (2001). Agricultural use of organophosphate pesticides and the risk of non-Hodgkin's lymphoma among male farmers (United States). Cancer Causes and Control 12: 509–517.
11. Merhi M, Raynal H, Cahuzac E, Vinson F, Cravedi JP, et al. (2007) Occupational exposure to pesticides and risk of hematopoietic cancers: meta-analysis of case-control studies. Cancer Causes Control. 18: 1209–1226.
12. IARC (1991) International Agency for Research on Cancer (IARC). Occupational Exposures in Insecticide applications and some pesticides. Lyon, France: IARC, 1991. Monographs on the Evaluation of Carcinogenic Risk to Humans, volume 53.
13. Morton LM, Slager SL, Cerhan JR, Wang SS, Vajdic CM, et al. (2014) Etiologic Heterogeneity Among NHL Subtypes: The InterLymph NHL Subtypes Project. J Natl Cancer Institute 48: 130–144.
14. Alavanja MCR, Sandler DP, McMaster SB, Zahm SH, McDonnell CJ, et al. (1996) The Agricultural Health Study. Environ Health Perspect 104: 362–369.
15. Alavanja MC, Samanic C, Dosemeci M, Lubin J, Tarone R, et al. (2003) Use of agricultural pesticides and prostate cancer risk in the Agricultural Health Study Cohort. Am J Epidemiol 157(9): 800–814.
16. SEER Program, National Cancer Institute. Available: http://seer.cancer.gov/lymphomarecode Accessed September 15, 2013.
17. Morton LM, Turner JJ, Cerhan JR, Linet MS, Treseler PA, et al. (2007) Proposed classifciation of lymphoid neoplasms for epidemiologic research from the Pathology Working Group of the International Lymphoma Epidemiology Consortium (InterLymph). Blood 110(2): 695–708.
18. Percy C, Fritz A, Ries L (2001) Conversion of neoplasms by topography and morphology from the International Classification of Disease for Oncology, second edition, to International Classification of Diseases for Oncology, 3rd ed. Cancer Statistics Branch, DCCPS, SEER Program, National Cancer Institute; 2001.
19. Heltshe SL, Lubin JH, Koutros S, Coble JB, Ji B-T, et al. (2012) Using multiple imputation to assign pesticide use for non-respondents in the follow-up questionnaire in the Agricultural Health Study J. Exp Sci Environ Epidemiol 22(4): 409–416.
20. Coble J, Thomas KW, Hines CJ, Hoppin JA, Dosemeci M, et al. (2011) An updated algorithm for estimation of pesticide exposure intensity in the Agricultural Health Study. Int J Environ Res Public Health. 8(12): 4608–4622.
21. ATSDR (2005) Agency for Toxic Substances and Disease Registry. Toxicological profile for Alpha-, Beta-, Gamma- and Delta- Hexachlorcyclohexane, August, 2005. Available: http://www.atsdr.cdc.gov/Toxprofiles/tp43.pdf. Accessed 2013 Sep 15.
22. EPA (2006a) US Environmental Protection Agency (2006). Lindane; Cancellation Order December 13, 2006. Federal Register/Vol 71, number 239, page 74905.
23. Stockholm Convention Report (2009) Report of the conference of the Parties of the Stockholm Convention on Persistent Organic Pollutants on the work of its fourth meeting. Convention on Persistent Organic Pollutants. Fourth Meeting, Geneva, 4–8 May 2009. Available: http;//chm.pops.int/Portals/0/Repository/COP4/UNEP-POPS-COP.4–38.English.pdf.
24. IARC (1987) International Agency for Research on Cancer (IARC). Overall evaluation of carcinogenicity: an updating of IARC monographs volume 1 to 42. Lyon, France: IARC, 1987. Monographs on the Evaluation of Carcinogenic Risk to Humans, Supplement 7.
25. Piskac-Collier AL, Smith MA (2009) Lindane-induced generation of reactive oxygen species and depletion of glutathione do not result in necrosis in renal distal tube cells. J Toxicol and Environ Health, Part A. 72: 1160–1167.
26. Rupa DS, Reddy PP, Reddi OS (1989). Genotoxic effect of benzene hexachloride in cultured human lymphocytes. Hum Genet 83: 271–273.
27. Blair A, Cantor KP, Zahm SH (1998) Non-Hodgkin's lymphoma and agricultural use of the insecticide lindane. Am J Ind Med 33: 82–87.
28. McDuffie HH, Pahwa P, Mclaughlin JR, Spinelli JJ, Fincham S, et al. (2001) Non-Hodgkin's lymphoma and specific pesticide exposures in Men: Cross-Canada Study of Pesticides and Health. Cancer, Epidemiology, Biomarkers & Prevention 10: 1155–1163.
29. Purdue MP, Hoppin JA, Blair A, Dosemeci M, Alavanja MCR (2007) Occupational exposure to organochlorine insecticides and cancer incidence in the Agricultural Health Study. Int J Cancer. 120(3): 642–649.
30. EPA (2012) US Environmental Protection Agency 2012. DDT-A Brief History and Status. Available: http://www.epa.gov/pesticides/factsheets/chemicals/ddt-brief-history-status.htm. Accessed 2013 Sep 15.
31. NTP (2011) National Toxicology Program, Report on Carcinogen- Twelfth Edition. Available: http://ntp.niehs.nih.gov/go/roc12. Accessed 2013 Sep 15.
32. Baris D, Zahm SH, Cantor KP, Blair A (1998). Agricultural use of DDT and the risk of non-Hodgkin's lymphoma: pooled analysis of three case-control studies in the United States, Occup Environ Med 55: 522–527.
33. Cocco P, Satta G, Dubois S, Pilli C, Pillieri M, et al. (2013) Lymphoma risk and occupational exposure to pesticides: results of the Epilymph study. Occupational Environ Med 70(2): 91–98.
34. Nanni O, Amadori D, Lugaresi C, Falcini F, Scarpi E, et al. (1996) Chronic lymphocytic leukaemia and non-Hodgkin's lymphomas by histological type in farming-animal breeding workers: a population case-control study based on a priori exposure matrics Occup Environ Med 53(10): 652–657.
35. Rothman N, Cantor KP, Blair A, Bush D, Brock JW, et al. (1997) A nested case-control study of non-Hodgkin lymphoma and serum organochlorine residues. The Lancet 350: 240–244.
36. Quintana PJE, Delfino RJ, Korrick S, Ziogas A, Kutz FW, et al. (2004) Adipose tissue levels of organochlorine pesticides and chlorinated biphenyls and the risk of non-Hodgkin's lymphoma. Environ Health Perspect 112: 854–861.
37. Brauner EV, Sorensen MA, Gaudreau E, LeBlanc A, Erikson KT, et al. (2012) A prospective study of organochlorines in adipose tissue and risk of non-Hodgkin lymphoma. Environ Health Perspect. 120(1): 105–111.
38. Spinelli JJ, Ng CH, Weber JP, Connors JM, Gascoyne RD, et al. (2007) Organochlorines and risk of non-Hodgkin lymphoma. Int J Cancer. 121(12): 2767–2775.
39. Cocco P, Brennan P, Ibba A, de Sanjose Llongueras S (2008) Plasma polychlorobiphenyl and organochlorine pesticide level and risk of major lymphoma subtypes. Occup Environ Med 65: 132–140.
40. EPA 2006(b). U.S. Environmental Protection Agency. 2006. Re-registration Eligibility Decision for Permethrin: Available: http://www.epa.gov/oppsrrd1/REDs/permethrin_red.pdf. Accessed 2013 Sep 15.
41. NPIC (2012). National Pesticide Information Center. Chemicals Evaluated for Carcinogenic Potential. Office of Pesticide Programs. U.S. Environmental Protection Agency. November 2012. Available: http://npic.orst.edu/chemicals_evaluated.pdf. Accessed 2013 Sep 15.
42. De Roos AJ, Zahm SH, Cantor KP, Weisenburger DD, Holmes FF, t al. (2003) Integrative assessment of multiple pesticides as risk factors for non-Hodgkin's lymphoma among men Occup Environ Med 60: E11.
43. Rusiecki JA, Patel R, Koutros S, Beane Freeman LE, Landgren O, et al. (2009) Cancer incidence among pesticide applicators exposed to permethrin in the Agricultural Health Study. Environ Health Perspect 117: 582–586.
44. EPA 2006(b). U.S. Environmental Protection Agency. 2006. Re-registration Eligibility Decision for Terbufos: Available: http://www.epa.gov/pesticides/reregistration/REDs/terbufos_red.pdf Accessed 2013 Nov 18.
45. Environmental Protection Agency, 2004. Interim registration eligibility decision: diazinon. Available: http://www.epa.gov/pesticides/reregistration/REDs/diazinon_red.pdf. Accessed 2013 Nov 18.
46. Donalson D, Kieley T, Grube A (2002) Pesticide industry sales and usage: 1998 and 1999 market estimates. Washington, DC.: US Environmental Protection Agency, 2002 (EPA-733-R-02-001).

47. Beane Freeman LE, Bonner MR, Blair A, Hoppin JA, Sandler DP, et al. (2005). Cancer incidence among male pesticide applicators in the Agricultural Health Study cohort exposed to diazinon. Am J Epidemiol. 162: 1070–1079.
48. Blair A. Tarone R, Sandler D, Lynch CF, Rowland A, et al. (2002) Reliability of reporting on life-style and agricultural factors by a sample of participants in the Agricultural Health Study from Iowa. Epidemiology 13(1): 94–99.
49. Blair A, Thomas HT, Coble J, Sandler DP, Hines CJ, et al. (2011). Impact of pesticide exposure on misclassification on estimates of relative risks in the Agricultural Health Study. Occup Environ Med 68: 537–541.
50. Thomas KW, Dosemeci M, Coble JB, Hoppin JA, Sheldon LS, et al. (2010) Assessment of a pesticide exposure intensity algorithm in the Agricultural Health Study. J Expo Sci Environ Epidemiol 20(6): 559–569.

14

The Parasitic Eyeworm *Oxyspirura petrowi* as a Possible Cause of Decline in the Threatened Lesser Prairie-Chicken (*Tympanuchus pallidicinctus*)

Nicholas R. Dunham, Steven T. Peper, Catherine E. Baxter, Ronald J. Kendall*

The Wildlife Toxicology Laboratory, The Institute of Environmental and Human Health, Texas Tech University, Lubbock, Texas, United States of America

Abstract

Lesser prairie-chickens (*Tympanuchus pallidicinctus*) have been declining range wide since the early 1900's despite efforts to establish conservation and improve their habitat. In early 2014, the lesser prairie-chicken was listed as a threatened species under the U.S Endangered Species Act and the need to find out why they are declining is more important than ever. Nine hunter shot lesser prairie-chickens were donated and sampled for the presence or absence of the eyeworm *Oxyspirura petrowi*, a known parasite that can cause damage to the eye of its host, and common environmental contaminants. Eyeworm infection was found in 7 of 9 birds (78% infection rate) with an infection range between 0–16 *O. petrowi* per bird. Breast, liver, and fat tissue samples from the lesser prairie-chickens were analyzed for the frequency of 20 organochlorine pesticides. Femurs and livers were also tested on these birds for metal contaminants. Pesticides were found in several samples above the detection limits but were still in the low ng/g range. Notable was the ubiquitous presence of endrin aldehyde across all tissues. One femur showed 5.66 μg/g of lead (Pb) but this is still relatively low. No liver samples had elevated mercury (Hg) above detection limits. The presence of these organochlorines is consistent with the historic use of pesticides in this region. With pesticide and metals found in such low levels and parasitic nematode infections at rather high levels, it is recommended that these parasites be further evaluated as a contributing factor to the decline of the lesser prairie-chicken.

Editor: Gregorio Moreno-Rueda, Universidad de Granada, Spain

Funding: The authors have no support or funding to report.

Competing Interests: The authors have declared that no competing interests exist.

* Email: ron.kendall@ttu.edu

Introduction

Historically, lesser prairie-chickens (*Tympanuchus pallidicinctus*) have thrived throughout much of the southern United States but since the early 1900's their population and range have been diminished by over 90% [1]. Arguably much of their decline has been blamed on anthropogenic factors including habitat loss due to agriculture or habitat fragmentation [2]. In early 2014, the lesser prairie-chicken was placed as a threatened species under the U.S Endangered Species Act which means if proper conservation and management isn't established the lesser prairie-chicken will become an endangered species in the foreseeable future under the law [3]. Recent reports of lesser prairie-chickens flying into stationary objects and other anecdotal reports of these birds flying into objects, as large as vehicles and barns, have led us to wonder if these birds have vision problems or other neurological problems. These problems are often caused through either parasitic infections, contamination by organochlorine pesticides, or metal toxicity which will be examined in this manuscript. With their recent listing as threatened under the Endangered Species Act, the need to find out why they are declining and increased conservation measures are needed now more than ever.

Oxyspirura petrowi has been receiving increased attention due to its potential role in negatively impacting gamebirds [4]. This parasite, known as the eyeworm, is a nematode that lives on the surface of and/or behind the eyeball in the lacrimal duct and its associated glands. Eyeworm sizes can range from microscopic egg to well over 15 mm in length [5]. While the lifecycle is not completely known, it has been suggested that the lifecycle of *O. petrowi* is likely similar to *Oxyspirura mansoni* which is known to infect poultry [6]. The lifecycle of *O. mansoni* starts when a gravid female deposits embryonated eggs in the eyes of the host, which are then washed down the naso-lacrimal ducts to the mouth, swallowed, and finally excreted into the feces where they are then ingested by an intermediate host [7]. Research is still underway to determine which arthropods are intermediate hosts for the eyeworm.

Research by Dunham et al. (2014a) on northern bobwhites (*Colinus virginianus*) revealed damage to tissues behind the eyeball and in the nasal-lacrimal glands causing localized hemorrhaging and swelling by these blood-feeding parasites. Increased eyeworm infections could cause severe hemorrhaging and swelling behind the eye, which applies pressure to the optic nerve, probably compromising the bird's vision. Visual impairments could impact their ability to forage, fly efficiently when escaping predators, and reproduce successfully, all of which could ultimately decrease their survivability in the wild.

The eyeworm *O. petrowi* has been previously found in lesser prairie-chickens in Texas [8] and Kansas [1]. A related species of eyeworm (*O. lumsdeni*) has been documented in lesser prairie-chickens in Oklahoma [9]. However, little research has been conducted on the impact that eyeworms could have on the decline of the lesser prairie-chicken.

Most pesticides of the organochlorine class have been banned or restricted for years; however, they are often still very persistent in soil and can be harmful to birds when they are exposed [10]. Organochlorines in avian species can lead to toxic effects such as lethargy, convulsions, and emaciation with more commonly known effects like eggshell thinning and reproductive inhibition [11]. Additionally both lead (Pb) and mercury (Hg) are commonly found throughout the environment and have well established detrimental effects on wildlife health. Lead and mercury are released into the environment by anthropogenic processes such as spent lead ammunition or by-products of the combustion of fossil fuels, and can lead to behavioral and neurological abnormalities.

This manuscript examines the influence of parasitic infections, organochlorine pesticides, and metal toxicity as a potential contributor to the lesser prairie-chicken decline.

Methods

The bodies of nine hunter-harvested lesser prairie-chickens from Kansas were donated to the Wildlife Toxicology Laboratory at The Institute of Environmental and Human Health (TIEHH), Texas Tech University for extensive evaluation for the presence of eyeworms, organochlorine pesticides, and toxic metals. These lesser prairie-chickens were harvested by hunters in Kansas during a limited hunting season. Specimens were salvaged from a registered taxidermist in the taxidermy specimen preparation process. Prior to processing by the taxidermists, each eyeball and its associated lacrimal ducts and tissues were removed and put into separate 70% ethanol vials. The eye sockets were examined for any remaining eyeworms and they were placed into their respective vials. In coordination with samples supplied by registered taxidermists, we were allocated the eyes and their associated ducts and glands, femur, liver, and breast muscle.

Eyeworm Examination

Each bird was thawed and examined for eyeworms. The examination started by removing the lacrimal duct, gland, and tissue from the eyeball and teasing them apart to look for eyeworms. Any eyeworms that were found during the examination process were placed in a physiological saline holding media. When all the examinations were complete, eyeworms were then transferred into a 70% ethanol+8% glycerin vial for preservation. Voucher parasite specimens of *O. petrowi* (USNPC No. 108249) were deposited in the U.S. National Parasite Collection, Beltsville, Maryland. Prevalence refers to the number of lesser prairie-chickens infected with *O. petrowi* in the sample divided by total number of lesser prairie-chickens examined in the sample, and mean is the number of *O. petrowi* found in the lesser prairie-chickens sampled by the total number of lesser prairie-chickens examined [12].

Contaminant Analysis

Samples of breast muscle, liver, and fat from the lower breast were taken from all nine donated birds. These tissues were extracted using a general QuEChERS method and analyzed by GC-ECD for 20 organochlorine pesticides and metabolites: hexachlorocyclohexane (HCH; alpha, beta, gamma (lindane), and delta isomers), alpha and gamma chlordane, heptachlor and heptachlor epoxide, DDT, DDE, DDD, methoxychlor, aldrin, dieldrin, endrin, endrin aldehyde, endrin ketone, endosulfan I, endosulfan II, and endosulfan sulfate [13]. The QuEChERS method started by taking approximately 3 g wet weight tissue that was then freeze dried and added to a 50 mL centrifuge tube containing 4,000 mg anhydrous magnesium sulfate and 1,000 mg anhydrous sodium chloride (United Chemical Technologies, Bristol, PA, USA). To this tube 15–20 mL acetonitrile was added with tetrachloro-*m*-xylene as an internal standard. After vortexing and centrifugation (3,000 rpm for 10 min), the extract was decanted and transferred into a 15 mL QuEChERS cleanup centrifuge tube containing 900 mg anhydrous magnesium sulfate, 300 mg primary-secondary amine (PSA) exchange material, and 150 mg endcapped C18 (United Chemical Technologies, Bristol, PA, USA).

Samples were analyzed on a Hewlett-Packard 6890 gas chromatograph equipped with two Agilent columns (primary: DB-17 ms: 30 m×0.32 mm×0.25 μm; secondary: DB-XLB: 30 m×0.32 mm×0.50 μm) and two μECD detectors. Inlet and detector temperatures were 220°C and 300°C (both detectors), respectively. A volume of 2 μL was injected in pulsed splitless mode with a pulse pressure of 40.0 psi, pulse time of 0.20 min, purge flow of 44.8 mL/min and purge time of 1.00 min. Each standard was made by spiking chicken breast extract with a certified OC pesticide mixture (Restek, Belafonte, PA, USA) and TCMX (Accustandard, New Haven, CT, USA).

Additionally, livers and a portion of each femur were extracted from all lesser prairie-chickens and sent to Trace Analysis Inc. (Lubbock, TX, USA). Livers were analyzed for elevated mercury levels and femurs were analyzed for elevated lead levels. Livers were tested for mercury using modified EPA method SW-846 7471B. [14]. First 5 mL of DI water was added to a 0.5–0.6 g portion of well-homogenized, freeze-dried sample. This mixture was heated for 2 minutes at 95±3°C. After cooling, an additional 50 mL of DI water was added, followed by 15 mL of potassium permanganate solution, and each sample was allowed to sit for a minimum of 15 minutes. Samples were mixed thoroughly and then heated at 95±3°C for 30 minutes. After cooling, 6 mL of sodium chloride hydroxylamine sulfate solution was added to each sample. These samples were then analyzed using a CETAC M-6100 Cold Vapor Atomic Absorbance mercury analyzer. Absorbance for Hg was measured at 253.7 nm. Femur samples were prepared for Pb analysis using a slightly modified version of modified EPA method SW-846 3050B [15]. Freeze dried femurs were digested for one hour at room temperature using 5 mL of HNO_3. An additional 5 mL of HNO_3 was added to each sample; samples were then heated to 100°C. The heating process continued until samples were completely digested and the digestant appeared clear. Once the digestion was complete, the samples were cooled and 10 mL of 30% H_2O_2 was added. Samples were then heated again until foaming subsided. At this point, 10 mL of concentrated HCl was added and samples were heated until the volume was reduced to 5 mL. Samples were subsequently diluted to 50 mL using DI water. Analysis of femurs for lead followed EPA method SW-846 6010C using a Perkin Optima 8300 Duel View Inductively Coupled Plasma-Optical Emission Spectrometer. Lead was measured at a wavelength of 220.353 nm in axial mode.

Method detection limits (MDLs) ranged from 8.0 ng/g for heptachlor epoxide to 70 ng/g for methoxychlor. In general, most MDLs are approximately10 ng/g. Generally speaking, organochlorine pesticide residues in the tissues of lesser prairie-chickens were relatively low to barely detectable. Standard detection limits for lead was 0.263 μg/g and mercury's was 0.00354 μg/g.

Results

Of the 9 lesser prairie-chickens examined, 7 were found to be infected with a total of 49 eyeworms. Prevalence of infection was 78% with a mean abundance of 5.44±6.02 [range:0–16]. Endrin aldehyde was detected but could not be quantitated due to partial removal in the QuEChERS cleanup step. The poor recovery of endrin aldehyde could have been a result of removal by PSA in the extraction step. Nevertheless, endrin aldehyde was detected on the primary column. Figure 1 shows the detection frequencies of each pesticide by tissue. Most residues were below method detection limits. The range and averages for pesticides above detection limits was recorded in Table 1. Even those pesticides found above detection limits are in the low ng/g range, which means it's unlikely that these levels pose a direct threat to lesser prairie-chicken health. Notable is the ubiquitous presence of endrin aldehyde across all tissues. Endrin aldehyde as well as DDE are common metabolites found in many birds including lesser prairie-chickens [16]. The presence of these organochlorines in low concentrations is consistent with the historic (and not recent) use of pesticides in the region.

Lead levels in the nine femur samples averaged 0.86 µg/g and only one of the femur samples was found to have lead above detection limits at 5.66 µg/g. This amount of lead is relatively low and not consistent with toxic lead exposure ranging between 2–8 µg/g [17,18]. Liver samples averaged 0.04 µg/g of mercury and no samples had mercury above the 0.263 µg/g detection limit.

Discussion

While our sample size is relatively low, it appears that lesser prairie-chickens in Kansas are not being exposed to a large number of organochlorines or two metals which are common throughout the environment. Since organochlorine pesticides have been banned or restricted, the likelihood of them exerting toxicity is being reduced daily. These pesticides are constantly being degraded in the environment and exposure is unlikely. However, this study only measured a subsample of potential organochlorine pesticides and didn't look at both organophosphate and carbamate pesticides, which means pesticides shouldn't be completely discarded as potential dangers to lesser prairie-chickens.

But, this is not the case when it comes to being exposed to potential eyeworm infection. Our eyeworm infection findings in the present study are considerably high and comparable to Robel et al. (2003) infection rates of *O. petrowi* in lesser prairie-chickens (95% infection), considering a further reduced population today.

A continuing presence of eyeworms is being reported in lesser prairie-chicken which could indicate that this parasite could potentially be a contributing factor to lesser prairie-chicken decline and have deeper implications than what was previously known. Previous research on the red grouse (*Lagopus lagopus scoticus*) in

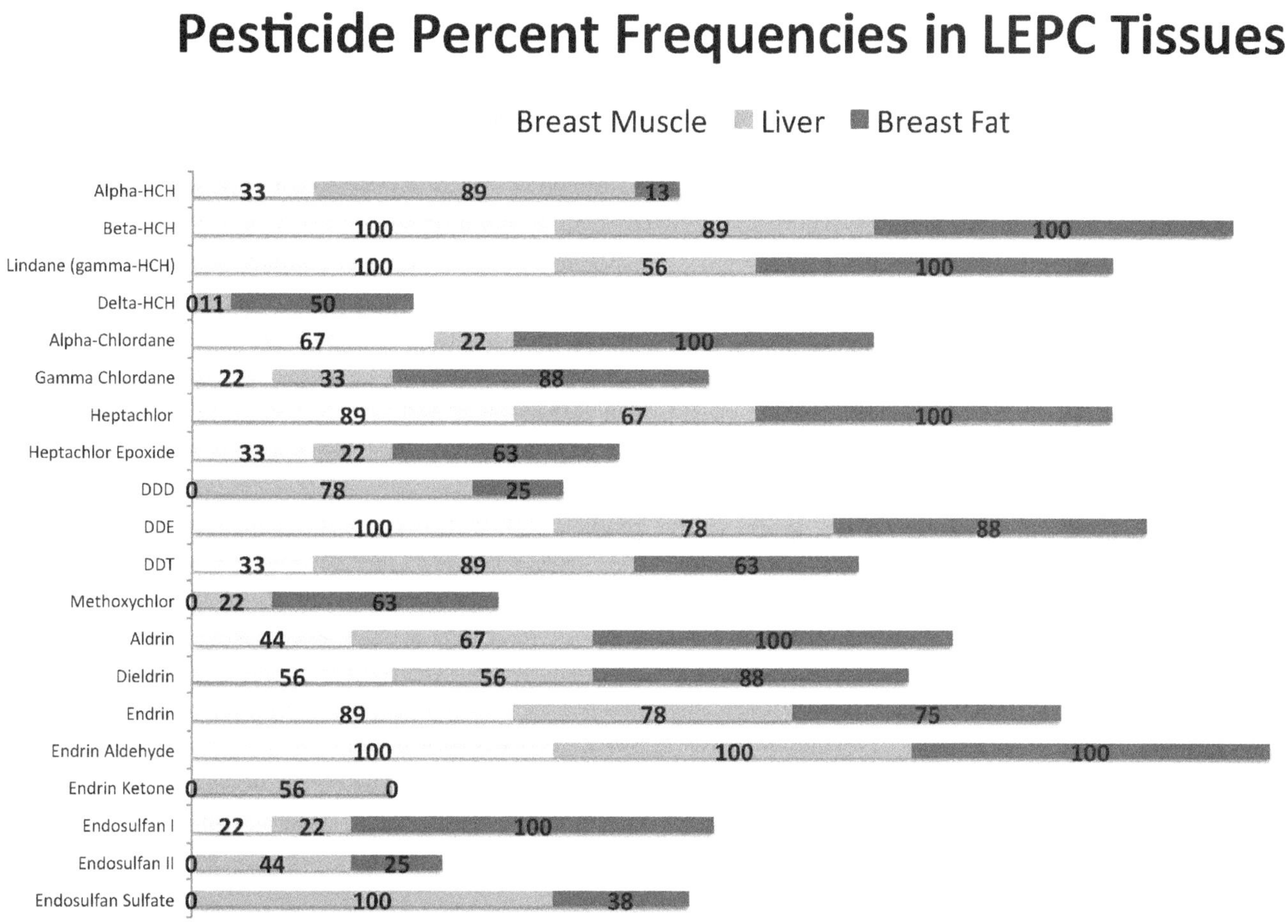

Figure 1. Percent frequencies of organochlorine pesticide detections, including pesticides above and below method detection limits, in lesser prairie-chickens (*Tympanuchus pallidicinctus*) by tissue type, Kansas, USA. *Method detection limits for heptachlor epoxide were 8 ng/g, 70 ng/g for methoxychlor, and 10 ng/g for all other organochlorines.

Table 1. Average and range of select pesticides found in lesser prairie-chickens (*Tympanuchus pallidicinctus*) tissues (ng/g wet weight)*, from Kansas, USA.

Pesticide	Breast Muscle (n = 9)	Liver (n = 9)	Breast Fat (n = 8)
Alpha-HCH		14 (n = 1)	
Beta-HCH	10 (n = 1)	21 (11–32) (n = 3)	17 (n = 1)
Gamma-HCH			32 (13–64) (n = 3)
Alpha Chlordane			17 (n = 1)
Gamma Chlordane			12 (n = 1)
Heptachlor	11 (9.9–13) (n = 4)	18 (9–39) (n = 6)	43 (16–120) (n = 8)
Heptachlor Epoxide			20 (8–33) (n = 2)
DDD		35 (n = 1)	
DDE			27 (n = 1)
DDT		35 (13–74) (n = 3)	
Aldrin		42 (n = 1)	
Endrin		21 (20–21) (n = 2)	43 (n = 1)
Endrin Ketone		35 (16–57) (n = 5)	
Endosulfan I			27 (n = 1)
Endosulfan Sulfate		31 (24–38) (n = 2)	

*Values and pesticides shown are only those above detection limits. Endrin aldehyde could not be quantitated.
Note: Method detection limits for heptachlor epoxide were 8 ng/g, 70 ng/g for methoxychlor, and 10 ng/g for all other organochlorines.

Scotland has shown that high parasite burdens can ultimately make a host more susceptible to predation [19]. Additional studies have also shown that increased parasite loads can correlate with the host population decline and reduced parasite loads throughout the populations correlate with an upward trend in the host population [20]. Eyeworm infection rates in northern bobwhites of the Rolling Plains region of Texas have been reported at greater than 90% in sampled birds and there is concern that the population of bobwhites is declining [4].

The infection rate throughout this region was much higher than what was previously thought which suggests that either the infection may have increased or was underestimated and this could be similar to what is happening in Kansas. However, because these birds are hunter shot, there may be some bias that these lesser prairie-chickens were more susceptible to being shot because of their increased parasitism than a lesser prairie-chicken that was eyeworm infection free.

The erratic behavior of lesser prairie-chickens flying into stationary objects is very similar to that of other gallinaceous or ground feeding birds, which have been found to be infected with the parasitic eyeworm *Oxyspirura petrowi* [21,22]. Eyeworms are known blood feeding nematodes that can cause inflammation and edema behind the eyes of other gallinaceous birds. A heavy infection of eyeworm may render lesser prairie-chickens susceptible to vision problems which may impair their ability to find food or escape from predators, and a heavy infection likely causes increased energy expenditure.

Despite efforts to restore habitat and limit hunting in areas where lesser prairie-chickens are commonly found, much of the population is steadily declining to the point where they are now threatened. Parasites like the eyeworm, which we believe has been underestimated as a factor, could silently be impacting the population by compromising vision which is a necessity for predator avoidance and finding/securing food. Additionally, new conservation practices by biologists and landowners have been to add fence signs and/or colored marking tape on fences to help reduce bird/fence collisions. The idea of this practice is to enable the lesser prairie-chicken to see the fence lines; however, this conservation practice may be counterintuitive. Marking the fences with signs provides no net benefit to the lesser prairie-chickens by being there and may actually provide targets with increased probability of impact by visually impaired birds. Further research needs to be conducted on the eyeworm and the impact they have on their infected host, including lesser prairie-chickens, so better management practices can be established.

Acknowledgments

We would like to thank all of the hunters who donated lesser prairie-chickens in support of this project. We thank Trace Analysis Inc. for their support in analyzing our samples for metal contamination. We would like to thank all of the Wildlife Toxicology Lab members for their laboratory support. Lastly, we would like to thank all of the reviewers of this manuscript for all of their time and guidance in getting this manuscript published.

Author Contributions

Conceived and designed the experiments: ND RK. Performed the experiments: ND SP CB. Analyzed the data: ND RK. Contributed reagents/materials/analysis tools: ND SP CB RK. Contributed to the writing of the manuscript: ND SP CB RK.

References

1. Robel RJ, Walker TL, Hagen CA, Ridley RK, Kemp KE, et al. (2003) Helminth parasites of Lesser Prairie-chicken *Tympanuchus pallidicinctus* in southwestern Kansas: Incidence, burdens and effects. Wildl Biol 9:341–349.
2. Hagen CA, Jamison BE, Giesen KM, Riley TZ (2004) Guidelines for managing lesser prairie-chicken populations and their habitats. Wildl Soc Bull 32:69–82.
3. United States Fish and Wildlife Service (2014) Endangered and Threatened Wildlife and Plants; Determination of Threatened Status for the lesser prairie-chicken; Final Rule. Federal Register 79: 10 April 2014. Available: http://www.gpo.gov/fdsys/pkg/FR-2014-04-10/pdf/2014-07302.pdf. Accessed 15 May 2014.

4. Dunham NR, Soliz LA, Fedynich AM, Rollins D, Kendall RJ (2014a) Evidence of an *Oxyspirura petrowi* epizootic in Northern Bobwhites (*Colinus virginianus*). J Wildl Dis 50: 552–558.
5. Dunham NR, Soliz LA, Brightman A, Rollins D, Fedynich AM, Kendall RJ (2014b) Live eyeworm (Oxyspirura petrowi) extraction, in-vitro culture, and transfer for experimental studies. J Parasitol: Published Online.
6. Ruff MD (1984) Nematodes and acanthocephalans. In: Diseases of poultry. 8th Ed, Hofstand MS, Calnek BW, Helmboldt CF, Reid WM, Yoder Jr HW, editor. Iowa State University Press, Ames, Iowa, pp. 614–648.
7. Schwabe CW (1951) Studies on *Oxyspirura mansoni*, the tropical eyeworm of poultry. II. Life history. Pac Sci 5:18–35.
8. Pence DB, Sell DL (1979) Helminths of the lesser prairie chicken, *Tympanuchus pallidicinctus* (Ridgway) (Tetraonidae), from the Texas panhandle. P Helm Soc Wash 46: 146–149.
9. Addison EM, Anderson RC (1969) *Oxyspirura lumsdeni* n. sp. (Nematoda:Thelaziidae) from Tetraonidae in North America. Can J of Zool 47:1223–1227.
10. Hellou J, Lebeuf M, Rudi M (2012) Review on DDT and metabolites in birds and mammals of aquatic ecosystems. Environ Rev 21:53–69.
11. Fimreite N (1984) Effects of lead shot ingestion in willow ptarmigan. Bull of Environ Contam Toxicol 33:121–126.
12. Bush AO, Lafferty KD, Lotz JM, Shostak AW (1997) Parasitology meets ecology in its own terms: Margolis et al. revisted. J Parasitol 83:575–583.
13. Cieślik E, Sadowska-Rociek A, Ruiz JMM, Surma-Zadora M (2011) Evaluation of QuEChERS method for the determination of organochlorine pesticide residues in selected groups of fruits. Food Chem 125:773–778.
14. United States Environmental Protection Agency (US EPA) (2007) SW-846, Method 7471B, Mercury in Solid or Semisolid Waste (Manual Cold Vapor Technique). Washington, DC.
15. United States Environmental Protection Agency (US EPA) (1996) SW-846, Method 3050B, Acid Digestion of Sediments, Sludges, and Soils. Washington, DC.
16. Mackay D, Shiu W-Y, Ma KC (1997) Illustrated handbook of physical-chemical properties and environmental fate for organic chemicals: Pesticide Chemicals. CRC press Boca Raton, FL, USA.
17. Friend M, Franson CJ (1999) Field manual of wildlife disease: general field guide procedures and disease of birds. United States Geological Survey, Washington, DC, pp.295–305.
18. Kendall RJ, Lacker TE, Bunck C, Daniel B, Driver C, et al. (1996) An ecological risk assessment of lead shot exposure in non-waterfowl avian species: Upland game birds and raptors. Environ Toxicol Chem 15:4–20.
19. Hudson PJ, Dobson AP, Newborn D (1992) Do parasites make prey vulnerable to predation? J Anim Ecol 61:681–692.
20. Cattadori IM, Haydon DT, Hudson PJ (2005) Parasites and climate synchronize red grouse populations. Nature 433:737–741.
21. Erickson AB, Highby PR, Carlson CE (1949) Ruffed Grouse populations in Minnesota in relation to blood and intestinal parasitism. J Wildl Manage 13:188–194.
22. McClure HE (1949) The eyeworm, *Oxyspirura petrowi*, in Nebraska pheasants. J of Wildl Manage 13:304–307.

15

Double-Stranded RNA Uptake through Topical Application, Mediates Silencing of Five CYP4 Genes and Suppresses Insecticide Resistance in *Diaphorina citri*

Nabil Killiny[1]*, Subhas Hajeri[2], Siddharth Tiwari[1], Siddarame Gowda[2], Lukasz L. Stelinski[1]

1 Department of Entomology and Nematology, Citrus Research and Education Center, IFAS, University of Florida, Lake Alfred, Florida, United States of America,
2 Department of Plant Pathology, Citrus Research and Education Center, IFAS, University of Florida, Lake Alfred, Florida, United States of America

Abstract

Silencing of genes through RNA interference (RNAi) in insects has gained momentum during the past few years. RNAi has been used to cause insect mortality, inhibit insect growth, increase insecticide susceptibility, and prevent the development of insecticide resistance. We investigated the efficacy of topically applied dsRNA to induce RNAi for five Cytochrome P_{450} genes family 4 (*CYP4*) in *Diaphorina citri*. We previously reported that these *CYP4* genes are associated with the development of insecticide resistance in *D. citri*. We targeted five *CYP4* genes that share a consensus sequence with one dsRNA construct. Quantitative PCR confirmed suppressed expression of the five *CYP4* genes as a result of dsRNA topically applied to the thoracic region of *D. citri* when compared to the expression levels in a control group. Western blot analysis indicated a reduced signal of cytochrome P_{450} proteins (45 kDa) in adult *D. citri* treated with the dsRNA. In addition, oxidase activity and insecticide resistance were reduced for *D. citri* treated with dsRNA that targeted specific *CYP4* genes. Mortality was significantly higher in adults treated with dsRNA than in adults treated with water. Our results indicate that topically applied dsRNA can penetrate the cuticle of *D. citri* and induce RNAi. These results broaden the scope of RNAi as a mechanism to manage pests by targeting a broad range of genes. The results also support the application of RNAi as a viable tool to overcome insecticide resistance development in *D. citri* populations. However, further research is needed to develop grower-friendly delivery systems for the application of dsRNA under field conditions. Considering the high specificity of dsRNA, this tool can also be used for management of *D. citri* by targeting physiologically critical genes involved in growth and development.

Editor: Immo A. Hansen, New Mexico State University, United States of America

Funding: This project was supported by a grant to LS from the Citrus Research and Development Foundation. The funders had no role in study design, data collection and analysis, decision to publish, or preparation of the manuscript.

Competing Interests: The authors have declared that no competing interests exist.

* Email: nabikilliny@ufl.edu

Introduction

RNA interference (RNAi) is a promising tool for studying functional genomics in eukaryotes and insects in particular [1,2]. Anti-sense (nonsense) RNA strand transcription has been used for over three decades to inhibit gene activity [3]. The efficacy of anti-sense silencing depends on hybridization between the injected RNA and an endogenous messenger. Since the discovery of double-stranded RNA (dsRNA) mediated gene-specific silencing in the nematode, *Caenorhabditis elegans* (Maupas) [4], dsRNA-mediated RNAi has been employed with various insects to silence specific genes [2]. RNAi has been widely used in various insect orders, including Coleoptera, Dictyoptera, Diptera, Hemiptera, Hymenoptera, Isoptera, Lepidoptera, Neuroptera, and Orthoptera [5,6,7]. The systemic nature of dsRNA-mediated RNAi has allowed this tool to be used in the management of various insect pests [8–13].

The cytochrome P_{450} monooxygenases are an important group of enzymes that are involved in the metabolism of xenobiotic compounds in insects. This group of enzymes is associated with insecticide resistance and metabolism of a wide range of endogenous and exogenous compounds that includes hormones, pheromones, insecticides, and plant secondary compounds in insects [14–18]. Overtranscription of families 4, 6, 9, and 12 has been frequently linked to insecticide metabolism and resistance [19–22].

The Asian citrus psyllid, *Diaphorina citri* Kuwayama (Hemiptera: Psyllidae), is perhaps the most destructive pest of citrus, mainly because it is a vector for the putative causal agent of huanglongbing (HLB), *Candidatus* Liberibacter asiaticus (*C*Las) [23]. HLB is a deadly citrus disease with no known cure [23,24]. Currently, the main tools that limit the spread of the disease are insecticides to manage the vector [18,25,26]. *D. citri* are susceptible to several insecticide classes, which includes the pyrethroids, organophosphates, carbamates, neonicotinoids, insect growth regulators, horticultural oils, and lipid synthesis inhibitors [27]. Foliar treatments may suppress populations for 3 weeks following application [27]. Broad-spectrum insecticides (pyrethroids, organophosphates, and neonicotinoids) are more effective against *D. citri* than IGRs or oils, and insecticide use against *D. citri* is most effective when populations are not actively reproducing [28]. Systemic soil-applied insecticides provide a much longer duration of population control (months) than foliar insecticides (weeks) [27]. The neonicitonoids have been the main class of effective systemic insecticides for *D. citri* control during the past decade [27]. Systemic neonicotinoids are particularly effective in protecting young trees as they mature into production [29].

Intense insecticide use has led to the development of varying levels of insecticide resistance in populations of *D. citri* in Florida, USA [26]. This is particularly concerning for the neonicotinoid class, since these are the main current tools for protecting young trees from *C*Las infection [29]. A metabolic mechanism for the evolution of insecticide resistance in populations of *D. citri*, particularly for neonicotinoids, is supported by increased activities of detoxifying enzymes and overexpression of Cytochrome P_{450} genes family 4 (*CYP4*) [17,21,26,30,31]. In the present study, we targeted the abovementioned *CYP4* genes for silencing by topical application of specific dsRNA to the thorax of newly emerged *D. citri* adults. Additionally, we tested the effect of dsRNA treatment on insecticide resistance by comparing mortality of known susceptible and resistant populations of *D. citri*.

Materials and Methods

Insect populations

A laboratory susceptible population (LS) of *D. citri* was maintained in a greenhouse at the Citrus Research and Education Center, Lake Alfred, Florida. The culture was established in 2000 using field populations from Polk County, Florida and maintained on sweet orange (*Citrus sinensis* (L.) Osbeck) without exposure to insecticides in a greenhouse at 27–28°C, with 60–65% relative humidity and a 14:10 (light:dark) photocycle hours. Three field populations of *D. citri* were collected from commercial citrus groves in Florida during 2013. The populations were collected with permissions from private groves. Name of groves, counties, and GPS coordinates are as the following: *i)* GapWay Groves (Private managed grove), Polk County (PL) (28° 05′ 40.14″ N; 81° 43′ 19.03″ W); *ii)* Winter Garden, Conserve II (Private managed grove), Lake County (LA) (28° 27′ 52.17″ N; 81° 39′ 31.69″ W); and *iii)* Uncle Matt's Organic (Organically managed grove), Lake County (OG) (28° 31′ 00.88″ N; 81° 40′ 01.90″ W). Adults were collected using sweep nets and aspirators, transferred to the laboratory, released onto citrus plants within Plexiglas cages (40×40×40 cm), and used in bioassays shortly thereafter.

Constructing dsRNA

A consensus sequence, derived from five previously published *CYP4* sequences [21], was used to design *CYP4*-specific primers (Table 1). The *CYP4*-specific primers were tailed with a T7 promoter sequence to generate sense and antisense transcripts separately.

Total RNA isolation was performed on groups of 40–50 psyllids using the SV total RNA isolation kit (Promega, Madison, WI, USA). One microgram of RNA was used to synthesize cDNA using the *CYP4*-specific reverse primers and iScript cDNA synthesis kit (Bio-Rad, Hercules, CA, USA). Sense and antisense PCR products were generated in separate PCR reaction using specific combination of primers (Table 1). To generate plus-sense

Table 1. Primers used in this investigation.

Purpose	Gene	Sequence	Reference
ds-RNA synthesis			
	CYP4	Forward CACG*ttaattaa*ACGTTCATGTTCGAGGGGCACGATACAACAAC	This
	Sense	Reverse GAGC*aggcct*GAAGGGTACATAGGAGTAAGGATGACGTTTCTG	investigation
	CYP4 Antisense	Forward GCAGCA*TAATACGACTCACTATAGGGAGA*GAAGGGTACATAGGAGTAAGGATGACGTTTCTG Reverse ACGTTCATGTTCGAGGGGCACGATACAACAAC	This investigation
	GFP	Forward GCGAACTAGTATGGCTAGCAAAGGAGAAGAACTTTTCACTG	El-Shesheny
		Reverse GAGACCGCGGCTACCCCTCGAGTTATTTGTAGAGCTCATC	et al. [32]
Gene expression			
	CYP4 genes		
	CYP4C67	Forward TGGAACGTGTCATCAAGGAG	Tiwari et al.
		Reverse CCGGATTGAAACTGTTAGGC	[21]
	CYP4DA1	Forward AGTGGTGTCGGAAATTGAGG	
		Reverse GTTCGAGCCACCTGGAGATA	
	CYPC68	Forward CTAGCCTGGACCCTCTTCCT	
		Reverse ACCCTCCCTATGAACGGAAC	
	CYPG70	Forward GCCGGAAGTTCTTTCTTCCT	
		Reverse TAACGGGTACTGGTGGGAAC	
	CYPDB1	Forward CTGTACGCTCTGGGACATCA	
		Reverse TTGAGCGGTGCATAGAGTTG	
	Reference genes		
	α-tubulin	Forward CAGGTCTTGTGTGGGACGTA	Tiwari et al. [21]
		Reverse GGCCACAGTTTGTTTCTTGC	
	Actin	Forward CCCTGGACTTTGAACAGGAA	
		Reverse CTCGTGGATACCGCAAGATT	

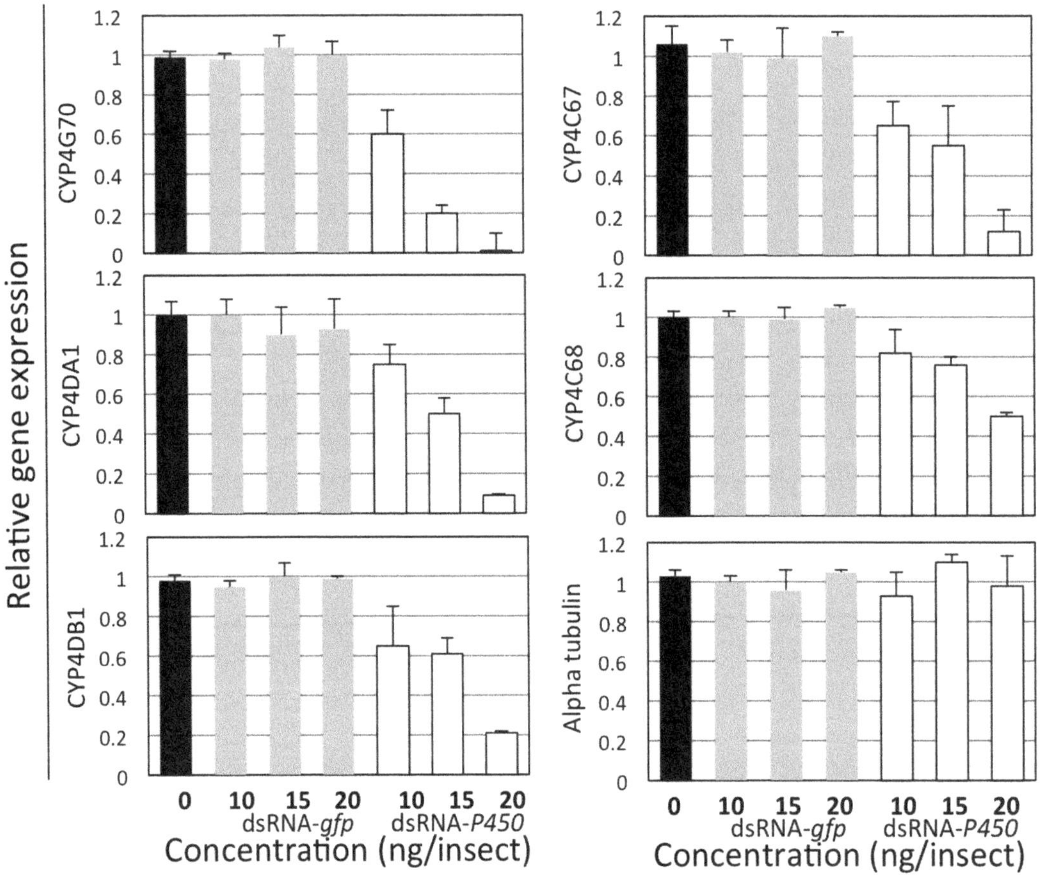

Figure 1. Relative expression levels of the five *CYP4* genes targeted by RNAi in *Diaphorina citri* adults 72 h after treatment with dsRNA. Ct values were first normalized to the endogenous control gene *Actin* followed by normalization to the treatment giving the lowest gene expression using the $2^{-\Delta\Delta CT}$ method. Standard deviations were calculated based on three independent experiments, each with three technical replicates. Alpha-tubulin was used as a non-target gene control. dsRNA-*gfp* treatment was used as a control targeting an irrelevant gene.

transcripts, sense primer with T7 promoter sequence and regular antisense primer were used; while to generate antisense transcripts, regular sense primer and antisense primer with T7 promoter sequence were used. Sense and antisense transcripts were annealed by denaturing at 70°C for 10 min, followed by slowly cooling to room temperature for 20 min. To eliminate DNA template and single-stranded RNA, dsRNA was treated with DNase I and RNase A. The dsRNA was then purified of proteins and free nucleotides using the phenol-chloroform purification method. The amount of purified dsRNA was measured with a NanoDrop Spectrophotometer. We used dsRNA-*gfp* as an irrelevant dsRNA (control). dsRNA-*gfp* was produced as described above. Green fluorescent protein (GFP) mRNA is 732 bp in length. Specific primers (Table 1) were used to amplify the full-length of GFP gene by using TMV-30BGFP according to El-Shesheny et al. [32].

D. citri treatment with dsRNA

Purified dsRNA was serially diluted using RNase-free water to obtain desired concentrations of dsRNA. Three concentrations of dsRNA (50, 75, and 100 ng/µl) and a control (0 ng/µl) were used to treat *D. citri* adults. *D. citri* adults were anaesthetized under CO_2 within a few hours of eclosion. A 0.2 µl droplet containing10, 15, or 20 ng of dsRNA was topically applied to the ventral side of the thorax using a 10 µl Hamilton syringe. To investigate the effect of dsRNA- P_{450} on gene and protein expression and enzymatic activity, treated adults were placed into 60 mm plastic disposable Petri dishes that were lined with citrus leaf disks, as a food source, over agar beds as described in Tiwari et al. [17]. Petri dishes with treated adults were kept at 25±1°C and 50±5% RH, with a 14:10 h light:dark photoperiod, in a growth chamber for 72 h. Insects were collected and stored in −20°C until use. dsRNA-*gfp* was used as a non-relevant dsRNA control.

Cytochrome P_{450} (general oxidase) assay

The activity of cytochrome P_{450} was quantified and expressed in terms of general oxidase level, which is an indirect measure of cytochrome P_{450} by using heme peroxidation as described in Tiwari et al. [31]. This method has been considered a reliable tool for comparing differences in general oxidase levels based on hemoprotein levels. Because heme constitutes the majority of cytochrome P_{450} in non-blood-fed insects, quantification of heme activity has been used to compare the levels of cytochrome P_{450} on the basis of general oxidase levels [33]. In brief, heme peroxidase activity was measured using 3,3′5,5′-tetra-methylbenzidine (TMBZ) (Sigma Aldrich) as the substrate. Five replicates, each consisting of three insects, were performed for each treatment.

Gene expression analysis

Live adult *D. citri* from each treatment were subjected to RNA isolation and cDNA synthesis. RNA isolations were performed in

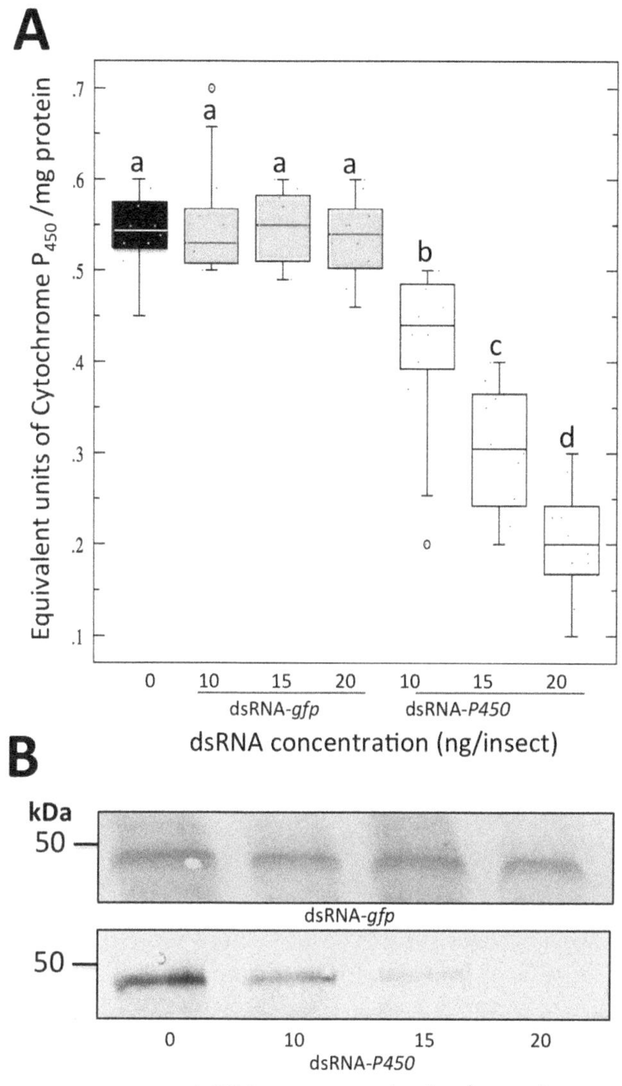

Figure 2. Cytochrome P_{450} (general oxidase) activity and protein expression in dsRNA-treated *D.citri*. A) Box blot representing general oxidase activity. Boxes indicate the interquartile range, including 50% of results, and the mid-horizontal lines represent the median; different letters above deviation error bars represent significant differences between treatments ($P<0.05$). B) Protein analysis of dsRNA-treated and control *D. citri* using Western blot. Western blot was performed using the microsomal proteins prepared from adults treated with three quantities of dsRNA and a control. dsRNA-*gfp* treatment was used as a control targeting an irrelevant gene.

three biological replicates using the SV total RNA isolation kit (Promega, Madison, WI, USA). The quantity and quality of RNA from each sample was measured on a NanoDrop 1000 Spectrophotometer using the absorbance at 260 nm and the A260/A280 ratio, respectively. Subsequently, cDNA was synthesized with the iScript cDNA synthesis kit (Bio-Rad, Hercules, CA, USA) for each replicate within each treatment. Quantitative real-time PCR (qPCR) was performed using iQ SYBR Green Supermix with an iCycler iQ real-time PCR detection system (Bio-Rad). Primers for five *CYP4* genes, *Alpha-tubulin* and the endogenous gene, *Actin*, were used to measure the gene expression of cytochrome P_{450}, as described in Tiwari et al. [21] (Table 1). Six biological replicates were performed for each treatment. The production of gene-specific products and absence of 'primer dimers' was verified by 1% agarose electrophoresis in TAE buffer with ethidium bromide staining.

The $2^{-\Delta\Delta CT}$ method was used to compare the relative expression of the consensus sequence among PCR products derived from the three dsRNA concentrations and control treatments [34]. This was done by first normalizing the expression level of dsRNA treated samples to *Actin* [21] gene expression, followed by normalization to the treatment giving the lowest gene expression. *Alpha-tubulin* was used as a non-targeted gene (control). Five biological replicates, each consisting of three insects and three technical replicates, were performed for each treatment.

Western blot assay

Subcellular protein fractions were extracted using the methods described by Wheeler et al. [35] from adults in each treatment. The protein concentration was determined by the Bradford method [36] using a protein assay kit (Bio-Rad Laboratories, Hercules, CA, USA) with ovalbumin as the standard. Since cytochrome P_{450} proteins were clearly detected in the microsomal protein [22], we used the microsomal fractions to perform the Western blot analysis, as described by Tiwari et al. [22].

Survival assay

The survival assay was carried out on *D. citri* treated with dsRNA-*gfp*, dsRNA- P_{450}, or RNase-free water as a control. Insects were placed on an autoclaved clear plant tissue culture container (75×75×100 mm) lined with 0.5 mm filter paper saturated with 20% sucrose. Fifty insects were placed per container and five replicates were performed for each treatment. Live insects were counted daily.

Residual activity of dsRNA- P_{450}

In order to assess the duration of the RNAi effect, 70 insects treated with RNase-free water or dsRNA-*P450* (20 ng/insect) were placed into plant tissue culture containers with filter paper saturated with 20% sucrose. Samples consisting of three insects were taken daily and kept at −20°C. Cytochrome P_{450} (general oxidase) activity was measured in all samples as described above. Five replicates were performed for both treatments.

Pesticide application

To investigate the effect of treatment with dsRNA- P_{450} on insecticide resistance, *D. citri* adults were treated with the dsRNAs (20 ng/insect) as described above. *D. citri* were initially maintained on Petri dishes with untreated citrus leaf disks for 72 h and thereafter transferred to new Petri dishes that contained leaf discs treated with insecticide solution. Briefly, the leaf disks (60 mm diameter) were excised, dipped in the insecticide solution made in acetone for 30 s, and allowed to air dry in a fume hood for 1 h prior to placement into the Petri dishes as described by Tiwari et al. [26]. We used analytical-grade imidacloprid at the LD_{50} dosage (0.02 ng AI/µl acetone) previously determined by Tiwari et al. [26]. The mortality of *D. citri* adults was assessed after 24 h. dsRNA-*gfp* was used as a negative control for dsRNA- P_{450}. Fives replicates (Petri dishes), with five insects each, were performed for each of the four *D. citri* populations tested. Each population was subjected to four different treatments: *D. citri* treated with RNase-free water on 1) RNase-free water- or 2) imidacloprid-treated disks, as well as *D. citri* treated with dsRNA- P_{450} on 3) RNase-free water- or 4) imidacloprid-treated leaf disks.

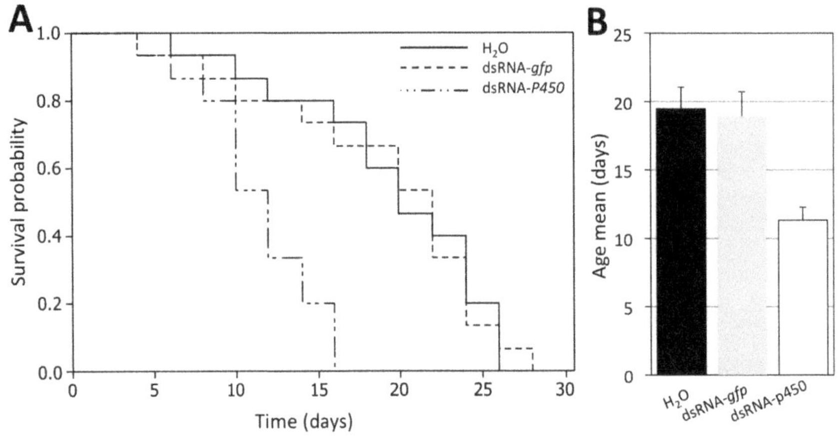

Figure 3. Effect of dsRNA- P_{450} on *D. citri* survival. A) Kaplan-Meier survival curve showing the effect of dsRNA- P_{450} treatment on lifespan of *D. citri*. B) Average life span of *D. citri*. Bars represent the standard deviations; different letters above error bars represent significant differences between treatments ($P<0.05$). dsRNA-*gfp* treatment was used as a control targeting an irrelevant gene.

Statistical analysis

All analyses were performed using SPSS version 19.0. Survival was calculated during the interval from initial treatment to when all insects died. Overall survival (OS) curves were obtained using the Kaplan-Meier method and comparisons were made using log rank and Wilcoxon tests. Analysis of variance (ANOVA) was used to compare: *i*) Calculated *D. citri* lifespans between various treatments, *ii*) The effect of dsRNA treatments on *CYP4* activity and insecticide resistance, and *iii*) The duration of the RNAi effect. Post hoc pairwise comparisons between treatments were performed with the Tukey honestly significant difference test. Statistical significance was established as $P<0.05$.

Results

Treatment with dsRNA-P_{450} causes down regulation of five *CYP4* genes

Relative expression levels for the five *CYP4* genes were compared between dsRNA- P_{450}, dsRNA-*gfp*-treated and control psyllids (Figure 1). Treatment with dsRNA- P_{450} caused reduced expression of the five *CYP4* genes. The effect of dsRNA- P_{450} was positively correlated with the quantity applied per treatment. The expression level of α-tubulin (non-target gene) remained constant among all treatments indicating the specificity of dsRNA- P_{450} to *CYP4* genes. In contrast, there was no effect of the dsRNA-*gfp* control treatments targeting irrelevant psyllid genes on expression levels of *CYP4* genes. The greatest reduction in expression level was found with *CYP4G70*, while the lowest effect was with *CYP4C68*. This observation may help in evaluating gene candidates for RNAi technology for *D. citri*.

Treatment with dsRNA-P_{450} reduces the protein expression and the enzymatic activity of *CYP4*

We investigated the effect of treating *D. citri* with dsRNA on general oxidase activity. The activities were similar for all doses of dsRNA-*gfp* treatment, while reduced when *D. citri* were treated with dsRNA-P_{450}. The activity decreased as the concentration of applied dsRNA- P_{450} was increased. Additionally, Western blots performed using the microsomal fractions revealed the presence of a band corresponding to a 45 kDa protein that cross-reacted with

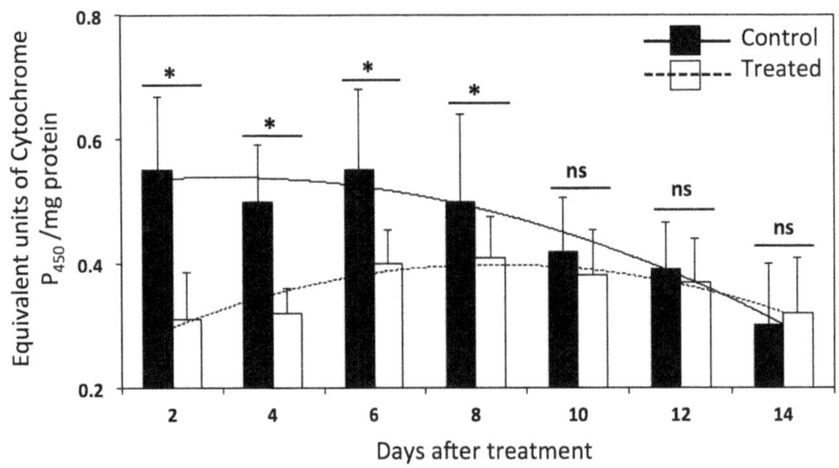

Figure 4. P_{450} as measured by general oxidase activity. Asterisks indicate significant differences between dsRNA- P_{450} treated and non-treated *D. citri*, while ns indicates no significant differences($P<0.05$).

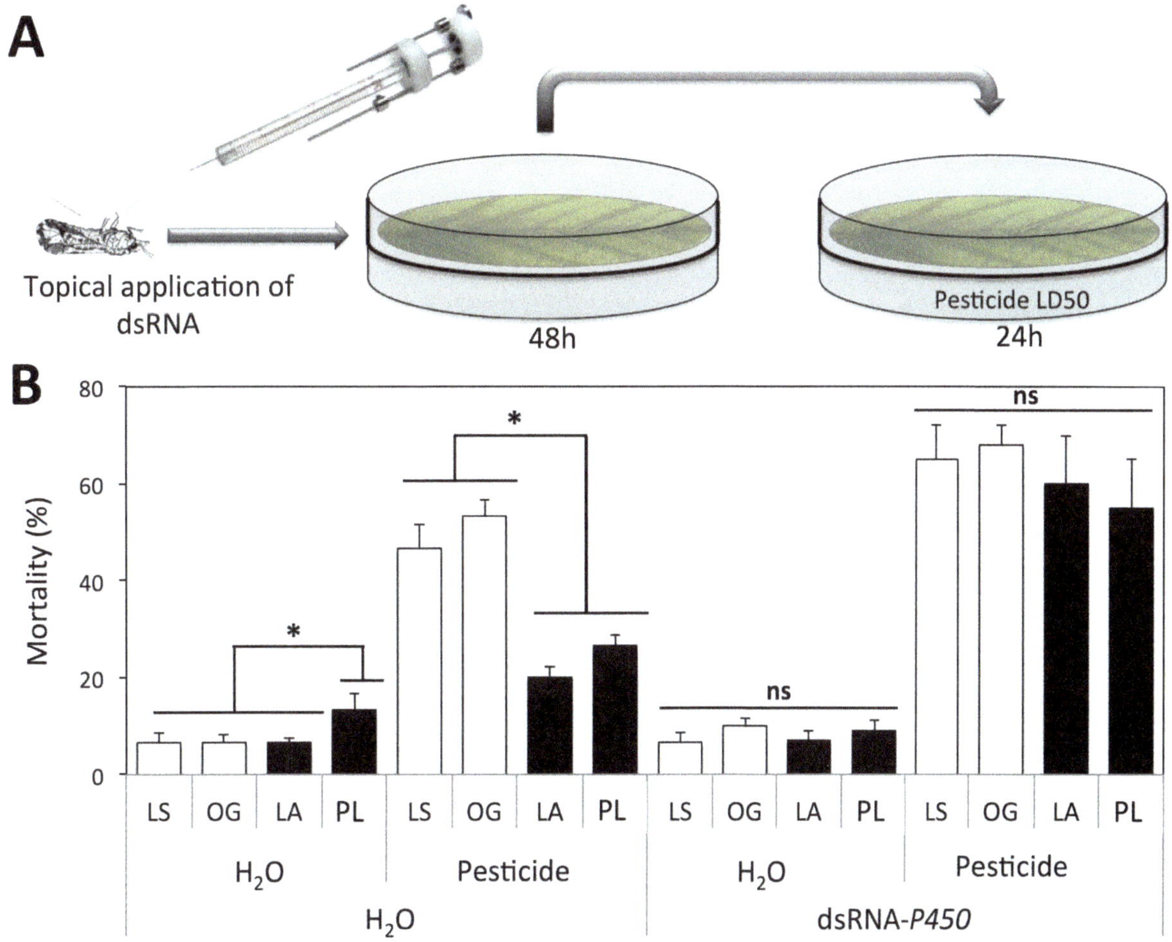

Figure 5. Effect of dsRNA- P_{450} treatment on *D. citri* susceptibility to imidacloprid using the LD_{50} dosage. A) Illustration of the protocol used to test the effect. B) Percent mortality of *D. citri* after exposure to imidacloprid for 24 h. Insects were maintained on non-treated leaf discs for 48 h after the dsRNA- P_{450} treatment prior to exposure to imidacloprid. LS: Laboratory susceptible population. OG: Susceptible population collected from an organic grove. PL: Resistant population collected from commercially managed Polk County, Florida. LA: Resistant population collected from commercially managed Lake County, Florida. Asterisks indicate significant differences, while ns indicates no significant differences ($P<0.05$).

the primary antibody of cytochrome P_{450} protein. Twenty-five micrograms of microsomal proteins for each treatment were used to perform the Western blot which indicated the highest amount of cytochrome P_{450} proteins (detoxifying enzymes) in *D. citri* treated with 0 ng/µl of dsRNA, followed by *D. citri* treated with 10, 15, and 20 ng/insect of dsRNA (Figure 2). There was no signal detected in *D. citri* treated with 20 ng/insect of dsRNA. Expression levels of target *CYP4* and oxidase activity, as a result of dsRNA treatment, directly correlated with the protein expression in adults treated with dsRNA.

Silencing of *CYP4* reduced the lifespan of *D. citri*

Survival of *D. citri* was quantified following treatment with dsRNA-*gfp*, dsRNA- P_{450}, and the control (water). The experiment was conducted under the conditions described earlier (Figure 3A). In this experiment, a 20 ng/insect concentration was used for both dsRNA-*gfp* and dsRNA- P_{450}. A Kaplan-Meier survival plot indicated significant differences among all treatments (log rank = 154.63, $P<0.001$). No significant differences in survival were found between the control and dsRNA-*gfp*-treated *D. citri* (log rank = 1.54, $P = 0.64$). The lifespan in dsRNA- P_{450}-treated *D. citri* was significantly shorter than that observed for other treatments. Mean lifespans for the treatments are presented in Figure 3B. This suggests that reduced expression of *CYP4* genes shortens the lifespan of *D. citri*.

Residual activity of dsRNA- P_{450} treatment

The residual activity of dsRNA- P_{450} was measured at 20 ng/insect concentration. We used general oxidase activity as an indicator for the residual of dsRNA- P_{450}. We compared oxidase activity between the control and dsRNA- P_{450}-treated *D. citri* daily after the treatment application. Oxidase activity was significantly reduced following application of dsRNA- P_{450} for up to 8 days (Figure 4).

Silencing of *CYP4* increased insecticide susceptibility

In order to determine the effect of dsRNA- P_{450} on insecticide susceptibility of *D. citri*, we used imidacloprid at the LD_{50} dosage. Two susceptible and two resistant populations were used in this experiment (Figure 5). *D. citri* that were treated with RNase-free water and then exposed to leaf discs treated with imidacloprid exhibited differing susceptibilities, depending on the population tested (Figure 5). Specifically, the two populations from commercially managed citrus groves that had received imidacloprid treatment over the previous several years (LA (Lake County) and PL (Polk County)) were less susceptible to imidacloprid at the LD_{50} dosage than *D. citri* collected from our Laboratory Susceptible culture (LS) and from the Organic Grove (OG) where imidacloprid had not been used previously (Figure 5). Mortality of *D. citri* exposed to imidacloprid after treatment with dsRNA- P_{450} was increased for each of the four populations as compared with the water control (Figure 5). Mortality of *D. citri* from the two resistant populations (LA, PL), at the LD_{50} dosage of imidacloprid, was significantly higher after treatment with dsRNA- P_{450} as compared with the water control (Figure 5). Given that dsRNA-P_{450} also increased mortality of *D. citri* from the two susceptible populations (LS and OG) further suggests that cytochrome P_{450} is implicated in imidacloprid resistance in *D. citri*.

Discussion

Most insect RNAi studies have relied on the delivery of specific dsRNA through either microinjections [4] or ingestion through feeding [10,13]. Each of these methods has advantages and disadvantages. The microinjection method requires intense training and is a notably time-consuming technique. In addition, optimization is required for volume selection, place of injection, and needle size for successful dsRNA injection into the insect body [7]. Delivery of dsRNA through ingestion also has limitations, such as reduced effectiveness for inducing RNAi [37], reduced efficacy of dsRNA due to the unfavorable gut environment [38], and difficulties in quantifying the amount of dsRNA ingested [39]. The current work highlights a novel method of dsRNA delivery through topical microapplication to the abdomen of adult *D. citri*. The ventral microapplication allows dsRNA uptake through the exoskeleton of insect. The uptake occurs via the intersegmental membranes. This investigation describes a relatively easy and efficient method for delivering and allowing the dsRNA to enter the insect's body to induce RNAi. This method has also been described for *D. citri* nymphs with high efficiency of activity [32]. Also, a similar delivery method was reported to induce RNAi in *Ostrinia nubilalis* larvae [11].

A dosage as low as 50 ng/µl of dsRNA down-regulated the expression of the consensus sequence derived from five *CYP4* genes from *D. citri*, as verified by qPCR and Western blot. The lifespan of *D. citri* following dsRNA treatment was statistically shorter as compared with untreated controls in the current investigation. In another example, mortality in *O. nubilalis* larvae ranged between approximately 40–50% following topical treatment with dsRNA [11]. The mortality, coupled with the lack of any other abnormality observed in the dsRNA-treated adult *D. citri*, suggests that the *CYP4* specific dsRNA are highly target specific. Target specificity of dsRNA is also useful considering the potential for dsRNA exposure to non-target organisms under field conditions. Designing target specific dsRNA is not uncommon; species-specific dsRNA has been shown to work like an insecticide by killing specifically targeted insect pests [40]. The low concentrations needed for induction of RNAi and the highly specific nature of dsRNA suggest it might be a tool for managing insecticide resistance in *D. citri*. Our results indicate that dsRNA-P_{450} reduced oxidase activity, which presumably increased insecticide susceptibility in both resistant and susceptible populations of *D. citri*. In comparison, dsRNA-*gfp* (our negative control treatment) did not affect *CYP4* gene expression or oxidase activity. These findings indicate specificity of RNAi for *D. citri* with the genes targeted in the present investigation.

An important challenge for the application of dsRNA for practical pest control is developing a delivery method for commercial field deployment. Another practical limitation of RNAi that needs to be addressed is that large quantities of dsRNA are expensive to produce. Currently, we are working on inserting the previously described dsRNA into citrus plants for direct ingestion by *D. citri* during feeding. Delivery of dsRNA through transgenic plants (Plant mediated RNAi) has been achieved in *Helicoverpa armigera* and *Diabrotica vergifera vergifera* [41,42]. The absence of interferon-regulated innate immunity pathways in insects allows the possibility of employing longer dsRNA for maximal RNAi [43]. Another potentially feasible way of delivering dsRNA would be to incorporate target-specific dsRNA into bacteria with an appropriate transfection reagent and then spraying the transformed bacteria onto citrus trees. However, future work is needed to evaluate the most efficient transfection reagents and bacterial formulations to prevent the breakdown of dsRNA under field conditions. Once the entire genome of *D. citri* is sequenced, this delivery method could be a convenient way to conduct high-throughput loss-of-function research for determining gene functions. In addition, the current results suggest that further work is needed to understand the mechanism of dsRNA entry into cells following topical application of dsRNA onto *D. citri* to induce RNAi.

Acknowledgments

We acknowledge Ian Jackson for *D. citri* collection and technical assistance.

Author Contributions

Conceived and designed the experiments: NK SH ST SG LLS. Performed the experiments: NK SH SG. Analyzed the data: NK SH ST SG LLS. Contributed reagents/materials/analysis tools: NK SH ST SG LLS. Contributed to the writing of the manuscript: NK SH ST SG LLS.

References

1. Scharf ME, Zhou X, Schwinghammer MA (2008) Application of RNA interference in functional genomics studies of a social insect. In: Barik S, editor. Methods in molecular biology, siRNA, shRNA and miRNA protocols. Vol. 442. Totowa, New Jersey: Humana Press. p. 205–229.
2. Huvenne H, Smagghe G (2010) Mechanisms of dsRNA uptake in insects and potential of RNAi for pest control: A review. J Insect Physiol 56: 227–235.
3. Izant JG, Weintraub H (1984) Inhibition of thymidine kinase gene expression by anti-sense RNA: A molecular approach to genetic analysis. Cell 36: 1007–1015.
4. Fire A, Xu S, Montgomery MK, Kostas SA, Driver SE, et al. (1998) Potent and specific genetic interference by double-stranded RNA in *Caenorhabditis elegans*. Nature 391: 806–811.
5. Tomoyasu Y, Miller SC, Tomita S, Schoppmeier M, Grossmann D, et al. (2008) Exploring systemic RNA interference in insects: A genome wide survey for RNAi genes in Tribolium. Genome Biol 9: R10.
6. Wuriyanghan H, Rosa C, Falk BW (2011) Oral delivery of double-stranded RNAs and siRNAs induces RNAi effects in the potato/tomato psyllid, *Bactericerca cockerelli*. PLoS ONE 6: e27736.
7. Yu N, Christiaens O, Liu J, Niu J, Cappelle K, et al. (2012) Delivery of dsRNA for RNAi in insect: An overview and future directions. Insect Sci 20: 4–14.
8. Gordon KH, Waterhouse PM (2007) RNAi for insect-proof plants. Nat Biotechnol 25: 1231–1232.
9. Price DRG, Gatehouse JA (2008) RNAi-mediated crop protection against insects. Trends Biotechnol 26: 393–400.
10. Zhou X, Wheeler MM, Oi FM, Scharf ME (2008) RNA interference in the termite *Reticulitermes flavipes* through ingestion of double-stranded RNA. Insect Biochem Molec Biol 38: 805–815.
11. Wang Y, Zhang H, Li H, Miao X (2011) Second-generation sequencing supply an effective way to screen RNAi targets in large scale for potential application in pest insect control. PLoS ONE 6(4): e18644.
12. Zhu F, Xu J, Palli R, Ferguson J, Palli SR (2011) Ingested RNA interference for managing the populations of the Colorado potato beetle, *Leptinotarsa decemlineata*. Pest Manag Sci 67: 175–182.

13. Rangasamy M, Siegfried BD (2012) Validation of RNA interference in western corn rootworm *Diabrotica virgifera virgifera* LeConte (Coleoptera: Chrysomelidae) adults. Pest Manag Sci 68: 587–591.
14. Hodgson E (1985) Microsomal monooxygenases. In: Kerkut GA, Gilbert LI, editors. Comprehensive insect physiology, biochemistry, and pharmacology. Vol. 11 Pharmacology. Elmsford, New York: Pergamon. p. 225–322.
15. Feyereisen R (1999) Insect P_{450} enzymes. Annu Rev Entomol 44: 507–533.
16. Scott JG (1999) Cytochromes P_{450} and insecticide resistance. Insect Biochem Mol Biol 29: 757–777.
17. Tiwari S, Stelinski LL, Rogers ME (2012) Biochemical basis of organophosphate and carbamate resistance in Asian citrus psyllid. J Econ Entomol 105: 540–548.
18. Tiwari S, Clayson PJ, Kuhns EH, Stelinski LL (2012) Effects of buprofezin and diflubenzuron on various developmental stages of Asian citrus psyllid, *Diaphorina citri*. Pest Manag Sci 68: 1405–1412.
19. Feyereisen R (2006) Evolution of insect P_{450}. Biochem Soc Trans 34: 1252–1255.
20. Li X, Schuler MA, Berenbaum MR (2007) Molecular mechanisms of metabolic resistance to synthetic and natural xenobiotics. Annu Rev Entomol 52: 231–253.
21. Tiwari S, Gondhalekar A, Mann RS, Scharf ME, Stelinski LL (2011) Characterization of five *CYP4* genes of Asian citrus psyllid and their expression levels in *Candidatus* Liberibacter asiaticus infected and uninfected adults. Insect Mol Biol 20: 733–744.
22. Tiwari S, Killiny N, Mann RS, Wenninger EJ, Stelinski LL (2012) Abdominal color of the Asian citrus psyllid, *Diaphorina citri*, is associated with susceptibility to various insecticides. Pest Manag Sci. 7 p. (wileyonlinelibrary.com) DOI 10.1002/ps.3407.
23. Halbert SE, Manjunath KL (2004) Asian citrus psyllids (Sternorrhyncha: Psyllidae) and greening disease of citrus: A literature review and assessment of risk in Florida. Fla Entomol 87: 330–353.
24. Manjunath KL, Halbert SE, Ramadugu C, Webb S, Lee RF (2008) Detection of '*Candidatus* Liberibacter asiaticus' in *Diaphorina citri* and its importance in the management of citrus Huanglongbing in Florida. Phytopathology 98: 387–396.
25. Sétamou M, Rodriguez D, Saldana R, Schwarzlose G, Parlang D, et al. (2010) Efficacy and uptake of soil-applied imidacloprid in the control of Asian citrus psyllid and a citrus leafminer, two foliar feeding citrus pests. J Econ Entomol 103: 1711–1719.
26. Tiwari S, Mann RS, Rogers ME, Stelinski LL (2011) Insecticide resistance in field populations of Asian citrus psyllid in Florida. Pest Manag Sci 67: 1258–1268.
27. Grafton-Cardwell E, Stelinski LL, Stansly PA (2013) Biology and management of Asian citrus psyllid, vector of huanglongbing pathogens. Annu Rev Entomol 58: 413–432.
28. Qureshi JA, Stansly PA (2010) Dormant season foliar sprays of broad spectrum insecticides: An effective component of integrated management for *Diaphorina citri* (Hemiptera: Psyllidae) in citrus orchards. Crop Prot 29: 860–866.
29. Serikawa RH, Backus EA, Rogers ME (2012) Effects of soil-applied imidacloprid on Asian citrus psyllid (Hemiptera: Psyllidae) feeding behavior. J Econ Entomol 105: 1492–1502.
30. Tiwari S, Pelz-Stelinski K, Stelinski LL (2011) Effect of *Candidatus* Liberibacter asiaticus infection on susceptibility of Asian citrus psyllid, *Diaphorina citri*, to selected insecticides. Pest Manag Sci 67: 94–99.
31. Tiwari S, Pelz-Stelinski K, Mann RS, Stelinski LL (2011) Glutathione S-transferase and cytochrome P_{450} (general oxidase) activity levels in *Candidatus* Liberibacter asiaticus-infected and uninfected Asian citrus psyllid (Hemiptera: Psyllidae). Ann Entomol Soc Am 104: 297–305.
32. El-Shesheny I, Hajeri S, El-Hawary I, Gowda S, Killiny N (2013) Silencing abnormal wing disc gene of the Asian Citrus Psyllid, *Diaphorina citri* disrupts adult wing development and increases nymph mortality. PLoS ONE 8(5): e65392.
33. Brogdon WG, McAllister JC, Vulule J (1997) Heme peroxidase activity measured in single mosquitoes identifies individuals expressing an elevated oxidase for insecticide resistance. J Am Mosquito Contr 13: 233–237.
34. Livak KJ, Schmittgen TD (2001) Analysis of relative gene expression data using realtime quantitative PCR and the $2^{-\Delta\Delta C}{}_T$ method. Methods 25: 402–408.
35. Wheeler MM, Tarver MR, Coy MR, Scharf ME (2010) Characterization of four esterase genes and esterase activity from the gut of the termite *Reticulitermes flavipes*. Arch Insect Biochem Physiol 73: 30–48.
36. Bradford MM (1976) A rapid and sensitive method for the quantitation of microgram quantities of protein utilizing the principle of protein-dye binding. Anal Biochem 72: 248–254.
37. Hunter CP (1999) Genetics: A touch of elegance with RNAi. Curr Biol 9: R440–R442.
38. Rajagopal R, Sivakumar S, Agrawal N, Malhotra P, Bhatnagar RK (2002) Silencing of midgut aminopeptidase N of *Spodoptera litura* by double-stranded RNA establishes its role as *Bacillus thuringiensis* toxin receptor. J Biol Chem 277: 46849–46851.
39. Surakasi VP, Mohamed AAM, Kim Y (2011) RNA interference of beta 1 integrin subunit impairs development and immune responses of the beet armyworm, *Spodoptera exigua*. J Insect Physiol 57: 1537–1544.
40. Whyard S, Singh AD, Wong S (2009) Ingested double-stranded RNAs can act as species-specific insecticides. Insect Biochem Molec Biol 39: 824–832.
41. Mao YB, Cai WJ, Wang JW, Hong GJ, Tao XY, et al. (2007) Silencing a cotton bollworm P_{450} monooxygenase gene by plant-mediated RNAi impairs larval tolerance of gossypol. Nature Biotechnol 25: 1307–1313.
42. Baum JA, Bogaert T, Clinton W, Heck GR, Feldmann P, et al. (2007) Control of coleopteran insect pests through RNA interference. Nature Biotechnol 25: 1322–1326.
43. Clemens MJ, Elia A (1997) The double-stranded RNA-dependent protein kinase PKR: Structure and function. J Interferon Cytokine Res 17: 503–524.

Modeling Dynamics of *Culex pipiens* Complex Populations and Assessing Abatement Strategies for West Nile Virus

Kasia A. Pawelek[1]*, Patrick Niehaus[1]§, Cristian Salmeron[1]§, Elizabeth J. Hager[2], Gregg J. Hunt[2]

1 Department of Mathematics and Computational Science, University of South Carolina Beaufort, Bluffton, South Carolina, United States of America, **2** Beaufort County Mosquito Control, Beaufort, South Carolina, United States of America

Abstract

The primary mosquito species associated with underground stormwater systems in the United States are the *Culex pipiens* complex species. This group represents important vectors of West Nile virus (WNV) throughout regions of the continental U.S. In this study, we designed a mathematical model and compared it with surveillance data for the *Cx. pipiens* complex collected in Beaufort County, South Carolina. Based on the best fit of the model to the data, we estimated parameters associated with the effectiveness of public health insecticide (adulticide) treatments (primarily pyrethrin products) as well as the birth, maturation, and death rates of immature and adult *Cx. pipiens* complex mosquitoes. We used these estimates for modeling the spread of WNV to obtain more reliable disease outbreak predictions and performed numerical simulations to test various mosquito abatement strategies. We demonstrated that insecticide treatments produced significant reductions in the *Cx. pipiens* complex populations. However, abatement efforts were effective for approximately one day and the vector mosquitoes rebounded until the next treatment. These results suggest that frequent insecticide applications are necessary to control these mosquitoes. We derived the basic reproductive number ($\Re_0$) to predict the conditions under which disease outbreaks are likely to occur and to evaluate mosquito abatement strategies. We concluded that enhancing the mosquito death rate results in lower values of $\Re_0$, and if $\Re_0<1$, then an epidemic will not occur. Our modeling results provide insights about control strategies of the vector populations and, consequently, a potential decrease in the risk of a WNV outbreak.

Editor: Tian Wang, University of Texas Medical Branch, United States of America

Funding: This work is partially supported by the grants from the University of South Carolina Beaufort Sea Islands Institute (KAP), University of South Carolina Magellan Scholar Program (KAP,PN), and the SC EPSCoR/IDeA Scientific Advocate Network (KAP,CS). The funders had no role in study design, data collection and analysis, decision to publish, or preparation of the manuscript.

Competing Interests: The authors have declared that no competing interests exist.

* Email: kpawelek@uscb.edu

§ These authors contributed equally to this work.

Introduction

West Nile virus (WNV) is a vector-borne enzootic pathogen amid a bird-mosquito-bird cycle throughout small to large regional areas, with humans and horses as incidental hosts [1–3]. A portion of mosquitoes potentially associated with WNV may rest and/or breed in underground stormwater systems, such as catch basins. These stormwater structures are engineered to drain excess rain water from streets, parking lots, and other non-pervious areas [4,5]. Mosquitoes may develop in large numbers in the nutrient-rich standing water among the catch basins and adjacent pipes. The most important mosquito species adopting this underground habitat in the United States is the *Culex pipiens* complex (*Cx. pipiens* Linnaeus, *Cx. quinquefasciatus* Say, and/or hybrids), primary vectors for WNV in the northern and southern regions, including Beaufort County, South Carolina [6].

Forty-eight states reported at least 39,000 human cases of WNV infection with 1,663 fatalities as a major health concern in the U.S. during 1999–2013 [7,8]. In SC, 73 human cases of WNV, including 5 deaths, were confirmed from 2002–2013 [9]. Anyone living in or visiting an area where WNV is prevalent is at risk of becoming infected. [1]. In humans, WNV infection may be asymptomatic or may cause febrile illness, encephalitis, or meningitis [1,10]. Although the majority (about 80%) of WNV-infected people are asymptomatic [1], adults at least 50 years of age have the greatest risk of developing the more severe symptoms of WNV upon infection [11]. Approximately 20% of the people infected with WNV will show flu-like symptoms whereas less than 1% of infected individuals will develop neurologic illness causing severe or life-threatening symptoms [1].

Public health insecticides (for immature and adult mosquitoes) and source reduction (elimination of mosquito breeding sites) represent important strategies to reduce the risk of WNV outbreaks throughout the U.S. [12]. The application of larvicides and adulticides must be conducted in a timely manner. Other risk-reducing approaches include the use of insect repellents and protective clothing as well as avoidance of peak mosquito feeding hours [1].

Public health plans developed before and during a disease epidemic are typically influenced by previous outbreaks. Further, mathematical models are needed to test abatement strategies based on current and past epidemics. For example, mathematical models represented an important tool in preparation for an influenza outbreak [13,14]. A number of models have also been developed to study within-host dynamics of virus infections (for instance, [15,16]). Susceptible, Exposed, Infected, Recovered (SIR/SEIR) models have been studied [17,18] and various mathematical models have been developed to study the spread of WNV [19–24]. Also, models have examined insecticide treatments to reduce mosquito and Triatomine bug populations, such as during malaria [25] and Chagas disease outbreaks, respectively [26,27]. However, only a few models have studied the abatement strategies used for WNV intervention [28,29]. Mathematical models for WNV transmission and calculations of the basic reproductive ratio, which is used to predict disease outbreaks and evaluate abatement strategies, have been compared to each other [21]. The study concluded that model assumptions are the key features when finding the basic reproductive ratio and varying assumptions may lead to contradictory results [21]. The feeding preferences of *Cx. pipiens* were evaluated for preferred and alternative bird species [22]. The study suggested that inclination of mosquitoes to feed on the American robin (*Turdus migratorius* Linnaeus) was the key parameter influencing the timing of the WNV infection peak [22]. Mosquito reduction strategies and personal protection were evaluated using a single-season deterministic model during a WNV outbreak [28]. The analysis of equilibria was conducted to determine the conditions for the WNV persistence [28]. None of the aforementioned articles examined the insecticide effectiveness based on surveillance data and simultaneously estimate the disease burden during a WNV outbreak, which we investigated in this study.

We developed a mathematical model established on previous modeling techniques [28,30] and compared the model with *Cx. pipiens* complex surveillance data to estimate the parameters associated with treatment effectiveness, as well as mosquito birth, maturation, and death rates. We employed the obtained estimates in an epidemiological model with various severity stages of the disease to test the mosquito abatement strategies during a WNV outbreak. This allowed us to obtain more precise predictions, which are crucial to develop and implement control strategies during a disease outbreak. Further, we obtained possible disease burden estimates by organizing the human population into various disease severity stages. These scenarios could be essential when various control strategies are used, such as drug administration, during limited medication supply and/or resistance.

Materials and Methods

Study area

Beaufort County is located within the Lower Coastal Plain in the southern portion of SC. The topography is generally low and flat with vast wetlands, including tidal salt and brackish waters as well as bottomland hardwood swamps [31]. Beaufort County consists of 923 square miles (576 square miles for land and 347 square miles for water) with about 168,000 residents [32]. Various habitats for birds, animals, mosquitoes, and other insects are located throughout Beaufort County [31]. Approximately 39.7% of the Beaufort County population consists of adults who are 50 years and older [33]. Sun City Hilton Head (located near the Town of Bluffton in Beaufort County) is the largest senior adult community in SC, with over 13,000 residents [34].

Study sites

The study sites consisted of the Sams Point neighborhood on Ladys Island (Figure 1) and the Battery Point neighborhood on Port Royal Island (Figure 2). The GPS coordinates were 32.423717–80.644145 and 32.422526–80.711176, respectively. These neighborhoods share similarities: urban residential communities, commercial properties, underground stormwater systems, roadside ditches, moderate to large tree canopies, navigable waterways, pest and/or vector mosquitoes, and feral birds.

Mosquito collections

As part of the surveillance program, Beaufort County Mosquito Control (BCMC) operated 20 strategically located mosquito traps (which included the two study sites) within the underground stormwater systems throughout Beaufort County. Encephalitis Vector Survey (EVS) traps (BioQuip Products, San Dominguez, CA) were used for the weekly collections. The traps were suspended underground below the catch basin lids and were operated for about 24 continuous hours each week. Each EVS trap included a blue LED light bulb and CO2 (approximately 3 pounds of dry ice) as attractants whereas a down-draft fan captured the mosquitoes into a collection net. From 2006 to 2012, weekly collections were identified to male and female species using a stereo microscope. We evaluated the *Cx. pipiens* complex for modeling because the mosquito species are the most abundant vectors for WNV in the study area.

Public health insecticide treatments

As part of the integrated mosquito management (IMM) program, BCMC used three abatement strategies (as needed) in 2006–2012 to control immature and adult pest and/or vector mosquitoes throughout Beaufort County, including the two study sites (except aerial spraying). For the first strategy, BCMC applied a larvicide (Altosid XR briquets) to about 20,000 catch basins, including those stormwater structures at Sams Point and Battery Point neighborhoods, at the beginning of each mosquito season. The application rate was one briquet per catch basin. This product was used to control mosquito breeding and was effective for up to 5 months (according to the manufacturer label) or longer depending on the frequency of rain events. BCMC used mapping software (ESRI ArcGIS ArcMap 10.1, Redlands, CA) to geocode this large inventory of catch basins. The treatment of aboveground breeding sites for the *Cx. pipiens* complex is mostly nonexistent because this type of breeding occurs among various water-holding containers (bird baths, waste tires, etc.) on private properties. For the second strategy, BCMC operated up to 7 spray trucks using various Ultra-Low Volume (ULV) adulticiding products during the 7-year-long study (Table 1). Before the start of each mosquito season, BCMC evaluated the efficacy rate (via 24-hour bioassays) of the adulticiding product selected for use. The efficacy rates ranged from 91–100% (Table 1). These ground-dispersed products are effective as contact insecticides during the night in which most of the active ingredients break down within an hour after sunlight exposure. For the third strategy, BCMC operated a fixed wing aircraft using another adulticide, Anvil 10+10 ULV (0.48 ounces per acre with a 90% efficacy rate), during mostly sunrise. However, aerial spraying did not occur at the two study sites because beehives and/or fish-bearing waters existed at both sites.

The mosquito surveillance data was examined to determine the relationships between the *Cx. pipiens* complex populations and the effects of the insecticide treatments during June-September when the highest average temperature and rainfall data were typically recorded (temperature data from [35]).

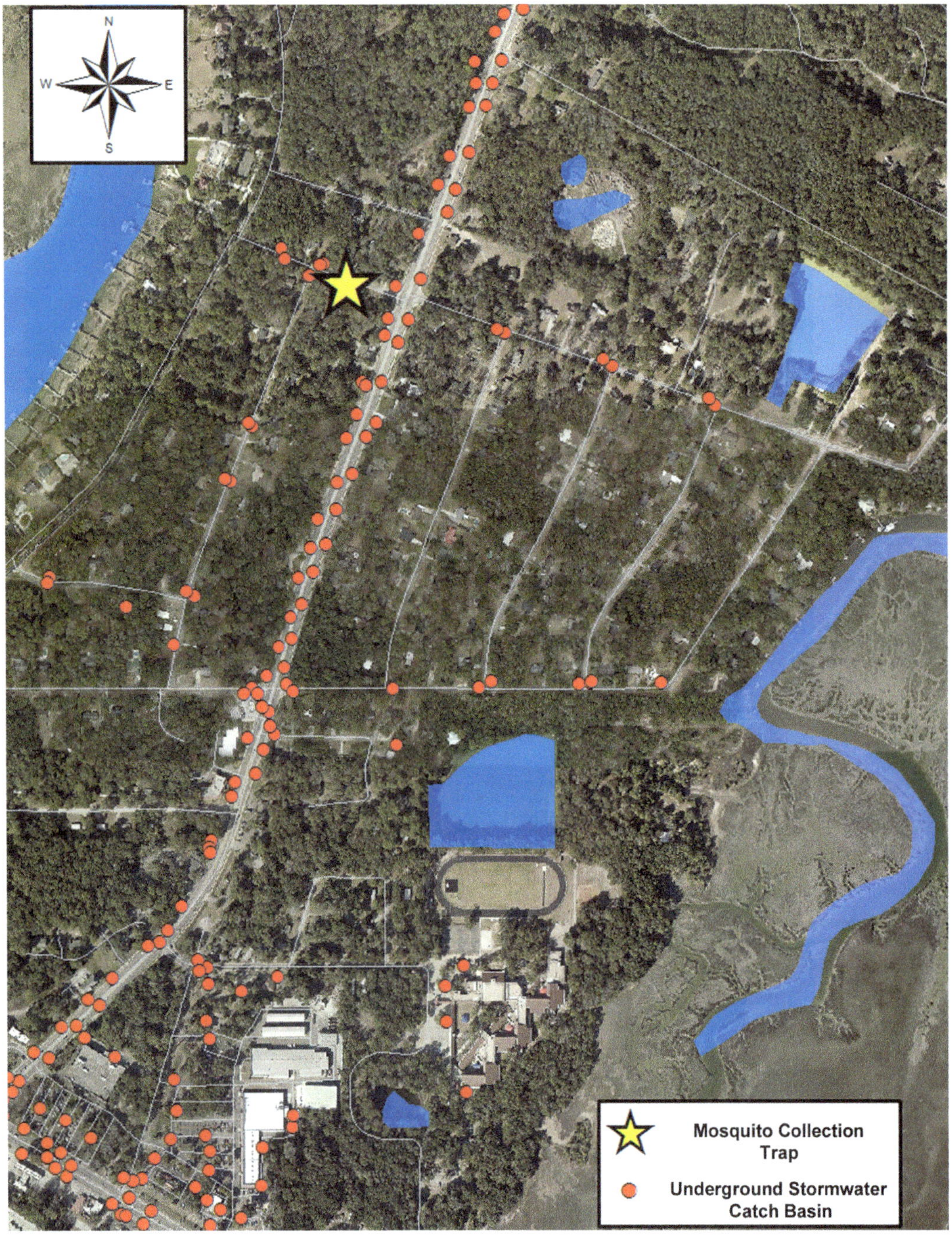

Figure 1. Aerial view of Sams Point neighborhood on Ladys Island, Beaufort County, SC depicting the location of the mosquito collection trap and various underground stormwater catch basins.

Ethics statement

A permit to conduct the study within Beaufort County was not required because the underground stormwater systems (including catch basins) are owned and maintained by several governmental entities, such as SC Department of Transportation, Beaufort County, and/or various municipalities. Specific permission was not required to conduct surveillance of mosquitoes collected at the various catch basins. However, the Director of BCMC coordinated activities with the Director of Beaufort County Public Works (who is responsible for the drainage infrastructure systems). BCMC applicators possess SC Non-Commercial Applicator Licenses as certified by Clemson University Department of Pesticide Regulation. County Council of Beaufort County has approved an annual budget for BCMC since 1974 and continues to endorse the control of mosquitoes and mosquito-borne diseases throughout its political boundary. Our field study did not involve endangered or protected species.

Mathematical model

Mathematical modeling is frequently used to study the dynamics of vector populations and disease outbreaks [17–30,36]. In this study, we designed a mathematical model based on Ordinary

Figure 2. Aerial view of Battery Point neighborhood on Port Royal Island, Beaufort County, SC depicting the location of the mosquito collection trap and various underground stormwater catch basins.

Differential Equations (model equations are listed in the Appendix S1 in File S1) to predict larval and adult mosquito populations and study the potential spread of WNV among birds and humans under various insecticide application scenarios and parameter estimates.

In the model, similar to [30,37], we assume that WNV transmission is dependent on the abundance of bird reservoirs and humans. The model consists of three populations: bird reservoirs, potential *Cx. pipiens* complex vectors, and human populations, which are organized into subpopulations based on the development and/or severity of WNV infection. Naïve adult *Cx. pipiens* complex mosquitoes (M_S), exposed (M_E), and infectious (M_I) adult mosquitoes lay eggs and, afterward, larvae (L) are born at the rate b. Larvae (L) originating from non-infected, exposed, and infectious mosquitoes mature to become susceptible adult mosquitoes (M_S) at the rate m. The natural death rates of larvae and three populations of adult mosquitoes are denoted by δ_L and δ_M, respectively. The naïve, exposed, and infectious mosquitoes have the same natural death rate (δ_M) because it is difficult to determine the cause of death.

We take the average number of mosquito bites on both birds and humans to be dependent on the total sizes of these populations

Table 1. Public health insecticides applied from spray trucks to control adult mosquitoes throughout Beaufort County, SC (including the two study sites) during 2006–2012.

Year	Public Health Insecticide	Application Rate (ounces per acre)	Efficacy Rate (%)*
2006	Aqua-Reslin	0.69	100
2007	Aqua-Reslin	0.69	100
2008	Aqualuer 20-20	0.55	91
2009	Aqualuer 20-20	0.55	91
2010	Zenivex E20	0.57	94
2011	Evoluer 30-30	0.83	93
2012	Evoluer 30-30	0.83	93

*After 24-hour bioassays.

because mosquitoes feed on birds and humans. The term $\beta\alpha_H M_I H_S/N_{Total}$ represents the rate of reservoir-frequency dependent infection when an infectious mosquito (M_I) takes a blood meal from a susceptible human (H_S), in which β is the average biting rate per day, α_H is the probability of virus transmission to human per infectious bite, and N_{Total} is the total blood supply of the entire populations of humans $N_H = H_S+H_E+H_F+H_N+R$, and birds $N_B = B_S+B_I+B_R$. In the rate of infection, the fraction of mosquito bites on susceptible humans is represented by the ratio of susceptible humans to the total populations of humans and birds, H_S/N_{Total}. The disease incubation period in humans is denoted by $1/\delta_E$. After $1/\delta_E$ days, a portion of the population (κ) develops WNV neuroinvasive disease (acute flaccid paralysis, encephalitis, or meningitis) (H_N). Another segment of the population (γ) is asymptomatic/recovered (R) and the remaining portion of the infected individuals (1-κ-γ) develops WNV fever (H_F). We did not distinguish between asymptomatic and recovered individuals because both populations do not exhibit any disease symptoms. After $1/\delta_E$ days, all of the individuals with WNV fever (H_F) recover and they are reclassified in the recovered category (R), because it is uncommon to die from the WNV fever [10]. Due to the severity of the neuroinvasive disease, a fraction of these individuals (ω) will die after $1/\delta_N$ days and they are repositioned in the deceased individual category (D).

Infected mosquitoes (M_I) may transmit the disease to susceptible birds (B_S) at the reservoir-frequency dependent rate $\beta\alpha_B M_I B_S/N_{Total}$, where α_B is the probability of virus transmission to a bird per infectious bite and fraction of mosquito bites on susceptible birds is represented by the ratio of susceptible birds to the total populations of humans and birds, B_S/N_{Total}. Due to the severity of the disease, a fraction of infected birds (σ) will die after $1/\delta B$ days and the rest of the birds (1-σ) will recover. The infected, recovered, and dead birds due to the infection with WNV categories are represented by B_I, B_R, and B_D, respectively. For simplicity, the birth or immigration of susceptible birds is represented by the recruitment rate Λ. Recruitment of infected birds could be taken into consideration by assuming that the fraction of the recruited birds $f\Lambda$ is infected and the remaining proportion $(1$-$f)\Lambda$ is susceptible. The natural death rates of birds in all categories are represented by τ.

Finally, susceptible mosquitoes (M_S) become exposed to the infection at the reservoir-frequency dependent rate $\beta\alpha_M M_S B_I/N_{Total}$ when these insects bite an infected bird (B_I), where α_M is the probability of virus transmission to a mosquito per infectious bite and the fraction of mosquito bites on infected birds is represented by the ratio of infected birds to the total populations of humans and birds, B_I/N_{Total}. The viral incubation rate in mosquitoes is denoted by η. We omitted the natural death, birth, and recruitment rates of human populations because of the short duration of the considered disease outbreak (one-season) and for model simplicity.

In the model, the population of adult mosquitoes (M_S, M_E, and M_I) is decreased by insecticide interventions. The insecticide treatment is represented by the step function $S(t)$, which is equal to the insecticide treatment effectiveness (s) for the duration of its activity (α) and 0 when the treatment is inactive (insecticide treatment function is listed in the Appendix S1 in File S1). In this study, we investigate the effect of varying treatment effectiveness and frequency of its applications.

The schematic representation of the model including the spread of WNV among the populations is given in Figure 3, whereas model variables and parameters are listed in Tables 2 and 3, respectively. The model equations are shown in the Appendix S1 in File S1.

Basic reproductive number

We determined the basic reproductive number ($\Re_0$) for the model (Appendix S2 in File S1). The threshold $\Re_0$ is one of the most influential tools developed to analyze and interpret models [38–41]. Basic reproductive ratio ($\Re_0$) is described as the average number of new infections caused by a single case in a fully susceptible population. If $\Re_0>1$, then an epidemic will arise whereas if $\Re_0<1$, then an epidemic will not occur. The Next Generation Method [42] was employed to determine $\Re_0$ (see Appendix S2 in File S1 for derivation). The dependence of $\Re_0$ on the mosquito death rate (δ_M) and the mosquito biting rate per day (β) was investigated in this study.

Data fitting and parameter values

We compared the model with the number of *Cx. pipiens* complex mosquitoes collected in traps located at the Sams Point and Battery Pont neighborhoods from 2006 to 2012 (Figure 4 and Table 4). We considered the months of June-September separately for each year and the corresponding insecticide applications.

The model was fitted to the *Cx. pipiens* complex surveillance data using Berkeley Madonna Version 8.3.18 software to estimate the model parameters. In the data fitting, we incorporated the actual insecticide treatment schedules reported by the BCMC for each location. Estimated values were established through the best nonlinear least squares fit of the model for the surveillance data,

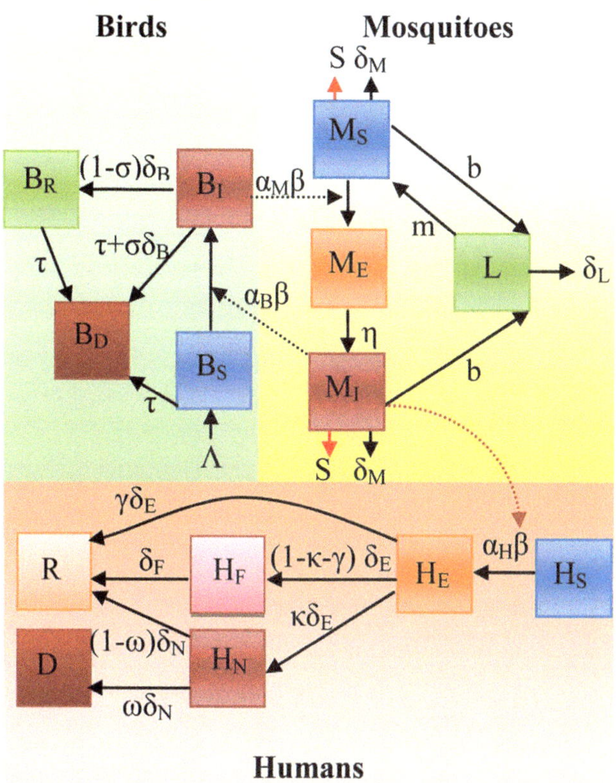

Figure 3. Schematic representation of the model. Model equations, variables, and parameters are given in the Appendix S1 in File S1, Tables 2 and 3, respectively.

(i.e., program minimized the root mean square (RMS) between the recorded data and the analogous model predictions) (Table 4 and Table S1 in File S1).

Lower and upper bounds for the selected parameters were employed in the data fitting procedure from the previous modeling studies (listed in Table 3). We scaled the model predictions, which are for the entire neighborhood, by a factor (p) to compare with the mosquito surveillance data from each trap. We took the lower bound of the treatment effectiveness to be 0.7 day^{-1}.The duration of the treatment effectiveness is assumed to last approximately 1 day since the ground-dispersed products are effective as contact insecticides during the night because most of the active ingredients break down within an hour after sunlight exposure.

Results

Collection and identification of mosquitoes

The *Cx. pipiens* complex represented 95.1% of all mosquito species collected and identified in the EVS trap located at Sams Point from 2006 to 2012 (Table 5). For the Battery Point site, the *Cx. pipiens* complex signified 86.8% of the total collections during the same 7 years (Table 6).

Overview of the comparison with surveillance data and numerical simulations

Our data fitting results suggest declines in the number of mosquitoes after insecticide treatments, which agree with the trend of the majority of the surveillance data from both trap locations in the absence of WNV infection (Figure 4). The best-fit parameter estimates are listed in Table 4 for both trap locations separately for each year from 2006 to 2012. Root Mean Square (RMS) for the best fit of the model to the surveillance data resulted in the highest values for years 2008 and 2009 (see Table S1 in File S1) and in the lowest values for years 2006, 2011, and 2012 for both trap locations. However, due to the low numbers of mosquitoes in the Sams Point trap in 2012, our model did not capture a minor increase of mosquitoes in August-September (Figure 4). We evaluated various treatment scenarios based on the obtained parameter estimates (Figures 5 and 6). We used the best-fit parameter estimates in the model with the WNV infection, to predict the changes of human, mosquito, and bird populations during a WNV outbreak (Figure 7). Further, we conducted additional sensitivity analysis of the model parameters in Figure 8 and Figures S1–S4 in File S1.

Table 2. Definitions of variables used in the model.

Variable	Definition	Initial Condition
L	Female larval mosquito density	$L_0 = L(0) = 1{,}000$
M_S	Susceptible female adult mosquito density	$M_0 = M_S(0) = 1{,}000$
M_E	Exposed female adult mosquito density	$M_E(0) = 0$
M_I	Infectious female adult mosquito density	$M_I(0) = 1/M_0$
B_S	Susceptible bird density	$B_0 = B_S(0) = 1{,}000$
B_I	Infectious bird density	$B_I(0) = 0$
B_R	Recovered bird density	$B_R(0) = 0$
B_D	Dead birds due to the infection with WNV density	$B_D(0) = 0$
H_S	Number of susceptible humans	$N = H_S(0) = 1{,}000$
H_E	Number of exposed humans	$H_E(0) = 0$
H_F	Number of humans with WNV fever	$H_F(0) = 0$
H_N	Number of humans with neuroinvasive disease	$H_N(0) = 0$
R	Number of recovered humans	$R(0) = 0$
D	Number of dead humans	$D(0) = 0$

Sensitivity analysis of initial conditions is illustrated in Figure S3 in File S1.

Table 3. Parameter definitions, values, and references.

Symbol	Definition, Units	Range, Reference	Value, Reference
s	Treatment effectiveness, Day^{-1}	0–1	fitted
α	Treatment duration, Days	0–1	1[I]
b	Birth rate, Larvae Day^{-1} $Adults^{-1}$	0.02–0.15, calculated[II]	fitted
δ_M	Natural death rate of adult mosquitoes, Day^{-1}	0.02–0.07, [54,55]	fitted
δ_L	Natural death rate of larvae, Day^{-1}	0.01–0.06, [54,55]	fitted
m	Maturation rate of larvae, Adults $Larvae^{-1}$ Day^{-1}	0.05–0.09, [37,56,57]	fitted
β	Biting rate per day	1–5[III]	mean
α_B	Probability of virus transmission to bird per infectious bite	0.27–1.00, [18,39,58–60]	mean
α_M	Probability of virus transmission to mosquito per infectious bite	0.23–1.00, [18,28,38,39,59,61]	mean
α_H	Probability of virus transmission to human per infectious bite	-	0.88, [28]
$1/\delta_B$	Duration of viremia in birds, Days	3.8–6.0[IV], [43]	4.5[IV], [43]
$1/\delta_E$	Incubation period for humans, Days	2–6	mean
$1/\delta_F$	Duration of the WN fever, Days	-	14, [28]
$1/\delta_N$	Duration of the neuroinvasive disease, Days	33–42, [62]	mean
κ	Fraction of the human population that can develop neuroinvasive disease	-	<0.01[V], [1]
γ	Fraction of the human population that is asymptomatic	0.7–0.8, [1]	mean
ω	Fraction of the human population with the neuroinvasive disease dying from the disease	-	0.1, [1]
σ	Fraction of the WNV-infected bird population dying from the disease	0.5–1.0[IV], [43]	0.72[IV], [43]
η	Virus incubation in mosquitoes, Day^{-1}	0.09–0.12, [63]	0.1, [63]
Λ	Recruitment rate of susceptible birds, Birds Day^{-1}	-	1.5[VI]
τ	Natural death rate of birds, Day^{-1}	0.001–0.002, [22,37,64]	0.0015, [22,37,64]

[I]Value based on BCMC field observations and discussions.
[II]The lower and upper bounds for the birth rate were calculated based on the steady state, $\delta_M(m+\delta_L)/m$ of the model without the infection, and the upper and lower bounds of the parameters utilized in the formula.
[III]It was estimated that *Culex quinquefasciatus* has 1–5 gonotrophic cycles [65], which are directly related to the number of blood meals taken by the mosquito [66]. We have also varied the values of mosquito biting rate in Figure 8 and Figure S4 in File S1.
[IV]Values taken from [43] for the following bird species: blue jay, house finch, American crow, house sparrow, and fish crow, which are mainly tested by SCDHEC [67].
[V]In simulations we assume $\kappa = 0.01$ to account for the worst case disease outbreak scenario.
[VI]The bird recruitment rate was calculated based on the steady state, $\Lambda = \tau B_0$ of the model without the infection.

Declines in the number of mosquitoes after insecticide treatments

According to our modeling predictions, the adulticide treatments produced significant reductions in the *Cx. pipiens* complex populations in specific neighborhoods (Figure 4). However, the abatement outcomes were effective for about one day and then the mosquito populations rebounded rapidly after the treatment became ineffective (Figure 4–6).

Insecticide treatment scenarios

Our model shows that the number of larval and adult *Culex pipiens* complex vary as we change the frequency of adulticide applications between every 1, 2, and 4 weeks (Figure 5A and B). Insecticide applied every 2 and 4 weeks results in a higher number of larval and adult mosquitoes than the weekly abatement scenario. Our results suggest that two months of weekly insecticide treatments will suppress the larval and adult mosquito populations to minor (acceptable) levels (Figure 5A and B). When the mosquito mortality due to the treatment is decreased from 0.7 day^{-1} to 0.4 day^{-1} or 0.2 day^{-1}, the vector populations are not as effectively controlled (Figure 5C and D). In particular, a mosquito mortality of 0.2 day^{-1} will result in approximately 3.5 times larger mosquito populations than the mortality of 0.7 day^{-1} (Figure 5C). The contour plot (Figure 6) for treatment effectiveness varying from 0–1 day^{-1} in which the adult mosquito population is shown as a function of the treatment effectiveness versus time provides a greater detail on the mosquito mortality needed to reduce the mosquito populations to the specific levels.

Control strategies during a WNV outbreak

In the absence of control, the number of mosquitoes remains elevated throughout the summer (Figure 7A). In particular, the number of infected mosquitoes quickly rises to a peak about a month after the beginning of the WNV infection (Figure 7A). When the biweekly insecticide treatment is applied, the number of susceptible and infectious vectors decreases significantly after each treatment (Figure 7B). The number of infectious mosquitoes is considerably reduced during the months of June–September and the population peak is approximately 5 times lower than without insecticide control (Figure 7B). However, more frequent weekly treatments result in an insignificant number of infectious mosquitoes (Figure 7C). The total number of vectors declines to an insignificant level in August and September, only two months after the start of the weekly treatments (Figure 7C). This scenario creates a minor number of disease cases (Figure 7F and I).

Our model predicts that the number of WNV fever cases will reach a peak in the mid-July in the non- insecticide treatment scenario (Figure 7G). The significant difference in the non-

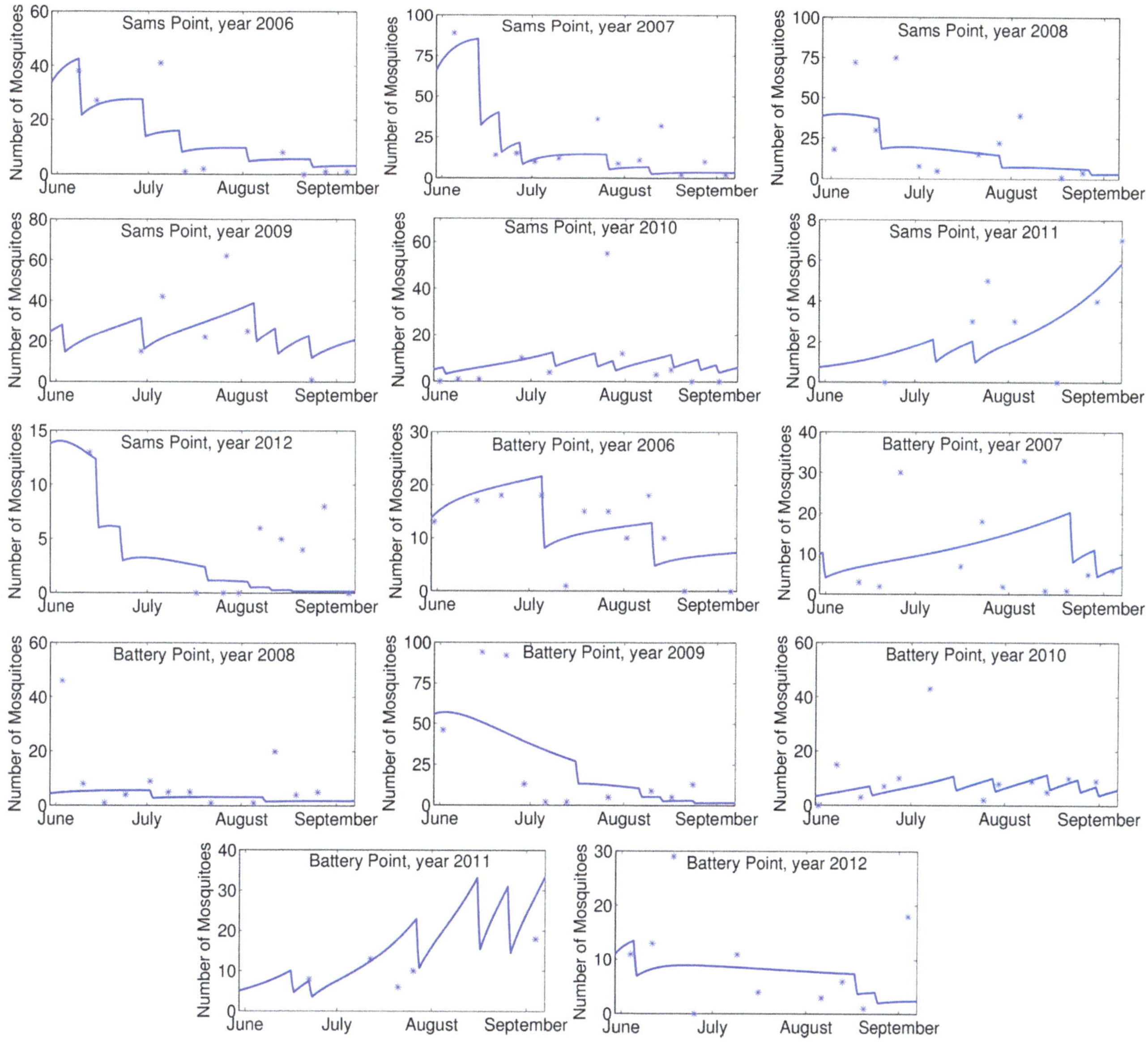

Figure 4. Model comparison with mosquito surveillance data. Model predictions of adult mosquito populations (blue lines) based on the best-fit to *Culex pipiens* complex surveillance data collected during the summer months at the Sams Point and Batter Pont trap locations (blue dots) during 2006–2012. The best-fit parameters estimates are listed in Table 4.

treatment versus biweekly application is the number of individuals with WNV fever and time of their amplification. The number of WNV fever cases is delayed by approximately 1.5 weeks and decreased by approximately a factor of 1.5. The number of neuroinvasive cases is low in the non-treatment and biweekly application and it is negligible when a weekly treatment is applied (Figure 7G–I).

Dead birds as disease indicators

The majority of people infected with WNV are asymptomatic or have flu-like symptoms [1]. Thus, it is difficult to detect a WNV disease outbreak based on the infrequent examination of these individuals by health care providers. However, an indication of a WNV disease outbreak may become obvious after numerous dead birds are discovered in a community. Residents can submit and/or report dead birds to the local and/or state government health department or a similar agency that monitors WNV activity. For instance, South Carolina Department of Health and Environmental Control (SCDHEC) requests the submission of a freshly dead blue jay (*Cyanocitta cristata* Linnaeus), house finch (*Haemorhous mexicanus* Muller), American crow (*Corvus brachyrhynchos* Brehm), house sparrow (*Passer domesticus* Linnaeus), and/or fish crow (*Corvus ossifragus* Wilson) from mid-March to the end of November [5]. Our modeling predictions indicate that such an occurrence of dead birds becomes noticeable in mid-June and reaches high numbers at the beginning of July in the absence of treatments (Figure 7D) and in the beginning of August when the biweekly treatments are applied (Figure 7E).

Basic Reproductive Number

The basic reproductive number ($\Re_0$) for the model is given by (see Appendix S2 in File S1 for derivation):

Table 4. Best fit parameter values to the *Culex pipiens* complex surveillance data.

	Location: Sams Point					
Year	δ_M	δ_L	*b*	*m*	*s*	*p*
2006	3.6×10^{-2}	1.6×10^{-2}	3.8×10^{-2}	9.0×10^{-2}	7.0×10^{-1}	3.4×10^{-2}
2007	2.1×10^{-2}	6.0×10^{-2}	7.7×10^{-2}	5.0×10^{-2}	1.0×10^{0}	6.6×10^{-2}
2008	2.2×10^{-2}	4.5×10^{-2}	3.0×10^{-2}	5.0×10^{-2}	7.0×10^{-1}	3.9×10^{-2}
2009	3.9×10^{-2}	1.0×10^{-2}	7.4×10^{-2}	7.8×10^{-2}	7.1×10^{-1}	2.4×10^{-2}
2010	3.1×10^{-2}	1.0×10^{-2}	8.7×10^{-2}	9.0×10^{-2}	7.0×10^{-1}	5.0×10^{-3}
2011	7.0×10^{-2}	1.9×10^{-2}	1.5×10^{-1}	9.0×10^{-2}	8.3×10^{-1}	7.3×10^{-4}
2012	5.2×10^{-2}	3.0×10^{-2}	3.0×10^{-2}	6.4×10^{-2}	7.2×10^{-1}	1.4×10^{-2}
Average	3.9×10^{-2}	2.7×10^{-2}	6.9×10^{-2}	7.3×10^{-2}	7.7×10^{-1}	2.6×10^{-2}
	Location: Battery Point					
2006	2.0×10^{-2}	4.9×10^{-2}	4.9×10^{-2}	6.4×10^{-2}	1.0×10^{0}	1.3×10^{-2}
2007	2.4×10^{-2}	6.0×10^{-2}	9.8×10^{-2}	5.0×10^{-2}	9.6×10^{-1}	1.0×10^{-2}
2008	2.0×10^{-2}	1.0×10^{-2}	2.0×10^{-2}	5.0×10^{-2}	7.0×10^{-1}	4.4×10^{-3}
2009	4.9×10^{-2}	1.0×10^{-2}	2.0×10^{-2}	6.0×10^{-2}	7.0×10^{-1}	5.6×10^{-2}
2010	2.0×10^{-2}	1.1×10^{-2}	7.5×10^{-2}	8.9×10^{-2}	7.1×10^{-1}	3.4×10^{-3}
2011	5.2×10^{-2}	1.9×10^{-2}	1.5×10^{-1}	9.0×10^{-2}	8.5×10^{-1}	5.1×10^{-3}
2012	3.1×10^{-2}	3.3×10^{-2}	3.7×10^{-2}	9.0×10^{-2}	7.0×10^{-1}	1.1×10^{-2}
Average	3.1×10^{-2}	2.7×10^{-2}	4.5×10^{-2}	7.0×10^{-2}	8.0×10^{-1}	2.1×10^{-2}

The fittings are displayed in Figure 4.

$$\Re_0=\sqrt{\frac{\alpha_B\beta\eta}{\delta_M(\delta_M+\eta)}\cdot\frac{\tilde{B}_S}{\tilde{B}_S+\tilde{H}_S}\cdot\frac{\alpha_M\beta\tilde{M}_S}{(\delta_B+\tau)(\tilde{B}_S+\tilde{H}_S)}}.$$

Similarly to [37], the first and second factors under the square root correspond to the number of secondary bird infections caused by a single infectious mosquito and the probability that the blood meal will be from a susceptible bird. The third factor denotes the number of secondary mosquito infections caused by one infected bird. The basic reproductive number does not depend on the secondary human infections because humans cannot infect mosquitoes with WNV. However, $\Re_0$ depends on the total blood supply (N_{Total}) evaluated at the disease free equilibrium, which consists of the populations of birds and humans ($\tilde{B}_S+\tilde{H}_S$). If $\Re_0$ is above the threshold value of 1, a disease outbreak is possible. The goal of mosquito abatement is to lower $\Re_0$ to a value below 1. In particular, $\Re_0$ depends on the mosquito death rate (δ_M) and the mosquito biting rate per day (β), which can impact the value of $\Re_0$. For the death rate value of 0.15 day^{-1} and for the mosquito biting rate per day of 3, we estimated $\Re_0$ to be 3.02, which suggest a rapid spread of WNV. A higher death rate and a lower biting rate result in much lower $\Re_0$. In particular, the death rate of $\geq$0.45 day^{-1} and the biting rates ranging from 1 to 4, will decrease $\Re_0$ to below the threshold value of 1, which will prevent a WNV outbreak (Table 7). Our model predicts $\Re_0$ to decline below the threshold value when the mosquito biting rate per day is 1 and the mosquito death rate is>1.15 day^{-1} (Table 7 and Figure 8). Also, increasing the mosquito death rate to 0.35 day^{-1} and decreasing the mosquito biting rate per day to below 4 results in a ratio below 1. Various scenarios that decrease the ratio to below the threshold value and increase the probability to control the disease are summarized in Table 7 and Figure 8.

Sensitivity tests

We also conducted sensitivity tests of the model parameters on the larval and adult mosquito populations (Figures S1 and S2 in File S1). We varied the parameters (described in each figure legend); specifically, we examined the upper and lower bounds of the parameters and their means (listed in Table 3). The remaining parameters were fixed at the mean of the upper and lower bounds. The variations in the initial number of larvae (L_O) and adults (M_O) show that these values increased proportionally with the final set point of the vector populations (Figure S1 in File S1). Sensitivity analysis of the maturation (m), birth (b), larval mosquito death (δ_L), and adult mosquito death (δ_M) rates on the resulting larval and adult mosquito populations show the mean values of these parameters maintain a consistent level of mosquito populations (Figure S2 in File S1). Lower bound parameter values of m and b cause the decline of these populations whereas upper bound estimates result in an increase. Conversely, lower bound parameter values of δ_L and δ_M cause the increase of these populations whereas upper bound estimates result in a decrease. This outcome suggests that decreasing m and/or b and/or increasing δ_L and/or δ_M will reduce the mosquito populations. For example, residents and visitors could support mosquito abatement efforts by removing *Cx. pipiens* complex breeding sites, such as water-filled containers, thus increasing δ_L, which will result in a decrease of the larvae and, subsequently, the adult mosquito populations.

In addition, we have also conducted sensitivity tests of predicted infected mosquitoes (M_I), infected birds (B_I), and humans with WNV fever (H_F) populations to the initial values in the model with WNV infection (Figure S3 in File S1) and sensitivity tests of predicted infected mosquitoes (M_I), infected birds (B_I), dead birds due to the infection with WNV (B_D), humans with WNV fever (H_F) populations to the mosquito biting rate per day (β) (Figure S4 in File S1). These sensitivity tests depict the differences in the time

Table 5. Summary of adult female mosquitoes (with mean females per trap night) collected in an EVS trap located at Sams Point neighborhood in Beaufort County, SC from 2006–2012.

Mosquito Species	2006	2007	2008	2009	2010	2011	2012	Totals	% Totals
Aedes albopictus		4 (0.08)	2 (0.04)		2 (0.04)		17 (0.33)	25	0.59
Aedes taeniorhynchus	18 (0.35)	4 (0.08)	6 (0.12)	4 (0.08)	55 (1.06)	56 (1.08)	16 (0.31)	159	3.73
Aedes vexans							1 (0.02)	1	0.02
Culex pipiens complex	288 (5.54)	1,409 (27.10)	1,019 (19.60)	498 (9.58)	470 (9.04)	128 (2.46)	236 (4.54)	4,048	95.05
Culex restuans		2 (0.04)	7 (0.13)				1 (0.02)	10	0.23
Culex salinarius			2 (0.04)		1 (0.02)		6 (0.12)	9	0.21
Culex territans	2 (0.04)							2	0.05
Uranotaenia sapphirina						5 (0.10)		5	0.12
Totals	308	1,419	1,036	502	528	189	277	4,259	100.00

We fitted the model to the *Culex pipiens* complex surveillance data from both traps separately during 2006–2012.

Table 6. Summary of adult mosquitoes (with mean females per trap night) collected in an EVS trap located at Battery Point neighborhood in Beaufort County, SC from 2006–2012.

Mosquito Species	2006	2007	2008	2009	2010	2011	2012	Totals	% Totals
Aedes albopictus	3 (0.06)	18 (0.35)	8 (0.15)	4 (0.08)	5 (0.10)	4 (0.08)		42	0.93
Aedes taeniorhynchus		2 (0.04)	1 (0.02)		2 (0.04)	5 (0.10)	3 (0.06)	13	0.29
Aedes vexans	16 (0.31)	6 (0.12)	19 (0.37)	2 (0.04)	4 (0.08)	38 (0.73)	20 (0.38)	105	2.34
Anopheles crucians	3 (0.06)					1 (0.02)		4	0.09
Culex erraticus				1 (0.02)				1	0.02
Culex nigripalpus	8 (0.15)	3 (0.06)	5 (0.10)	13 (0.25)				29	0.65
Culex pipiens complex	415 (7.98)	735 (14.13)	534 (10.27)	728 (14.00)	547 (10.52)	442 (8.50)	498 (9.58)	3,899	86.80
Culex restuans	35 (0.67)	17 (0.33)	70 (1.35)	38 (0.73)		16 (0.31)	9 (0.17)	185	4.12
Culex salinarius	18 (0.35)	8 (0.15)	163 (3.13)	6 (0.12)	3 (0.06)		6 (0.12)	204	4.54
Culex territans	1 (0.02)		4 (0.08)					5	0.11
Uranotaenia sapphirina		1 (0.02)	2 (0.04)		2 (0.04)			5	0.11
Totals	499	790	806	792	563	506	536	4,492	100.00

We fitted the model to the *Culex pipiens* complex surveillance data from both traps separately during 2006–2012.

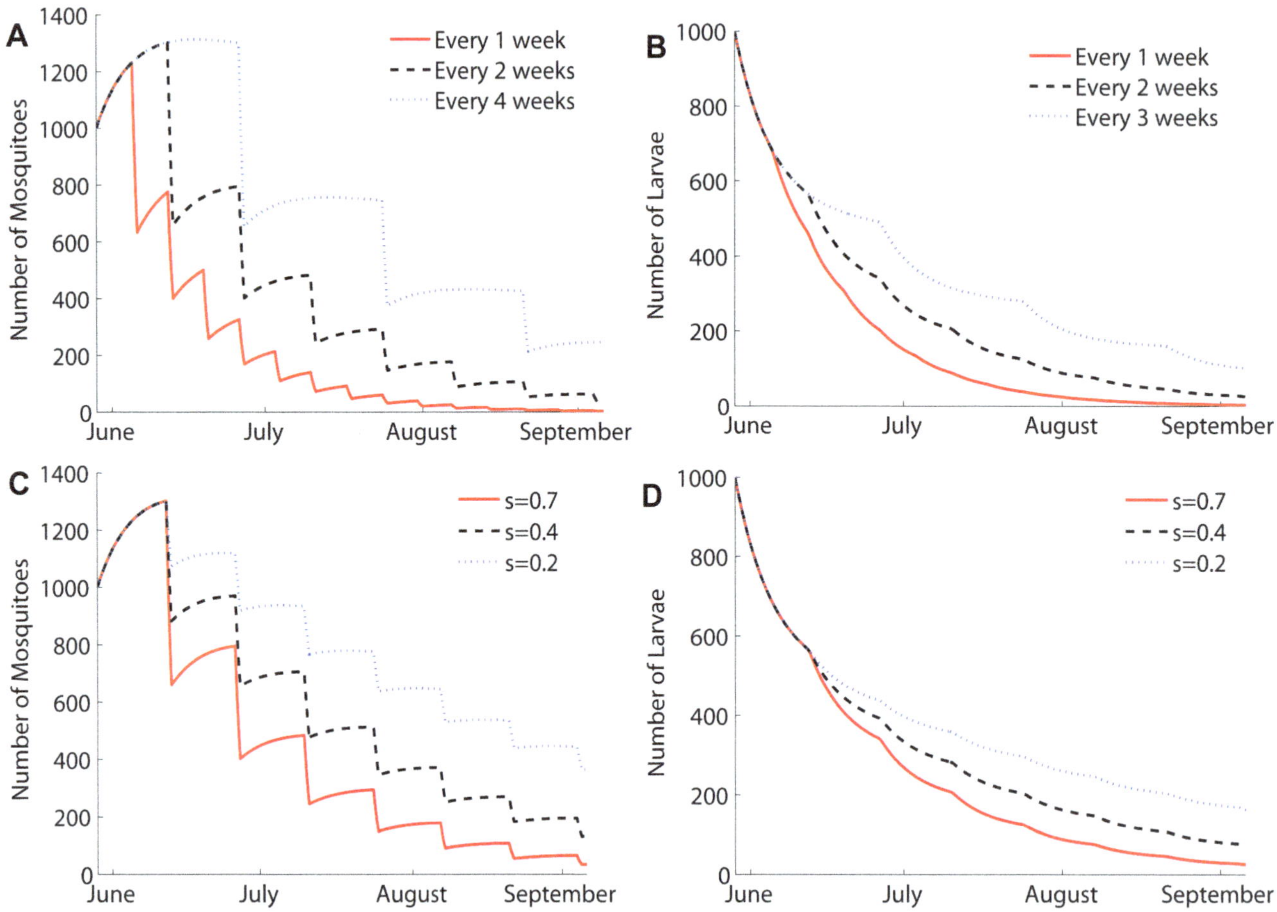

Figure 5. Effectiveness of various treatment scenarios predicted by the model in the absence of the WNV infection. Demonstration of how various insecticide applications impact the adult mosquito (A and C) and larval populations (B and D). Dates of the insecticide treatments (A and B) and treatment effectiveness were varied (C and D) as indicated in the legends in each figure. The remaining parameter estimates were taken from the Tables 3 and 4 (best-fit parameter values are taken for the year 2006 from Sams Point trap location).

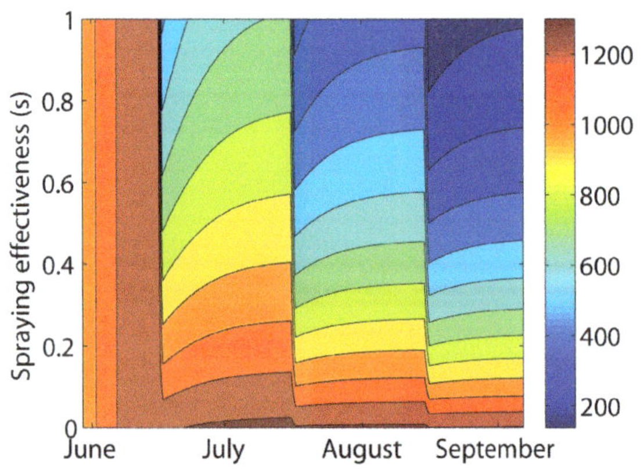

Figure 6. Sensitivity test of the predicted adult mosquito population to the model parameter representing treatment effectiveness in the absence of the WNV infection in the model. Contour plot of the adult mosquito populations as a function of the treatment effectiveness (*s*) versus time. The remaining parameters were fixed and chosen from the Tables 3 and 4 (best-fit parameter values are taken for the year 2006 from Sams Point trap location).

and height of the infection peak as the initial conditions and the mosquito biting rate are varied.

Discussion

Mosquito surveillance data was evaluated for a specific mosquito trap used to collect *Cx. pipiens* complex mosquitoes, primary vectors of WNV in regions of the U.S. [6]. While comparison to the field data is vital to estimate the effectiveness of various mosquito control strategies, previous modeling efforts rarely utilized surveillance data to estimate the parameters and validate the models. Our modeling predictions and the trend of the majority of the surveillance data depict declines in the number of mosquitoes after each insecticide treatment (Figure 4). The data and curves generated by our model show significant declines in the number of mosquitoes when insecticide treatments are applied; however, the vector populations rebound rapidly (Figure 4–6). These results suggest that frequent applications of public health insecticides are necessary to control adult vectors and, ultimately, the spread of WNV. As part of a multidisciplinary approach before and during a WNV outbreak, this control strategy should be supplemented by: 1) source reduction or the elimination of breeding sites, 2) treatment of catch basins and other suitable breeding sites, and 3) initiation of community outreach activities.

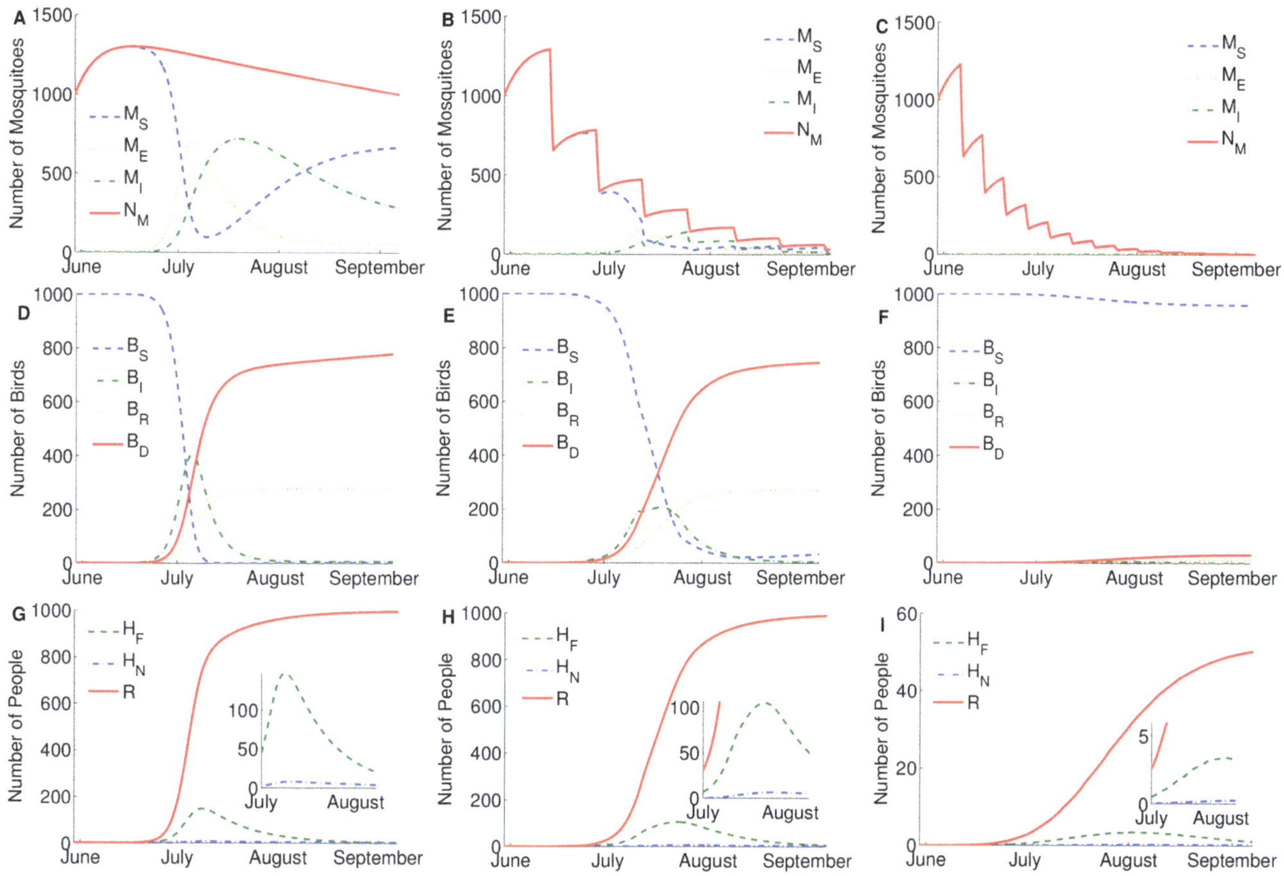

Figure 7. Model predictions of the mosquito, bird, and human populations in the presence of the WNV infection. The changes of mosquito (A–C), bird (D–F), and human (G–I) populations predicted by model (equations are listed in the Appendix S1 in File S1). The insecticide treatments scenarios, non-treatment (A, D, and G), biweekly (B, E, and H), and weekly (C, F, and I) are represented in each column. Model variables and parameter values are given in Tables 2–4 (best-fit parameter values are taken for the year 2006 from Sams Point trap location).

Through the data fitting procedure, we obtained parameter estimates (Table 4) that enabled us to attain a more reliable evaluation of the treatment effectiveness and predictions during a WNV outbreak at a specific location (Figure 7). We identified the treatment scenarios necessary to control WNV transmission and conducted sensitivity analysis of the selected model parameters (Figures 5–8 and Figures S1–S4 in File S1). Our model predicts a minimal number of mosquitoes and, subsequently, negligible WNV transmission when weekly insecticide treatments are conducted (Figure 7C, F, and I). Further, we investigated the impact of mosquito biting rate and their death rate on the basic reproductive ratio (Table 7 and Figure 8). We concluded that lower values of mosquito biting rate and higher mosquito death rate results in lower values of $\mathfrak{R}_0$ (Table 7 and Figure 8) and if we can lower it enough to achieve $\mathfrak{R}_0<1$, then an epidemic can be avoided.

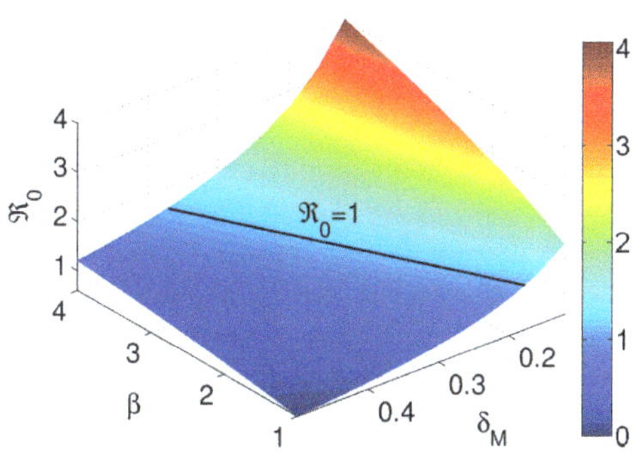

Figure 8. Sensitivity test of the predicted basic reproductive ratio to the selected model parameters. The basic reproductive ratio ($\mathfrak{R}_0$) as the function of the mosquito biting rate per day (β) and adult mosquito death rate per day (δ_M). The remaining parameters in the formula are taken from the Tables 3 and 4 (best-fit parameter values are taken for the year 2006 from Sams Point trap location). The ratio formula and its derivation are provided in the Appendix S2 in File S1.

We also demonstrated that in the biweekly treatments and non-treatment scenarios an increase in the number of dead birds would be observed, which would be an indicator of a WNV outbreak (Figure 7D and E). Not all birds have the same mortality rate due to WNV infection. For example, an experimental study reported that the common grackle (*Quiscalus quiscula* Linnaeus) had a mortality rate of 33% in contrast to the American crow, red-billed gull (*Larus scopulinus* Forster), and the house finch, which depicted nearly 100% mortality rates [43]. A future modeling study could assess more classes of birds depending on their susceptibility and morality due the infection. However, more data is necessary to adequately estimate the unknown parameters in a more complex model with these features.

A large occurrence of American crow die-offs preceded the 1999 laboratory confirmation of WNV among various bird species

Table 7. Sensitivity test of the predicted basic reproductive ratio to the mosquito biting rate and the death rate of the susceptible adult mosquitoes.

	δ_M=0.15	δ_M=0.20	δ_M=0.25	δ_M=0.30	δ_M=0.35	δ_M=0.40	δ_M=0.45	δ_M=0.50
β=1	1.007	0.689	0.511	0.400	0.322	0.267	0.226	0.195
β=2	2.013	1.379	1.021	0.796	0.643	0.533	0.452	0.340
β=3	3.020	2.068	1.532	1.194	0.965	0.801	0.679	0.585
β=4	4.027	2.757	2.042	1.592	1.286	1.068	0.905	0.780

The impact of varying mosquito biting rate (β) and their death rate per day (δ_M) on the ratio, given by the formula: $\Re_0 = \sqrt{\frac{\alpha_B \beta \eta}{\delta_M(\delta_M+\eta)} \cdot \frac{\tilde{B}_S}{\tilde{B}_S+\tilde{H}_S} \cdot \frac{\alpha_M \beta \tilde{M}_S}{(\delta_B+\tau)(\tilde{B}_S+\tilde{H}_S)}}$ (see Appendix S2 in File S1 for derivation). The remaining parameter values in the formula are taken from the Tables 3 and 4 (best-fit parameter values are taken for the year 2006 from Sams Point trap location). The parameter definitions and units are listed in Table 3.

in New York [44]. Also, abnormal bird deaths (primarily American crows) were observed in 2003 at Riverside, CA and nearby areas prior to a WNV outbreak [45]. The number of birds reported by the second month of the outbreak was about 5,000. This number represented a small portion of the bird mortalities because of the discontinued use of the residential hotline when dead bird sightings became common [45]. Other areasin California (Kern and Los Angeles counties) reported bird die offs during the summer of 2004 [46].

A recent study reported that the blood feeding behavior of *Cx. quinquefasciatus* significantly shifted with the change of season [47]. While another study noted that the feeding preference varied completely with location [48]. Further, information of the local regional preference is necessary to obtain reasonable estimates of the host feeding preference percentages. Although *Cx. quinquefasciatus* may have a preferred host for blood feeding, however availability of hosts is a key factor for an ultimate choice. We have added a variation of the model that includes feeding preferences in the Supplementary Materials (see Appendix S3 in File S1).

A limitation in our study is the impact of various natural factors, such as the seasonality of the vector populations. Many factors affect the rate at which mosquitoes reproduce, including larval rearing conditions, adult size, age, and the quality and quantity of the blood meal [49]. In a recent study of larval abundance in stormwater catch basins, low precipitation and high mean temperature were associated with high larval abundance in urban and suburban areas [50]. A study of spatiotemporal dynamics of the spread of WNV demonstrated a key role of the temperature on the seasonality and emergence of WNV [51]. Therefore, we have considered only a short duration (June–September) for multiple years in our study to avoid the dependence of the model on the aforementioned parameters.

Another limitation of the study is the treatment effectiveness that may vary with time and space. The insecticide treatment in our study is very effective for a short period of time and we have reviewed data from two locations in which treatments were applied throughout the neighborhoods on specific dates. Such changes in the treatments may be considered upon availability of other data.

There is a need to study temperature and rain events for evaluating the development of the *Cx. pipiens* complex throughout the underground stormwater systems. In past studies, the design of underground management systems and water temperature significantly impacted the development cycles of *Culex* mosquitoes [52,53]. Furthermore, other relationships between natural factors and mosquito populations can be reviewed by analyzing data and constructing mathematical models, which are essential to optimize control methods for a particular mosquito breeding habitat.

Surveillance strategies can be improved by increasing the frequency of samples collected throughout the year as well as increasing the number of strategically located mosquito trap sites. These changes may reduce uncertainties among the collected data and would assist to quantify the effects of inconsistent weather conditions, such as floods and droughts, on the various mosquito populations. Upon the establishment of these relationships, it would be possible to propose improved surveillance and abatement strategies to monitor and control the mosquito populations. Our modeling results support the long-standing importance of mosquito control and surveillance activities.

Supporting Information

File S1 Supplementary appendices, table, and figures. Appendix S1, Model Equations. **Appendix S2**, Basic Reproductive Number. **Appendix S3**, Mosquito Feeding Preferences. **Table S1, Root Mean Square generated by the best fit of the model to the *Culex pipiens* complex surveillance data collected in EVS traps located in Sams Pont and Battery Point.** The model was fitted to the surveillance data using Berkeley Madonna Version 8.3.18 software to estimate the model parameters. Estimated values were established through the best nonlinear least squares fit of the model for the surveillance data, (i.e. program minimized the root mean square (RMS) between the recorded data and the analogous model predictions). **Figure S1, Sensitivity tests of predicted larva and adult mosquito populations to their initial values in the absence of the WNV infection in the model.** Plots illustrating the sensitivity of the initial number of larval (L_0) and adult mosquitoes (M_0) on the subsequent larval and adult mosquito populations, respectively. The initial condition parameter in the legend was varied while the remaining parameters were fixed and chosen from the Tables 3 and 4 (best-fit parameter values are taken for the year 2006 from the Sams Point trap location). **Figure S2, Sensitivity tests of predicted larva and adult mosquito populations to the model parameters in the absence of the WNV infection in the model.** Plots illustrating the sensitivity of the maturation (m), birth (b), larval mosquito death (δ_L), and adult mosquito (δ_M) rates on the subsequent larval and adult mosquito populations, respectively. The parameter in the legend was varied while the remaining parameters were fixed and chosen from the Tables 3 and 4 (best-fit parameter values are taken for the year 2006 from the Sams Point trap location). **Figure S3, Sensitivity tests of predicted infectious mosquitoes, infectious birds, and humans with WNV fever populations to the initial values in the model with WNV infection.** Plots illustrating the sensitivity of the initial susceptible populations of female adult mosquitoes (M_0), birds (B_0), and humans (N) on the infected mosquitoes (M_I), infectious birds (B_I), and humans with WNV fever (H_F) populations. The initial condition parameter in the legend was varied while the remaining parameters were fixed and chosen from the Tables 3 and 4 (best-fit parameter values are taken for the year 2006 from the Sams Point trap location). **Figure S4, Sensitivity tests of predicted infectious mosquitoes, infectious birds, dead birds due to the infection with WNV, humans with WNV fever populations to the mosquito biting rate per day.** Plots illustrating the sensitivity of the mosquito biting rate per day (β) on the infectious mosquitoes (M_I), infectious birds (B_I), dead birds due to the infections with WNV (B_D), and humans with WNV fever (H_F) populations. The mosquito biting rate per day in the legend was varied while the remaining parameters were fixed and chosen from the Tables 3 and 4 (best-fit parameter values are taken for the year 2006 from the Sams Point trap location).

Acknowledgments

Disclaimer: The findings and conclusions of this report are those of the authors and do not necessarily represent the views of Beaufort County, SC. The results correspond to research only. Mention of a proprietary product does not constitute an endorsement or a recommendation for its use by Beaufort County, SC.

T. Crosby (BCMC), D. Davis (BCMC), C. Evans (SCDEHC), A. Jenkins (BCMC), J. Lightsey (BCMC), J. Loverin (BCMC), G. Millidge (BCMC), and A. Washington (BCMC) provided technical support for the BCMC surveillance program.

Author Contributions

Conceived and designed the experiments: KAP. Performed the experiments: KAP PN. Analyzed the data: KAP PN. Contributed reagents/materials/analysis tools: EJH GJH. Wrote the paper: KAP CS GJH.

References

1. CDC (2013) West Nile Virus. Available: http://www.cdc.gov/ncidod/dvbid/westnile/index.htm.
2. Bernard KA, Maffei JG, Jones SA, Kauffman EB, Ebel G, et al. (2001) West Nile virus infection in birds and mosquitoes, New York State, 2000. Emerg Infect Dis 7: 679–685.
3. Gubler DJ, Campbell GL, Nasci R, Komar N, Petersen L, et al. (2000) West Nile virus in the United States: guidelines for detection, prevention, and control. Viral Immunol 13: 469–475.
4. Howard Country (2013) Stormwater Management Structures. Available: http://www.howardcountymd.gov/displayprimary.aspx?id=354.
5. Environmental Protection Agency (2012) Stormwater Management. Available: http://www.epa.gov/greeningepa/stormwater/.
6. BCMC (2012) West Nile Virus. Available: http://www.co.beaufort.sc.us/departments/Public-Safety/mosquito-control/west-nile-virus.php.
7. CDC (2014) West Nile virus disease cases and deaths reported to CDC by year and clinical presentation, 1999–2012. Available: http://www.cdc.gov/westnile/resources/pdfs/cummulative/99_2012_CasesAndDeathsClinicalPresentationHumanCases.pdf.
8. CDC (2014) West Nile Virus Disease Cases and Presumptive Viremic Blood Donors by State – United States, 2013 (as of January 7, 2014). Available: http://www.cdc.gov/westnile/statsMaps/preliminaryMapsData/histatedate.html.
9. South Carolina Department of Health and Environmental Control Bureau of Laboratories (2012) 2012 West Nile Virus Activity. Available: http://www.scdhec.gov/health/lab/micro/medical_entomology/docs/WNV_Cases_2012.pdf.
10. MedlinePlus (2013) West Nile Virus. Available: http://www.nlm.nih.gov/medlineplus/ency/article/007186.htm.
11. Doheny K, WebMD (2012) West Nile Virus: Who's at Risk? www.webmd.com: WebMD.
12. Peterson RKD, Macedo PA, Davis RS (2006) A Human-Health Risk Assessment for West Nile Virus and Insecticided Used in Mosquito Management. Environment Health Perpectives 114: 366–372.
13. Ferguson NM, Cummings DA, Cauchemez S, Fraser C, Riley S, et al. (2005) Strategies for containing an emerging influenza pandemic in Southeast Asia. Nature 437: 209–214.
14. Ferguson NM, Cummings DA, Fraser C, Cajka JC, Cooley PC, et al. (2006) Strategies for mitigating an influenza pandemic. Nature 442: 448–452.
15. Pawelek KA, Huynh GT, Quinlivan M, Cullinane A, Rong L, et al. (2012) Modeling within-host dynamics of influenza virus infection including immune responses. PLoS Comput Biol 8: e1002588.
16. Rong L, Guedj J, Dahari H, Perelson AS (2014) Treatment of hepatitis C with an interferon-based lead-in phase: a perspective from mathematical modelling. Antivir Ther.
17. Brauer F, Castillo-Chavez C (2000) Mathematical Models in Population Biology and Epidemiology, Texts in Applied Mathematics Series. New York: Springer.
18. Diekmann O, Heesterbeek JAP (2000) Mathematical Epidemiology of Infectious Diseases. Model Building, Analysis and Interpretation. Chichester: Wiley.
19. Jiang J, Qiu Z, Wu J, Zhu H (2009) Threshold conditions for west nile virus outbreaks. Bull Math Biol 71: 627–647.
20. Lewis M, Renclawowicz J, van den Driessche P (2006) Traveling waves and spread rates for a West Nile virus model. Bull Math Biol 68: 3–23.
21. Lewis MA, Renclawowicz J, van den Driessche P, Wonham M (2006) A comparison of continuous and discrete-time West Nile virus models. Bull Math Biol 68: 491–509.
22. Simpson JE, Hurtado PJ, Medlock J, Molaei G, Andreadis TG, et al. (2012) Vector host-feeding preferences drive transmission of multi-host pathogens: West Nile virus as a model system. Proc Biol Sci 279: 925–933.
23. Cruz-Pacheco G, Esteva L, Vargas C (2009) Seasonality and outbreaks in West Nile virus infection. Bull Math Biol 71: 1378–1393.
24. Fan G, Liu J, van den Driessche P, Wu J, Zhu H (2010) The impact of maturation delay of mosquitoes on the transmission of West Nile virus. Math Biosci 228: 119–126.
25. Al-Arydah M, Smith R (2011) Controlling malaria with indoor residual spraying in spatially heterogenous environments. Math Biosci Eng 8: 889–914.
26. Spagnuolo AM, Shillor M, Kingsland L, Thatcher A, Toeniskoetter M, et al. (2012) A logistic delay differential equation model for Chagas disease with interrupted spraying schedules. J Biol Dyn 6: 377–394.
27. Spagnuolo A, Shillor M, Stryker G (2011) A model for Chagas disease with controlled spraying. Journal of Biological Dynamics 5: 299–217.
28. Bowman C, Gumel AB, van den Driessche P, Wu J, Zhu H (2005) A mathematical model for assessing control strategies against West Nile virus. Bull Math Biol 67: 1107–1133.
29. Ghosh D, Guha R (2011) Using a neural network for mining interpretable relationships of West Nile risk factors. Soc Sci Med 72: 418–429.

30. Wonham MJ, Lewis MA (2008) A comparative analysis of models for West Nile virus. Mathematical Epidemilogy: Lecture Notes in Mathematics: Springler. pp. 365–390.
31. Beaufort County South Carolina (2012) Beaufort County Nature Facts. Available: http://www.co.beaufort.sc.us/County-Projects/calendar-contest/wow-facts.php.
32. United States Census (2012) 2012 U.S. Gazetteer Files. Available: http://www2.census.gov/geo/gazetteer/2012_Gazetteer/2012_gaz_counties_45.txt.
33. United States Census (2010) Interactive Population Map. Available: http://www.census.gov/2010census/popmap/.
34. Sun City Hilton Head (2014) Welcome. http://www.suncityhiltonhead.org/Home.aspx.
35. South Carolina Department of Natural Resources (2014) South Carolina State Climatology Office. Available: http://www.dnr.sc.gov/climate/sco/ClimateData/countyData/county_beaufort.php.
36. Coffield DJ, Jr., Spagnuolo AM, Shillor M, Mema E, Pell B, et al. (2013) A model for Chagas disease with oral and congenital transmission. PLoS One 8: e67267.
37. Wonham MJ, Lewis MA, Renclawowicz J, van den Driessche P (2006) Transmission assumptions generate conflicting predictions in host-vector disease models: a case study in West Nile virus. Ecol Lett 9: 706–725.
38. Anderson RM, May RM (1991) Infectious Diseases of Humans. Oxford: Oxford University Press.
39. Dobson A, Foufopoulos J (2001) Emerging infectious pathogens of wildlife. Philos Trans R Soc Lond B Biol Sci 356: 1001–1012.
40. Heesterbeek H (2002) A brief history of R0 and a recipe for its calculation. 189–204 p.
41. Hethcote HW (2000) The mathematics of infectious diseases. SIAM Review 42: 599–653.
42. van den Driessche P, Watmough J (2002) Reproduction numbers and sub-threshold endemic equilibria for compartmental models of disease transmission. Math Biosci 180: 29–48.
43. Komar N, Langevin S, Hinten S, Nemeth N, Edwards E, et al. (2003) Experimental infection of North American birds with the New York 1999 strain of West Nile virus. Emerg Infect Dis 9: 311–322.
44. Eidson M, Kramer L, Stone WB, Hagiwara Y, Schmit K, et al. (2001) Dead Bird Surveillance as an Early Warning System for West Nile Virus. Emerg Infect Dis 7: 5.
45. University of California Riverside (2013) West Nile Watch. Available: http://westnile.ucr.edu/index.php?content=newsletters/newsletter071904.html.
46. Reisen WK, Fang Y, Lothrop HD, Martinez VM, Wilson J, et al. (2006) Overwintering of West Nile Virus in Southern California. Journal of Medical Entomology 43: 344–355.
47. Molaei G, Andreadis TG, Armstrong PM, Bueno RJ, Dennett JA, et al. (2007) Host feeding pattern of Culex quinquefasciatus (Diptera: Culicidae) and its role in transmission of West Nile Virus in Harris County, Texas. Am J Trop Med Hyg 77: 73–81.
48. Zinser M, Ramberg F, Willott E (2004) Culex quinquefasciatus (Diptera: Culicidae) as a potential West Nile virus vector in Tucson, Arizona: Blood meal analysis indicates feeding on both humans and birds. Journal of Insect Science 4: 3.
49. Hurd H, Hogg JC, Renshaw M (1995) Interactions between bloodfeeding, fecundity and infection in mosquitoes. Parasitology today (Personal ed) 11: 411–416.
50. Gardner AM, Hamer GL, Hines AM, Newman CM, Walker ED, et al. (2012) Weather variability affects abundance of larval Culex (Diptera: Culicidae) in storm water catch basins in suburban Chicago. J Med Entomol 49: 270–276.
51. Hartley DM, Barker CM, Le Menach A, Niu T, Gaff HD, et al. (2012) Effects of temperature on emergence and seasonality of West Nile virus in California. Am J Trop Med Hyg 86: 884–894.
52. Henn JB, Metzger ME, Kwan JA, Harbison JE, Fritz CL, et al. (2008) Deveme of Culex Mosquitoes in Stormwater Management Structures in California. The American Mosquito Control Association 24: 8.
53. Harbison JE, Metzger ME, Hu R (2010) Association Between Culex quinquefasciatus (Diptera: Culicidae) Oviposition and Structural Features of Belowground Stormwater Treatment Device. Journal of Medical Entomology 47: 7.
54. Daszak P, Cunningham AA, Hyatt AD (2000) Emerging infectious diseases of wildlife–threats to biodiversity and human health. Science 287: 443–449.
55. Cruz-Pacheco G, Esteva L, Montano-Hirose JA, Vargas C (2005) Modelling the dynamics of West Nile Virus. Bull Math Biol 67: 1157–1172.
56. Castillo-Chavez C, Blower S, van den Driessche P, Kirschner D, Yakubu AA (2002) Mathematical Approaches for Emerging and Reemerging Infectious Diseases: An Introduction. New York: Springer.
57. Chowell G, Castillo-Chavez C, Fenimore PW, Kribs-Zaleta CM, Arriola L, et al. (2004) Model parameters and outbreak control for SARS. Emerg Infect Dis 10: 1258–1263.
58. Blower SM, Dowlatabadi H (1994) Sensitivity and Uncertainty Analysis of Complex Models of Disease Transmission: An HIV Model, as an Example. International Statistical Review/Revue Internationale de Statistique 62: 229–243.
59. Dobson A (2004) Population dynamics of pathogens with multiple host species. Am Nat 164 Suppl 5: S64–78.
60. Enserink M (2004) Emerging infectious diseases. A global fire brigade responds to disease outbreaks. Science 303: 1605.
61. Getz WM, Pickering J (1983) Epidemic models: thresholds and population regulation. Am Nat 121: 892–898.
62. Flores Anticona EM, Zainah H, Ouellette DR, Johnson LE (2012) Two case reports of neuroinvasive west nile virus infection in the critical care unit. Case Rep Infect Dis 2012: 839458.
63. Darensburg T, Kocic VL (2004) On the discrete model of West Nile-like epidemics. Proc Dyn Sys Appl 4: 358–366.
64. Apperson CS, Hassan HK, Harrison BA, Savage HM, Aspen SE, et al. (2004) Host feeding patterns of established and potential mosquito vectors of West Nile virus in the eastern United States. Vector Borne Zoonotic Dis 4: 71–82.
65. Gerberg EJ, Barnard DR, RA W, 61–62. N (1994) Manual for Mosquito Rearing and Experimental Techniques. American Mosquito Control Association Bulletin 5: 61–62.
66. Elizondo-Quiroga A, Flores-Suarez A, Elizondo-Quiroga D, Ponce-Garcia G, Blitvich BJ, et al. (2006) Gonotrophic cycle and survivorship of Culex quinquefasciatus (Diptera: Culicidae) using sticky ovitraps in Monterrey, northeastern Mexico. J Am Mosq Control Assoc 22: 10–14.
67. The South Carolina Department of Health and Environmental Control Bureau of Laboratories (2014) Report and Submit Dead Birds to Help DHEC Track West Nile Virus. Available: http://www.scdhec.gov/HomeAndEnvironment/ReportIt/ReportDeadBirds/#other.

17

A Small-Scale Field Trial of Pyriproxyfen-Impregnated Bed Nets against Pyrethroid-Resistant *Anopheles gambiae* s.s. in Western Kenya

Hitoshi Kawada[1]*, Gabriel O. Dida[1,2], Kazunori Ohashi[3], Emiko Kawashima[1], George Sonye[4], Sammy M. Njenga[5], Charles Mwandawiro[6], Noboru Minakawa[1,7]

1 Department of Vector Ecology and Environment, Institute of Tropical Medicine, Nagasaki University, Nagasaki, Japan, **2** School of Public Health, Maseno University, Kisumu, Kenya, **3** Health and Crop Sciences Research Laboratory, Sumitomo Chemical Co. Ltd., Hyogo, Japan, **4** Springs of Hope, Mbita, Kenya, **5** Eastern and Southern Africa Centre of International Parasite Control, Nairobi, Kenya, **6** Kenya Medical Research Institute, Nairobi, Kenya, **7** The Global Center of Excellence Program, Nagasaki University, Nagasaki, Japan

Abstract

Pyrethroid resistance is becoming a major problem for vector control programs, because at present, there are few suitable chemical substitutes for pyrethroids, as when used on bed nets the insecticide must have low mammalian toxicity as well as high activity to mosquitoes. Pyriproxyfen (PPF) is one of the most active chemicals among the juvenile hormone mimic (JHM) group. Sterilizing mosquitoes by using PPF could be a potential control measure for pyrethroid-resistant malaria vectors. We investigated the sterilizing effects of two types of PPF-impregnated bed nets – a 1% PPF-impregnated net and a 1% PPF +2% permethrin-impregnated net (Olyset Duo) – to pyrethroid-resistant wild population of *Anopheles gambiae* s.s. in western Kenya. High mortality of blood-fed mosquitos was observed 3 days post-collection, in the houses where PPF-impregnated nets were used, indicating the effect of PPF on the longevity of mosquitos that came in contact with the net. Reduction in the number of ovipositing females, number of eggs, and number of progeny per female were also observed in the houses in which both Olyset Duo and PPF-impregnated nets were used. This is the first field study showing the high sterilizing efficacy of PPF against wild pyrethroid-resistant *An. gambiae* s.s. population. In addition, we recognized the necessity of combined use of permethrin with PPF, in order to reduce the risk of mosquito bites and provide a level of personal protection. Further studies on wild pyrethroid-resistant mosquito populations such as *An. arabiensis* and *An. funestus* s.s. would provide more information on the practical use of the PPF-impregnated bed nets.

Editor: Takafumi Tsuboi, Ehime University, Japan

Funding: This study was funded as a joint research between Nagasaki University and Sumitomo Chemical Co. Ltd in creating the field and semi-field bioassay system for insecticide delivering technology. The funders had no role in study design, data collection and analysis, decision to publish, or preparation of the manuscript. KO is an employee of funder company, Sumitomo Chemical Co., Ltd. KO gave an advisory opinion at the intervention of the bed nets as one of the developer of Olyset Duo.

Competing Interests: KO is an employee of Sumitomo Chemical Co., Ltd., but he is not the board member, has no stocks of the company. No patent was applied by the study.

* Email: vergiss@nagasaki-u.ac.jp

Introduction

After more than 30 years since the introduction of permethrin, synthetic pyrethroids have emerged as the newest class of insecticides in the vector control program [1]. On a global level, pyrethroids constitute approximately 81% of the spray utility, among which 68% is used for residual spraying and 24% for space spraying, and 100% of WHO-recommended insecticides for the treatment of long lasting mosquito nets (LNs) [2]. The resistance of vector mosquitoes is one of the main obstacles in effective vector control, because currently, there are no suitable substitutes for pyrethroids [1]. More than 90% of the current annual malaria incidence occurs in Africa. *Anopheles gambiae* Giles (*An. gambiae* s.s.), *Anopheles arabiensis* Patton and *Anopheles funestus* Giles (*An. funestus* s.s.) are the primary malaria vectors in sub-Saharan Africa. Recently, a causal relationship between the high coverage of LNs due to mass campaigns and the increase in the *kdr* frequency in *An. gambiae* s.s. have been reported [3,4,5]. In contrast, such *kdr* mutations do not seem to be common in *An. arabiensis* [6–8], but the metabolic resistance seems to be most common in this species [8,9]. Fortunately, *kdr* mutations have not been reported in African malaria vectors except in these two species. *Anopheles funestus* s.s. is the major malaria vector in southern Africa. Re-emergence or increase of this species as well as the development of metabolic pyrethroid resistance in eastern and southern Africa have been reported in this species [9–15].

Juvenile hormone mimics (JHMs) developed from natural origins are one of the commercially successful chemicals in the insect growth regulators (IGRs) category, with a unique mode of action that is insect-specific, stage-specific, slow acting, and non-neurotoxic [16]. Methoprene [17] and pyriproxyfen [18] are the most successful JHMs. Such chemicals have been commonly used as larvicides, since most of the effects of JHMs are on the last instar larvae, which become deformed or die at the pupal stage [18].

Pyriproxyfen was reported to cause vacuolation, inhibited development of imaginal buds of *Aedes aegypti* (L) larvae, and histolysis such as disrupted mitochondria, abundant vacuoles and poorly structured cytoplasmic organelles [19]. Adult *An. balabacensis* Baisas, which survived 48 h of immersion in 0.005 ppb (one-eighth of LC_{50}) of pyriproxyfen during the last larval instar, was found to show considerable reduction in sperm and egg production, as well as in blood feeding and mating activity [20].

JHMs also acted as a "sterilant" in adult insects. Methoprene was reported to affect ovarian development and adult longevity in *Ae. aegypti* [21,22]. The number of eggs per female and hatchability decreased with the application of pyriproxyfen to female *Ae. aegypti* [23,24], housefly (*Musca domestica* L.), and German cockroach (*Blattella germanica* L.) [25]. Aiku et al. [26] first demonstrated the use of pyriproxyfen-impregnated bed net as a sterilizing device for Anopheline mosquitoes. Ohashi et al. reported that the adult females of insecticide-susceptible *An. gambiae* strain were completely sterilized after exposure to 0.01% pyriproxyfen-treated nets (3.5 mg active ingredient [AI]/m^2) before and after consuming blood. It was also seen that the adult longevity decreased by exposure to pyriproxyfen in a dose-dependent manner [27]. Similarly, Harris et al. reported that the laboratory colony of *An. arabiensis*, which blood-fed one day prior to pyriproxyfen exposure (3 mg [AI]/m^2), produced no viable offspring, while other treatments (blood-fed 3 days before exposure, 1 day after exposure, and 3 days after exposure) had no significant effect [28].

The above findings suggested that sterilizing technique using pyriproxyfen could be one of the most effective measures for controlling pyrethroid-resistant malaria vectors. However, the field conditions in which people use bed nets and the wild population of malaria vectors come into contact with the bed nets voluntarily, have not been thoroughly studied. Here, we assessed the sterilizing effects of two types of pyriproxyfen-impregnated bed nets −1% pyriproxyfen-impregnated net and 1% pyriproxyfen +2% permethrin-impregnated net (Olyset Duo) – on a pyrethroid-resistant wild population of *An. gambiae* s.s. in western Kenya.

Materials and Methods

Bed nets

Three kinds of bed nets were provided by Sumitomo Chemical Co. Ltd. (Tokyo, Japan). The first one was impregnated with 1% pyriproxyfen (PPF), the second with 1% pyriproxyfen and 2% permethrin (Olyset Duo), and the third with 2% permethrin (Olyset Net). All bed nets were of the same size (130 cm [W]×180 cm [L]×150 cm [H]) and were made of polyethylene netting material (mesh 20 holes/cm^2), with active ingredients incorporated into the polymer before monofilament yarn extrusion.

Study location

The study was conducted in Ragwe village, in the Suba south sub-county in Homabay county of Nyanza Province in Western Kenya. The Suba south sub-county has been identified as a high malaria transmission area in Kenya, with more than 50% of the population exposed to ≥40% *Plasmodium falciparum* parasite rate corrected to a standard age range of 2 years old to <10 years old [29]. The main malaria vectors in the area are *An. gambiae* s.s., *An. arabiensis*, and *An. funestus* s.s. *Anopheles rivulorum* Leeson, which belongs to the Funestus Group is a minor vector in that area [12]. These four species were recently reported to have developed multimodal pyrethroid resistance [9,12]. *Anopheles gambiae* s.s. mainly dominated in Ragwe village [30].

Selection of houses and intervention with bed nets

Preliminary mosquito collection was done at the residential houses in Ragwe village and 15 houses, each with a relatively high mosquito density were selected. The houses had standard structures and sizes with eaves or gaps between the top of the wall and the roof, which was common in African houses. The number of people sleeping in bedrooms and living rooms, the number of bed nets (Olyset Net and Permanet) used before intervention, and the number of Olyset Net, Olyset Duo, and PPF-impregnated net after intervention are shown in Table 1.

New bed nets were supplied and hung in houses in sufficient numbers to cover all residents while sleeping. The distance between each house was 100–500 m. For each type bed net intervention, 5 houses were used. The house residents were informed about the study and their written consent was obtained before conducting the interventions on April 27 and 28, 2013.

Mosquito collection

Pre- and post-intervention mosquito collections were done in the morning (07:00–09:00) by three groups consisting of three staffs each, for ca. 20 min per house. Three clusters consisting of five houses each in the vicinity were arranged and three collection groups rotated the clusters daily. Mosquito collections were performed four times prior to intervention (April 22, 24, 25, and 26, 2013) and four times after intervention (April 29, 30, May 1, and 2, 2013). Anopheline mosquitoes resting on the walls, under furniture, or anywhere else in the house were collected using a battery-powered aspirator (C-cell aspirator, BioQuip Products, CA, USA). The nozzles and containers of the aspirators were changed for the collection in the houses with Olyset Duo and PPF-impregnated net to avoid contamination. The collected mosquitoes were examined microscopically in a laboratory to distinguish the *Anopheles gambiae* complex from other anopheline species, based on the identification keys of Gillies and Coetzee [31]. Individual species within *An. gambiae* s.l. were identified using the multiplex polymerase chain reaction (PCR) method described by Scott et al. [32].

Observation of egg production, hatchability, and longevity of blood-fed females

After the collection of mosquitoes, freshly blood-fed female mosquitoes were individually held in a 20-mL glass vial with a small amount of water (1–2 mL) and damp filter paper as an oviposition substrate until egg production or death occurred. The presence and the number of eggs oviposited were recorded daily and the whole egg batch was transferred to a paper cup (ca. 50 mL) with water to allow hatching. A daily record of egg batch hatching was conducted. It required 7–21 days to complete the observation, since egg hatching did not take place simultaneously. After collection of eggs and bodies, mosquitoes were identified by PCR as previously described. Ohashi et al. reported that the mean longevity of *An. gambiae* s.s. females exposed to the 0.1% PPF-treated nets were 5.6 days [27]. The mean longevity of the blood-fed females contacted 1% PPF-treated nets was thought to be less than 5.6 days. On the other hand, most blood-fed mosquitoes normally lay eggs in less than 4 days after blood meals and they sometimes die after oviposition. Therefore, in the present trial, the deaths of field-collected blood-fed females within 3 days were judged as the deaths caused by the effect of PPF and the comparison of the mortality between PPF-exposed mosquitoes (Olyset Duo and PPF-impregnated net) and unexposed ones (Olyset Net) was done.

Table 1. Information on the houses in which bed net interventions were conducted.

House No.	No. of people sleeping[a)]		No. of bed nets used before intervention		No. of bed nets used after intervention		
	Bedroom	Living room	Olyset Net	Permanet	Olyset Net (Oly)	Olyset Duo (Duo)	Pyriproxyfen (PPF)
RAG002	2A	3C	0	3	2	-	-
RAG004	2A	1A+1C	2	0	-	3	-
RAG005	2A	0	2	0	-	-	2
RAG007	2A	3C	2	0	2	-	-
RAG008	2A	2C	2	0	-	2	-
RAG009	3A	3C	4	0	-	-	3
RAG010	1A	1C	0	2	2	-	-
RAG011	3A	4C	2	0	-	4	-
RAG014	1A	2C	1	0	-	-	2
RAG020	1A	1A+1C	0	2	1	-	-
RAG023	1A	1C	1	0	-	2	-
RAG027	2A	1C	2	0	-	-	2
RAG028	2A	4C	2	0	2	-	-
RAG029	1A	6C	1	0	-	1	-
RAG030	2A	0	1	1	-	-	2

[a)]A, adults (age>15 years); C, children (age ≤15 years).

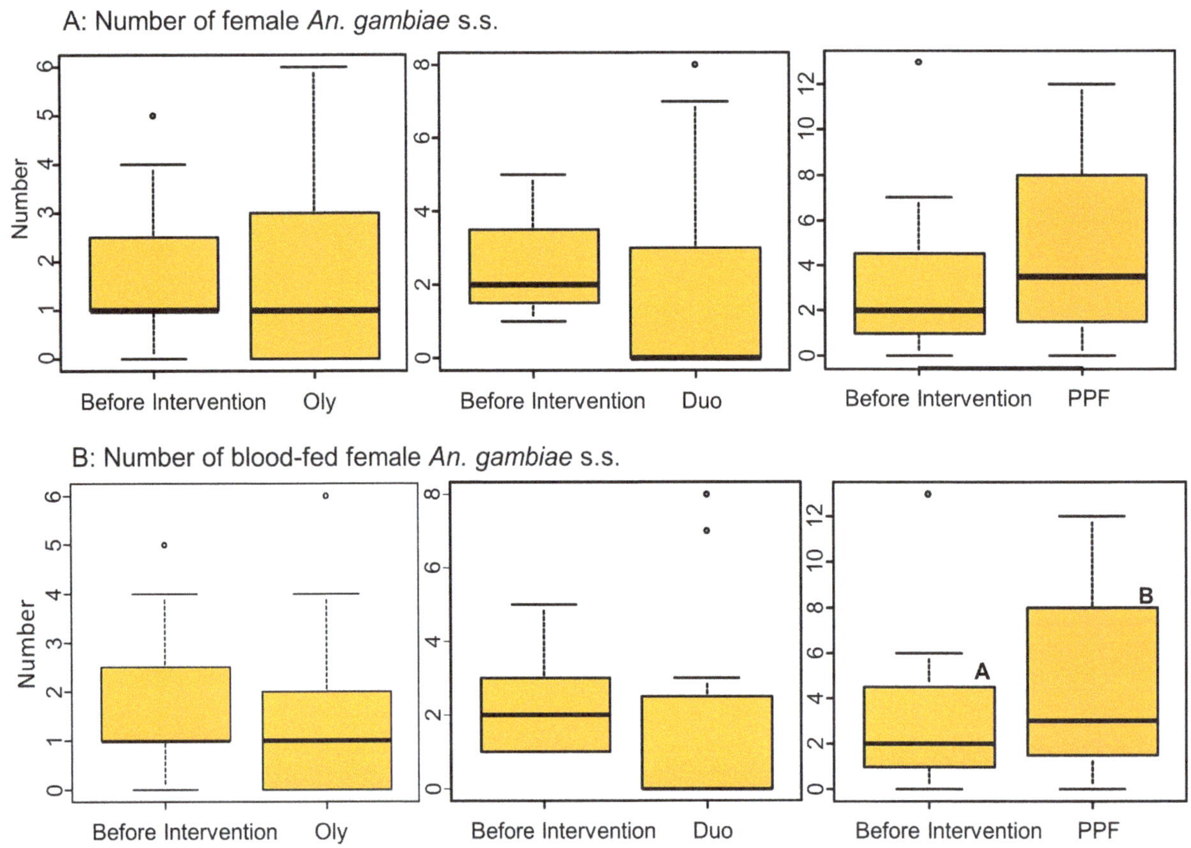

Figure 1. Box-and-whisker plots (medians: thick lines, with inter-quartile ranges) of the number of *Anopheles gambiae* s.s. females collected in the houses with Olyset Net (Oly), Olyset Duo (Duo), and PPF-impregnated net (PPF), before and after intervention. A, Average number of female *An. gambiae* s.s. females collected/house/day; B, Average number of blood-fed *An. gambiae* s.s. females collected/house/day. Different letters indicate significant difference by GLMM analysis.

Detection of Point Mutations in the Voltage-Gated Sodium Channel in *An. gambiae* s.s.

PCR and direct DNA sequencing were performed to identify point mutations at 1014L in the field-collected mosquitoes according to the method by Kawada et al. [10,11]. Direct DNA sequencing was performed using the 3730 DNA Analyzer (Applied Biosystems). The electropherogram of the targeted amino acid replacement was analyzed using MEGA 4.0 public domain software. The unique DNA haplotype sequences were deposited into the GenBank.

Data analysis

A generalized linear mixed model (GLMM) using the Poisson distribution was used to examine the number of mosquitoes collected before and after bed net intervention by using the package lme4 in R (http://www.R-project.org). The dates of collection and the houses were used as random effects. Comparisons of the total number of blood-fed and unfed females, the total number of dead and alive blood-fed females within 3 days after collection, and the total number of oviposited and non-oviposited blood-fed females before and after intervention of bed nets were performed with χ^2 test using JMP 7.0J (SAS Institute Japan Inc., Tokyo, Japan). Kruskal–Wallis one-way analysis of variance was used to compare the number of oviposited eggs, hatchability, and the number of progenies before and after intervention.

Ethics Statement

The protocol for the study (Case No. 2522) was approved by the Scientific Steering Committee and the National Ethics Review Committee of the Kenya Medical Research Institute. All necessary permits were obtained for the described field studies. No mosquito collection was done without the approval of the head of the village and the owner and occupants of the collection house.

Results

Effect of Olyset Net, Olyset Duo, and PPF-impregnated net on the number of collected female *An. gambiae* s.s.

One hundred and forty-five female *An. gambiae* s.s. and 5 female *An. arabiensis* were collected before the intervention, while 153 *An. gambiae* s.s. and 6 *An. arabiensis* were collected afterward. The allelic frequency of the point mutation in the voltage-gated sodium channel (L1014S) in *An. gambiae* s.s. collected was 98.4% (n = 96, Accession AB776705, AB776706). Average numbers of *An. gambiae* s.s. females collected in the houses with Olyset Net, Olyset Duo, and PPF-impregnated net were 1.90, 2.37, and 3.26 per house/day before intervention, and 1.56, 1.55, and 4.70 per house/day after intervention, respectively. Generalized linear mixed model for the number of collection revealed that net-by-intervention (before and after) interaction ($P = 0.0047$) was significant, while net ($P = 0.083$) and intervention ($P = 0.740$) were not significant. Significant difference was not observed in the number of mosquitoes collected in the houses with

Olyset Net, Olyset Duo, and PPF-impregnated net (P>0.1 for Olyset Net and Olyset Duo, and P = 0.092 for PPF-impregnated net; Figure 1). Average numbers of blood-fed *An. gambiae* s.s. females collected in the houses with Olyset Net, Olyset Duo, and PPF-impregnated net were 1.85, 2.26, and 3.16 per house/day before intervention, and 1.33, 1.45, and 4.35 per house/day after intervention, respectively. Generalized linear mixed model for the number of collection revealed that both net (P<0.001) and net-by-intervention (before and after) interaction (P = 0.0089) were significant, while intervention (P = 0.891) was not significant. Decreased number of blood-fed females were observed in the houses with Olyset Net and Olyset Duo, although these differences were not significant (P>0.1), while a significant increase was observed in the houses with PPF-impregnated net (P<0.0001; Figure 1).

Effect of Olyset Net, Olyset Duo, and PPF-impregnated net on blood feeding, longevity of blood-fed females, and oviposition of *An. gambiae* s.s.

Mortality (3 days after collection) of the females in the houses with Olyset Net, Olyset Duo, and PPF-impregnated net were 10.8%, 4.7%, and 8.3% before intervention, and 8.3%, 17.2%, and 33.3% after, respectively. The difference in the proportion of the total number of dead blood-fed females against the total number of live blood-fed females of *An. gambiae* s.s. before and after intervention of PPF-impregnated net was significant (P = 0.00041), while those in the houses with Olyset Net (P>0.1) and with Olyset Duo (P>0.05) were not significant (Figure 2).

Percentage of the blood-fed females ovipositing in the houses with Olyset Net, Olyset Duo, and PPF-impregnated net were 78.4%, 76.7%, and 75.0% before intervention, and 66.7%, 44.8%, and 26.4% after, respectively. Significant differences in the proportion of the total number of ovipositing blood-fed females vs. the total number of non-ovipositing blood-fed females of *An. gambiae* s.s. before and after intervention were observed in the houses with Olyset Duo (P = 0.0057) and PPF-impregnated net (P<0.0001), while that in the houses with Olyset Net was not significant (P>0.1) (Figure 3).

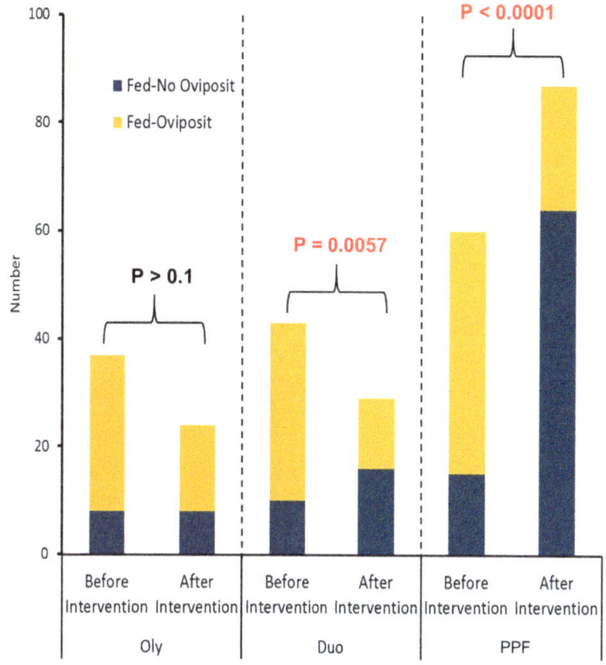

Figure 3. Total number of oviposited (Fed-Oviposit) and non-oviposited (Fed-No Oviposit) blood-fed females of *Anopheles gambiae* s.s. before and after intervention with Olyset Net (Oly), Olyset Duo (Duo), and PPF-impregnated net (PPF). Figures indicate the P-value by χ^2 test.

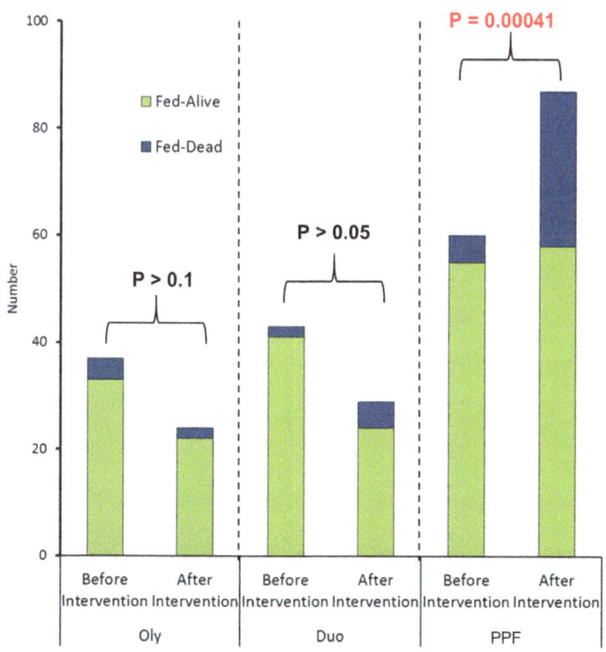

Figure 2. Total number of dead (within 3 days after collection) (Fed-Dead) and alive (Fed-Alive) blood-fed females of *Anopheles gambiae* s.s. before and after intervention with Olyset Net (Oly), Olyset Duo (Duo), and PPF-impregnated net (PPF). Figures indicate the P-value by χ^2 test.

Effect of Olyset Net, Olyset Duo, and PPF-impregnated net on the fecundity of *An. gambiae* s.s.

Average numbers of eggs oviposited before intervention in the houses with Olyset Net, Olyset Duo, and PPF-impregnated net were 124.6, 126.9, and 118.1/female and those after intervention were 111.8, 66.1, and 39.5/female, respectively. The difference in the number of oviposited eggs among the pre-intervention and post-intervention houses was significant (P = 0.0008), and higher and significant reduction in the number of eggs were observed in the houses where Olyset Duo (P = 0.0083) and PPF-impregnated nets (P<0.0001) were used compared to that in the pre-intervention houses (Figure 4).

Average hatchability of eggs oviposited by the blood-fed *An. gambiae* s.s. females collected before intervention in the houses with Olyset Net, Olyset Duo, and PPF-impregnated net was 60.9%, 59.3%, and 57.9% and those in the houses with Olyset Net, Olyset Duo, and PPF-impregnated net after intervention were 42.6%, 51.3%, and 42.9%, respectively. Hatchability in the pre-intervention houses and post-intervention houses was not significant (P = 0.852) (Figure 4).

Average numbers of progeny produced by the blood-fed *An. gambiae* s.s. females collected before intervention in the houses with Olyset Net, Olyset Duo, and PPF-impregnated net was 79.9, 77.2, and 72.6/female and those in the houses with Olyset Net, Olyset Duo, and PPF-impregnated net after intervention were 54.7, 38.4, and 19.6/female, respectively. Difference in the

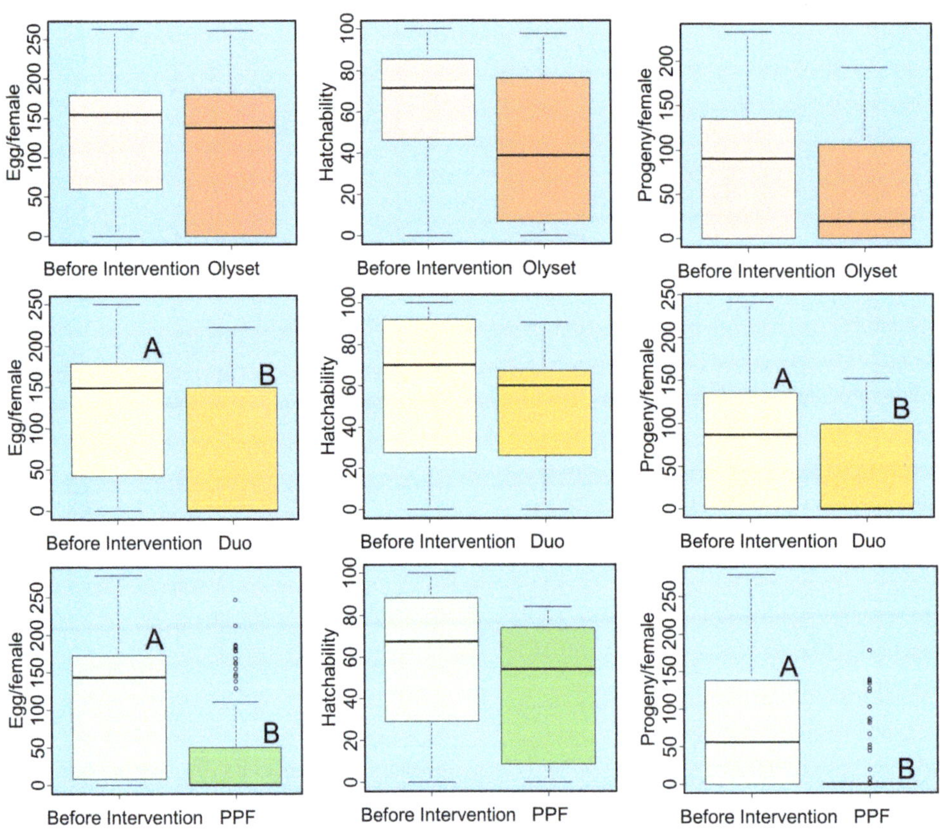

Figure 4. Box-and-whisker plots (medians: thick lines, with inter-quartile ranges) of the number of eggs oviposited, hatchability of eggs, and number of progenies produced by *Anopheles gambiae* s.s. females collected from houses with Olyset Net (Oly), Olyset Duo (Duo), and PPF-impregnated net (PPF), before and after intervention. Different letters indicate significant difference by Kruskal-Wallis one-way analysis of variance.

number of progenies between the pre-intervention houses and post-intervention houses was significant (P = 0.0021), and considerably higher reduction was observed in the houses in which Olyset Duo (P = 0.0195) and PPF-impregnated nets (P<0.0001) were used compared to that in the pre-intervention houses (Figure 4).

Discussion

The *An. gambiae* s.s. population in the study area was pyrethroid-resistant governed by a point mutation of the voltage-gated sodium channel (L1014S) at high frequency, as previously reported [9,33]. Additionally, a reduced repellency to pyrethroids has been reported in *An. gambiae* s.s. population in the study area as compared to the other malaria vectors, *An. arabiensis* and *An. funestus* s.s. [33]. This population attacked humans mainly at midnight irrespective of the use of LNs [34]. Therefore, we hypothesized that almost all blood-fed females had contact with the bed nets before or after the blood feeding, although the frequency of contact or total duration of contact was not measured in the present study.

An increase in the number of collected mosquitoes were observed in the houses in which PPF-impregnated nets were used. Further, some blood-fed mosquitoes were observed resting inside, as well as outside the PPF-impregnated bed nets, while no mosquitoes were collected from inside the permethrin-incorporated Olyset Net and Olyset Duo. This indicated that permethrin played a role in reducing the mosquito numbers inside the houses. No significant difference was noted in the number of blood-fed and unfed females of *An. gambiae* s.s. before and after intervention in the houses with Olyset Net, Olyset Duo, and PPF-impregnated nets were used. Since the present study was performed in houses where the residents had no restrictions on the sleeping times, some might have been outside the bed nets during midnight or might have slept under the bed nets in a manner that could have resulted in receiving mosquito bites.

Mortality (within 3 days after collection) of the blood-fed females in the houses with Olyset Duo and PPF-impregnated net increased from 4.7% to 17.2% and from 8.3% to 33.3%, respectively, after intervention. Significant differences were noted in the proportion of the number of dead blood-fed females against the number of live blood-fed females before and after intervention in the houses with PPF-impregnated nets. This increase in mortality clearly indicated the effect PPF had on the longevity of blood-fed females as previously reported [27]. Similarly, methoprene was reported to cause mortality to *Ae. aegypti* at 24 h following blood consumption [22]. Damages to internal tissues such as Malpighian tubules, midguts, and salivary gland cells have been observed in PPF-treated cat flea *Ctenocephalides felis* (Bouché), as reported by Meola et al. [35], and similar damages might have caused the deaths in the mosquitoes. Additionally, the percentage of the blood-fed females oviposited in the houses with Olyset Duo and PPF-impregnated nets reduced from 76.7% to 44.8% and from 75.0% to 26.4%, respectively, after intervention. Significant differences in the proportion of ovipositing, blood-fed females before and after intervention were observed in the houses with Olyset Duo and PPF-impregnated nets. The mortality rate in 3 days and the oviposition of the blood-fed females might be one of the means to determine the efficacy of PPF-impregnated bed nets in a field study, since the observation of egg production and hatchability requires time, special apparatus such as binoculars, is costly, and requires technical skill.

The number of oviposited eggs and progenies were significantly reduced in Olyset Duo and PPF-impregnated nets as compared to those before intervention. There was no difference in hatchability. In the current study, the same materials were used for Olyset Net and bed nets in which 1% PPF was impregnated. In the laboratory study using the laboratory-bred colony of *An. gambiae* s.s. (insecticide-susceptible ICIPE strain having 6.3% allelic L1014S mutation [9]), more than 90% reduction in offspring/blood-fed female was observed by the 3-min exposure to the same net materials (unpublished data). Ohashi et al. reported 100% reduction in fecundity of *An. gambiae* s.s. upon 3-min contact to the nets dipped in 0.01% isopropyl alcohol solution of PPF [27]. Kawada et al. reported that the minimum dosage of PPF, which causes sterility in female houseflies, was 5 µg per female [25]. On the other hand, 0.05–0.49 µg per female of PPF was thought to be effective in *Ae. aegypti*, since these amounts were detected in the mosquitoes that had reduced fecundity [24]. Even though the amount of PPF picked up by the mosquito after a short contact time with the PPF-impregnated net surface was not known, we believe it could be within the above range.

PPF has been reported to be effective against DDT, dieldrin, organophosphate, and carbamate-resistant anopheline mosquitoes as a larvicide, indicating the absence of cross-resistance of PPF to these insecticides [36]. However, only few studies have shown the relationship between pyrethroid-resistance and efficacy of PPF in mosquitoes. Kasai et al. suggested cross-resistance between etofenprox and insect growth regulators (IGRs) such as diflubenzuron and PPF in *Culex pipiens pipiens*, which were found in Japan [37]. There does not seem to be no other reports studying the above issue. Absence in cross resistance of PPF to other insecticides, especially pyrethroids, is essential for the selection of PPF as an active ingredient for LNs because of the aforementioned serious resistance problems in malaria vectors. The present study showing that the high sterilizing efficacy of PPF against pyrethroid-resistant *An. gambiae* s.s. might provide a solution to the above issue.

The current study also demonstrated the impact of permethrin and PPF when used in combination. The use of only PPF in a bed net might increase the risk of mosquito bites even though the mosquitoes that came in contact with the net might later have died (due to the impact of PPF on longevity) or have been sterilized. However, permethrin has an excito-repellent activity to mosquitoes that might reduce the contact chance of mosquitoes to bed net surface. Siegert et al. reported in a recent study that Olyset Net reduced the landing attempts of mosquitoes and elevated their flight frequency, resulting in low incidence of mortality [38]. Combined use of permethrin and PPF, therefore, might be antagonistic. The current study, however, demonstrated the effectiveness of permethrin incorporated PPF bed net (Olyset Duo) in reducing the number of ovipositing females and the number of eggs laid, and suppressing progeny through these mechanisms. The fact that a combination of a repellent pyrethroid and PPF was effective in sterilizing mosquitoes might be partly due to the reduced repellency of permethrin to *An. gambiae* s.s. in the study area [33]. Further, high anthropophily and midnight feeding habits [34] might have contributed positively to the effect of PPF. Recently, Ngufor et al. conducted the experimental hut trial of Olyset Duo against pyrethroid resistant laboratory colonies of *An. gambiae* s.s. [39]. The authors concluded that the pyrethroid-resistant mosquitoes which fail to be killed by permethrin could be sterilized if the net also contains PPF, resulting in greater reductions in the abundance of pyrethroid resistant malaria vectors.

The observation of oviposition and hatchability of mosquitoes requires skills, apparatus (binoculars and other equipment), and long term incubation of eggs, since it takes 1–3 weeks for the completion of egg hatching because egg hatching does not take place simultaneously [40]. We could not use large number of houses in the present study because of the limited manpower and apparatus. Larger scale field studies with sufficient number of skilled staffs and apparatus will be required for the further comprehension of the effectiveness of PPF. Further studies on the other wild pyrethroid-resistant mosquito populations such as *An. arabiensis* and *An. funestus* s.s. will provide a more convincing explanation to the above suggestions. Selection of PPF concentration, which would give the maximum efficacy while being cost-effective, as well as selection of the optimum concentration of permethrin or selection of another insecticide that can substitute for permethrin are issues of interest for future studies.

Acknowledgments

The protocol for the study was reviewed and approved by the Scientific Steering Committee (SSC) and the National Ethics Review Committee (ERC) of the Kenya Medical Research Institute (KEMRI). We are grateful to all the staff at the Thomas Odhiambo campus of the International Center of Insect Physiology and Ecology (ICIPE), for providing facilities and experimental insects. We also thank B. Awuor, F. Sonye, A. Oreng' Obielo, R. Muga, S. Omito, C. Opora, J. Oreng', M. Shikuku, D. Otieno, P. Oboyi, Springs of Hope, Mbita, Kenya, and C. Kimura, Institute of Tropical Medicine, Nagasaki University, Nagasaki, Japan, for providing technical support and assisting with the study, and J. Invest and J. Lucas, Sumitomo Chemical (UK), London, UK, for reviewing the manuscript.

Author Contributions

Conceived and designed the experiments: HK KO. Performed the experiments: HK GOD EK. Analyzed the data: HK. Contributed reagents/materials/analysis tools: GOD GS. Wrote the paper: HK. Reviewed the protocol for the study and criticized on the discussion in the manuscript: SMN CM NM.

References

1. Nauen R (2007) Insecticide resistance in disease vectors of public health importance. Pest Manag Sci 63: 628–633.
2. van den Berg, Zaim M, Yadav RS, Soares A, Ameneshewa B, et al. (2012) Global trends in the use of insecticides for vector-borne disease control. Environmental Health Perspectives. http://dx.doi.org/10.1289/ehp.1104340.
3. Stump AD, Atieli FK, Vulule JM, Besansky NJ (2004) Dynamics of the pyrethroid knockdown resistance allele in western Kenyan populations of *Anopheles gambiae* in response to insecticide-treated bed net trials. Am J Med Trop Hyg 70: 591–596.
4. Mathias D, Ochomo EO, Atieli F, Ombok M, Bayoh MN, et al. (2011) Spatial and temporal variation in the *kdr* allele L1014S in *Anopheles gambiae* s.s. and phenotypic variability in susceptibility to insecticides in Western Kenya. Malar J 10: 10.
5. Bayoh MN, Mathias DK, Odiere MR, Mutuku FM, Kamau L, et al. (2010) *Anopheles gambiae*: historical population decline associated with regional distribution of insecticide-treated bed nets in western Nyanza Province, Kenya. Malar J 9: 62.
6. Himeidan YE, Chen H, Chandre F, Donnelly MJ, Yan G (2007) Short report: permethrin and DDT resistance in the malaria vector *Anopheles arabiensis* from eastern Sudan. Am J Trop Med Hyg 77: 1066–1068.
7. Kulkarni M, Rowland M, Alifrangis M, Mosha FW, Matowo J, et al. (2006) Occurrence of the leucine-to-phenylalanine knockdown resistance (*kdr*) mutation

in *Anopheles arabiensis* populations in Tanzania, detected by a simplified high-throughput SSOP-ELISA method. Malar J 5: 56.
8. Kawada H, Futami K, Komagata O, Kasai S, Tomita T, et al. (2011) Distribution of a knockdown resistance mutation (L1014S) in *Anopheles gambiae* s.s. and *Anopheles arabiensis* in Western and Southern Kenya. PLoS One 6: e24323.
9. Kawada H, Dida GO, Ohashi K, Komagata O, Kasai S, et al. (2011) Multimodal pyrethroid resistance in malaria vectors, *Anopheles gambiae* s.s., *Anopheles arabiensis*, and *Anopheles funestus* s.s. in western Kenya. PLoS One 6: e22574.
10. Wondji CS, Irving H, Morgan J, Lobo NF, Collins FH, et al. (2009) Two duplicated P450 genes are associated with pyrethroid resistance in *Anopheles funestus*, a major malaria vector. Genome Res 19: 452–459.
11. Hargreaves K, Koekemoer LL, Brooke BD, Hunt RH, Mthembu J, et al. (2000) *Anopheles funestus* resistant to pyrethroid insecticides in South Africa. Med Vet Entomol 14: 181–189.
12. Kawada H, Dida GO, Sonye G, Njenga SM, Mwandawiro C, et al. (2012) Reconsideration of *Anopheles rivulorum* as a Vector of *Plasmodium falciparum* in Western Kenya: some evidence from biting time, blood preference, sporozoite positive rate, and pyrethroid resistance. Parasit Vect 5: 230.
13. Lwetoijera DW, Harris C, Kiware SS, Dongus S, Devine GJ, et al. (2014) Increasing role of *Anopheles funestus* and *Anopheles arabiensis* in malaria transmission in the Kilombero Valley, Tanzania. Malar J. 13: 331.
14. McCann RS, Ochomo E, Bayoh MN, Vulule JM, Hamel MJ, et al. (2014) Reemergence of *Anopheles funestus* as a vector of *Plasmodium falciparum* in western Kenya after long-term implementation of insecticide-treated bed nets. Am J Trop Med Hyg. 90: 597–604.
15. Hunt R, Edwardes M, Coetzee M (2010) Pyrethroid resistance in southern African *Anopheles funestus* extends to Likoma Island in Lake Malawi. Parasit Vectors. 3: 122.
16. Miyamoto J, Hirano M, Takimoto Y, Hatakoshi M (1993) Insect growth regulators for pest control, with emphasis on juvenile hormone analogs. Present status and future prospects. In: Duke, SO, Menn, JJ, Plimmer JR editors. Pest control with enhanced environmental safety (ACS symposium series 524). Washington, DC: American Chemical Society. 144–168.
17. Henrick CA, Staal GB, Siddall JB (1973) Alkyl-3,7,11-trimethyl-2,4-dodecadienoates, a new class of potent insect growth regulators with juvenile hormone activity. J Agr Food Chem 21: 354–359.
18. Hirano M, Hatakoshi M, Kawada H, Takimoto Y (1998) Pyriproxyfen and other juvenile hormone analogues. Rev Toxicol 2: 357–394.
19. Syafruddin, Arakawa R, Kamimura K, Kawamoto F (1990) Histopathological effects of an insect growth regulator, 4-phenoxyphenyl (*RS*)-2-(2-pyridyloxy)-propyl ether (pyriproxyfen), on the larvae of *Aedes aegypti*. Jpn J Sanit Zool 41: 15–22.
20. Iwanaga K, Kanda T (1988) The effects of a juvenile hormone active oxime ether compound on the metamorphosis and reproduction of an Anopheline vector, *Anopheles balabacensis*. Appl Entomol Zool 23: 186–193.
21. Judson CL, Lumen HZ (1976) Some effects of juvenile hormone and analogs on ovarian follicles of the mosquito *Aedes aegypti* (Diptera: Culicidae). J Med Entomol 13: 197–201.
22. Klowden MJ, Chambers GM (1989) Ovarian development and adult mortality in *Aedes aegypti* treated with sucrose, juvenile hormone and methoprene. J Insect Physiol 35: 513–517.
23. Kawada H, Itoh T, Abe Y, Horio M (1993) Can mosquito be a carrier of larvicides? In: Proceedings of the 1st International Congress on Insect Pests in the Urban Environment, Cambridge, UK, 497.
24. Itoh T, Kawada H, Abe Y, Eshita Y, Rongsriyam Y, et al. (1994) Utilization of blood-fed females of *Aedes aegypti* as a vehicle for the transfer of the insect growth regulator pyriproxyfen to larval habitats. J Am Mosq Control Assoc 10: 344–347.
25. Kawada H, Senbo S, Abe Y (1992) Effects of pyriproxyfen on the reproduction of the housefly, *Musca domestica*, and the German cockroach, *Blattella germanica*. Jpn J Sanit Zool 43: 169–175.
26. Aiku AO, Yates A, Rowland M (2006) Laboratory evaluation of pyriproxyfen treated bednets on mosquito fertility and fecundity. A preliminary study. West Afr J Med 25: 22–6.
27. Ohashi K, Nakada K, Ishiwatari T, Miyaguchi J, Shono Y, et al. (2012) Efficacy of pyriproxyfen-treated nets in sterilizing and shortening the longevity of *Anopheles gambiae* (Diptera: Culicidae). J Med Entomol 49: 1052–1058.
28. Harris C, Lwetoijera DW, Dongus S, Matowo NS, Lorenz LM, et al. (2013) Sterilising effects of pyriproxyfen on *Anopheles arabiensis* and its potential use in malaria control. Parasit Vect 6: 144.
29. Noor AM, Amin AA, Akhwale WS, Snow RW (2007) Increasing coverage and decreasing inequity in insecticide-treated bed net use among rural Kenyan children. PLoS Med 4: e255.
30. Futami K, Dida GO, Sonye G, Lutiali PA, Mwania MS, et al. (2014) Impacts of insecticide treated bed nets on *Anopheles gambiae* s.l. populations in Mbita district and Suba district, Western Kenya. Parasit Vect 7: 63.
31. Gillies MT, Coetzee M (1987) A supplement to the Anophelinae of Africa south of the Sahara (Afrotropical region). South African Institute for Medical Research No. 55.
32. Scott JA, Brogdon WG, Collins FH (1993) Identification of single specimens of the *Anopheles gambiae* complex by the polymerase chain reaction. Am J Trop Med Hyg 49: 520–529.
33. Kawada H, Ohashi K, Dida GO, Sonye G, Njenga SM, et al. (2014) Insecticidal and repellent activities of pyrethroids to the three major pyrethroid-resistant malaria vectors in western Kenya. Parasit Vect 7: 208.
34. Kawada H, Ohashi K, Dida GO, Sonye G, Njenga SM, et al. (2014) Preventive effect of permethrin-impregnated long-lasting insecticidal nets on the blood feeding of three major pyrethroid-resistant malaria vectors in western Kenya. Parasit Vect 7: 383.
35. Meola R, Pullen S, Meola S (1996) Toxicity and histopathology of the growth regulator pyriproxyfen to adults and eggs of the cat flea (Siphonaptera: Pulicidae). J Med Entomol 33: 670–679.
36. Kawada H, Shono Y, Itoh T, Abe Y (1993) Laboratory evaluation of an insect growth regulators against several species of Anopheline mosquitoes. Jpn J Sanit Zool 44: 349–353.
37. Kasai S, Shono T, Komagata O, Tsuda Y, Kobayashi M, et al. (2007) Insecticide resistance in potential vector mosquitoes for West Nile virus in Japan. J Med Entomol 44: 822–829.
38. Siegert PY, Walker E, Miller JR (2009) Differential behavioral responses of *Anopheles gambiae* (Diptera: Culicidae) modulate mortality caused by pyrethroid-treated bednets. J Econ Entomol 102: 2061–2071.
39. Ngufor C, N'Guessan R, Fagbohoun J, Odjo A, Malone D, et al. (2014) Olyset Duo (a pyriproxyfen and permethrin mixture net): An experimental hut trial against pyrethroid resistant *Anopheles gambiae* and *Culex quinquefasciatus* in Southern Benin. PLoS One 9: e93603.
40. Ebrahimi B, Shakibi S, Foster WA (2014) Delayed egg hatching of *Anopheles gambiae* (Diptera: Culicidae) pending water agitation. J Med Entomol 51: 580–90.

18

A Multi-Criteria Index for Ecological Evaluation of Tropical Agriculture in Southeastern Mexico

Esperanza Huerta[1]*, Christian Kampichler[2,3], Susana Ochoa-Gaona[4], Ben De Jong[4], Salvador Hernandez-Daumas[1], Violette Geissen[5,6]

1 El Colegio de la Frontera Sur, Unidad Campeche, Dpto. Agroecología, Campeche, México, **2** Universidad Juárez Autónoma de Tabasco, División de Ciencias Biológicas, Villahermosa, Tabasco, México, **3** Sovon Dutch Centre for Field Ornithology, Natuurplaza (Mercator 3), Nijmegen, The Netherlands, **4** El Colegio de la Frontera Sur, Unidad Campeche, Dpto. Sustainability Sciences, Campeche, México, **5** University of Bonn - INRES, Bonn, Germany, **6** Wageningen University and Research Center – Alterra, Wageningen, Gelderland, Netherlands

Abstract

The aim of this study was to generate an easy to use index to evaluate the ecological state of agricultural land from a sustainability perspective. We selected environmental indicators, such as the use of organic soil amendments (green manure) versus chemical fertilizers, plant biodiversity (including crop associations), variables which characterize soil conservation of conventional agricultural systems, pesticide use, method and frequency of tillage. We monitored the ecological state of 52 agricultural plots to test the performance of the index. The variables were hierarchically aggregated with simple mathematical algorithms, if-then rules, and rule-based fuzzy models, yielding the final multi-criteria index with values from 0 (worst) to 1 (best conditions). We validated the model through independent evaluation by experts, and we obtained a linear regression with an $r^2 = 0.61$ ($p = 2.4e\text{-}06$, $d.f. = 49$) between index output and the experts' evaluation.

Editor: Yong Deng, Southwest University, China

Funding: Financial support was obtained from the Conacyt-SEMARNAT Project "Uso sustentable de los recursos naturales en la frontera sur de México" (Sustainable use of natural resources on the southern border of Mexico) (code SEMARNAT-2002-C01-1109). El Colegio de la Frontera Sur provided infrastructural resources. The funders had no role in study design, data collection and analysis, decision to publish, or preparation of the manuscript.

Competing Interests: The authors have declared that no competing interests exist.

* Email: ehuerta@ecosur.mx

Introduction

In the past 60 years, degradation and deforestation of tropical forests worldwide have occurred much faster and more extensively than in any other period in history [1], [2]. Furthermore, countries like Mexico have been undergoing drastic land use changes. In the tropics of Mexico large parts of its lowland rainforest areas have been converted into pasture and cropland. In the state of Tabasco, for example, only 3.4% of the state is covered with original forest [3], whereas 76.4% of the surface was used for cattle production in 2000 [4] and 15.6% was used for agriculture, principally sugarcane and fruit plantations [3]. The ecological consequences of these land-use changes in Tabasco are well documented. Soil losses in hilly regions are very high up to 200 t ha^{-1} $year^{-1}$ [5]; high pesticide and fertilizer inputs to crops that have replaced forests have caused considerable environmental contamination [6], [7], and soil fertility is decreasing [8], [5], [9].

It is of the utmost importance to identify sustainable land use strategies which are economically attractive for the region's farmers and which may also reconcile the need for food production with that of soil conservation. In order to assist a variety of stakeholders at the local and regional level in making land use decisions, simple evaluation tools are needed. This is even more needed since a high percentage of the population consists of immigrants from other Mexican states, who are unfamiliar with the conditions of the humid tropics and use intensive techniques for farming the land. This is mostly due to large-scale agricultural development projects, such as "Plan Balancán-Tenosique," named after the two municipalities involved, in which, in the 1970 s, over 1100 km^2 of lowland rain forest was destroyed and converted into crop and pasture land. Ecological values of the 1970 s were very different from those of the present, and government representatives were willing to deforest in order to grant land to farmers [10].

The direct impact of farming is difficult to measure due to methodological difficulties (impossibility of measurement, complexity of the system) or practical reasons (time, costs) [11]. Therefore, the use of indicators appears to be an alternative way of guiding land use decisions [12], [13]. However, the "indicator explosion" [14], that is, the use of an exaggerated number of indicators aimed at assessing environmental impacts of agricultural activities, has been of little use to local decision-makers. Particularly in the tropics, land use decisions are still based on the informal opinion of local experts rather than on implementation of Decision Support Systems for environmentally sound resource management [15]. On the one hand, this is due to farmers' restricted access to modern communications and information technologies; on the other, application of indicators is often beyond the capability of local farmers. For example, the agricultural sustainability index proposed by Nambiar et al. [16], which aims to measure agricultural sustainability as a function of biophysical, chemical, economic, and social indicators, would require considerable training of government stakeholders and farmers in order to be applied, and such training is rarely available.

Thus, any tool which local farmers or regional decision makers may use to support their decision making must be as simple as possible. Agroecosystems (like any other ecosystem) are too complex to be precisely measured and evaluated [17]. We agree with the view of Darnhofer et al. [18] in favour of developing less precise rules of thumb which may be used by farmers as well as to guide local land-use decisions toward a more environmentally friendly system of agriculture.

This index may provide farmers and others involved with a tool to evaluate the sustainability of their management of crop land which is oriented toward diminishing soil damage and conserving soil fertility. The index, exclusively based on terms which describe the environmental conditions of the crop system, is accessible to most farmers, and may be calculated using a simple internet application. In this paper we present a simple, easy-to-use index in order to evaluate the ecological state of farms in south-eastern Mexico. We applied the indicator system to 52 crop production systems in south-eastern Mexico and compared the results of the indicator system with expert opinions.

Human knowledge of how to efficiently and sustainably manage complex systems (including agricultural systems) is incomplete and much of what is thought to be known about this topic is actually incorrect. Yet, decisions must be made by policy makers, agricultural extension agents, and farmers despite uncertainty and knowledge gaps [19]. Therefore, tools to support local decision-makers must be flexible, should not enter into too much detail or precision, and should allow for an adaptive strategy which promotes "learning through management" [19]. Consequently, our rationale for developing an index which aids farmers in making environmentally friendly land-use decisions is based on basic, simplified ecological concepts, i.e. the presence of trees, since trees within an agroecosystem enhance soil microclimate in terms of radiation partitioning (shading), evapotranspiration partitioning, and rain interception/redistribution [20]. These factors all help to retain soil moisture. Branches, bark, roots, and living and dead leaf surfaces provide shelter [21] for soil micro-, meso-, and macro-invertebrates. Tree cover for instance, enhances above- and below-ground diversity, serving to support agricultural sustainability [22].

Materials and Methods

1. Rationale of index composition

We define conventional agriculture as a cropping system, typically promoted by government development programs, that is "capital-intensive, large-scale, highly mechanized agriculture with monocropping and extensive use of synthetic fertilizers, and pesticides" [23], [24]. Furthermore, we acknowledge that farming systems are sustainable only if "they minimize the use of external inputs and maximize the use of internal inputs already present on the farm" [25], [26]. The strategy most frequently linked to sustainability is reduction or elimination of agrochemicals, particularly chemical fertilizers and pesticides [25], [27], [28], [29], [30], [31]. Another key to sustained productivity of agricultural systems is the maintenance of soil functions, such as organic matter and nutrient cycling [32], based on organic inputs [33], above-and below-ground biodiversity [22], and diversifying crop systems with nitrogen-fixing legumes [34]. The principal role of the index we propose is to characterize methods of tillage, external inputs, and crop structure.

2. Primary indicators

We chose 12 field variables as primary indicators related to the above mentioned aspects of ecologically sound agricultural land use based on farmer's practices. These are easy to evaluate in the field and characterize plot structure (primary indicators: tree cover, tree density, tree diversity), crop structure and crop conditions (primary indicators: crop type, crop rotation, crop density, crop colour), tillage (primary indicators: type of tillage, timing of tillage), the use of fertilizers (chemical versus organic) and pesticide application.

2.1 Tree cover. Tree cover is defined as the canopy of trees, measured in the field, and recorded as percentage classes of tree cover in three height classes (trees >15 m, 10–15 m, <10 m). Thus, this variable characterized one aspect of agroecosystem management: the farmers' decision to maintain the canopy of the trees in his or her agroecosystem.

2.2 Tree density. Tree density is defined as the number of trees per area. To measure this, we distinguished three categories of tree density: high density (abundant), medium density, and low density (isolated or no trees). A high number of trees per area guarantee carbon sequestration [35] while a stable microclimate is maintained. This variable, measured in the field, is one the variables that characterize the effect of trees in the agroecosystem.

2.3 Tree diversity. This variable was measured in the field by counting the number of trees species within the plot. In agroecosystems, biodiversity may; (i) contribute to constant biomass production and reduce the risk of crop failure in unpredictable environments, (ii) restore disturbed ecosystem services such as water and nutrient cycling, and (iii) reduce risks of pests and diseases through enhanced biological control or direct pest control [36], [20], [22].

2.4 Crop type. This variable indicates whether the crop is annual, seasonal, or perennial. We obtained this information by observing the type of crop. Annual crops in general have higher environmental impacts, ie: greenhouse emissions, and nutrient leaching, than perennial crops [37].

2.5 Crop rotation. In sustainable farm systems leguminous crops are increasingly used in crop rotations as a source of nutrients, particularly nitrogen for crop growth [38], [39], nitrogen-fixing legumes, contribute to maintaining biodiversity above and in the soil, contribute nitrogen to the soil/plant system, and help avoid the build-up of pest populations [34]. In this study, we asked farmers whether they planted another crop before planting the main crop and whether they practice crop rotation, as crop rotation may assist with weed and pest control [40], [34]. According to Bellon [41], an activity which leads into the maintenance or increase of renewable resources in agroecosystems, is considered as an ecological technology. This variable helped us to characterize the technology used in the agroecosystem.

2.6 Crop density. Crop density is defined as the number of plants (individuals) per area. Three categories were recorded in the field: abundant (high density: 3,000 plants/ha), medium density (1000–1600 plants/ha), and sparse (<1000 plants/ha). This variable also indicated the level of technology applied to the crop, as less intensive techniques typically yield lower densities [42].

2.7 Crop colour. The colour of a crop indicates the nutritional status of the plants; green plants generally have sufficient nutrients, while yellow plants lack nitrogen [43].

2.8 Type of tillage. Type of tillage was categorized into no tillage, manual tillage, and mechanical tillage using machinery, the latter of which generally indicates high disturbance of the soil surface and rapid loss of soil organic carbon and other nutrients

[44]. In the field, we asked the farmers how they prepared the land for crop planting.

2.9 Frequency of tillage. This variable indicates frequency of tillage - every year or every 2 years. With this information, it was possible to estimate the frequency of soil disturbance due to tillage.

2.10 Chemical input. Searching for sustainable productions, it is recommended no or low or use of inorganic fertilizers and pesticides [45], [46]. Long term use of some pesticides, as glifosate can decrease earthworm species number, density and biomass [47], [48]. This information allowed us to estimate the amount of chemical fertilizers and pesticides applied within a given area. These variables (chemical fertilizers and pesticides) are included in the indicator for chemical disturbance.

2.11 Green manure. Is defined as the presence or absence of leguminous crops mixed with the principal crop, generally used to increase total soil nitrogen content. Green manures should always be intercropped, as it has been proven that growing legumes with cereal crops decreases N_20 emissions [49], and therefore is a sustainable, environmentally friendly practice. Examples of green manure use have been observed in traditional Mesoamerican cultures, for example, intercropping beans, as well as other edible plants, within the *milpa* or traditional maize cropping system [50].

3. Index development

Primary indicators were hierarchically aggregated into higher levels, forming intermediate variables, which in turn are structured into a single index that evaluates the ecological condition of a given plot on a scale from 0 (worst) to 1 (best). An index close to zero either would mean that a more environmentally friendly farming techniques need to be implemented or that the plot should be subjected to a fundamental change in land-use, e.g., reforestation, in order to return to a more ecologically sound state. An index close to one, on the other hand, would indicate an ecologically sound land-use. Methods applied in indicator aggregation were (i) simple mathematical operations, ii) sets of if-then rules, and (iii) sets of if -then rules combined with fuzzy logic.

3.1. Aggregation through mathematical operations. Mathematical operations include calculating averages, weighed averages, minimum values, and maximum values. For example, if primary indicator A has the value a and primary indicator B has the value b, then the value x of the intermediate variable X is determined as $x = (a+b)/2$, $x = (w_1*a+w_2*b)$, where $w1$ and $w2$ are weights, or $x = \min(a, b)$ or $\max(a, b)$.

3.2. Aggregation by IF-THEN rules. If primary indicator A can have the discrete values a_1, a_2, and a_3, and primary indicator B can have the discrete values b_1, b_2, and b_3, then the value x of the intermediate variable X is determined by a set of nine (number of levels of A x number of levels of B) rules. For example, IF $A = a_1$ and $B = b_1$ THEN $X = x_1$.

3.3. Aggregation by IF-THEN rules and fuzzy logic. We used small fuzzy rule-based models for aggregation in the case of non-linear interactions among indicators using a continuous numerical scale or an ordinal scale with a large number of possible values. In classic set theory, an object can either be a member (membership = 1) or not (membership = 0) of a given set. The central idea of fuzzy set theory is that an object may have a partial membership of a set, which consequently may possess all possible values between 0 and 1. The closer an element is to 1, the more it belongs to the set; the closer the element is to 0, the less it belongs to the set. To apply the fuzzy set theory, three steps are involved in calculating the model's output. First, for any observed value of the primary indicators, its corresponding membership value in the fuzzy set domain is calculated (*fuzzification*); second, the memberships of the intermediate variable X are calculated, applying the rules in the fuzzy set theory (*fuzzy inference*); third, the fuzzy results are converted into a discrete numerical output (*defuzzification*; see Wieland 2008 for an introduction to fuzzy models). Fuzzy rule-based models have become popular in ecological modelling [51], [52], and several examples exist of its usefulness in the context of ecosystem evaluation, bioindication, and sustainable management [15], [53], [54]. Here, if both primary indicators A and B can have numerical values from 0 to 1, then the value x of the intermediate variable X is determined by a series of fuzzy set rules representing the linguistic variables "low A", "medium A", "high A", and "low B", "medium B", and "high B", as well as the output "low X", "medium X", and "high X". The value of the intermediate variable X is determined by nine levels of A x the number of levels of B; for example: If A = low and B = low Then X = [low, medium, or high].

To maintain the number of rules as well as their complexity as low as possible, we aggregated only two variables at a time. A simple example shows the reasoning behind this decision; if there are three primary variables, A, B, and C with three categories for each variable, a single rule node requires 3*3*3 = 27 If-Then rules of the type "If $A = a_1, a_2,$ or a_3 and If $B = b_1, b_2,$ or b_3 and if $C = c_1, c_2,$ or c_3 Then...", whereas if an additional intermediate variable X is introduced to the model, only 18 rules are needed: 3*3 = 9 rules to aggregate A and B to X (If $A = a_1, a_2,$ or a_3 and If $B = b_1, b_2,$ or b_3), and 3*3 = 9 rules to aggregate X and C (If $X = x_1, x_2,$ or x_3 and If $C = c_1, c_2,$ or c_3).

4. Study area and application of the index

The state of Tabasco in south-eastern Mexico is characterized by a humid tropical climate with a mean annual rainfall between 1200 and 4000 mm and a mean annual temperature of 27°C [55]. Predominant soils are Gleysols and Fluvisols over alluvial sediments in the plains, Vertisols, Cambisols, Luvisols, and Acrisols over Miocene or Oligocene sediments, and Leptosols and Regosols over limestone mountains [9], [56]. We chose the municipalities Balancán and Tenosique in western Tabasco (17°81′50″–18°81′00″ N, 91°80′10″–91°84′60″ W) as a study area (Figure 1), we worked with private, *ejidal* (multipurpose land, where owners can or cannot sell the property according to their legal status in the National Agrarian File), and communal lands (coordinates of each plot are shown in Table 1). The region is mainly a plain towards the North (67% of area has an elevation <20 m. a. s. l.) with hills (29%, 20–200 m. a. s. l.), and mountains (4%, max. 640 m. a. s. l.) in the South, comprising a total area of 5474 km^2. These municipalities have undergone a high degree of land use change over the past 40 years. Until the early 1970 s, this region was still covered by lowland rain forest. The principal form of land use is pastureland, and covers 60% of the land [57]. An additional 30% is cropland, mainly cultivated under small-medium holder systems with seasonal conventional agriculture using high levels of agrochemical inputs (Table 2 & 3). Common crops are maize (*Zea mays*), a variety of hot peppers (*Capsicum* sp), cucumbers (*Cucurbita argyrosperma*), watermelon (*Citrullus lanatus*), perennial fruit crops such as papaya (*Carica papaya*), and biannual crops such as sugar cane (*Saccharum officinarum*) [58].

We chose 52 farms in the study area (Table 1), and selected those farms whose main economic activity (100%) is agriculture and chose one agricultural plot from each farm for evaluation. There were annual and biannual crops with or without trees (Table 2). Average plot size was 32.4±55.1 ha, and the average time that the plot had been used for a given crop system was 2.5±3.0 years.

Table 1. Characterization of plots.

Plot #	Municip	Lat	Long	Altitud (mls)	Plot size (ha)	Type of property	Original-vegetation
1	Bal	2007880	652445	43	18	Ejidal	A
2	Bal	1964368	642505	10	0.5	Private	TRF
3	Bal	1970936	647701	11	160	Private	TRF
4	Bal	1957459	667496	14	2	Ejidal	TRF
5	Bal	1964107	660656	14	7	Private	PT
6	Bal	1960297	658989	22	15	Private	TRF
7	Bal	1976300	676424	40	30	Private	TRF
8	Bal	1990326	665947	40	28	Private	TRF
9	Bal	1964590	667264	11	3	Private	RV
10	Bal	1972260	672382	28	2.5	Ejidal	TRF
11	Bal	1950339	666409	49	40	Private	TRF
12	Bal	1966112	682994	40	80	Private	PT
13	Ten	1931754	704474	65	14	Private	TRF
14	Ten	1912951	690381	130	2	Communal	TRF
15	Ten	1932644	665521	15	3	Ejidal	TRF
16	Ten	1926163	678608	56	68	Private	TRF
17	Ten	1933663	674227	59	35	Private	A
18	Ten	1926570	669486	35	8.5	Ejidal	TRF
19	Ten	1942431	663885	21	1.5	Ejidal	TRF
20	Ten	1942068	671156	31	7.25	Ejidal	TRF
21	Ten	1943167	650736	82	14	Ejidal	A
22	Ten	1923838	669841	118	0. 25	Private	TRF
23	Bal	1973301	695548	105	40	Ejidal	TRF
24	Bal	1982392	701383	74	20	Ejidal	A
25	Bal	1965354	709053	44	11	Ejidal	A
26	Bal	1960496	706464	51	44	Ejidal	TRF
27	Bal	1942484	709731	66	10	Ejidal	A
28	Ten	1944148	673673	16	320	Ejidal	TRF
29	Ten	1930350	659650	29	20	Private	TRF
30	Ten	1924549	656910	41	53	Ejidal	TRF
31	Ten	1926226	652978	65	3	Ejidal	TRF
32	Bal	1955194	688843	45	44	Private	TRF
33	Bal	1960980	679265	49	29	Ejidal	TRF
34	Bal	1942552	696222	48	10	Private	TRF
35	Ten	1938120	686470	49	5	Private	TRF

Table 1. Cont.

Plot #	Municip	Lat	Long	Altitud (mls)	Plot size (ha)	Type of property	Original-vegetation
36	Bal	1968750	695595	60	2	Private	A
37	Bal	1977546	699417	53	200	Ejidal	TRF
38	Bal	1949770	708854	53	23	Ejidal	TRF
39	Bal	1951216	697890	46	2	Ejidal	TRF
40	Ten	1931041	667544	31	17	Private	TRF
41	Ten	1927447	664223	3	3	Ejidal	TRF
42	Ten	1928226	664551	31	3	Ejidal	TRF
43	Ten	1930579	691785	54	15	Communal	TRF
44	Ten	1931219	677895	59	63	Private	TRF
45	Ten	1931525	676500	60	6.75	Ejidal	TRF
46	Ten	1908445	675040	140	40	Ejidal	TRF
47	Ten	1910875	669750	222	45	Ejidal	TRF
48	Ten	1924222	704702	42	30	Ejidal	A
49	Ten	1935076	712227	88	20	Ejidal	TRF
50	Ten	1914876	686285	113	4	Ejidal	TRF
51	Ten	1911365	684878	389	22	Ejidal	TRF
52	Ten	1911503	698606	139	10	Ejidal	TRF

Municip: municipality, Bal: Balancan, Ten: Tenosique. TRF: tropical rain forest, PA: pasture, A: acahual (secondary vegetation). Ejidal: multipurpose land, where owners can or cannot sell the property according to their legal status in the National Agrarian File.

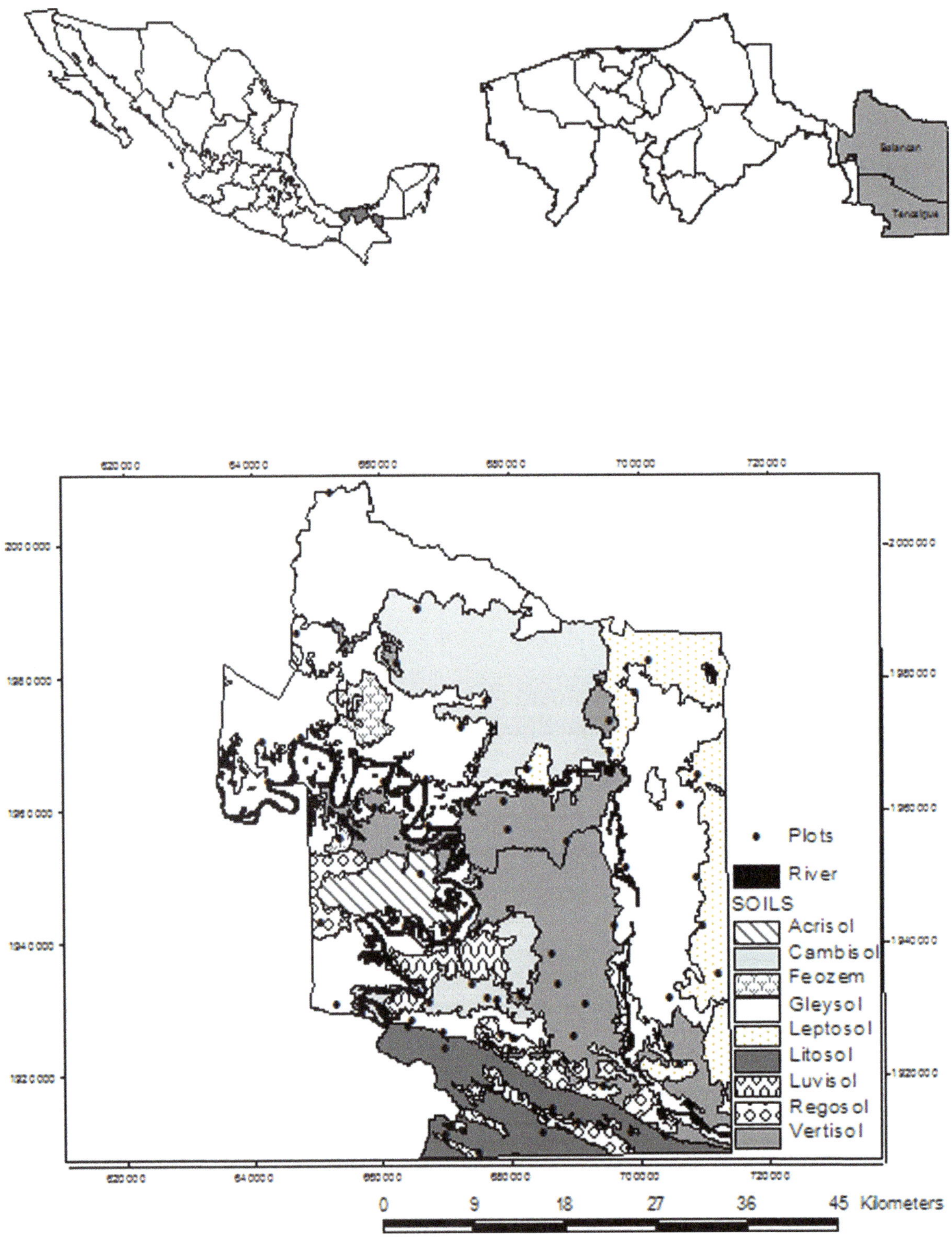

Figure 1. Distribution of 52 evaluated agricultural plots in tropical South-East Mexico.

We questioned farmers as to frequency, amount and type of chemical fertilizers and pesticides applied per area of cropland. After their verbal consent, farmers gave us their complete name and signed next to their name on the record sheet, validating all the information. Due to the fact that we only asked about the use of fertilizers and management of the land, the procedure approval from the Ethics Committee was not required.

Between March and October 2004, for each plot, the values of the primary indicators were determined and the index of ecological condition was calculated. Prior to index calculation, the plots were also evaluated by experts (2 scientists, each with

Table 2. Crop characterization.

Plot	Cycle	Main seasonal or perennial crop	Trees
1	s-s	*Zm, Pv, Cs, Cl*	*Tr, Pa, Cpa*
2	s-s	*Zm, Pv, Cs, Cl*	*Sh, Dg*
3	w-s	*Sv*	*Sa*
4	a-w	*Zm, Pv, Cpa*	*0*
5	Y	*Zm, Pv, Cl, Cm*	*0*
6	Y	*Zm*	*Cpa, Eg, Tr*
7	s-a	Ca	Tr
8	Y	*Cp*	*Cp*
9	Y	*Zm*	*0*
10	Y	*Zm*	*0*
11	w-s	*Zm, Pv, Cl*	*Sm, Cpa*
12	Y	*Co*	*Co*
13	s-s	*Zm*	*0*
14	s-s	*Le, Se*	*0*
		Cpa, Zm, Pv	
15	s-s	*Cl, Ta, Mp,*	*Fruit trees*
16	Y	*Jj*	*Gs, Hc, Cc, Tr*
17	Y	*So*	*0*
18	Y	*So*	*0*
19	Y	*So*	*0*
20	Y	*Zm*	*Mi*
21	s-s	*Zm*	*0*
22	Y	*Csi, Cli, Js*	*Co, Pa, Bc, Cn, Mi*
23	s-s	*Zm, Cpa*	*Cpe*
24	s-s	*Zm*	*Cpa*
25	s-s	*Zm*	*0*
26	s-s	*Zm, Cpa*	*0*
27	s-s	*Zm, Ca*	*Sm*
28	Y	*So*	*0*
29	s-s	*Zm, Pv, Cs, Cl*	*0*
30	Y	*Zm, Pv, Cs, Cl, Ta*	*0*
31	Y	*Zm, Pv, Cs*	*0*
		Cl, Os, Le	
32	s-s	*Zm, Pv, Cpa, Cl, Os*	*Tr*
33	Y	*Zm, Pv*	*Sm, Bg*
34	s-s	*Zm, Pv, Os*	*Cpa*
			Tr
35	Y	*Zm, Cl*	*Tr*
36	Y	*Zm, Pv, Cpa, Cl, Ca*	*0*
37	s-s	*Zm, Pv, Ca*	*Sm, Pa, Sh*
38	Y	*Zm, Pv, Cpa, Cl, Ta*	*Co, Sh*
39	Y	*Zm, Pv, Cpa, Cl, Ca*	
		Cs, Cm	
40	Y	*So*	*Tr, Cpe*
			Co, D a
41	Y	*So*	*Tr, Cpe*
			Co, D a
42	Y	*So*	*Tr, Cpe*
			Co, D a

Table 2. Cont.

Plot	Cycle	Main seasonal or perennial crop	Trees
43	Y	*So*	*Tr, Cpe*
			Co, Da
44	Y	*So*	*Tr, Cpe*
			Co, Da
45	Y	*So*	*Tr, Cpe*
			Co, D a
46	Y	*Zm, Pv, Cpa, Me*	*Sm*
47	Y	*Zm, Pv, Me, Ib*	*Bc, Mi, Dg, Cs, Ll*
48	Y	*Zm, Pv, Cpa, Cl*	*0*
49	Y	*Zm, Pv, Ca*	*Tr, Gs*
50	Y	*Zm, Pv, Cpa, Cl*	*Sm, Co, Ma*
		Cs, Ib, Me	0
51	Y	*Zm, Pv, Cpa, Cs*	*Fruit trees*
52	Y	*Zm, Pv, Cpa, Cl, Cs*	*Co, Sh, Bc*
		Mp, Ca, Me	0

Cycle: s–s: spring-summer, w-s: winter-summer, s-a: summer-autumn, Y: all the year. **Main seasonal or perennial crop**: Ca: *Cucurbita argyrosperma* (cushaw pumpkin); Cp: *Carica papaya* (papaya); Cl: *Citrullus lanatus* (watermelon); Cli: *Citrus limon* (lemon); Cm: *Cucumis melo* (muskmelon); Cpa: *Cucurbita pepo* (squash); Cs: *Capsicum* sp. (pepper); Csi: *Citrus sinensis* (orange tree); Ib: *Ipomoea batatas* (sweet potato); Jj: *Jarthropha jurcas* (oil palm); Js: *Jobo spondia* (plum); Le: *Lycopersicon esculentum* (tomato); Me: *Manihot esculenta* (cassava); Mp: *Musa paradisiaca* (banana); Os: *Oryza sativa* (rice); Pv: *Phaseolus vulgaris* (bean), So: *Saccharum officinarum* (sugarcane); Se: *Sechium edule* (pear squash); Sv: *Sorghum vulgare* (milo); Ta: *Triticum aestivum* (wheat); Zm: *Zea mays* (maize). **Trees**: *Bc: Byrsonima crassifolia* (nanche); *Cc: Crescentia cujete* (calabash tree); *Co: Cedrela odorata* (Mexican cedar); *Cpa: Carludovica palmate* (toquilla palm); *Cpe: Ceiba pentandra* (kapok); Cn: *Cocos nucifera* (Coconut); *Dg: Dialium guianense* (wild tamarind); Eg: *Eucalyptus grandis* (Eucalyptus); *Gs: Gliricidia sepium* (Cocoite, gliricidia, cacao de nance); *Ll: Leucaena leucocephala* (white lead tree, jumbay); *Ma: Mammea americana* (mamey apple); *Mi: Mangifera indica* (mango); *Pa: Persea americana* (avocado); *Sa: Sterculia apetala* (camoruco, manduvi tree); *Sh: Swietenia humilis* (small mahagoni); *Sm: Spondias mombin* (Yellow plum, bai makok); *Tr: Tabebuia rosea* (savannah oak, macuilis).

over 15 years of experience in agroecology) on a scale from 0 (poor condition) to 1 (good condition) in order to test the correlation between the experts' opinions and the index.

Data was normalized for carrying out multiple regression with plot size, altitude of plot site, or previous vegetation cover and the ecological index.

Finally we located the index in this public web address: http://201.116.78.102/~modelo/Index.html, where farmers in the near future can introduce independent data sets and evaluate or monitor their own agro ecosystems.

Results

1. Index structure

Ten nodes were used to aggregate primary indicators to the final index of ecological conditions of agricultural systems (Figure 2). Six of these used simple mathematical operations; two were based on rule sets, and two on fuzzy rule-based models (Table 4).

2. Index application

2.1. General characterization of sampled plots. A total of 67% of the plots were cultivated after cutting of primary lowland forest, 24% after cutting secondary forests of various ages, and 1.5% after cutting riparian vegetation, the rest conversion from pastureland (livestock production) to crop system. On 63% of the farms, maize, beans and pumpkins were cultivated, 13% sugar cane, 12% watermelon, 9% rice and 3% pepper. On 68.5% of the plots, trees were scattered among the crops. Conventional tillage was used on 55.5% of the plots, and pesticides were used on 79.4%, green manure was used on 21% of the plots and chemical fertilizers on 67%. All farmers understood the meaning of the variables evaluated for their plots (for a complete list of plots, see Table 5).

2.2. Plot evaluation with the index. Ecological condition of the plots ranges from 0.0 to 0.8125 (Figure 3; Table 5). One plot which was intercropped with timber and fruit trees presented the highest value. This site was characterized by an absence of agrochemical use, use of green manure, presence of annual crop rotation, and high tree diversity. We found 18 plots with index values of 0–0.25, 7 plots with values of 0.3–0.5, and 26 plots with values of 0.5–0.7. Thus, the majority of the plots evaluated in this study had an intermediate index value. 50% of the plots with this intermediate index are *ejidal* property, 38% private land, and 12% is under another type of ownership (smallholder or communal). 80% of these intermediate plots were lowland rain forest before being converted into agricultural land, 12% were secondary vegetation, and 8% were pastureland (Table 1). One might believe that a prior forest condition implies a more environmentally friendly agroecosystem that allows for preserving a more diverse system. However, plots with low index values were also lowland rain forest before being turned into agricultural land (Table 5). In this case, the land managers or owners decided to deforest the land to subsequently plant annual crops. 13 plots were larger than 40 ha (Table 1). All of these were previously covered with lowland rain forest; the ecological index ranges from 0.5–0.56 for 46% of these plots, and another 23% have an ecological index of 0.37–0.43. 10 plots fell under the smallest size category (0.25–3 ha) and had an ecological index of 0.5–0.68 (4 plots), 0 to 0.18 (5 plots), and 0.39 (1 plot).

Carrying out multiple regression with normalized data, we did not observe significant correlations between plot size, altitude of

Table 3. Crop land preparation and inputs characterization.

Plot	Tillage	Chemical Fertilization	Pesticides Use	Gm	Oa
1	Ma & Me	NPK 17:17:17 & U	Endosulfan**	0	0
2	Ma	0	Chlorpyrifos*	0	0
3	Me	Pholiar	0	0	c s m
4	Ma	0	0	0	0
5	Ma & Me	NPK 17:17:17	Chlorpyrifos**	Y	Y
6	Ma & Me	NPK 17:17:17	Chlorpyrifos*	0	0
7	Me	0	ID	0	0
8	Me	ID	ID	Y	0
9	Ma & Me	0	Carbofuran*	0	0
10	Me	NPK 18-46-0	ID	0	0
11	Me	NPK 18- 46- 0, U & P	ID	Y	0
12	Me	NPK 18-46-0	ID	0	0
13	Me	NPK 17:17:17 & U	Chlorothalonil**	0	0
			Chlorpyrifos*		
14	Ma	0	Endosulfan*	0	0
15	Me	0	0	Y	0
16	Ma	Urea	0	0	0
17	Me	NPK 19-19-19	ID	0	0
18	Me	NPK 19-19-19	ID	0	0
19	Me	Urea	ID	0	bg
20	Me	0	Methilic*	0	0
21	Me	NPK 17:17:17 & U	Zeta**	0	0
22	Ma	0	0	0	0
23	Me	NPK 17:17:17 & U	Chlorpyrifos**	0	lcf
24	Me	Urea & P	Zeta*	Y	lcf
25	Me	Urea	(2,4-D AMINA)*	Y	0
26	Me	NPK 17:17:17 & U	(Z)-(1R,3R)**	0	0
27	Me	0	Zeta*	N	0
28	Me	NPK 17:17:17	(RS)*	0	0
29	Me	NPK 17:17:17 & U	Chlorpyrifos**	0	0
30	Ma	0	Chlorpyrifos**	Y	0
31	Me	NPK 17:17:17 & U	Chlorpyrifos*	0	0
32	Me	0	0	Y	0
33	Me	0	0	0	0
34	Me	0	0	0	0
35	Me	NPK 17:17:17 & U	0	0	0
36	Me	NPK 17:17:17 & U	Chlorpyrifos*	Y	0
37	Me	0	Urea	0	0
38	Me	NPK 17:17:17 & U	Carbofuran*	Y	0
39	Me	NPK 17:17:17 & U	Chlorpyrifos**	0	0
40	Me	NPK 17:17:17	(Z)-(1R,3R)*	0	0
41	Me	NPK 17:17:17 & U	(Z)-(1R,3R)*	0	0
42	Me	NPK 17:17:17 & U	(Z)-(1R,3R)*	0	0
43	Me	NPK 17:17:17 & U	(Z)-(1R,3R)*	0	0
44	Me	NPK 17:17:17 & U	(Z)-(1R,3R)*	0	0
45	Me	NPK 17:17:17 & U	(Z)-(1R,3R)*	0	0
46	Ma	0	0	Y	0
47	Me	U	0	0	0
48	Ma &Sb	Urea	Chlorpyrifos**	0	0

Table 3. Cont.

Plot	Tillage	Chemical Fertilization	Pesticides Use	Gm	Oa
49	Me	Urea & P	Zeta*	0	0
50	Ma	Urea	Endosulfan*	0	0
51	Ma	0	ID	0	0
52	Me	NPK 17:17:17 & U	Chlorpyrifos*	0	0

Tillage: Ma: Manual; Me: Mechanized. **Pesticides**: Endosulfan: 6,7,8,9,10,10-Hexachlor- 1,5,5a,6,9,9a-Hexahidro-6,9-metane-2,4,3-benzodioxatiepin-3-oxide; Chlorpyrifos: 0,0-dimetil 0-(3,5,6-trichlore-2-piridinil) fosforotioate (33.8%), Permetrine: 3-fenoxibenzil (1RS)-cis, trans-3-(2,2 diclorovinil)-2,2 dimetil ciclopropane-carboxilate (4.8%); Carbofuran: 2,3 Dihidro-2,2-dimetil-7-benzofuranil metil carbamate; Chlorothalonil: Tetrachloroisoftalonitrile; Zeta: Zeta-cipermetrine a-ciano-3-(fenoxifenil) metil (±) cis-trans; (Z)-(1R,3R): (Z)-(1R,3R)-3-(2-chloro-3,3,3-trifluoroprop-1-enil)-2,2-dimetilcichlopropanecarboxilate de (S)-α-ciano-3-fenoxibencile & (Z)-(1S,3S)-3-(2-chloro-3,3,3-trifluoroprop-1-enil)-2,2-dimetilcichlopropanecarboxilate de (R)-α-ciano-3-fenoxibencile; (RS): (RS)- alfa- ciano-3-fenoxybencil(1RS)-cis-trans-3-(2,2-dichlorvinil)- 1,1-dimetilcichlopropanecasrboxilate. * Once per year, ** 2 per year. **Chemical Fertilization**: U: Urea, P: Phosphorus. **Gm**: Green manure, Y: Yes, 0: null application. **Oa**: Organic amendments: csm: cow, sheep manure, bg: burned grass, lcf: last crop fallow, 0: null application.

plot site, or previous vegetation cover and the ecological index (Kendall's Tau, T = 0.016 p = 0.98, T = 0.11 p = 0.90, T = −0.31 p = 0.75, respectively). Therefore, in this study, it seems that neither size nor location of the plot determines the type of plot management; rather, plot management is likely determined by government development programs and traditional farming techniques.

3. Correlation between index and expert opinion

We obtained a Pearson correlation coefficient of $r^2 = 0.61$ ($p = 2.4e\text{-}06$, $d.f. = 49$) between the index and the values determined by independent experts, indicating a satisfactory correspondence (Figure 4). However, the experts systematically awarded higher scores to the plots than did the index. Moreover, they suggested to include additional variables to the index which would yield better information regarding (i) type of organic inputs to the crop system, (ii) types of pest and disease control used, (iii) number of native plant species among the crop, (iv) origin of crop seeds, (v) vegetation surrounding the crop, (vi) presence of vertebrate fauna, and (vii) diversity of soil macroinvertebrates.

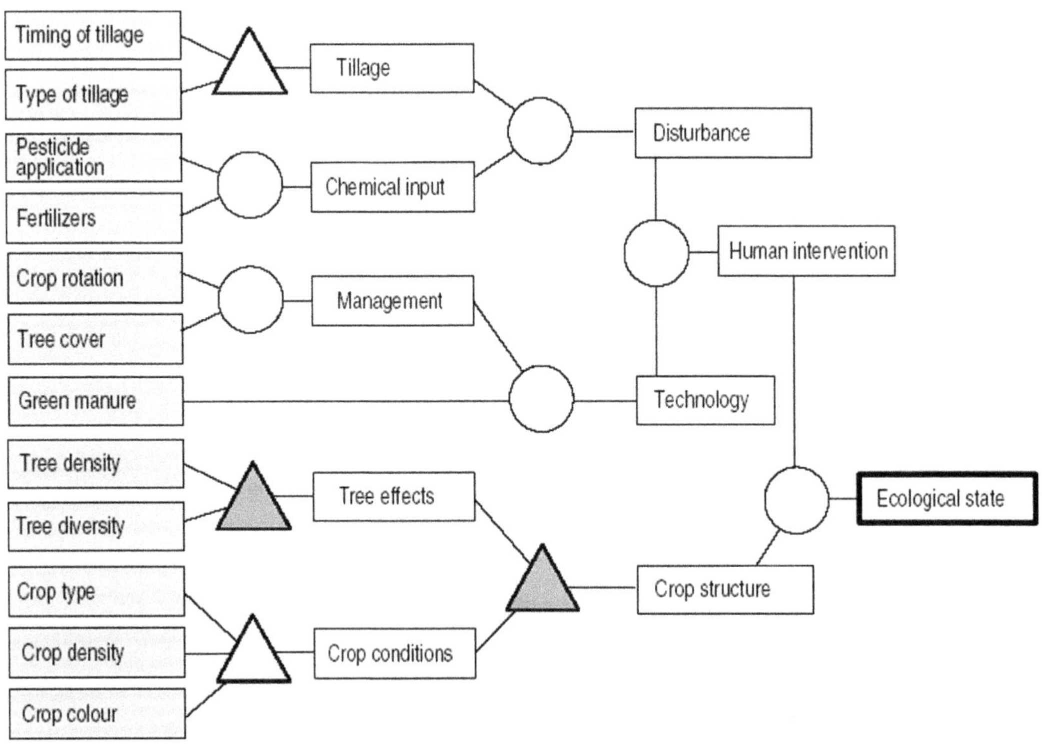

Figure 2. Structure of the Index of ecological condition of tropical agroecosystems. Primary indicators are shaded in grey. Circles represent simple mathematic algorithms, white triangles represent rule sets, and grey triangles represent rule sets, and grey triangles represent rule sets based on fuzzy logic. This index was presented together with other indexes within a frame of Indicators of environmentally sound land use in the humid tropics [15].

Table 4. Description of the measurement levels of the variables.

	Groupal variables	Field data variables	Best	Intermediate			Worst
				5 levels			
			1	0.75	0.5	0.25	0
	Management	Tree cover	all year	presence polyculture	present	rare	null
	Tree effect	Tree density	high	medium	low	isolated	null
				4 levels			
			1	0.66	0.33		0
		Tree diversity	>4 species	4 species	2 species		1 species
Crop structure	Crop conditions	Crop type	perennial polyculture	perennial monoculture	biannual		annual
				3 levels			
			1	0.33			0
	Plough (tillage)	Tillage time	null	frequent			constant
		Tillage form	null	manual			technical
Disturbance	Chemical inputs	Pesticides	null	frequent			constant
		Fertilizers	null	frequent			constant
	Management	Crop rotation	constant	frequent			null
Technology		Green manure	constant	frequent			null
				2 levels			
			1				0
	Crop conditions	Crop density	abundant				disperse
		Crop coleur	green				yellow

Table 5. Normalized Ecological Index for plots evaluated (52).

Plot	TT	TF	P	F	GM	RC	CA	DA	DivA	CT	CD	Ccol	Ev	Index
1	n	m	c	c	n	n	>1	l	>4	pm	a	g	0.24	0.56
2	n	n	c	n	n	n	>1	i	>4	a	a	g	0.35	0.5
3	c	m	n	c	c	n	>1	i	>4	a	a	g	0.26	0.56
4	n	n	n	n	n	n	n	n	1	a	a	g	0.01	0.37
5	n	n	c	c	c	c	>1	i	>4	a	a	g	0.09	0.62
6	n	m	c	c	n	n	>1	l	>4	a	a	g	0.22	0.5
7	c	t	c	n	n	n	>1	i	>4	a	d	b	0	0.15
8	c	t	c	c	c	c	n	n	1	pp	d	g	0.1	0.12
9	n	t	c	n	n	n	n	n	1	a	a	g	0.09	0.25
10	c	t	c	c	n	n	n	n	1	a	d	b	0	0
11	c	t	c	c	n	c	>1	i	>4	pm	ID	ID	0.03	0.21
12	c	t	c	c	n	n	>1	h	>4	ba	a	g	0.2	0.5
13	n	t	c	c	n	n	>1	i	>4	a	a	g	0.03	0.5
14	n	n	c	n	n	n	n	n	1	pp	a	g	0.29	0.37
15	n	t	n	n	n	c	>1	h	>4	pp	a	g	0.2	0.81
16	n	n	c	c	n	n	>1	l	>4	ba	a	g	0.25	0.5
17	c	t	c	c	n	n	n	n	1	a	a	g	0	0.25
18	c	t	c	c	n	n	n	n	1	a	d	g	0	0
19	c	t	c	c	c	n	n	n	1	a	d	g	0	0.06
20	n	t	c	n	n	n	>1	i	>4	a	a	g	0.1	0.5
21	c	t	c	c	n	n	n	n	1	a	a	g	0	0.25
22	n	n	n	n	n	n	>1	i	>4	pp	d	g	0.33	0.62
23	n	t	c	c	c	n	>1	i	>4	a	a	g	0.15	0.56
24	n	t	c	c	c	c	>1	l	>4	a	a	g	0.2	0.62
25	n	n	c	c	n	c	n	n	1	a	d	g	0.01	0.06
26	n	m	c	c	n	n	n	n	1	a	a	g	0.25	0.25
27	n	t	c	n	n	n	>1	i	>4	a	a	g	0.01	0.5
28	n	t	c	c	n	n	n	n	1	a	a	g	0	0.25
29	c	t	c	c	n	n	n	n	1	pm	a	g	0.01	0.37
30	n	m	c	n	n	c	n	n	1	pm	a	g	0.1	0.43
31	c	t	c	c	n	n	n	n	1	a	a	g	0.01	0.25
32	c	t	n	n	n	c	>1	i	>4	a	d	y	0.09	0.34
33	c	t	n	n	n	n	>1	h	>4	a	a	g	0.15	0.62
34	c	t	n	n	n	n	>1	i	>4	ba	a	g	0.13	0.62
35	c	t	n	c	n	n	>1	l	>4	a	a	g	0.01	0.5

Table 5. Cont.

Plot	TT	TF	P	F	GM	RC	CA	DA	DivA	CT	CD	Ccol	Ev	Index
36	c	t	c	c	n	c	n	n	1	a	a	g	0	0.31
37	c	t	c	n	n	n	>1	i	>4	pm	a	y	0.01	0.15
38	c	t	c	c	n	c	>1	h	>4	pm	a	g	0.2	0.68
39	c	t	c	c	n	n	>1	l	>4	a	a	b	0.06	0.18
40	c	t	c	c	n	n	>1	l	>4	a	a	g	0.19	0.5
41	c	t	c	c	n	n	>1	l	>4	a	a	g	0.19	0.5
42	c	t	c	c	n	n	>1	l	>4	a	a	g	0.17	0.5
43	c	t	c	c	n	n	>1	l	>4	a	a	g	0.2	0.5
44	c	t	c	c	n	n	>1	l	>4	a	a	g	0.2	0.5
45	c	t	c	c	n	n	>1	l	>4	a	a	g	0.2	0.5
46	n	n	n	n	n	c	>1	l	>4	pp	a	b	0.6	0.37
47	n	t	n	c	n	n	>1	l	>4	pp	d	g	0.18	0.5
48	c	n	c	c	n	n	n	n	1	a	a	b	0	0
49	c	t	c	c	n	n	>1	l	>4	a	a	b	0.07	0.18
50	c	n	c	c	n	n	>1	l	>4	pm	a	g	0.2	0.56
51	c	n	c	n	n	n	>1	l	>4	pm	a	g	0.15	0.56
52	c	t	c	c	n	n	>1	l	>4	pm	a	ID	0.24	0.18

TT: Tillage Time: c: >once per year; n: once per year; TF: Tillage Form: n: null, m: manual, t: technical; P. Pesticide use: c: constant, n: null, F: Use of chemical Fertilizers: c: constant, n: null; GM: Green Manure, c: constant, n: null; RC: Crop Rotation: c: constant, n: null; CA: tree cover: >1 year, n: null; DA: tree density: h: high, l: low, i: isolated, n:null; DivA: tree diversity: >4 species, 1 species, CT: crop type: a: annual, ba: biannual, pp: perennial polyculture, pm: perennial monoculture; CD: crop density: a: abundant, d: disperse; Ccol: crop colour: g: green, b: brown, y: yellow; EV: evaluation.

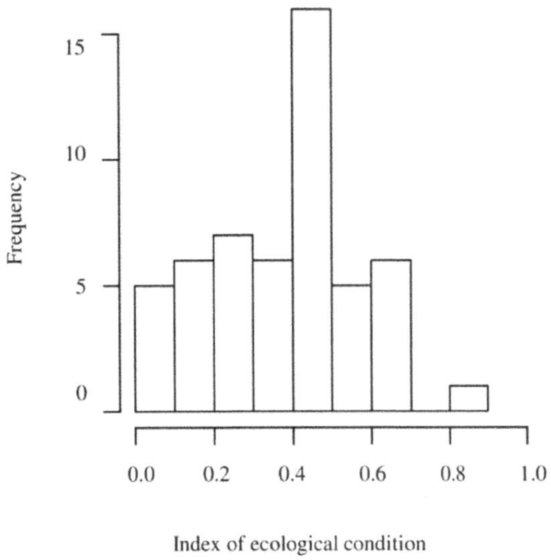

Figure 3. Frequency histogram of values of the index of ecological condition applied to 52 plots in South-East Mexico.

Discussion

Since the Rio Earth Summit, there has been a concerted effort to construct indicators to monitor progress toward sustainable development [59]. Most of these indicators have been developed in Europe (10) and Asia (2) [60]. In Latin America, sustainable indicators have been developed by Astier et al. [61]; this index mainly focuses on subsistence level agriculture, and evaluates the sustainability of a system.

Our index is geared toward small and medium scale producers who principally grow for market and have a fairly large crop area (32.4 ha on average). According to Bockstaller and Girardin [62], indicators must be elaborated according to a scientific approach, and one of the important steps in this elaboration is validation.

Our index was developed according to the consensus of a group of scientists, with knowledge in agroecology, and was validated independently by 2 scientists, each with over15 years of experience in agroecology. The evaluation included 3 important steps: design validation, output validation, and end use validation [62].

In previous studies, only seven indicators have been used to evaluate farm systems: crop diversity, crop succession, pesticide use, nitrogen level, phosphorus level, soil organic matter, and irrigation methods [63]. All of these practices depend on the farmer's decision, and to a large extent they impact the environment.

Our index is based on qualitative and quantitative concrete data and includes most of these 7 indicators, except nitrogen and phosphorus, both of which are observed indirectly via plant health, through the crop colour indicator (according to whether plants have a greenish or yellowish colour); organic matter, which is indirectly characterized by the technology applied in the system (green manure indicator, Figure 2); and irrigation, which in this study was not evaluated, given that all plots evaluated were only used for seasonal rainfed agriculture, according to local rainfall patterns.

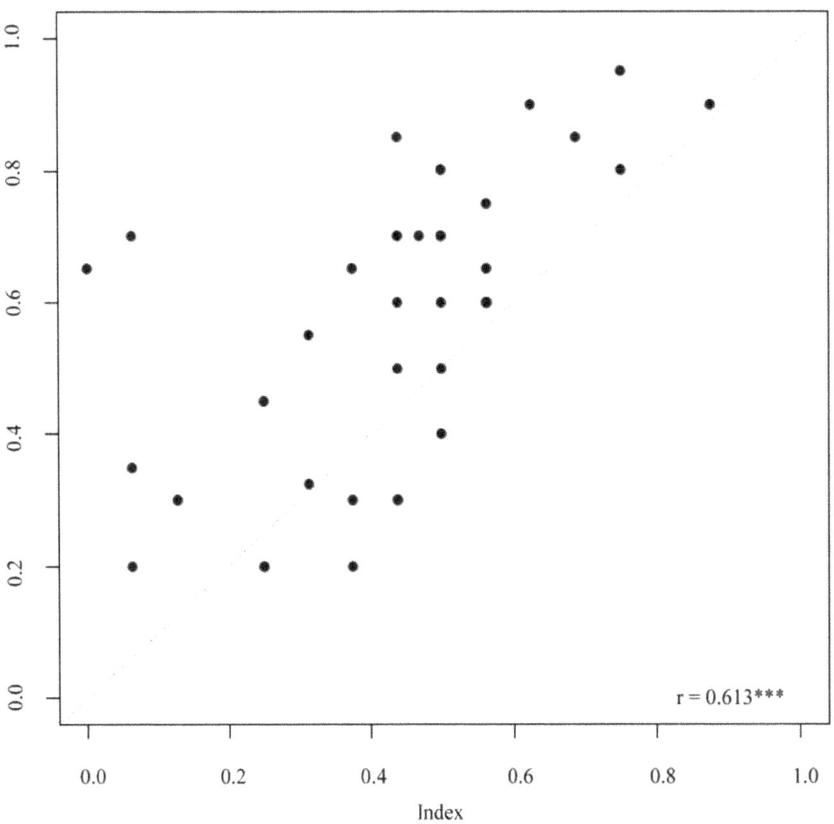

Figure 4. Scatterplot of the values of the index of ecological condition applied to 52 plots in South-East Mexico (x-axis) versus quality values between 0 (worst) and 1 (best) assigned by experts to the same plots (y-axis).

At the farm level there are indicators that evaluate the environmental impact of the agricultural practices and indicators that evaluate the effect of those practices at the local and global level [60].

Our index includes both types of indicators; evaluating agricultural practices: those variables taken in the field: type and frequency of tillage, pesticide application, fertilizer use, crop rotation, tree cover and density, green manure, crop density (see Table 2 & 3). Effect indicators used were disturbance (mainly soil disturbance) measured by tillage (frequency and type) and chemical inputs, technology used, and crop structure (see index, Figure 2). Some existing indexes focus on evaluation or evolution of environmental performance, thus encouraging environmentally sound practices [60], such as crop rotation, organic fertilizers, and no-tillage. Meanwhile, our index identifies the indicator that has the highest environmental impacts in each plot evaluated, with the idea that the farmer could potentially improve these with a given practice, ie. to monitor the soil ecological condition via the use of a soil macroinvertebrates index [64], where the lack of macroinvertebrates informs of a severe negative activity as pollution or conventional tillage.

In developing the variables to be included in our index, we reviewed bibliographic studies and carried out field work obtaining data which we hoped would reflect the negative and environmentally friendly practices commonly used in agro ecosystems of southeastern Mexico. The index can be used by farmers, using the following web address: http://201.116.78.102/~modelo/Index.html.

However, we did not evaluate certain indicators such as water use, and water quality, as did - for example - the index (monitoring tool) of ecological indicators used on a Flemish dairy farm [65]. Nor did we evaluate environmental impacts due to energy consumption [66]. Within a tropical framework, in south-eastern Mexico, the priorities were to identify those practices that were soil perturbing and environment polluting, practices that can be modified by the farmers, by an attitude changing. Nambiar et al. [16], proposed an agricultural sustainability index (ASI) to measure sustainability as a function of biophysical, chemical, economic, and social indicators, our index only measures the ecological state of the agroecosystem, and provides easy to use tools for improving those practices which negatively impact the environment. Van der Werf and Petit [60], state that indicators based on farmer practices cost less in data collection but do not allow for an actual evaluation of environmental impact. In the case of our study, the experts' evaluation correlated satisfactorily with the index, although the index rendered more penalizing scores than did the experts. Some improvements must be made to our index, relating quality and quantity of the applied inputs for instance; the index should specify the kind of manures used and then to evaluate their effect when added to the systems. The consulted experts found important to integrate this information into the index, they also found that the possible relations among different crops and environmental effects of using cow manure, vermicompost, or traditional compost have to be considered. Another variable which the experts suggested should be added to the index is the presence of natural vegetation surrounding the crop. Farmers see advantages of having crops surrounded by secondary forest, diversity in the agricultural area can be increased, ie when different pollinators arrive.

The advantages of using this index is that the common agricultural practices (mechanized land preparation and use of common pesticides ie. Carbofuran, Chlorpyrifos), evaluated in this study as indicators are used throughout the world; over time, through practice, the index may be improved. Farmers and other land owners may realize which of the practices they use are disturbing the environment, due to the fact that these practices generate a value in each of the evaluated indicators. The variables obtained in the field contribute to the information of each of the indicators, and the index is the compendium of all the indicators. Our index doesn't give a sustainability measure, because it does not include socioeconomic indicators of the farms. Further studies are required in order to observe the acceptance of this index by farmers in a regional scale.

Acknowledgments

We thank Lorena Soto and Guillermo Jimenez for participating in the evaluation of the plots. We thank Simon Hernandez de la Cruz, Alejandra Sepulveda Lozada, Lauritania Ibarra Hernandez, and Marcelo Rodriguez Ricardes, who helped to collect data in the field. We are also thankful to Lorena Reyes for bibliographic support. El Colegio de la Frontera Sur provided infrastructural resources. Language revision was done by Ann Greenberg.

Author Contributions

Conceived and designed the experiments: EH CK SOG BDJ SHD VG. Performed the experiments: EH CK. Analyzed the data: EH CK. Contributed reagents/materials/analysis tools: EH CK SOG BDJ SHD VG. Contributed to the writing of the manuscript: EH CK SOG BDJ VG. Figures elaboration: CK SOG.

References

1. Houghton RA (1994) The world wide extent of land-use change. BioScience 44: 305–313.
2. Tilman D, Lehman C (2001) Human-cause environmental change: impacts on plant diversity and evolution. Proceedings of the National Academy of Sciences 98: 5433–5440.
3. INEGI (2008) Cuaderno Estadístico Municipal de Tenosique. Villahermosa, Tabasco: Gobierno del Estado de Tabasco.
4. Grande D, de Leon F, Nahed J, Perez-Gil F (2010) Importance and Function of Scattered Trees in Pastures in the Sierra Region of Tabasco, Mexico. Research Journal of Biological Sciences 5: 75–87.
5. Geissen V, Morales-Guzman G (2006) Fertility of tropical soils under different land use systems-a case study of soils in Tabasco, Mexico. Appl Soil Ecol 841: 1–10.
6. Melgar C, Geissen V, Cram S, Sokolov M, Bastidas P, et al. (2008) Pollutants in drainage channels following long-term application of Mancozeb to banana plantations in southeastern Mexico. Journal of Plant Nutrition and Soil Science 171: 597–604.
7. Aryal DR, Geissen V, Ponce-Mendoza A, Ramos-Reyes RR, Becker M (2012) Water quality under intensive banana production and extensive pastureland in tropical Mexico. Journal of Plant Nutrition and Soil Science 175(4): 553–559.
8. Ortiz SM, Anaya G, Estrada BW (1994) Evaluación, Cartografía y Políticas Preventivas de la Degradación de la Tierra. Chapingo, Mexico.
9. Geissen V, Sanchez-Hernandez R, Kampichler C, Ramos-Reyes R, Sepulveda-Lozada A, et al. (2009) Effects of land-use change on some properties of tropical soils – An example from Southeast Mexico. Geoderma 151: 87–97.
10. Moreno-Unda AA (2011) Environmental effects of the National Tree Clearing Program, Mexico, 1972–1982 Cologne: Cologne University of Applied Sciences 119 p.
11. Bockstaller CGL, Keichinger O, Girardin P, Galan MB, Gaillard G (2009) Comparison of methods to assess the sustainability of agricultural systems. A review. Agron Sustain Dev 29: 223–235.
12. Mitchell GMA, McDonald A (1995) PICABUE: a methodological framework for the development of indicators of sustainable development. Int J Sust Dev World 104–123.

13. Bockstaller C, Guichard L, Makowski D, Aveline A, Girardin P, et al. (2008) Agri-Environmental Indicators to Assess Cropping and Farming Systems: A Review Sustainable Agriculture. In: Lichtfouse E, Navarrete M, Debaeke P, Véronique S, Alberola C, editors. Springer Netherlands. 725–738.
14. Riley J (2001) The indicator explosion: local needs and International challenges. Agr Ecosyst Environ 87: 119–120.
15. Kampichler C, Hernández-Daumás S, Ochoa-Gaona S, Geissen V, Huerta-Lwanga E, et al. (2010) Indicators of environmentally sound land use in the humid tropics: The potential roles of expert opinion, knowledge engineering and knowledge discovery. Ecological Indicators 10: 320–329.
16. Nambiar KK, Gupta AP, Fu Qinglin, Li S (2001) Biophysical, chemical and socio-economic indicators for assessing agricultural sustainability in the Chinese coastal zone Agriculture Ecosystems and Environment 87: 209–214.
17. Cabell JF, Oelofse M (2012) An indicator framework for assessing agroecosystem resilience. Ecology and Society 17: 18 http://dx.doi.org/10.5751/ES-04666-170118.
18. Darnhofer I, Bellon S, Dedieu B, Milestad R (2010) Adaptiveness to enhance the sustainability of farming systems. A review. Agronomy for Sustainable Development 30: 545–555.
19. Allen CR, Fontaine JJ, Pope KL, Garmestani AS (2011) Adaptive management for a turbulent future. Journal of Environmental Management 92: 1339–1345.
20. Malezieux E, Crozat Y, Dupraz C, Laurans M, Makowski D, et al. (2009) Mixing plant species in cropping systems: concepts, tools and models. A review. Agron Sustain Dev 29: 43–62.
21. Jones CG, Lawton J, Shachak M (1997) Positive and negative effects of organisms as physical ecosystem engineers. Ecology and Society 78: 1946–1957.
22. Brussaard L, Caron P, Campbell B, Lipper L, Mainka S, et al. (2010) Reconciling biodiversity conservation and food security: scientific challenges for a new agriculture. Current opinion in Environmental sustainability 2: 34–42.
23. Knorr D, Watkins TR (1984) Alterations in Food Production Knorr DW, Watkins, T.R., editor. New York: Van Nostrand Reinhold.
24. Seufert V, Ramankutty N, Foley JA (2012) Comparing the yields of organic and conventional agriculture. Nature 485: 229–232.
25. Carter H (1989) Agricultural sustainability: an overview and research assessment. Calif Agric 43: 1618–1637.
26. Tellarini V, Caporali F (2000) An input/output methodology to evaluate farms as sustainable agroecosystems: an application of indicators to farms in central Italy. Agriculture, Ecosystems and Environment 77: 111–123.
27. Stinner BR, House GJ (1987) Role of ecology in lower-input, sustainable agriculture: an introduction. Am J Alternative Agric 2: 146–147.
28. Lockeretz W (1988) Open questions in sustainable agriculture. Am J Alternative Agric 3: 174–181.
29. Hauptli H, Katz D, Thomas BR, Goodman RM (1990) Biotechnology and crop breeding for sustainable agriculture. In: Edwards CA, Lal R, Madden P, Miller RH, House G, editors. Sustainable Agricultural Systems Soil and Water Conservation Society: Ankeny, Iowa. 141–156.
30. Madden P (1990) The economics of sustainable low-input farming systems. In: Francis CA, Flora CB, King LD, editors. Sustainable Agriculture in Temperate Zones. New York: John Wiley & Sons. 315–341.
31. Dobbs TL, Becker DL, Taylor DC (1991) Sustainable agriculture policy analyses: South Dakota on-farm case studies. J Farming Systems ResExt 2: 109–124.
32. Blair GJ, Lefroy RD, Lisle L (1995) Soil carbon fractions based on their degree of oxidation, and the development of a carbon management index for agricultural systems. Aust J Soil Res 46: 1459–1466.
33. Ouédraogo E, Mando A, Zombré NP (2001) Use of compost to improve soil properties and crop productivity under low input agricultural system in West Africa. Agriculture, Ecosystems and Environment 84: 259–266.
34. Pretty J, Toulmin C, Williams S (2011) Sustainable intensification in African agriculture. International Journal of Agricultural Sustainability 9: 5–24.
35. Lal R (2004) Soil carbon sequestration impacts on global climate change and food security. Science 304: 1623–1627.
36. Gurr GM, Wratten SD, Luna JM (2003) Multi-function agricultural biodiversity: pest management and other benefits. Basic Appl Ecol 4: 107–116.
37. Börjesson P (1999) Environmental effects of energy crop cultivation in Sweden I: Identification and quantification. Biomass and Bioenergy 16: 137–154.
38. Rommelse R (2001) Economic assessment of biomass transfer and improved fallow trials in western Kenya In: ICRAF Natural Resource Problems PaPP, editor. Natural Resource Problems. Nairobi (Kenya): International Centre for Research in Agroforestry (ICRAF).
39. Nyende P, Delve R (2004) Farmer participatory evaluation of legume cover crop and biomass transfer technologies for soil fertility improvement using farmer criteria, preference ranking and logit regression analysis. Exp Agric 40: 77–88.
40. Koocheki A, Nassiri M, Alimoradi L, Ghorbani R (2009) Effect of cropping systems and crop rotations on weeds. Agronomy for Sustainable Development 29: 401–408.
41. Bellon M (1995) Farmers 'Knowledge and Sustainable Agroecosystem Management: An Operational Definition and an Example from Chiapas, Mexico. Human Organization 54: 263–272.
42. Belalcázar S, Espinosa J (2000) Effect of Plant Density and Nutrient Management on Plantain Yield. Better Crops International 14: 12–15.
43. Shaahan MM, El-Sayed AA, Abou El-Nour EAA (1999) Predicting nitrogen, magnesium and iron nutritional status in some perennial crops using a portable chlorophyll meter. Scientia Horticulturae 82: 339–348.
44. Agbede TM (2008) Nutrient availability and cocoyam yield under different tillage practices. Soil Tillage Research 99: 49–57.
45. Edwards CA (1989) The Importance of Integration in Sustainable Agricultural Systems. Agriculture, Ecosystems and Environment 27: 25–35.
46. Edwards CA, Grove TL, Harwood RR, Pierce Colfer CJ (1993) The role of agroecology and integrated farming systems in agricultural sustainability. Agriculture, Ecosystems and Environment 46: 99–121.
47. García-Pérez JA, Alarcón-Gutiérrez E, Perroni Y, Barois I (2014) Earthworm communities and soil properties in shaded coffee plantations with and without application of glyphosate. Applied Soil Ecology 83: 230–237.
48. Correia FV, Moreira JC (2010) Effects of Glyphosate and 2,4-D on Earthworms (*Eisenia foetida*) in Laboratory Tests. Bull Environ Contam Toxicol 85: 264–268.
49. Dick J, Kaya B, Soutoura M, Skiba U, Smith R, et al. (2008) The contribution of agricultural practices to nitrous oxide emissions in semi-arid Mali Soil. Use and Management 24: 292–301.
50. Morales H, Perfecto I (2000) Traditional knowledge and pest management in the Guatemalan highlands. Agriculture and Human Values 17: 49–63.
51. Li BL, Rykiel EJ (1996) Introduction. Ecological Modelling 90: 109–110.
52. Salski A (1996) Introduction Fuzzy Logic in Ecological Modelling. Ecological Modelling 85: 1–2.
53. Mendoza GA, Prabhu R (2003) Fuzzy methods for assessing criteria and indicators of sustainable forest management. Ecological Indicators 3: 227–236.
54. Kampichler C, Platen R (2004) Ground beetle occurrence and moor degradation: modelling a bioindication system by automated decision-tree induction and fuzzy logic. Ecological Indicators 4: 99–109.
55. INEGI (2000) Cuaderno Estadístico Municipal de Tenosique. Gobierno del Estado de Tabasco; INEGI, editor. Villahermosa, Tabasco.
56. INEGI (1985) Carta Edafológica, Villahermosa In: E15-8, editor. 1: 250,000. Aguascalientes, México: Instituto Nacional de Estadística, Geografía e Informática.
57. Manjarrez-Muñoz B (2008) Ordenamiento territorial de la ganadería bovina en Balancán y Tenosique, Tabasco. Villahermosa Tabasco: El Colegio de la Frontera Sur. 105 p.
58. Isaac-Márquez R (2008) Análisis del Cambio de Uso y Cobertura del Suelo en los Municipios de Balancán y Tenosique, Tabasco, México. Villahermosa, Tabasco, México: El Colegio de la Frontera Sur.
59. Rigby D, Woodhouse P, Young T, Burton M (2001) Constructing a farm level indicator of sustainable agricultural practice. Ecological Economics 39: 463–478.
60. van der Werf HMG, Petit J (2002) Evaluation of the environmental impact of agriculture at the farm level: a comparison and analysis of 12 indicator-based methods. Agriculture, Ecosystems and Environment 93: 131–145.
61. Astier M, Speelman S, López-Ridaura S, Masera O, Gonzalez-Esquivel CE (2011) Sustainability indicators, alternative strategies and trade-offs in peasant agroecosystems: analysing 15 case studies from Latin America. International Journal of Agricultural Sustainability 9: 409–422.
62. Bockstaller C, Girardin P (2003) How to validate environmental indicators. Agr Syst 76: 639–653.
63. Bockstaller C, Girardin P, Van der Werf HGM (1997) Use of agroecological indicators for the evaluation of farming systems. Eur J Agron 7: 261–270.
64. Huerta E, Kampichler C, Geissen V, Ochoa-Gaona S, de Jong B, et al. (2009) Towards an ecological index for tropical soil quality based on soil macrofauna. Pesq agropec bras 44 (8): 1056–1062.
65. Meul M, Nevens F, Reheul D (2009) Validating sustainability indicators: Focus on ecological aspects of Flemish dairy farms. Ecological Indicators 9: 284–295.
66. Pervanchon F, Bockstaller C, Girardin P (2002) Assessment of energy use in arable farming systems by means of an agro-ecological indicator: the energy indicator. Agricultural Systems 72: 149–172.

Xenopsylla cheopis (Siphonaptera: Pulicidae) Susceptibility to Deltamethrin in Madagascar

Sebastien Boyer[1]*, Adélaïde Miarinjara[1,2], Nohal Elissa[1]

1 Unité d'Entomologie Médicale, Institut Pasteur de Madagascar, Antananarivo, Madagascar, 2 Ecole doctorale Sciences de la vie et de l'environnement, Université d'Antananarivo, Antananarivo, Madagascar

Abstract

The incidence of bubonic plague in Madagascar is high. This study reports the susceptibility of 32 different populations of a vector, the flea *Xenopsylla cheopis* (Siphonaptera: Pulicidae), to the insecticide Deltamethrin. Despite the use of Deltamethrin against fleas, plague epidemics have re-emerged in Madagascar. The majority of the study sites were located in the Malagasy highlands where most plague cases have occurred over the last 10 years. *X. cheopis* fleas were tested for susceptibility to Deltamethrin (0.05%): only two populations were susceptible to Deltamethrin, four populations were tolerant and 26 populations were resistant. KD50 (50% Knock-Down) and KD90 (90% Knock-Down) times were determined, and differed substantially from 9.4 to 592.4 minutes for KD50 and 10.4 min to 854.3 minutes for KD90. Susceptibility was correlated with latitude, but not with longitude, history of insecticide use nor date of sampling. Combined with the number of bubonic plague cases, our results suggest that an immediate switch to an insecticide other than Deltamethrin is required for plague vector control in Madagascar.

Editor: Érika Martins Braga, Universidade Federal de Minas Gerais, Brazil

Funding: The authors have no support or funding to report.

Competing Interests: The authors have declared that no competing interests exist.

* Email: seboyer@pasteur.mg

These authors contributed equally to this work.

Introduction

Ectoparasites (including ticks and fleas) are both pests and vectors of various diseases of humans, livestock, pets, and wild animals. They can transmit diverse pathogens of medical and/or veterinary significance, including viruses, bacteria, protozoa, and helminthes [1]. Plague is one of these diseases and still remains a health problem with occasional epidemics occurring in the world. Over the last decade 82% of all cases worldwide (more than 2,000 annually, total 21,725) have been in the Democratic Republic of Congo (10,581 cases between 2000 and 2009) and Madagascar (7,182) [2,3]. In Madagascar, 2,409 cases were confirmed between 2007 and 2011, and the country declared 67% of the worldwide cases in 2012 [4,5].

Plague is a life-threatening infectious disease caused by the Gram-negative bacterium *Yersinia pestis* [6–8]. It primarily affects rodents, but can also cause outbreaks in human populations. The infection is classically transmitted in the murine population by infected fleas and the risk to human increases during epizootic periods that are associated with high rodent and flea densities. At least 80 flea species are known to carry the etiological agent of the plague, although their role in disease transmission varies [9]. For instance, the oriental rat flea *Xenopsylla cheopis* (Siphonaptera: Pulicidae), Rothschild, 1903, is considered to be the most efficient vector as well as the major vector to humans. Other flea species have been identified as vectors in East Africa, including the Islands of the South West Indian Ocean (*Ctenophtalmus bacopus*, *C. cabirus*, *Dinopsyllus lypusus*, *Pulex irritans*, *Xenopsylla brasiliensis*, *X. cheopis*) [10]. Plague was introduced into Madagascar in 1898, and then spread throughout the central highlands [4]. Two flea species are involved in the transmission of plague: *X. cheopis* and *Synopsyllus fonquerniei* (Siphonaptera, pulicidae), Wagner & Roubaud. 1932. Although *X. cheopis* is the main vector, the black rat is also frequently parasitized by the endemic flea *S. fonquerniei*, and this species contributes to the circulation of *Y. pestis* in the rural murine population [4].

Since 1947, chemical insecticides have been used to limit plague transmission during outbreaks. DDT was the first chemical insecticide used to control plague vectors in Madagascar. In 1956, the organophosphate Malathion and organochlorines were applied to plague control. At the same time, the use of the EV vaccine significantly decreased human plague cases [11]. However, the numbers of plague cases increased, particularly in the capital Antananarivo and in the coastal city of Mahajanga, after long periods of absence: 28 and 63 years, respectively [12,13]. One of the possible causes of the increase in the incidence of human cases despite an active control campaign against plague vectors is the emergence of resistance to insecticides.

The first cases of *X. cheopis* resistance to DDT were described in 1965 in Madagascar and were first demonstrated in 1981 [14,15]. In 1983, *X. cheopis* was reported to be resistant to Malathion, Fenitrothion and Propoxur in the field [16,17]. Thus, the National Plague Control Program (NPCP) used Deltamethrin, a pyrethroid, to control plague outbreaks in the 1990s. After six years of use, insecticide susceptibility tests revealed *X. cheopis* resistance to Deltamethrin [18,19]. In 2000, four populations in Madagascar were assayed for susceptibility to Deltamethrin (0.025%): one population was resistant and three populations were tolerant [20].

Pyrethroids interfere with the normal activities of nerve membranes by delaying the closing of activation gates for the sodium channel [21]. The knock-down (KD) effect describes the insect paralysis following the arrival of the molecule at the central lymph nodes. One of the major mechanisms of pyrethroid resistance involves the loss of sensitivity of the active site of the protein targeted by pyrethroids (the voltage-gated sodium channel); this is known as knockdown resistance (kdr) [22]. The existence of resistance mechanisms and their selection has been extensively studied in Diptera species (Drosophila and mosquito species) and in stored-product insects, but not in flea species. As already described for other insect vector species, *X. cheopis* in Madagascar undoubtedly displays multiple resistances against different insecticides [14,18,20].

Because Deltamethrin is the insecticide used by Malagasy NPCP, it is crucial to detect and evaluate the current status of flea population's susceptibility to this insecticide, in order to conduct appropriate and efficient vector control program. Here, we report an investigation of the susceptibility to Deltamethrin of *X. cheopis* in the highlands of Madagascar.

Materials and Methods

Ethics statements

This research in Madagascar was systematically made possible thanks to extant conventions between the IPM and local governments. The rats were caught either during an epidemic event at the request of the Ministry of Health and the World Health Organization on the basis of a National Health Priority, or during an investigative mission. In this last case, a letter was sent to the local authority and local general inspector, and to the national authority of the Ministry of Health, to explain the main objective of the field mission. The mission orders were authorized by the CSB (Centre de Santé de Base), SSD (Service de Santé de District) and Fokotany authorities. Additional authorizations were not required because *Rattus* species are considered to be pest species (especially *R. rattus* and *R. norvegicus*) and have no protected status (see IUCN and CITES lists). Traps were set only after agreement was explicitly obtained from both the village head and the field/house owner. In cultivated fields, traps were always placed at the edge of the farmed area, so that no damage was caused to crops. Rats were caught alive in wire-mesh and Sherman traps. All animals were killed by cervical dislocation. All members of the IPM involved in the fieldwork have been trained to handle and kill rodents. The study was conducted in accordance with the guidelines (http://www.pasteur.fr/ip/easysite/pasteur/en/institut-pasteur/ethics-charter) adapted to wild rodents. Animals were treated in a humane manner, and in accordance with guidelines of the American Society of Mammalogists [23]. No country-specific ethics agreement could be obtained because the country where sampling occurred and the IPM have no ethics committees relevant to animal experimentation.

Young mice used to feed flea population were from the Institut Pasteur de Madagascar (IPM) animal breeding facility. They were not purchased or donated, but were bred for this purpose.

Trapping rats and collection of fleas

The species *Xenopsylla cheopis* was collected in 32 localities in Madagascar (Figure 1). *X. cheopis* fleas were collected on rats trapped with Sherman traps (H.B. Sherman Trap. Inc, Tallahassee, Florida) or with wire-mesh BTS traps (Besancon Technical Service, Besancon, France). For public health and sanitary reasons, all rats trapped were killed. We used trapping protocols developed and routinely used by IPM for research on *R. rattus* [24,25].

To harvest fleas, the host was placed in a large and pale-colored basin, deep enough to prevent the fleas from escaping. Then, the fur was brushed systematically with an adapted hard-bristled brush until the fleas jumped out. The fleas were caught with a manual pump aspirator and transported alive in a box containing rice bran and food for the larvae. Fleas were identified to the species level under binocular magnifier with the Duchemin systematic key [26].

Rearing of fleas

Fleas collected in the field were reared under laboratory conditions (22–27°C; 75–80% RH). Eggs, larvae, nymphs and adults were maintained together in clear two-liter glass jars containing 200 ml sterilized rice bran. Sixty five grams of sterilized dried ox blood, 5.5 g of dried yeast and 200 g of laboratory animal diet powder were added to the litter. The litter was changed once a year. The fleas were removed to a clean jar containing new litter, counted and divided into several jars when there were more than 200 adults; this allowed the flea population to increase more quickly. To feed the adults, each jar was provided with one live young mouse for three days every week.

Deltamethrin bioassays

Adult fleas were bioassayed according to the WHO protocol [27]. Bioassays were carried out in 18 cm glass test tubes, covered with fine mesh cloth. Groups of ten fleas were exposed for 8 hours to paper impregnated with 0.05% Deltamethrin (1.5×6 cm; Vector Control Research Unit, Penang, Malaysia) [28]. Four replicates were carried out for each flea population. The number of dead or paralyzed fleas was counted after 10, 20, 30, 40, 60, 120, 180, 300 and 480 minutes. After the 8-hour exposure, the impregnated paper was removed (and replaced with non-impregnated filter paper). Flea mortality was scored after 24 hours. A tube with a paper impregnated with silicone oil (Vector Control Research Unit, Penang, Malaysia) was used as negative control (two replicates).

Knock Down 50 (KD50; the time by which 50% of fleas were knocked down) and KD90 (the time by which 90% of fleas were knocked down) were estimated from the counts during the 8 h exposure, and the susceptibility status was assessed after the 24 h observation [29,30]: mortality rates of 98 to 100% were considered to indicate susceptibility; 80 to 98%, tolerance; and less than 80%, resistance [31]. The test was not validated, and the data not included, if the negative control mortality rate was over 10%. The mortality rate was corrected with the Abbott formula when control values were between 5% and 10%.

Statistical analysis

The analysis was performed with R software (R version 2.15.3 2013), and Tinn-R environment (Tinn-R Editor Version 2.4.1.5 2013).

Mean KD50 and KD90 and the standard errors for each flea population were estimated with a binomial generalized linear model (glm) analysis. This glm including a probit function is a fitted model giving a prediction and a standard error at each response probability.

Analysis of Variance (ANOVA) and Tukey's b test were used to compare survival rates. Correlations between the mortality rate, KD50 and KD90 were calculated with Pearson tests. Correlations between the sampling date and the KD50, KD90 and the mortality after 24 h were also estimated with Pearson tests. ANOVA was used to test the influence of the latitude, the

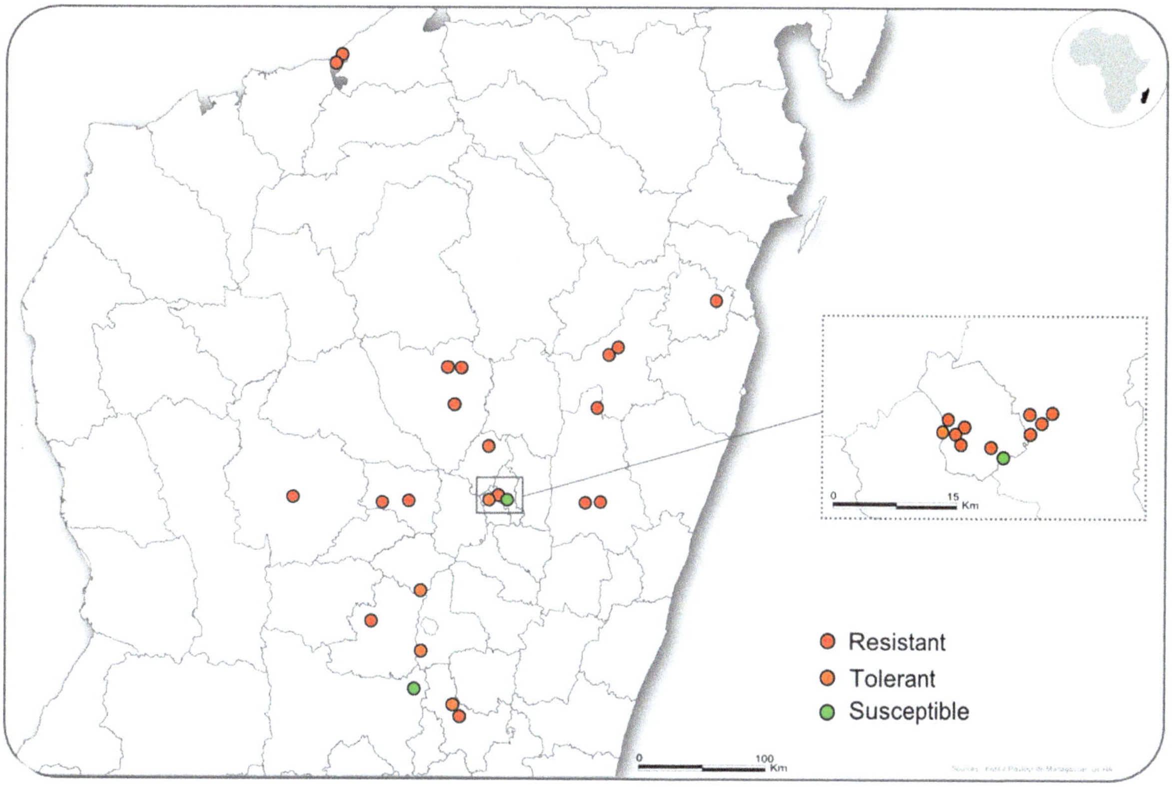

Figure 1. Sampling sites in Madagascar. The circles represent the villages where the flea populations were caught.

longitude, and the history of Indoor Residual Spraying (IRS) treatment on the three measures of sensitivity.

Results

Knock Down 50

KD50 values for the 32 populations were from 9.44 min to 592.35 min, and thus displayed a 63-fold difference between the flea population with the highest and that with the lowest KD50 (Table 1). Five of the 32 populations had a KD50 below 30 minutes, and six populations had KD50 values over 3 hours.

Knock Down 90

KD90 values for the 32 populations were between 10.44 min and 854.29 min (82 fold difference). The populations with the highest and the lowest KD90 values were the same as those with the highest and lowest KD50 values (Table 1) Four of the five populations with the lowest KD50 values had a KD90 value below 1 hour. Four populations had KD90 values greater than the 8 h-exposure time with a highest value with 854.29 min.

Mortality rate

The mortality rates of the 32 populations were between 2.5% and 100%, with an average of 50.55% (Figure 2). Two of the 32 populations were susceptible to Deltamethrin: the populations from Mandroseza and Soanierana showed 100% mortality. Four populations (Ambodirano Ampefiloha, Iarinoro Tsarasaotra, Amparhimboahangy Betafo and Sahatany) were tolerant to Deltamethrin (85–92.5% mortality), and the other 26 populations (81.25% of the populations) were resistant (2.5–77.5% mortality). The mortality rate was significantly correlated with the KD50 ($r = -0.588$, $p<0.001$) and KD 90 ($r = -0.549$, $p = 0.0011$) values.

Correlations

No correlation was detected between the times that fleas spent in the insectarium and their KD50, KD90 or mortality (Table 2). The effect of IRS treatment on the mortality rate, the KD50 and the KD90 was not significant (Table 3). They were significant correlations between latitude and mortality, and between longitude and the KD50.

Discussion

The KD50 and KD90 on one hand, and mortality on the other, are not indicators of the same mechanisms of resistance, and therefore not the same evolutionary selections; nevertheless, the correlations between these measures are strong enough to conclude about the Deltamethrin resistance of populations in the field. The results we report are unambiguous and raise serious concerns about flea control: 81.25% of 32 tested populations were resistant to 0.05% Deltamethrin. In 2000, only one *X. cheopis* population was found to be resistant to Deltamethrin (0.025%) [20]. Of the four populations tested in 2000, besides the one resistant population, three have become tolerant to a higher Deltamethrin concentration. Thus, the resistance has increased substantially. Analyses of pests with multiple resistances indicated that treatment with one insecticide can favor resistance to a second, different, insecticide; the example of DDT treatment influencing the resistance to other insecticides has been extensively studied and documented [32,33]. As the resistance of *X. cheopis* to DDT has been demonstrated in the central highland of Madagascar since the early 80's, our hypothesis is that *X. cheopis* may have acquired Deltamethrin resistance after exposure to DDT treatment.

Table 1. Mortality values of Xenopsylla cheopis populations to Deltamethrin in Madagascar.

Sampling places	KD50[a] (minute)		KD90[b] (minute)		% mortality (24 hours)		GPS			
Station	**Mean**	**SE**	**Mean**	**SE**	**Mean**	**SE**	**X**	**Y**	**IRS[c] 1993–2012**	**Months in insectarium**
Sahatany	*9,44*	*74,76*	*10,44*	*58,06*	*85,0*	*3,3*	20°02.145′	46°57.870′	9	1
Mandroseza	*11,43*	*0,89*	*18,51*	*1,48*	*100,0*	*3,9*	18°56′7,05″	47°33′1,25″	0	18
Ambanivolafotsy	*24,12*	*1,82*	*48,48*	*3,57*	47,5	1,8	20°32′24,94″	47°14′35,44″	2	2
Andavamamba	*24,66*	*4,97*	107,57	10,44	65,0	2,5	18°55′15,83″	47°30′41,35″	0	20
Ambodirano Ampefiloha	*28,59*	*2,05*	*57,14*	*4,22*	*92,5*	*3,6*	18°54′38,23″	47°29′58,69″	0	13
Soanierana	31,71	6,13	106,57	13,71	*100,0*	*5,4*	20°20′02,4″	46°54′47,7″	9	1
Ambohimiandra	33,49	3,46	91,04	7,76	65,0	2,5	18°55′44,62″	47°32′34,29″	0	17
Miadamanjaka	37,94	2,80	80,09	6,04	*67,5*	*2,6*	18°56′52,79″	46°51′57,75″	12	5
Amparihimboahangy Betafo	39,37	9,14	213,86	19,53	*85,0*	*3,3*	19°35.892′	46°38.059′	10	75
Andohatapenaka	49,52	5,29	147,03	12,36	47,5	1,8	18°54′19,63″	47°30′0,54″	0	76
Route de la gare	51,04	8,91	228,42	19,97	60,0	2,3	18°56′38,22″	48°13′49,76″	3	18
Ambohipananina	53,28	6,33	174,59	14,70	77,5	3,0	17°49′48	48°26′45,96″	4	19
Andoharano Ankazobe	56,96	3,85	115,87	8,24	37,5	1,4	18°15′0,00″	47°10′60,00″	5	3
Ankadindambo	62,80	6,61	190,83	15,41	57,5	2,2	18°54′11,08″	47°35′27,41″	10	2
Antsahatsaka	71,58	14,65	388,92	37,95	45,0	1,7	18°57′10,62″	48°16′40,84″	3	20
Ikianja	76,35	6,37	194,33	14,56	37,5	1,4	18°54′8,48″	47°34′51,38″	10	2
Tsaramasoandro	80,79	3,87	127,01	7,06	**12,5**	**0,5**	17°57′39,95″	47°13′1,20″	5	9
Tsinjorano	86,37	4,67	149,08	9,11	20,0	0,8	17°57′34,8″	47°14′52,7″	5	9
Ambatondrazaka	90,46	12,39	363,81	31,92	50,0	1,9	17°50′5,23″	48°25′29,31″	4	18
Andranofotsy	91,55	8,01	250,23	18,47	65,0	2,5	18°33′0,00″	47°28′0,00″	4	19
Tanambao	120,44	11,83	379,56	30,60	57,5	2,2	18°53′58,04″	47°35′38,65″	10	2
Iarinoro Tsarasaotra	122,02	7,60	251,45	15,76	*87,5*	*3,4*	20°26′37,76″	47°12′46,83″	6	11
Soavinarivo Betafo	125,79	8,55	284,53	18,62	35,0	1,3	19°49.452′	46°36.144′	11	77
Ambaniala	128,82	8,48	283,46	18,22	**12,5**	**0,5**	18°54′49,73″	47°34′45,24″	10	2
Andaingo	174,83	9,60	337,89	19,36	30,0	1,2	18°15′37,25″	48°15′56,15″	3	108
Andranomanalina	200,98	6,61	375,20	13,25	47,0	1,1	18°54′46,60″	47°30′34,95″	0	85
Amparaky	**230,02**	**13,62**	471,97	30,98	**2,5**	**0,1**	18°57′13,36″	46°40′20,29″	13	13
Vavatenina	**231,42**	**11,42**	415,60	22,80	17,5	0,7	17°28′3,77″	49°12′0,27″	0	78
Miandrarivo	**296,37**	**13,95**	**500,62**	**27,97**	32,5	1,3	18°56, 156′	45′59,329′	9	4
Abattoir Mahajanga	**359,34**	**27,76**	**739,93**	**69,86**	42,5	1,6	15°43′23,06″	46°19′22,75″	0	5
Tsena be Isotry	**375,39**	**22,83**	**853,58**	**63,70**	28,0	1,1	18°54′32,94″	47°31′0,65″	0	9
Tsararano Ambony	**592,35**	**51,78**	**854,29**	**93,51**	**7,5**	**0,3**	15°42′6,32″	46°19′49,48″	0	2

The most sensitive populations to Deltamethrin with the lowest KD50 and KD90 values are highlighted with italicized font. The most resistant populations to Deltamethrin with the highest KD KD50 and KD90 values are highlighted in bold. The same font code was used for the mortality percentage after 24 hours with the most sensitive population italicized, while the most resistant populations are in bold.

[a]KD50 represents the time, in minute, by which 50% of fleas were knocked down.

[b]KD90 represents the time, in minute, by which 90% of fleas were knocked down.

[c]IRS mean Indoor Residual Spraying, and this column represents the number of insecticide interventions the population have underwent (Max: 1 per year)

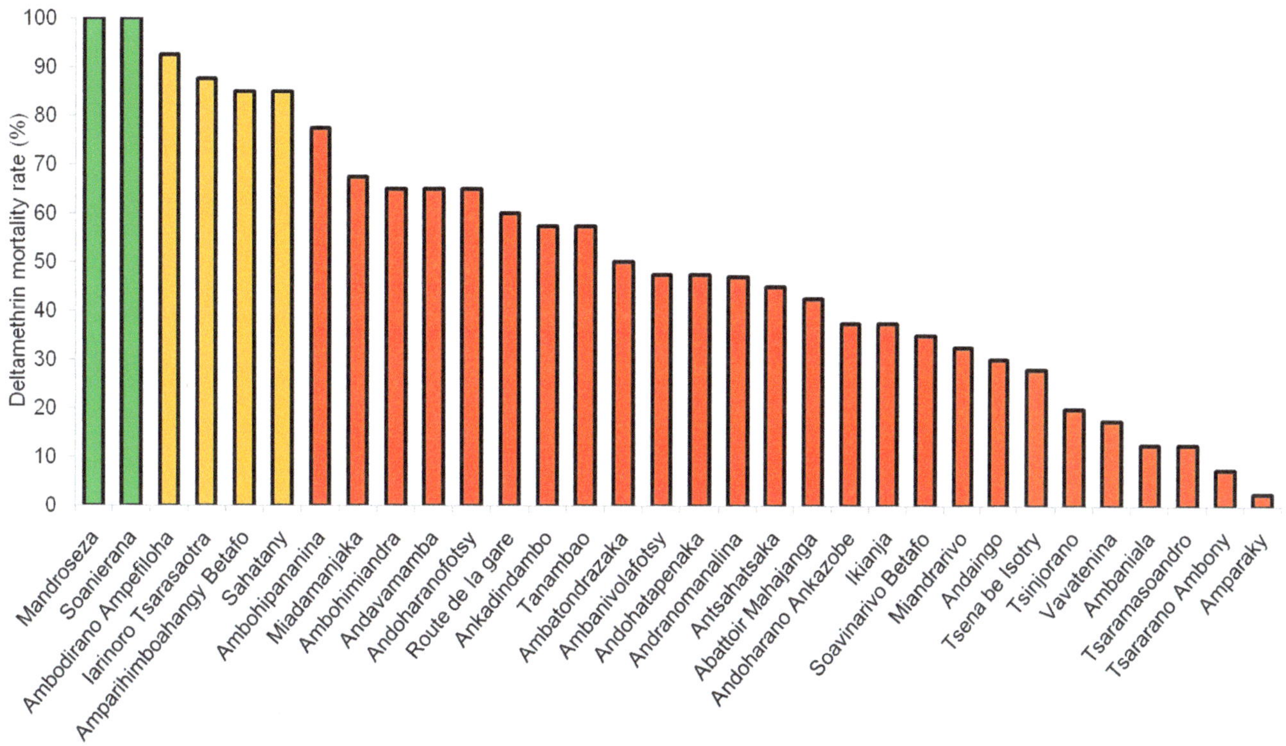

Figure 2. Deltamethrin mortality of flea populations sampling in Madagascar. Green bars represent sensitive populations, orange bars tolerant populations and red bars resistant populations (WHO definition).

The environmental effects and/or of the anthropic pressures undoubtedly have a role in the emergence of insecticide resistance. However, latitude was the only factor found to correlate with mortality. Note that the latitude reflects a complex combination of climatic conditions (precipitations, temperature), environmental conditions such as altitude and various ecological characteristics (plant species, human social and cultural behaviors, and agricultural practices). It is therefore difficult to determine whether the relationship between latitude and insecticide resistance is due to anthropic factors or natural selection. Anthropogenic selection pressure may result from the effects of IRS, pollution, agriculture, urbanization, insecticide spraying (public health, agriculture), and the individual or collective use of pesticides. Indeed, Gratz [34] reported that mosquito vector control program mainly with DDT IRS affected the susceptibility of *X. cheopis* to this insecticide. However, in this study we find that there is no relationship between IRS treatment history (Deltamethrin, Bendiocarb), at least over the past 17 years, and flea susceptibility to Deltamethrin. Hence, at this controversy, it is difficult to assess the main factor which induces *X. cheopis* resistance to Deltamethrin, and further studies must be conducted to understand the resistance mechanisms and the associated factors. Therefore, due to the ability insects to acquire cross resistance to insecticides of the same family, other pyrethroids should not be used, and the resistance of flea populations to α-Cypermethrin, Cyfluthrin, Etofenprox, λ-Cyhalothrin, or Permethrine insecticides should be assessed before any large-scale use. It would also be valuable to understand the

Table 2. Correlations results between the mortality values parameters and between the times spent by fleas in the insectariums (Pearson's correlation Test).

Tested parameters	r	95% confidence interval		t	df	p-value
24 h Survival×KD50	−0,588	−0,777	−0,301	−3,983	30	0,0004
24 h Survival×KD90	−0,549	−0,753	−0,247	−3,595	30	0,0011
KD50×KD90	0,933	0,867	0,967	14,197	30	<0.0001
KD50×Insectarium	0,010	−0,340	0,358	0,056	30	0,956
KD90×Insectarium	0,030	−0,323	0,374	0,162	30	0,8727
24 h Survival×Insectarium	−0,090	−0,425	0,267	−0,495	30	0,624

KD50 represents the time, in minute, by which 50% of fleas were knocked down, while KD90 represents the time, in minute, by which 90% of fleas were knocked down. 24 h Survival is the percentage of mortality to Deltamethrin after 24 h. And Insectarium represents the number of months during which fleas were present in the insectarium before the Deltamethrin bioassays occurred.

Table 3. The effect of parameters (X latitude and Y longitude: GPS localization of the sampling place; IRS: Indoor Residual Spraying which represent the number of insecticide interventions the population should underwent (Max: 1 per year) on KD50, KD90 and 24 hours survival rate of *Xenopsylla cheopis*.

	Knock-Down 50				Knock Down 90				24 h Survival rate			
	Df	F value	P		Df	F value	P		Df	F value	P	
X latitude	1	22,55	<0.0001	***	1	14,22	0,0009	***	1	9,27	0,006	**
Y longitude	1	6,91	0,015	*	1	1,78	0,194		1	0,22	0,644	
IRS	1	0,39	0,538		1	0,01	0,906		1	2,09	0,161	
latitude×longitude	1	0,03	0,870		1	0,28	0,604		1	0,46	0,505	
latitude×IRS	1	2,46	0,130		1	1,78	0,194		1	2,46	0,130	
longitude×IRS	1	2,22	0,149		1	2,60	0,120		1	2,13	0,158	
latitude×longitude×IRS	1	0,33	0,574		1	0,04	0,851		1	0,58	0,453	
Residuals	24				24				24			

Signification codes: 0 '***'0.001 '**'0.01 '*'0.05.

mechanisms of insecticide resistance in flea species, and in particular multiple resistance mechanisms.

Conclusions

We report that only two of the 32 flea populations sampled from different locations were susceptible to Deltamethrin. Consequently, in the current context of the re-emergence of plague and the increasing numbers of human plague cases in Madagascar, Deltamethrin is ineffective against fleas. Its use in Madagascar should be stopped and the control program for plague diseases needs to change to another insecticide. Twelve insecticides will be tested in our laboratory to identify which is the most appropriate for national flea control program.

Acknowledgments

We are very grateful to the entomological technicians which done a great job with all the sampling missions and the bioassays, particularly Tojo Ramihangihajason, Etienne Tata and Mandimby Andriatsiferana Rajaonarimanana.

Author Contributions

Conceived and designed the experiments: SB NE. Performed the experiments: AM. Analyzed the data: SB. Contributed reagents/materials/analysis tools: SB AM NE. Wrote the paper: SB AM NE.

References

1. Cleaveland S, Laurenson MK, Taylor LH (2001) Diseases of humans and their domestic mammals: pathogen characteristics, host range and the risk of emergence. Philosophical Transactions of the Royal Society of London Series B: Biological Sciences 356(1411): 991–999.
2. Butler T (2013) Plague gives surprises in the first decade of the 21st century in the United States and worldwide. The American Journal of Tropical Medicine and Hygiene 89(4): 788–793.
3. WHO (2010) Peste humaine: Examen de la morbidité et de la mortalité régionales, 2004–2009. Relevé pidémiologique hebdomadaire N°6(85): 40–45.
4. Andrianaivoarimanana V, Kreppel K, Elissa N, Duplantier J-M, Carniel E, et al. (2013) Understanding the Persistence of Plague Foci in Madagascar. PLOS Neglected Tropical Diseases 7(11): e2382.
5. WHO (2013) World health statistics 2013: a wealth of information on global public health.
6. Guiyoule A, Rasoamanana B, Buchrieser C, Michel P, Chanteau S, et al. (1997) Recent emergence of new variants of *Yersinia pestis* in Madagascar. Journal of Clinical Microbiology 35(11): 2826–2833.
7. Perry RD, Fetherston JD (1997) *Yersinia pestis*-etiologic agent of plague. Clinical microbiology reviews 10(1): 35–66.
8. Carniel E (2002) La peste. Comptes Rendus Biologies 325(8): 851–853.
9. Gage KL, Kosoy MY (2005) Natural History of plague: Perspectives from More than a Century of Research. Annual Review of Entomology 50: 505–528.
10. Eisen RJ, Gage KL (2012) Transmission of Flea-Borne Zoonotic Agents. Annual Review of Entomology 57: 61–82.
11. Brygoo ER (1966) Epidémiologie de la peste à Madagascar. Archives de l'Institut Pasteur de Madagascar 35.
12. Migliani R, Ratsitorahina M, Rahalison L, Rakotoarivony I, Duchemin JB, et al. (2001) Résurgence de la peste dans le district d'Ikongo à Madagascar en 1998. 1. Aspects épidémiologiques dans la population humaine. Bulletin de la Société de Pathologie Exotique 94(2): 115–118.
13. Ratsitorahina M, Chanteau S, Rosso ML, Randriambelosoa J, Ratsifasoamanana L, et al. (2002) Actualités épidémiologiques de la peste à Madagascar. Archives de l'Institut Pasteur de Madagascar 68 (1&2): 51–54.
14. Coulanges P, Clerc Y, Randrianantoanina E (1982) *X. cheopis* and *S. fonquerniei*, plague-carrying Malagasian fleas-demonstration of their resistance to DDT, dieldrin and malathion. Archives de l'Institut Pasteur de Madagascar 49(1): 171.
15. Fontenille D, Coulanges P (1987) Notes sur la sensibilité des puces *Xenopsylla cheopis* de la région d'Antananarivo à la Déltamethrine et au Propoxur. Archives de l'Institut Pasteur de Madagascar 53: 249–259.
16. Coulanges P, Randrianantoanina E (1984) Résistance exceptionnelle aux insecticides de puces pestigènes malgaches. Bulletin de la Société de pathologie exotique 77(5): 705–711.
17. Coulanges P, Randrianantoanina E (1984) Résistance des puces pestigènes malgaches aux insecticides organochlores, organophosphores et aux carbamates. Archives de l'Institut Pasteur de Madagascar 51(1): 253–260.

18. Ratovonjato J (1998) Sensibilité de *Xenopsylla cheopis* aux insecticides en milieu urbain à Madagascar. Archives de l'Institut Pasteur de Madagascar 64(1–2): 25–28.
19. Ratovonjato J, Duchemin JB, Duplantier J-M, Laventure S, Rabarison P, et al. (1998) Evaluation de la sensibilité des puces pestigènes malgaches aux insecticides en milieu urbain. Résultats et analyse préliminaires. Archives de l'Institut Pasteur de Madagascar 64(1–2): 29–33.
20. Ratovonjato J, Duchemin JB, Duplantier J-M, Chanteau S (2000) *Xenopsylla cheopis* (Siphonaptera: Xenopsyllinae), puces des foyers ruraux de peste des Hautes Terres malgaches: niveau de sensibilité au DDT, aux pyréthrinoïdes et aux carbamates après 50 années de lutte chimique. Arch Inst Pasteur Madagascar 66(1&2): 9–12.
21. Soderlund DM, Clark JM, Sheets LP, Mullin LS, Piccirillo VJ, et al. (2002) Mechanisms of pyrethroid neurotoxicity: implications for cumulative risk assessment. Toxicology 171(1): 3–59.
22. Kasai S, Shono T, Komagata O, Tsuda Y, Kobayashi M, et al. (2007) Insecticide resistance in potential vector mosquitoes for West Nile virus in Japan. Journal of Medical Entomology 44(5): 822–829.
23. Sikes RS, Gannon WL (2011) Guidelines of the American Society of Mammalogists for the use of wild mammals in research. Journal of Mammalogy 92(1): 235–253.
24. Rahelinirina S, Léon A, Harstskeerl RA, Sertour N, Ahmed A, et al. (2010) First isolation and direct evidence for the existence of large small-mammal reservoirs of *Leptospira sp.* in Madagascar. PloS one 5(11): e14111.
25. Tollenaere C, Brouat C, Duplantier JM, Rahalison L, Rahelinirina S, et al. (2010) Phylogeography of the introduced species Rattus rattus in the western Indian Ocean, with special emphasis on the colonization history of Madagascar. Journal of Biogeography 37(3): 398–410.
26. Duchemin J-B (2003) Biogéographie des puces de Madagascar. Thesis: Université de Paris XII - Val de Marne. Available: http://doxa.u-pec.fr:80/theses/th0212881.pdf
27. WHO (1970) Résistance aux insecticides et lutte antivectorielle: dix-septième rapport du Comité OMS d'experts des insecticides Genève: WHO. 306p p.
28. WHO (1981) Instructions for determining the susceptibility or resistance of fleas to insecticides. Genève: WHO. WHO/VBC/81.815 WHO/VBC/81.815. 1–6 p.
29. WHO (1976) Report of WHO Expert Committee: Resistance of Vectors and Reservoirs of disease to pesticide: twenty-second report of the WHO Expert Committee on Insecticides [meeting held in Geneva from 16 to 23 September 1975] Geneva: WHO.
30. WHO (1998) Test procedures for insecticide resistance monitoring in malaria vectors, bio-efficacy and persistence of insecticides on treated surfaces: report of the WHO informal consultation, Geneva, 28–30 September 1998. Genève: WHO. 43 p.
31. WHO (2011) The technical basis for coordinated action against insecticide resistance. 36 p.
32. Chandre F, Darriet Fdr, Manguin S, Brengues Cc, Carnevale P, et al. (1999) Pyrethroid cross resistance spectrum among populations of *Anopheles gambiae ss* from Cote d'Ivoire. Journal of the American Mosquito Control Association 15(1): 53–59.
33. Guedes RNC, Lima JOG, Santos JP, Cruz CD (1994) Inheritance of deltamethrin resistance in a Brazilian strain of maize weevil (*Sitophilus zeamais* Mots.). International Journal of Pest Management 40(1): 103–106.
34. Gratz NG, Traub R, Starcke H (1980) Problems and developments in the control of flea vectors of disease. AA Balkema. pp. 217–240.

20

Use-Exposure Relationships of Pesticides for Aquatic Risk Assessment

Yuzhou Luo*, Frank Spurlock, Xin Deng*, Sheryl Gill, Kean Goh

Department of Pesticide Regulation, California Environmental Protection Agency, Sacramento, California, United States of America

Abstract

Field-scale environmental models have been widely used in aquatic exposure assessments of pesticides. Those models usually require a large set of input parameters and separate simulations for each pesticide in evaluation. In this study, a simple use-exposure relationship is developed based on regression analysis of stochastic simulation results generated from the Pesticide Root-Zone Model (PRZM). The developed mathematical relationship estimates edge-of-field peak concentrations of pesticides from aerobic soil metabolism half-life (AERO), organic carbon-normalized soil sorption coefficient (KOC), and application rate (RATE). In a case study of California crop scenarios, the relationships explained 90–95% of the variances in the peak concentrations of dissolved pesticides as predicted by PRZM simulations for a 30-year period. KOC was identified as the governing parameter in determining the relative magnitudes of pesticide exposures in a given crop scenario. The results of model application also indicated that the effects of chemical fate processes such as partitioning and degradation on pesticide exposure were similar among crop scenarios, while the cross-scenario variations were mainly associated with the landscape characteristics, such as organic carbon contents and curve numbers. With a minimum set of input data, the use-exposure relationships proposed in this study could be used in screening procedures for potential water quality impacts from the off-site movement of pesticides.

Editor: Guy Smagghe, Ghent University, Belgium

Funding: The authors have no support or funding to report.

Competing Interests: The authors have declared that no competing interests exist.

* E-mail: yluo@cdpr.ca.gov (YL); xdeng@cdpr.ca.gov (XD)

Introduction

As part of the registration process, pesticides are evaluated for their potential to move off-site and impact non-target organisms. Surface runoff and tile flow are significant pathways for pesticides movement to surface waters. Monitoring-based surface water risk assessments of pesticides are usually conducted at the watershed scale using measured concentration data from river sites, especially at watershed outlets. For example, in-stream measurements of pesticides were assessed for U.S. watersheds with spatial scales across 14 orders of magnitude [1,2]. However, water flow from non-application areas and non-agricultural headwaters may significantly dilute pesticide concentrations in the river. For example, in California's Central Valley, one of the most productive agricultural areas in the world, pesticide concentrations are substantially higher in small creeks dominated by irrigation return flows, as compared to main streams where the majority of flow originates in Sierra Nevada mountains [3,4,5,6]. Because of the dilution effects, data in larger streams are associated with great spatial variability and thus not able to provide reliable and comparative information for pesticide management and mitigation. Therefore, assessments of aquatic risk now generally focus on smaller water bodies close to the edge-of-field.

Monitoring data are not always available and adequate for risk assessment, especially for pesticide products with new active ingredients. Environmental fate and transport models may be used to predict likely concentrations and associated risks of pesticides and to determine priorities for monitoring and regulatory assessments. Water quality modeling is a key component of pesticide management, as in the development of Best Management Practices (BMPs) and Total Maximum Daily Loads (TMDL). Compared to watershed-scale models, field-scale models better account for hydrologic processes within agricultural fields and have the capability to simulate agricultural management practices. Field-scale models, such as the Pesticide Root Zone Model (PRZM) and the Root Zone Water Quality Model (RZWQM) [7], provide dynamic simulation of pesticide fate and transport processes, from pesticide applications to edge-of-field discharge. However, these models usually require a large set of model input parameters for the full descriptions of landscape characteristics, climate conditions, and management practices [8]. Consequently parameterization and simulation using field-scale models could be complicated and time-consuming, particularly when batch simulations and post-data analysis are involved [9]. In addition, during the pesticide registration process, the required input data may be difficult to obtain, especially for new pesticides which have not been applied in field conditions. Therefore, there is a research need to develop simple mathematical relationships to determine the potential aquatic risks of pesticides based on a minimum set of input parameters. Such simple relationships may be used in a screening procedure to identify pesticides that require more refined studies. As an early modeling effort, U.S. Environmental Protection Agency (USEPA) developed the model for Generic Estimated Environmental Concentration (GENEEC) to mimic more sophisticated simulations of pesticide transport from crop field to a standard pond [10]. However, differences in climate, soil, topography or crop are not considered in estimating potential exposure, thus substantially limiting its applications to pesticide evaluation and registration.

This study develops "use-exposure relationships" in the form of linear regression equations that link pesticide application rate and physicochemical properties to a predicted exposure level (such as peak concentration) for specific environmental configurations. The relationship is developed from the results of more detailed field-scale model simulations, but use significantly fewer input parameters. Specific study objectives are [1] to identify governing parameters in pesticide fate and transport processes in canopy-soil system [2]; to establish empirical relationships between those parameters; and [3] to demonstrate the developed model with parameterizations in the crop scenarios of California. The approach enables a quick risk assessment based on limited input data, and yields accuracy comparable to the dynamic simulation of the selected field scale model. The parameterized use-exposure relationship provides useful information for decision making in pesticide registration and management.

Materials and Methods

PRZM Model and Aquatic Exposure Assessment

PRZM is a one-dimensional compartmental model developed by USEPA for predicting pesticide movement in unsaturated soils [11]. It is designed to evaluate the influence of climate, soil properties, and management practices on pesticide transport and transformation processes, e.g., surface runoff, plant uptake, leaching, erosion, and volatilization. PRZM generates daily pesticides fluxes in both dissolved and adsorbed forms at the edge of fields. The resulting fluxes are useful for further analyses, such as aquatic risk assessment [12], loading calculation [13], and water quality evaluation [14]. PRZM has undergone validation and testing to field-scale runoff and leaching studies [15,16]. An enhanced version of PRZM is being used for surface water and groundwater exposure assessments in the European Union [17].

PRZM was selected in this study based on its ability to simulate relevant governing processes of pesticide transport and because of its use by regulatory agencies in their pesticide exposure assessments [10,17]. USEPA has also developed crop scenarios to facilitate the application of PRZM in risk assessment [18]. Those scenarios specify the environmental configurations for typical crops in major agricultural regions of U.S., including weather conditions, landscape characteristics, crop growth parameters, and soil properties.

To assess pesticide risks to aquatic organisms, an exposure index (EI) was defined in this study as follows. First, the estimated environmental concentrations (EEC) of pesticide in surface runoff and soil erosion were predicted as daily time series by PRZM. For dissolved pesticides, the exposure index was then calculated as the peak concentration of 4-day moving averages in the 1-in-3 year return period. This definition is consistent with the current regulatory surface water criteria for two widespread pesticidal surface water contaminates chlorpyrifos and diazinon [19,20].

Simulation Design and Input Data

Stochastic PRZM simulations were conducted to develop crop-scenario-specific "use-exposure relationships", i.e., empirical equations for predicting edge-of-field pesticide runoff concentrations. The simulations were based on crop scenarios developed by USEPA for pesticide risk assessment. A single annual pesticide application, repeated every year during 1961–1990, was simulated for a specific scenario. Annually repeated applications were utilized to incorporate the effects of climatic and hydrologic variations on pesticide off-site movement. In addition, the 30-year simulation also accounted for the accumulation of persistent pesticides. For those pesticides, residues from previous applications may remain in the soil and add to the newly applied chemicals in the next year. A random application date was assigned to a PRZM simulation and pesticide was applied on the assigned date for each year in that simulation. The random date was generated within the pesticide's application season depending on the actual use pattern of the pesticide, such as dormant-season application, in-season application, and pre-emergent application. On each day of application, pesticide was applied at a fixed rate (a "base rate" or BASE, kg/ha as the active ingredient), which was an arbitrary small application rate for stochastic PRZM runs. A linear relationship was assumed between pesticide application rate and pesticide loadings from the field. A small base rate was used to avoid high predicted concentrations that exceed the water solubility (SOL) of the pesticide during the simulation. In this study, the base rate was set as 0.1 kg/ha. Preliminary simulations showed that, with base rate of 0.1 kg/ha, EECs were always lower than the corresponding SOL in all PRZM runs. Predicted concentrations should be compared to the SOL when applying the developed use-exposure relationships with actual label rates.

The chemical properties of aerobic soil metabolism half-life (AERO), organic carbon-normalized soil adsorption coefficient (KOC), and SOL are the governing factors for pesticide runoff potential in both dissolved and adsorbed phases [21,22]. KOC and SOL are direct input parameters in PRZM, and AERO is used in calculating the model inputs of decay rate constants [23]. In PRZM and most of other field-scale models, SOL is considered only as an upper limit on the dissolved concentration. In addition, significant association between the two properties of KOC and SOL has been reported in several previous studies. For example, linear correlation ($p<0.001$) was confirmed between log-transformed KOC and SOL [24]. Similar linear relationships were also used to estimate KOC from SOL [11,25]. Therefore, only the independent chemical properties of AERO and KOC were selected in this study for stochastic PRZM simulations. The two selected parameters were also used by other studies for estimating pesticide runoff potentials [21,26,27,28].

The probability distributions for AERO and KOC were derived from a database of physiochemical property and reaction half-life complied by Spurlock [24] for 172 pesticides. Spurlock [24] suggested that log normality is a reasonable assumption for AERO and KOC, and that the two properties are independent ($p = 0.551$). Maximum likelihood estimation (MLE) was applied to estimate the distribution parameters (Table 1). Latin Hypercube Sampling (LHS) was used to generate random input data of

Table 1. Parameters for the log-normal distribution of aerobic soil metabolism half-life (AERO) and organic carbon-normalized soil sorption coefficient (KOC).

Variable	μ	σ	E	SD
AERO	3.44	1.99	226.01	1613.14
KOC	6.51	2.52	1.61e4	3.82e5

Notes:
[1] the parameter estimation was based on the median fate properties derived from registration studies of 172 pesticides [24].
[2] μ and σ are the mean and standard deviation of the data's natural logarithm, respectively; E and SD are the mean and standard deviation of the data, respectively.

AERO and KOC within 95% of cumulative frequency of the corresponding log normal distribution as defined in Table 1. For each PRZM run, the exposure index from the single pesticide application at base rate, denoted as EI_BASE, was calculated from the predicted daily EECs. Finally, the general mathematical relationship between the EI_BASE and input chemical properties of AERO and KOC for the particular crop scenario was developed based on regression analysis. The built-in Monte Carlo simulation in PRZM does not report daily time series of edge-of-field pesticide concentrations. In addition, a deficiency has been reported for the built-in Monte Carlo module in PRZM [29]. Therefore, LHS algorithm was taken from our previous study [30]; and a batch program was developed for stochastic PRZM runs and post-data analysis.

Crop Scenarios in California

Crop scenarios developed by USEPA for California were used for simulations. Available scenarios include standard crop scenarios [18], crop scenarios developed for organophosphate pesticide cumulative risk assessment [31], and crop scenarios developed for effects determinations for California listed endangered and threatened species [32]. Combined, approximately 30 scenarios are available for California, some of which are associated with pesticide use patterns with high runoff potential. These include crops with flood or furrow irrigation, winter rain season application, and pre-emergent herbicide application. These scenarios were selected in this study to demonstrate the development of the pesticide use-exposure screening model. Results of a statewide survey of California irrigation methods [33] indicated that field crops and tomatoes are dominated by flood and furrow irrigation. Scenarios of almond and turf were selected to represent wet season application and pre-emergent herbicide application, respectively. Tables 2 and 3 summarize the selected scenarios for PRZM simulations for California. A non-California scenario (Florida tomatoes) was also included in this study to compare/contrast results with a wetter climate.

The crop scenarios specify the weather conditions, soil properties, and crop growth parameters used in the PRZM simulations. Other model inputs used in this study, including chemical property and pesticide application, are summarized in Table 4. To provide conservative estimation of pesticide residues at the edge of field, pesticides were assumed to be incorporated into the soil at application and all mass loss fluxes by interception, volatilization, and decay on the plant canopy were set as zero.

Results and Discussion

Use-Exposure Relationship for a Single Pesticide Application

The response of EI_BASE to random values of AERO and KOC was evaluated by stochastic PRZM simulation. For each crop scenario, the predicted dissolved or adsorbed EI_BASE were paired with corresponding inputs of AERO and KOC for further regression analysis. The logarithmic transformation was also applied to EI_BASE according to preliminary analyses on pesticide concentrations detected in surface water of California [34]. Finally, an $N \times 3$ matrix of (lnAERO, lnKOC, lnEI_BASE), with N denoting the number of stochastic PRZM runs, was generated from Monte Carlo simulation. Demonstrated in Figure 1a is an example plot of the matrix for dissolved pesticides for the standard crop scenario for cotton in California. Significant correlations were identified between dissolved EI_BASE vs. AERO and KOC, especially for pesticides with KOC higher than a certain value (e.g., about 5 for lnKOC as shown in Figure 1a). This correlation reflected the effects of degradation and partitioning of pesticides on the peak concentration predicted at the field edge. For pesticides with lower KOC, EI_BASE was generally invariant with KOC. With low KOC values, pesticides are mainly present in dissolved phase, thus the change in KOC do not have a great effect on the phase partitioning.

Stochastic PRZM simulations for other crop scenarios revealed similar relationships among the predicted EI_BASE and input parameters, i.e., the general linear relationship between EI_BASE and AERO and KOC, and the presence of an approximate lnKOC cutoff, below which EI_BASE is independent of KOC. Therefore, a conceptual model was developed for the use-relationships from a single pesticide application based on that general data structure (Figure 1b). First, a breakpoint for KOC (KOC*) was determined from the trend of dissolved EI_BASE with KOC for the given crop scenario. Multivariate linear regression with logarithmic transformations was conducted between EI_BASE vs. AERO and KOC on the data points with KOC>KOC*. A similar relationship was applied to pesticides with KOC≤KOC* by substituting KOC with KOC*, in order to provide conservative estimation of EI_BASE for those pesticides.

The general use-exposure relationship is:

$$\ln(EI\_BASE)=f(AERO,KOC)=b_1+b_2\ln(AERO)+b_3\ln(KOC) \quad (1)$$

Table 2. Overview of selected California crop scenarios developed by USEPA.

Crop scenario	Represented use pattern	Soil (hydrologic group)	Weather station
Alfalfa (OP)	Pasture, gravity irrigation	Sacramento clay (D)	Fresno
Almond (STD)	Dormant application	Manteca fine sandy loam (C)	Sacramento
Cotton (STD)	Field crop, gravity irrigation	Twisselman Clay (C)	Fresno
Sugar beet (OP)	Field crop, gravity irrigation	Ryde clay loam (C)	Fresno
Tomato (STD)	Tomato, gravity irrigation	Stockton clay (D)	Fresno
Turf (RLF)	Pre-emergent application	CapaySilty Clay Loam (D)	San Francisco
Wheat (RLF)	Grain, gravity irrigation	San Joaquin Loam (D)	Fresno
Tomato_FL (STD)	Tomato scenario in Florida	Riviera Sand (C)	West Palm Beach

Data source: USEPA Tier 2 crop scenarios for PRZM/EXAMS Shell [18,31,32]. "STD" = Standard crop scenarios, "OP" = scenarios developed for the cumulative risk assessment of organophosphate pesticides, and "RLF" = scenarios developed for the effects determinations for the California red-legged frog and other California listed species. "Tomato_FL" denotes the standard USEAP crop scenario for tomato in Florida, provided as an example of the crop scenarios in other states.

Table 3. Landscape characteristics and soil properties of selected California crop scenarios.

Crop scenario	CN	USLE K/LS/P	USLE C	OC1
Alfalfa	90/88/89	0.20/0.30/1.0	0.051–0.217	1.77%
Almond	84/79/84	0.28/0.30/1.0	0.034–0.221	0.81%
Cotton	89/86/89	0.21/0.37/1.0	0.054–0.412	0.29%
Sugar beet	89/86/89	0.28/0.30/1.0	0.015–0.769	3.48%
Tomato	91/87/91	0.24/0.13/1.0	0.035–0.255	0.95%
Turf	80/80/80	0.37/1.80/0.5	0.001	35.6%
Wheat	92/89/90	0.37/0.79/1.0	0.027–0.604	0.44%
Tomato_FL	91/87/91	0.03/0.20/1.0	0.177–0.938	1.16%

Parameters:
CN = Runoff curve numbers of antecedent moisture condition II for fallow, cropping, and residue, respectively;
USLE K = soil erodibility for the universal soil loss equation (USLE);
USLE LS = topographic factor for the USLE;
USLE P = practice factor for the USLE;
USLE C = cover management factor for the USLE;
OC1 = Organic carbon content in the surface soil.

And the relationship for dissolved pesticides is:

$$\ln(EI\_BASE) = f(AERO, KOC) = b_1 + b_2 \ln(AERO) + b_3 \ln(\max(KOC, KOC^*)) \quad (2)$$

where b_1, b_2, and b_3 are coefficients derived by regression. KOC* was determined by maximizing the coefficient of determination (R^2) in the regression analysis. Based on the linear assumption between pesticide application and exposure, the dissolved exposure index (EI, µg/L) from pesticide applications at the actual rate (RATE, kg/ha) was expressed as,

$$EI = \frac{RATE}{BASE} \cdot EI\_BASE \quad (3)$$

Derived Parameters for Dissolved Pesticides

For each selected scenario, 5 000 stochastic simulations of PRZM (N = 5 000) were conducted for the 30-year period of 1961–1990. Regression coefficients and other statistics for the use-exposure relationship for dissolved pesticides are summarized in Table 5.

Values of lnKOC* varied with different scenarios, ranging from 0.5 to 5.5. With higher organic carbon content in surface soil (OC1), such as for the California turf scenario (35.6%), lower KOC* values were observed relative to other scenarios with smaller OC1 ranging from 0.44% to 3.48% (Table 3). However, the product of KOC* and OC1, equivalent to the corresponding limiting distribution coefficient (KD*), was approximately constant among the scenarios, ranging from 0.5 to 0.7. Since the distribution coefficients indicate pesticide mobility in the soil, the KD* value determined from the PRZM simulations was considered as the critical KD value below which the transport process of dissolved pesticides with surface runoff was insensitive to their KOC values.

The empirical AERO-KOC based use-exposure relationships explained 90–95% of the variances on the predicated EI_BASE of dissolved pesticides. The predictive ability of the relationships was mainly attributable to lnKOC, which solely explained 85–90% of

Table 4. Chemical property and pesticide application in PRZM simulations.

Variable	Description	Values/notes
APPDAY	Application date	Random numbers (uniform) in the application season
APPEFF	Application efficiency	0.99 (ground application) [2]
CAM	Pesticide application method	4 (soil incorporation)
DAIR	Diffusivity in air (cm2/day)	4300 [1]
DEPI	Incorporation depth (cm)	4 [2]
DRFT	Drift fraction	0.01 [2]
DSRATE	Adsorbed phase decay rate (1/d)	= ln2/AERO, LHS sampling
DWRATE	Dissolved phase decay rate (1/d)	= DSRATE [2]
ENPY	Enthalpy of vaporization (kcal/mol)	20 [1]
FEXTRC	Washoff extraction (1/cm)	0.5 [1]
HENRYK	Henry's law constant (g/aq, dimensionless)	0 [3]
IPSCND	Disposition of foliar pesticide after harvest	1 (surface applied) [3]
PLDKRT	Decay rate on foliage (1/d)	0 [2]
PLVKRT	Volatilization rate on foliage (1/d)	0 [2]
KOC	Organic carbon-normalized soil adsorption coefficient (L/kg[OC])	LHS sampling
TAPP	Application rate (kg/ha)	0.1 (base application rate used in this study)
UPTKF	Pesticide uptake	0 [2]

Notes:
[1]Suggested value in the PRZM manual [11].
[2]USEPA-suggested model input parameter value [23].
[3]Assumptions made for conservative evaluation of pesticide exposure.

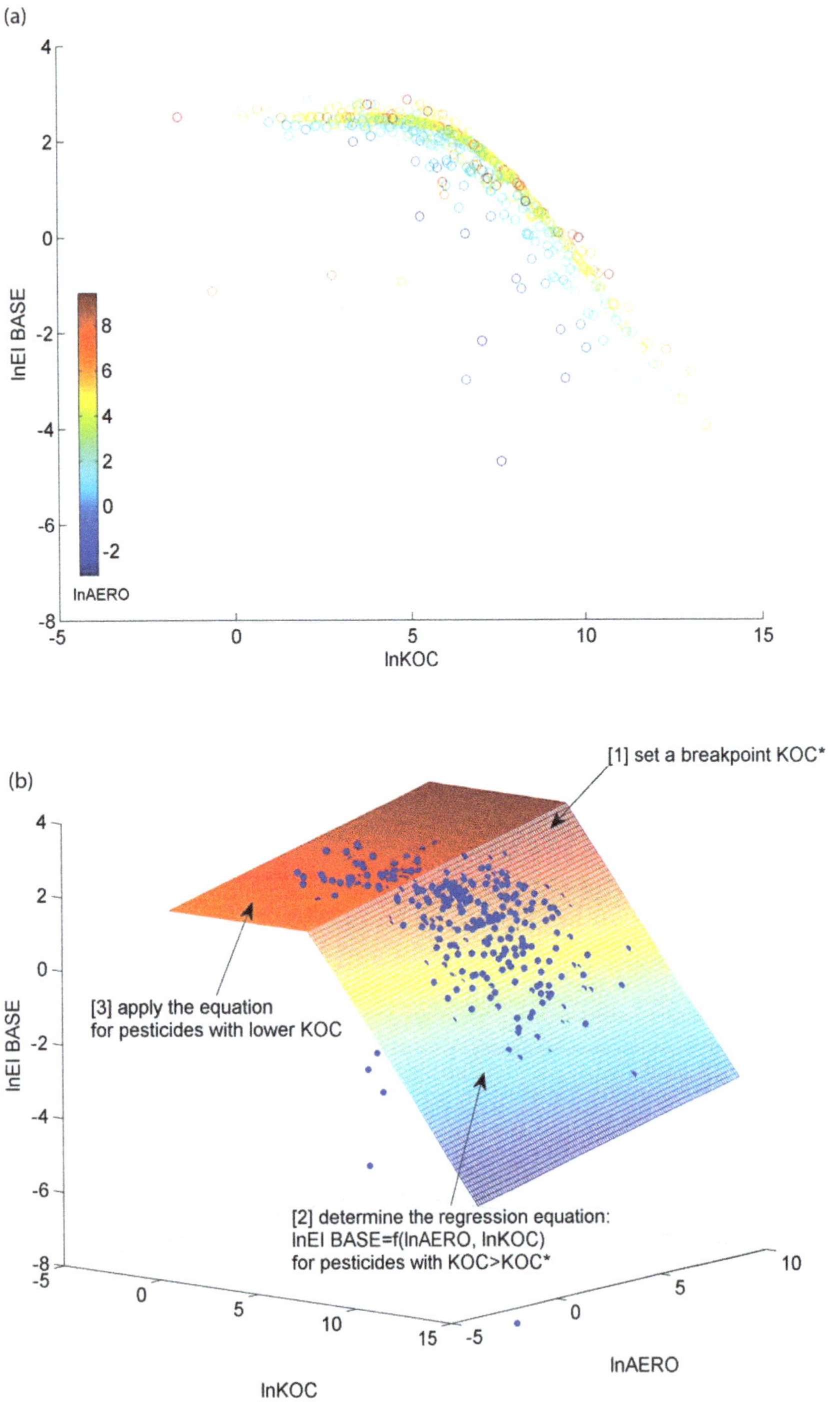

Figure 1. Use-exposure relationship for dissolved pesticides (EI_BASE in μg/L): (a) example results of Monte Carlo simulation and (b) conceptual model.

the total variances in lnEI_BASE, while AERO had only a limited contribution. Generally, the relative magnitudes of dissolved EI_BASE for pesticides with large KOC were mainly related to the regression coefficients of KOC (b3), while those for pesticides with lower KOC were determined by the intercepts (b1). This reflected the competition between phase partitioning and water runoff extraction on the pesticide yields from the applied field. Relatively higher regression coefficients for AERO (b2) were observed for scenarios with higher OC1 such as sugar beet (OC1 = 3.48%) and turf (OC1 = 35.6%) (Table 3). With elevated OC contents, pesticides are less mobile in the soil and could be accumulated for a longer period. Previous studies indicated that pesticide half-life in the soil is the key parameter in determining the total amount of pesticide residues discharged from fields [21,22,35]. However, the measure of exposure in risk characterization is estimated from peak concentrations, which are usually observed shortly after pesticide applications once surface runoff induced by precipitation or irrigation is available. Thus, soil

Table 5. Use-exposure relationships for dissolved pesticides in selected California crop scenarios.

Scenarios	Coefficients			R2	lnKOC*
	b1	b2	b3		
Alfalfa	5.2156	0.1907	−0.8288	0.9494	3.5
Almond	4.8131	0.1869	−0.7467	0.9335	4.5
Cotton	6.3173	0.1467	−0.7662	0.9102	5.5
Sugar beet	4.9105	0.2412	−0.8377	0.9193	3.0
Tomato	5.9979	0.1785	−0.7844	0.8970	4.0
Turf	3.3647	0.2821	−0.8248	0.9546	0.5
Wheat	6.0764	0.1853	−0.7954	0.9487	5.0
Tomato_FL	4.9362	0.2531	−0.8063	0.9422	4.0

Note: "Tomato_FL" denotes the standard USEAP crop scenario for tomato in Florida, which is provided as an example of the crop scenarios in other states.

metabolism might have only moderate effects on pesticide exposure at the edge of field as measured by peak concentrations.

The regression coefficients for AERO and KOC did not vary much over scenarios. For instance, the maximum regression coefficient for KOC was −0.7467 (almonds), while the minimum value was −0.8377 (sugar beet). The regression coefficients for AERO ranged from 0.1467 (cotton) to 0.2821 (turf) (Table 5). The use-exposure relationship derived for crops in other states, taking tomato in Florida as an example in this study, also showed similar regression coefficients as in California crops. This suggested that, for a specific pesticide, the difference of predicted EI_BASE over scenarios were mainly determined by the intercepts of b1 in Eq. (1). In another words, the effects of chemical fate processes such as partitioning and degradation on pesticide exposure were similar among scenarios, while the spatial variability was related to environmental parameters including climate condition, soil property, and landscape characteristics.

While the California scenarios are developed for areas with similar climate, they are associated with substantial variability in soil type and hydrologic group (Table 3). The intercept in the regression equation for the use-exposure relationship (b1 in Table 5) was significantly correlated to curve numbers for residue surface soil condition (with a p-value, $p = 0.008$), and moderately correlated to curve numbers for cropping surface condition ($p = 0.063$). In most of the crop scenarios, curve numbers for residue surface condition were implemented for winter months, or the rainfall season in California. In the use-exposure relationship, the intercept was associated with water runoff generation since the chemical fate processes such as partitioning and degradation were represented by the chemical properties. Therefore, the significant correlation between b1 and curve number for residue surface soil condition indicated that peak concentrations of pesticide at the field edges were most likely observed during the rainfall season in California. The dependence of pesticide runoff potential on curve number in the PRZM simulation has been reported in previous studies [22,36,37]. Findings in this study confirmed the effects of curve number on the predicted pesticide concentrations and loadings from the crop fields.

Derived Parameters for Sediment-Bound Pesticides

At present there are no surface water quality criteria at either federal or state level for sediment-bound pesticides. Water quality assessments for pesticides in sediment, such as those for Clean Water Act Section 303(d) listing [38], are based on 10-day *Hyalella azteca* sediment toxicity tests [39]. To mimic the sediment toxicity tests, 10-day averages were calculated as adsorbed exposure index from PRZM-predicted daily concentrations of pesticide associated with soil erosion. The same frequency as for dissolved pesticide, i.e., once every three years return period, was used in the development of use-exposure relationship for adsorbed pesticides. Median lethal concentration (LC50) values for sediment toxicity are usually reported on an OC-normalized basis. For example, *Hyalella azteca* 10-day LC50 values for pyrethroids are typically reported at 1% OC, as compiled by Domagalski et al. [40]. To match the toxicity data, PRZM-predicted concentrations of pesticide in eroded sediment were normalized by OC1 defined in each scenario (Table 3).

For pesticide associated with eroded soil, Figure 2 shows the results of stochastic PRZM runs based on the USEPA standard scenario for cotton in California. There was a general increasing trend of EI_BASE with increases of KOC and AERO, especially for pesticides with KOC lower than a certain value. For pesticides with extremely high values, such as those with lnKOC larger than about 11 as shown in Figure 2a, the predicted EI_BASE for adsorbed pesticides were associated with high uncertainty and might not be significantly correlated with their KOC values. For these pesticides, the majority of the residues have been partitioned into the particulate phase. Based on the similar equations for dissolved pesticides, the following use-exposure relationship was developed for adsorbed pesticides,

$$\ln(EI\_BASE) = f(AERO, KOC) = b_1 + b_2 \ln(AERO) + b_3 \ln(\min(KOC, KOC^*)) \quad (4)$$

where EI_BASE (ng/g) is the predicted exposure index associated with eroded soil from a single pesticide application at BASE rate of 0.1 kg/ha, b's are regression coefficients, and KOC* is a threshold value for KOC above which the EI_BASE was assumed to be independent the pesticide's KOC.

Table 6 shows the parameters of the use-exposure relationships for adsorbed pesticides under selected California scenarios. The R^2 values ranged from 60–85%, substantially lower than those for dissolved pesticides (Table 5). Similar to the equations for dissolved pesticides, the majority of the variance in the use-exposure relationships for adsorbed pesticides wasexplained by KOC. Although the KOC* varied greatly, the KD* values as the product of KOC* and OC1 for the modeled scenarios were generally invariant, ranging from 160–240. However, the regression coefficients varied greatly among scenarios. Observed uncertainty in predicted exposure of adsorbed pesticides might be related to soil erosion processes. PRZM simulates soil erosion based on themodified universal soil loss equation (MUSLE) with input parameters. These parameters are usually associated with seasonality and variability in soil properties and management practices (Table 3). With only input parameters of AERO and KOC, therefore, the proposed use-exposure relationship was inadequate for capturing the variability in soil erosion processes.

Use-Exposure Relationship for Multiple Applications

The use-exposure relationships in Eqs. (1)through(4) provide estimates for the exposure index from a single pesticide application. For multiple applications, the exposure could be conservatively estimated from the maximum season application rate. To refined the estimation, especially for pesticides with short field dissipation half-lives (FD), equivalent application rate at any given time, RATE_eq(t), is calculated as the total pesticide amount in the soil available for runoff and soil erosion processes. RATE_eq includes contributions from both applied pesticide on

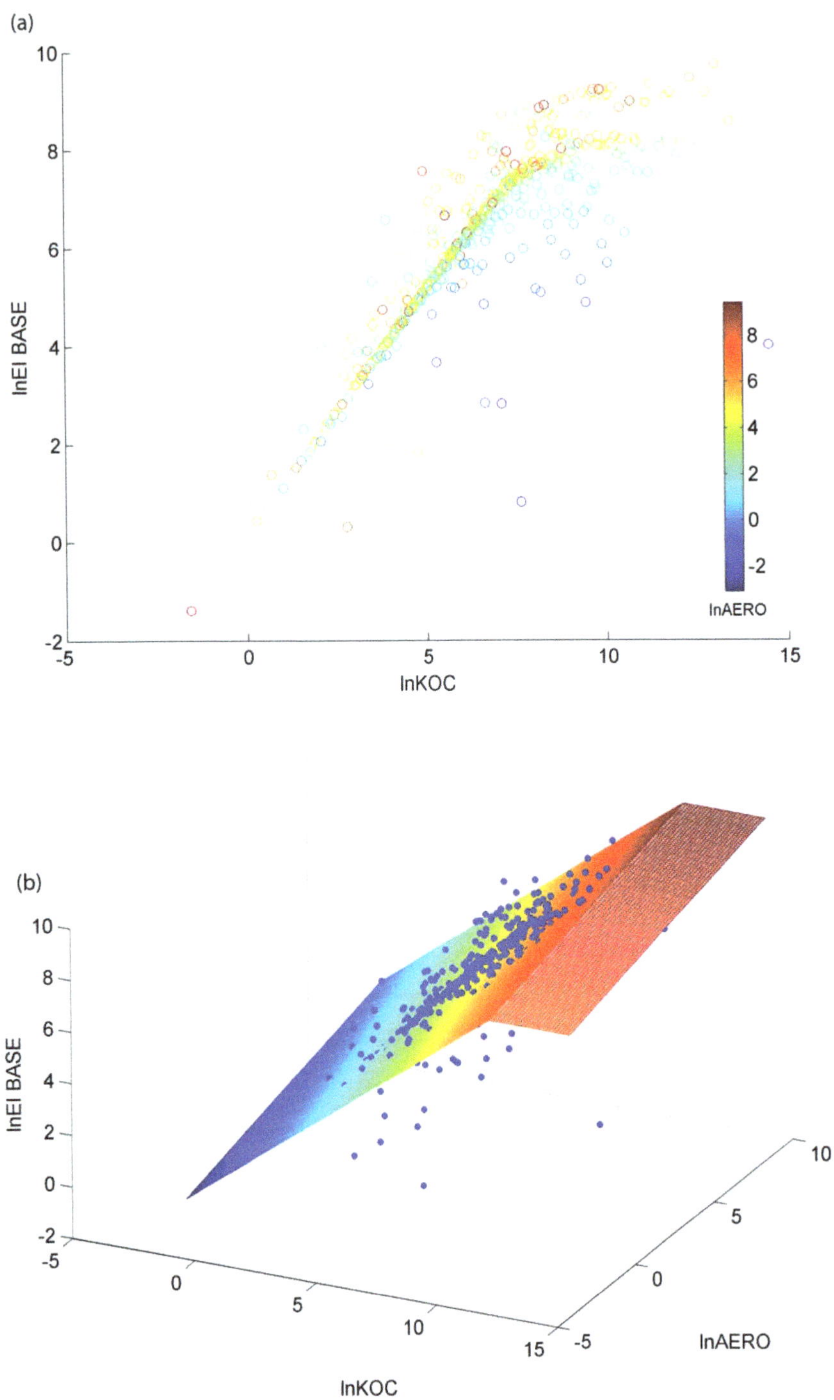

Figure 2. Use-exposure relationshio for adsorbed pesticides (EI_BASE in ng/g): (a) example results of Monte Carlo simulation and (b) conceptual model.

the given day and residues from previous applications, and could be calculated in the form of a convolution,

$$RATE\_eq(t) = \int_{t_0}^{t} [RATE(\tau) \cdot D(t-\tau)] d\tau$$
$$D(\Delta t) = \exp(-\frac{\ln 2}{FD}\Delta t) \quad (5)$$

where t0 is the first day of the application season, RATE(t) is the application rate at day t, and D(Δt) is the total fractional decay during Δt. The calculation of RATE_eq accounted for only the pesticide loss by degradation. Losses from surface runoff, soil erosion, and leaching were ignored to provide a conservative estimation of the amount of pesticide remaining in the soil.

Figure 3 presents a schematic of the RATE_eq calculation from multiple applications of carbaryl for tomatoes (maximum single application rate = 2.24 kg/ha, application interval = 7 day, and

Table 6. Use-exposure relationships for adsorbed pesticides in selected California crop scenarios.

Scenarios	Coefficients			R^2	ln(KOC*)
	b1	b2	b3		
Alfalfa	1.7756	0.3140	0.4936	0.6896	9.5
Almond	0.1179	0.2116	0.6937	0.7955	10.0
Cotton	0.9213	0.1890	0.7221	0.8466	11.0
Sugar beet	2.7386	0.3254	0.5118	0.6409	8.5
Tomato	3.2070	0.1912	0.6062	0.7770	10.0
Turf	2.7715	0.2832	0.4486	0.6106	6.5
Wheat	1.0782	0.3233	0.5848	0.7210	10.5
Tomato_FL	1.7065	0.4105	0.4809	0.7607	10.0

maximum number of applications = 4) [32]. The exposure index from multiple pesticide applications could be estimated by substituting RATE in Eq. (3) with the maximum value of RATE_eq during a year or a cropping season. For single pesticide application rate and application interval (INTERVAL, day) as fixed values, as usually documented in pesticide labels, the maximum RATE_eq could be directly calculated as,

$$\max[RATE\_eq(t)] = RATE \cdot \sum_{i=1}^{M-1} \exp[-\frac{\ln 2}{FD}(M-i) \cdot INTERVAL] \quad (6)$$

where M is the maximum number of applications.

Summary and Conclusions

Use-exposure relationships were developed as an alternative approach to field-scale modeling for pesticide runoff and associated aquatic risks. The relationships require a minimum set of input parameters to estimate exposure, which is defined herein as peak pesticide concentrations at the edge of field. The selected input parameters, half-life in the soil, adsorption coefficient, and recommended application rates, are generally available during pesticide registration. Thus the proposed approach is appropriate for quickly screening pesticide products for their potential adverse effects on the environment and human health. While the PRZM model was chosen to parameterize the weighting factors of the selected parameters for this study, the approach could be applied with other field-scale models.

The development of use-exposure relationships was demonstrated using crop scenarios developed by the USEPA for California. The relationships explained 90–95% of variations in the exposure index of dissolved pesticides as predicted by PRZM modeling. Regression coefficients for AERO and KOC for the simulated scenarios varied only in small ranges, suggesting that the effects of chemical property-related fate processes such as partitioning and degradation on the predicted exposure index were similar among scenarios. KOC was the governing factor in predicting pesticide exposures for all scenarios. Since aquatic risk analysis is mainly focused on the peak concentrations of pesticides, and these concentrations are usually observed shortly after pesticide applications, the half-life in the soil had limited influence on the exposure index defined in this study. The results of this study suggested that the selection of evaluation approaches for pesticide exposure could be dependent on the purpose of regulatory assessment and management planning. For instance, total pesticide loadings from agricultural fields might be very sensitive to chemical persistence, while the peak concentrations of a pesticide are mainly related to its mobility. For a particular

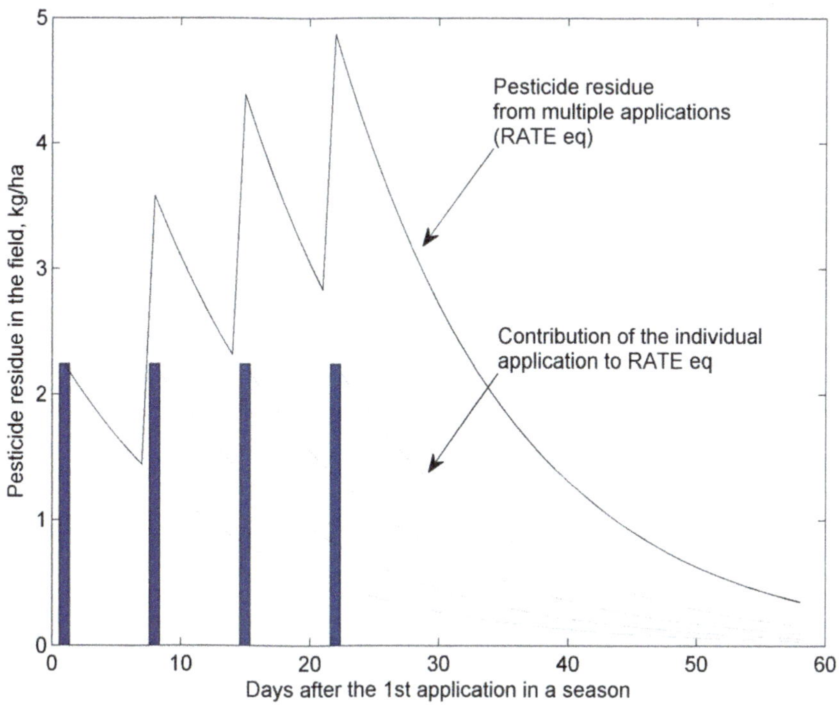

Figure 3. Equivalent application rate from multiple pesticide applications, illustrated with recommended application rates and intervals of carbaryl for tomatoes [32]. Bars represent four applications at 2.24 kg/ha and 7-day intervals.

pesticide, the spatial variability on its exposure indices across various scenarios is associated with landscape characteristics, such as OC and CN values. In addition, significant correlation was observed between intercept constants (b1) in the use-exposure relationships and the curve numbers for residue in the crop scenarios. Those findings indicated the possibility in developing generic equations of use-exposure relationships.

Methods to extend the capability of use-exposure relationship modeling are provided, including applications for assessing sediment-bound pesticides and provisions for multiple pesticide applications within a growing season. Since the use-exposure relationships were parameterized from regression analysis on the simulation results of existing field-scale models, the accuracy of risk assessment based on those relationships is associated with the model itself and the configuration of crop scenarios. Future work issuggested to implement the developed equations in real field conditions and to compare the predictions with measured pesticide data. An evaluation of measured data with corresponding field conditions would generate instructive information for developing crop scenarios for pesticide exposure assessment and risk characterization.

Acknowledgments

The authors acknowledge Dr. John Sanders, Michael Ensminger, and Keith Starner of California Department of Pesticide Regulation and W. Martin Williams of Waterborne Environmental, Inc. for valuable discussions and critical reviews.

Author Contributions

Conceived and designed the experiments: YL FS XD SG KG. Performed the experiments: YL FS XD. Analyzed the data: YL XD. Contributed reagents/materials/analysis tools: YL FS. Wrote the paper: YL FS XD SG KG.

References

1. Capel PD, Larson SJ, Winterstein TA (2001) The behaviour of 39 pesticides in surface waters as a function of scale. Hydrologic Processes 15: 1251–1269.
2. Capel PD, Larson SJ (2001) Effect of scale on the behavior of atrazine in surface waters. Environmental Science & Technology 35: 648–657.
3. Domagalski JL, Munday C (2003) Evaluation of diazinon and chlorpyrifos concentrations and loads, and other pesticide concentrations, at selected sites in the San Joaquin Valley, California, April to August, 2001. United States Geologic Survey. Water-Resources Investigations Report 03-4088 Water-Resources Investigations Report 03-4088.
4. Kratzer CR, Zamora C, Knifong DL (2002) Diazinon and chlorpyrifos loads in the San Joaquin River Basin, California, January and February 2000. United States Geological Survey Water Resources Investigation Report 02-4103 Water Resources Investigation Report 02-4103.
5. Dubrovsky NM, Kratzer CR, Brown LR, Gronberg JM, Burow KR (1998) Water quality in the San Joaquin-Tulare Basins, California, 1992–95. United States Geological Survey. Circular 1159 Circular 1159.
6. Ross LJ, Stein R, Hsu J, White J, Hefner K (1999) Distribution and mass loading of insecticides in the San Joaquin River, California. Sacramento: Environmental Hazards Assessment Program, Environmental Monitoring and Pest Management Branch, California Department of Pesticide Regulation, EH 99-01 EH 99-01.
7. Ahuja LR, Rojas KW, Hanson JD, Shaffer MJ, Ma L, eds. Root zone water quality model – Modeling management effects on water quality and crop production. Highlands Ranch, CO: Water Resources Publications LLC.
8. USGS (2005) Evaluation of unsaturated-zone solute-transport models for studies of agricultural chemicals, Open-File Report 2005-1196 (http://pubs.usgs.gov/of/2005/1196/ofr20051196.pdf, accessed 02/2011). Reston, VA: U.S. Geological Survey.
9. Nolan BT, Bayless ER, Green CT, Garg S, Voss FD, et al. (2005) Evaluation of Unsaturated-Zone Solute-Transport Models for Studies of Agricultural Chemicals, Open-File Report 2005-1196. Denver, CO: U.S. Geological Survey.
10. USEPA (2010) Water exposure models used by the Office of Pesticide Programs (http://www.epa.gov/oppefed1/models/water/models4.htm, accessed 10/2010). Washington, DC: U.S. Environmental Protection Agency, Office of Pesticide Programs.
11. USEPA (2006) PRZM-3, a model for predicting pesticide and nitrogen fate in the crop root and unsaturated soil zones: users manual for release 3.12.2. Washington, DC: Center for Exposure Assessment Modeling, U.S. Environmental Protection Agency, EPA/600/R-05/111 EPA/600/R-05/111.
12. Luo Y, Zhang M (2009) A geo-referenced modeling environment for ecosystem risk assessment: organophosphate pesticides in an agriculturally dominated watershed. Journal of Environmental Quality 38(32): 664–674.
13. Dasgupta S, Cheplick JM, Denton DL, Troyan JJ, Wiliams WM (2008) Predicted runoff loads of permethrin to the Sacramento River and its tributaries In: Gan J, Spurlock F, Hendley P, eds. Synthetic Pyrethroids, Occurrence and Behavior in Aquatic Environments Oxford University Press.
14. Cryer SA, Fouch MA, Peacock AL, Havens PL (2001) Characterizing agrochemical patterns and effective BMPs for surface waters using mechanistic modeling and GIS. Environmental Modeling and Assessment 6: 195–208.
15. Jones RL, Russell MH (2001) Final Report of FIFRA Environmental Model Validation Task Force (http://femvtf.com/femvtf/Files/FEMVTFbody.pdf, accessed 10/2010). Washington, DC: Federal Insecticide, Fungicide and Rodenticide Act (FIFRA) Environmental Model Validation Task Force, American Crop Protection Association.
16. Singh P, Jones RL (2002) Comparison of pesticide root zone model 3.12: Runoff predictions with field data. Environmental Toxicology and Chemistry 21: 1545–1551.
17. FOCUS (2001) FOCUS surface water scenarios in the EU evaluation process under 91/414/EEC (http://focus.jrc.ec.europa.eu/, accessed 10/2010). European Commission, Forum for the Co-ordination of Pesticide Fate Models and their Use (FOCUS), Document Reference SANCO/4802/2001-rev.2. 245 p.
18. USEPA (2008) USEPA Tier 2 crop scenarios for PRZM/EXAMS Shell (http://www.epa.gov/oppefed1/models/water/index.htm, accessed 09/2010). Washington, DC: U.S. Environmental Protection Agency, Office of Pesticide Programs.
19. Siepmann S, Finlayson B (2000) Water quality criteria for diazinon and chlorpyrifos. Administrative Report 00-3. SacramentoCA: California Department of Fish and Game. 59 p.
20. USEPA (2005) Aquatic Life Ambient Water Quality Criteria, Diazinon (EPA-822-R-05-006). U.S. Environmental Protection Agency, Office of Water, Office of Science and Technology, Washington, DC.
21. Goss EW (1992) Screening procedure for soils and pesticides for potential water quality impacts. Weed Technology 6: 701–708.
22. Luo Y, Zhang M (2010) Spatially distributed pesticide exposure assessment in the Central Valley, California, USA. Environmental Pollution 15: 1629–1637.
23. USEPA (2002) Guidance for selecting input parameters in modeling the environmental fate and transport of pesticides, version II Office of Pesticide Programs, U.S. Environmental Protection Agency.
24. Spurlock F (2008) Distribution and variance/covariance structure of pesticide environmental fate data. Environmental Toxicology and Chemistry 27: 1683–1690.
25. Chapra SC (1997) Surface water-quality modeling. WCB/McGraw-Hill, Boston.
26. Chen W, Hertl P, Chen S, Tierney D (2002) A pesticide surface water mobility index and its relationship with concentrations in agricultural drainage watersheds. Environmental Toxicology and Chemistry 21: 298–308.
27. Kellogg RL, Nehring R, Grube A, Goss DW, Plotkin S (2000) Environmental Indicators of Pesticide Leaching and Runoff from Farm Fields (http://www.nrcs.usda.gov/technical/NRI/pubs/eip_pap.html, accessed 11/2010). Washington, DC: U.S. Department of Agriculture, Natural Resources Conservation Service.
28. Larson SJ, Crawford CG, Gilliom RJ (2004) Development and application of Watershed Regressions for Pesticides (WARP) for estimating atrazine concentration distributions in streams, Water-Resources Investigations Report 03-4047. Sacramento, CA: U.S. Geologic Survey.
29. USEPA (2010) Pesticide Root Zone Model (PRZM) release notes (http://www.epa.gov/ceampubl/gwater/przm3/prz3reln.html, accessed 10/2010). Athens, GA: U.S. Environmental Protection Agency, Center for Exposure Assessment Modeling.
30. Luo Y, Yang X (2007) A multimedia environmental model of chemical distribution: fate, transport, and uncertainty analysis. Chemosphere 66: 1396–1407.
31. USEPA (2006) Organophosphate pesticides: revised cumulative risk assessment (http://www.epa.gov/pesticides/cumulative/rra-op/, accessed 10/2010). Washington, DC: U.S. Environmental Protection Agency.
32. USEPA (2010) Effects Determinations for the California Red-legged Frog and other California Listed Species (http://www.epa.gov/espp/litstatus/effects/redleg-frog/index.html, assessed 02/2011). Washington, DC: U.S. Environmental Protection Agency, Office of Pesticide Programs.
33. CDWR (2002) 2001 Statewide Irrigation Methods Survey (http://www.water.ca.gov/landwateruse/, accessed 09/2010). Sacramento, CA: California Department of Water Resources, Division of Statewide Integrated Water Management.
34. CEPA (2010) Surface Water Database. Sacramento, CA: California Environmental Protection Agency, Department of Pesticide Regulation, (http://www.cdpr.ca.gov/docs/sw/, accessed 10/2010).

35. Villeneuve J-P, Lafrance P, Banton O, Frechette P, Robert C (1988) A sensitivity analysis of adsorption and degradation parameters in the modeling of pesticide transport in soils. Journal of Contaminant Hydrology 3: 77–96.
36. Wolt J, Singh P, Cryer S, Lin J (2002) Sensitivity analysis for validating expert opinion as to ideal data set criteria for transport modeling. Environmental Toxicology and Chemistry 21: 1558–1565.
37. Don DF, Havens PL, Blau GE, Tillotson PM (1992) The Role of Sensitivity Analysis in Groundwater Risk Modeling for Pesticides. Weed Technology 6: 716–724.
38. CEPA (2010) 2010 Integrated Report (Clean Water Act Section 303(d) List/305(b) Report (http://www.waterboards.ca.gov/water_issues/programs/tmdl/integrated2010.shtml, accessed 11/2010). Sacramento, CA: California Environmental Protection Agency, State Water Resources Control Board.
39. USEPA (1999) Methods for Measuring the Toxicity and Bioaccumulation of Sediment-associated Contaminants with Freshwater Invertebrates, Second Edition (EPA-600/R-99/064) U.S. Environmental Protection Agency, Office of Research and Development, Duluth, MI; U.S. Environmental Protection Agency, Office of Water, Washington, DC.
40. Domagalski JL, Weston DP, Zhang M, Hladik M (2010) Pyrethroid insecticide concentrations and toxicity in streambed sediments and loads in surface waters of the San Joaquin Valley, California, USA. Environmental Toxicology and Chemistry 29: 813–823.

Pyrethroids Differentially Alter Voltage-Gated Sodium Channels from the Honeybee Central Olfactory Neurons

Aklesso Kadala[1], Mercedes Charreton[1,4], Ingrid Jakob[1], Thierry Cens[2], Matthieu Rousset[2], Mohamed Chahine[3], Yves Le Conte[1,4], Pierre Charnet[2], Claude Collet[1,4]*

1 INRA, UR 406 Abeilles et Environnement, Toxicologie Environnementale, Avignon, France, **2** CNRS, UMR 5237, Centre de Recherche de Biochimie Macromoléculaire, Université Montpellier 2, Montpellier, France, **3** Department of medicine, Laval University, Québec city, Canada, **4** UMT Protection des Abeilles dans l'Environnement, Avignon, France

Abstract

The sensitivity of neurons from the honey bee olfactory system to pyrethroid insecticides was studied using the patch-clamp technique on central 'antennal lobe neurons' (ALNs) in cell culture. In these neurons, the voltage-dependent sodium currents are characterized by negative potential for activation, fast kinetics of activation and inactivation, and the presence of cumulative inactivation during train of depolarizations. Perfusion of pyrethroids on these ALN neurons submitted to repetitive stimulations induced (1) an acceleration of cumulative inactivation, and (2) a marked slowing of the tail current recorded upon repolarization. Cypermethrin and permethrin accelerated cumulative inactivation of the sodium current peak in a similar manner and tetramethrin was even more effective. The slow-down of channel deactivation was markedly dependent on the type of pyrethroid. With cypermethrin, a progressive increase of the tail current amplitude along with successive stimulations reveals a traditionally described use-dependent recruitment of modified sodium channels. However, an unexpected decrease in this tail current was revealed with tetramethrin. If one considers the calculated percentage of modified channels as an index of pyrethroids effects, ALNs are significantly more susceptible to tetramethrin than to permethrin or cypermethrin for a single depolarization, but this difference attenuates with repetitive activity. Further comparison with peripheral neurons from antennae suggest that these modifications are neuron type specific. Modeling the sodium channel as a multi-state channel with fast and slow inactivation allows to underline the effects of pyrethroids on a set of rate constants connecting open and inactivated conformations, and give some insights to their specificity. Altogether, our results revealed a differential sensitivity of central olfactory neurons to pyrethroids that emphasize the ability for these compounds to impair detection and processing of information at several levels of the bees olfactory pathway.

Editor: David J Schulz, University of Missouri, United States of America

Funding: This work received financial support from the French National Research Agency (ANR-13-BSV7-0010, Bee-channels). The funders had no role in study design, data collection and analysis, decision to publish, or preparation of the manuscript.

Competing Interests: The authors have declared that no competing interests exist.

* Email: claude.collet@avignon.inra.fr

Introduction

In social bees, olfaction is a key function that underlies many activities such as nursing, defense against parasites and predators, foraging, and orientation. Antennae play a key role in olfaction as they house olfactory receptor neurons (ORNs) which are responsible for odors and pheromones detection [1,2]. The second stage of the olfactory pathway involves antennal lobe neurons (ALNs) which are responsible for the processing of olfactory information [3]. Residues of many pyrethroid insecticides (that are commonly used to protect fields from insects considered as pests from an agricultural point of view) have been detected in a number of hives and their outer environment [4,5]. In insects, the symptoms generally associated with type I pyrethroid (e.g., tetramethrin) poisoning are the absence of coordination, hyper-activity and prostration. In addition, for type II pyrethroids (e.g., cypermethrin), periods of convulsions followed by paralysis are also noticed [6]. It should be noted that the earlier classification as type I or type II initially relied on extreme symptoms of poisoning and thereafter the absence or presence of an alpha-cyano residue, but this structural dichotomy turned out to be simplistic since some compounds show intermediate properties [7]. From a toxicological point of view, the study of sublethal effects of pyrethroids is now a priority since subtle modifications can strongly affect highly complex organizations such as those exhibited by social bees. In the honeybee, these sublethal effects include impairment of olfaction and learning performances [8,9], behavioral changes such as disorientation and desertion from the hive [10,11] that would globally lead to colony disturbance. Whereas some of these sublethal effects (especially the so called 'knockdown' effect) have been ascribed to their deleterious action on the peripheral nervous system, little direct evidence has been so far obtained in honeybees [12]. Considering their peripheral location, antennal ORNs are likely to be primarily exposed to these neurotoxic insecticides during various activities, especially foraging. In accordance with the peripheral hypothesis, pyrethroids change the action potential activity in ORNs of moths, as demonstrated in electroantennogram and single sensilla recording assays [13]. Some sublethal effects of pyrethroids (e.g., decrease of queen's egg-laying) have also been attributed to their action on the central nervous system [14], and in particular on neurons from the olfactory pathway, such as ALN neurons.

In neurons, pyrethroids primarily target the voltage-gated sodium channels responsible for action potential generation [15,16]. Most of these studies however rely on voltage-clamp experiments on vertebrate voltage-sensitive channels either in their native environment [17,18] or heterologously-expressed in *Xenopus* oocytes [19,20]. To our knowledge, the direct analysis of the effects of pyrethroids on insects sodium channels in their native neuronal environment are very rare under voltage-clamp conditions (for example *Heliothis virescens* [21]) and do not take into account use-dependent processes that are typical of this insecticide class. A large part of the data on insect channels susceptibility came from expression studies (*Xenopus oocytes*) on sodium channels from insects that are considered as pests, such as *Blatella germanica*, *Musca domestica* or *Drosophila melanogaster* [22,23,24]. Unfortunately, in the case of honey bee, no cloning and no heterologous expression of the voltage-gated sodium channels have been reported yet. Our group has already studied the effects of pyrethroids on peripheral ORN neurons from the honeybee antenna [12]. The present paper reports on the differential modifications of voltage-gated sodium channels functions by 3 types of pyrethroids in central neurons from identified honeybee brain regions involved in the processing of olfactory information, the antennal lobes, with an emphasis on use-dependent channels modifications.

Materials and Methods

Cell culture

Antennal lobe neurons (ALNs) were isolated from the brains of domestic honeybee *A. mellifera* pupae (at stages between four and six days before emergence). Pupae were first dipped in alcohol for few seconds and then rinsed in sterile distilled water for sterility purposes. The brain was dissected out of the insect's forehead in a sterile Ca^{2+} and Mg^{2+} -free Tyrode (400 mOsm/l, see Solutions). The brain sheath was then removed and the antennal lobes were isolated. After a hyperosmotic non-enzymatic dissociation in Ca^{2+} - Mg^{2+} free Tyrode, (500 mOsm/l, 4°C; 15 min) and centrifugation (0.3 g, 3 min), the pellet was suspended (one antennal lobe per 15 µl) in culture medium (see Solutions). Fragments of the antennal lobes were gently triturated through the disposable tip of a p100 pipette. Isolated neurons were plated on poly-L-lysine coated plastic Petri dishes, and cultured within a liquid space formed by a coverslip supported by two pieces of glass spacers attached with a non-cytotoxic silicone grease to the bottom of the Petri dish. Dishes were thus kept upside down in an incubator (29°C, high humidity). All experiments were done on 2 to 5 days-old cell cultures and performed at room temperature (20–22°C).

Electrophysiology

Membrane currents were measured in the whole-cell configuration using a patch-clamp amplifier (RK400, Bio-Logic, Claix, France). Voltage pulse generation and data acquisition were done using WinWCP software (John Dempster, Strathclyde University, UK) driving an A/D, D/A converter (PCI-6014 board, National Instruments Corp. Austin, TX, USA). Patch-clamp pipettes were pulled from borosilicate glass capillaries on a vertical pipette puller (P30, Sutter Instruments Co, Novato, AS, USA) and then sylgard-coated to minimize pipette capacitance. The resistance of the recording electrode filled with intracellular solution (see Solutions below) ranged between 3 and 5 MΩ in standard extracellular solution. The liquid junction potential (<3 mV) was not corrected. Microelectrode offset potential was nulled before seal formation. After gigaseal formation and before membrane patch rupture, the resting holding potential was established at −80 mV and the residual microelectrode capacitance was nulled with a fast analog compensation circuit available on the amplifier. In the whole cell configuration, capacitance and series resistance were maximally compensated for. Passive leak currents and residual linear capacitative currents were subtracted using a P/4 protocol. Current-voltage relationships were generated by step depolarizations to test potentials from −80 to +70 mV (10 mV increment) from −80 mV. Individual curves of the sodium current voltage dependence were fitted with equation 1: $I(V) = G_{max}(V-V_{rev})/(1+\exp[V_{0.5}-V)/k])$, where I(V) is the peak density of the current for a depolarization to a membrane potential V, G_{max} is the maximum conductance, V_{rev} is the apparent reversal potential, $V_{0.5}$ is the half activation voltage and k is a steepness factor. To explore steady-state inactivation, the membrane potential was stepped from à −80 mV holding potential to a series of 100 ms conditioning prepulses in the range of −100 to +20 mV, followed by a 0 mV test potential. Experimental data were fitted with equation 2: $h = 1/(1+\exp[(V-V_h)/K_h])$, where V_h is the potential for half inactivation and K_h a slope parameter.

To assess pyrethroid effects, a single stimulation protocol consisted of a 3 ms depolarization from a holding potential of −80 mV to −10 mV. Trains of repetitive depolarizations at 13 Hz were also used and consisted of ten short stimulations (3 ms to −10 mV) with an interpulse duration (time lapse between the initiation of two successive pulses) of 78 ms. Three milliseconds pulses were selected since short pulses durations allow for the development of the maximal tail currents as demonstrated in other neuronal preparations [25,26]. The percentage of channels modified by pyrethroids was calculated using Equation 3: $M = (I_{tail}/(E_h-E_{rev}))/(I_{Na}/(E_t-E_{rev}))$ x 100 where I_{tail} is the maximal tail current amplitude, E_h is the potential to which the membrane is repolarized, E_{rev} is the reversal potential for the sodium current, I_{Na} is the amplitude of the current during depolarization (measured in control conditions) and E_t is the membrane potential reached during the test pulse [25]. Patch-clamp data were analyzed with OriginPro software. Values are given as mean ± S.E.M. The Student's *t*-test was used to compare means, with a significance level set at $P<0.05$.

Solutions

The Ca^{2+} and Mg^{2+} -free Tyrode used for dissection contained (in mM): 140 NaCl, 5 KCl, 10 HEPES, 90 sucrose (pH 7.2, adjusted with NaOH, 400 mOsm/l). The hyperosmotic Tyrode contained 190 instead of 90 mM sucrose. Culture medium was made of a commercial liquid L15 medium (with L-glutamine) supplemented with 5.5 mM D-Glucose, 3.3 mM L-proline, 75 mM sucrose, 10% fetal bovine serum, 1% penicillin/streptomycin (pH 7.2, adjusted with HCl, 400 mOsm/l). The standard extracellular solution used for patch-clamp in order to isolate the sodium current contained (in mM): 120 NaCl, 20 TEA-Cl, 2 $MgCl_2$, 2 $BaCl_2$, 0.1 $CdCl_2$, 1 4-aminopyridine, 10 HEPES, 90 sucrose (pH 7.2 ajusted with HCl, 400 mOsm/l). A concentration of 0.1% DMSO was also added to the extracellular solution during the experiments with pyrethroids, to match the concentration present in the insecticide-containing solution. Intracellular (pipette) solution contained (in mM): 135 CsCl, 5 NaCl, 1 $MgCl_2$, 1 $CaCl_2$, 10 EGTA, 10 HEPES, 90 sucrose (pH 7.2, adjusted with CsOH, 380 mOsm/l). Pyrethroid compounds were purchased from Sigma-Aldrich Co (St-Louis, MO, USA). Stock solutions

(10 mM) were prepared in DMSO (thoroughly vortexed) and dissolved in the perfusion solution (concentration in DMSO never exceeded 0.1%, a concentration that did not have any detectable effect on the sodium current). We used a single use glass capillary device as described by Tatebayashi and Narahashi [25] to expose olfactory neurons to pyrethroids. Only one neuron was recorded per dish.

Modeling

The state model of the sodium channels was constructed using a formalism derived from previous model as well as from available structural and functional data (see Results and Discussion). Four Closed and one Open states represent the gating process that opens the channel. The Closed states refer to the successive voltage-dependent activation (transition rates α and β) of the 3 similar S4 voltage-sensors necessary for channel opening. Transition to the Open state also requires an additional voltage dependent mechanism (rates KFwr, and Kbck, see Results and Discussion). Fast inactivation (hinged-lid -IFM inactivation due to the loop connecting domains III and IV) is also voltage-dependent since it relies on the activation of the 4th voltage sensor (DIVS4, transition rates KiF, KiB, [27]). The deep, slow, pore-dependent inactivation is not dependent on voltage and therefore transition rates are set as constants (Kin and Kout). Voltage-dependent transition rates are set as $A.\exp^{(V/k)}$, with A the value at V = 0, k the voltage dependency, and V the membrane potential. The set of differential equations describing the system is solved numerically to fit the experimental data recorded in control conditions, giving a set of values able to reproduce most of the channel properties (see Figure S1). For the effects of pyrethroids, drug binding has been limited to the Open state, as suggested from their use-dependent effects described in a number of articles, and the location of the pyrethroid binding site within the channel pore [28,29]. We are aware of the possible binding to Closed or intermediate states during the gating process [30] through lateral fenestrations of the channel pore within the plasma membrane for example [31]. In our conditions, this possibility was also explored (see Supplementary Informations), but not presented in the results section because (i) the use-dependent inhibition and the location of the binding site suggest Open state binding, as stated above, (ii) data with tetramethrin that could suggest Closed state binding are better fitted with a model that does not required it, (iii) this pyrethroid is also the most hydrophylic one, which clearly does not favor Closed-state binding on a buried binding site and (iv) using Closed state binding gave qualitatively the same results on the kinetics modifications that affect transitions to Open and Inactivated states (see Supplementary Informations), but changed the Kd values to less sensitive values from tens of nanomoles to micromoles. However, binding to the Open state does not mean therefore that the drugs accesses its binding sites by the hydrophilic pathway of the open pore, but just that the open conformation is required, which, in this case is functionally equivalent. Once bound, the kinetics parameters that can be affected are those that depend on the pore module where the binding site is located (DIS4-S5 linker, DIS5, DIIS6, DIIS4-S5linker, DIIS5, and DIIIS6 [28,29]) i.e. transition rates to and from Open and Inactivated states. We have limited the possible changes to the amplitude only, leaving voltage-dependency unaffected. The p, q, r, s, t and u parameters are therefore the factors affecting KFwr, KBck, KiF, KiB, Kin and Kout, respectively, giving the drug-bound values KFwrb, KBckb, KiFb, KiBb, Kinb and Koutb. This model was used to fit the experimental data in the presence of cypermethrin, permethrin and tetramethrin, and values of p, q, r, s, t and u were then displayed on a radargraph with a log scale for visualization (see Results and Discussion). The binding and unbinding of the drugs (KPyrF, and KPyrB) were also evaluated and used to calculate the Kd values for each drug. Simulation and fitting were done using the Berkeley-Madonna software.

Results

Electrophysiological properties of the sodium current in ALNs

In ALNs, the mean current-voltage curve peaks at 0 mV (Figures 1A and B) and at this potential, the mean maximal current amplitude is -121 ± 12 pA/pF (n = 24). Individual I/V curves were fitted (Eqn 1) and on average, the potential for half activation (V_m) and the slope factor (K_m) are -18.9 ± 1.0 mV and 5.2 ± 0.2 respectively (n = 24). The mean parameters of the steady-state inactivation curve (Figure 1C) were obtained by fitting a Boltzmann equation to experimental data from individual ALNs (Eqn 2). On average, the potential for half inactivation (V_h) and the slope factor (K_h) are -46.9 ± 1.7 mV and 7.2 ± 0.3 respectively (n = 12). The time course of the sodium current was characterized by measuring the time to peak (t_p) and the time of half inactivation (t_h). On average, t_p decreases from 0.81 ± 0.03 ms at -20 mV to 0.46 ± 0.02 ms at +30 mV (n = 24). On average t_h decreases from 0.77 ± 0.08 ms at -20 mV to 0.17 ± 0.01 ms at +30 mV (n = 24). The recovery from inactivation was explored using a two-pulse protocol (Figure 1D) and data from individual ALNs were fitted with a two exponential function. A fast time constant ($\tau_1 = 2.6\pm0.2$ ms, n = 12) accounts for 75% of recovery and a slow time constant ($\tau_2 = 69\pm18$ ms, n = 12) is responsible for the rest of the recovery. We next assessed the effect of repeated short depolarizations (ten 3ms-pulses, Figure 1E), mimicking those encountered by neurons during trains of action potentials. On average (Figure 1F), the sodium current amplitude is significantly reduced to 77 ± 2% of its control value after ten pulses at 13 Hz (n = 36). Under a higher stimulation frequency i.e. 35 Hz, the peak is significantly more reduced and reaches 70 ± 2% of its control value (n = 7).

Effects of pyrethroids on ALNs

The effects of pyrethroids were quantified using repetitive depolarizations (3 ms, 13 Hz) to mimic trains of action potentials encountered during neuronal activity. Figure 2 shows sodium currents recorded in the presence of 10 µM of cypermethrin, permethrin or tetramethrin. The membrane potential was stepped to -10 mV from a holding potential of -80 mV and the effects of pyrethroids were assessed 3 minutes after the beginning of exposure. Peak current as well as tail current amplitudes were measured. All pyrethroids induced the appearance of a prominent tail current upon repolarization from each step of the train (compare with control currents in Figure 1E), revealing a slowing-down of channels deactivation (Figure 2). Cypermethrin elicits a progressive tail current summation in all neurons (n = 7). With permethrin, the majority of neurons (7 out of 10) showed also a progressive summation of the tail current, at least for the first pulses. However, a stationary value (2 neurons out of 10) as well as a decreasing value (1 neuron out of 10) could also be recorded along with successive steps. Interestingly, in the presence of tetramethrin, the tail current amplitude already reached its maximum after the first or second pulse and amplitude then decayed to a smaller value in all neurons (n = 9). In single neurons, the fraction of channels modified by pyrethroids out of the total active channel population is traditionally calculated from tail current amplitudes using Eqn 3. The percentage of modified channels significantly increases from 3 ± 1 to 5 ± 1% ($n = 7$, $p < 0.05$) with cypermethrin (Figure 3A). With permethrin, the

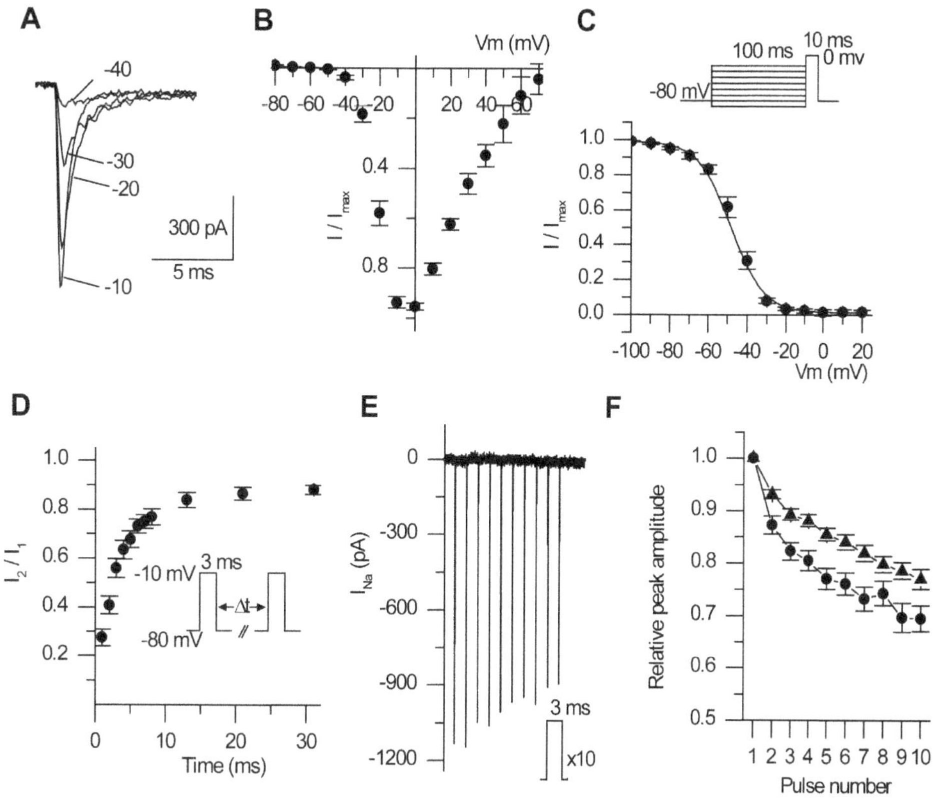

Figure 1. Properties of the voltage-gated sodium current from honeybee ALNs. A, Sodium current traces obtained from an ALN for voltage steps to the indicated potential from a holding potential of −80 mV. B, Mean relative I/V curve (n = 24). C, Mean steady-state inactivation curve (n = 12). D, Recovery from fast inactivation in ALN (n = 12). E, Use-dependent decrease in the amplitude of the sodium current observed in an ALNs submitted to a 10-pulses train at 13 Hz. I, Average decrease in the amplitude of the sodium current in ALNs submitted to a 10-pulse train at 13 Hz (filled triangles, n = 36) or 35 Hz (filled circles n = 7).

percentage of modified channels is 6±1% after the first pulse and 11±3% after the tenth pulse (n = 10 neurons) but these values are not significantly different. With tetramethrin, the average percentage of modified channels decreases by a factor of three between the first and the tenth pulse (the mean value drops from 43±4 to 15±5% n = 9, $p<0.001$). Therefore while significantly different percentages of modified channels are obtained at the first pulse for the three compounds (tetramethrin> permethrin> cypermethrin), this difference tended to attenuate with activity. We then estimated the decay-rate of the tail currents induced by each compound by measuring the remaining tail current amplitude 600 ms after the end of the tenth pulse (R600). After that delay, the residual tail current is decreased to 43±10% of its initial value for cypermethrin (n = 7, $p<0.05$, Figure 3B), 34±11% for permethrin (n = 10, $p<0.05$) and 3±2% of its initial value for tetramethrin (n = 10, $p<0.05$), Tetramethrin thus produced significantly faster decaying tail currents than permethrin and cypermethrin ($p<0.05$ and $p<0.01$ respectively).

Not only a prominent effect of pyrethroids was observed on tail currents, but use-dependent effects were also observed on the peak of the sodium current during depolarization. While incubation with cypermethrin and tetramethrin significantly reduce the peak sodium current amplitude obtained in response to the first depolarization by 33±4% ($p<0.05$, n = 13), and 6±11% ($p<0.05$, n = 10), respectively, this effect was not significant with permethrin. Moreover, all pyrethroids accelerated the cumulative inactivation of the sodium current peak seen in control conditions (Figure 1E and F). As shown in Figure 4, the peak current decreases to 62, 69 and 77% of its initial values, levels that are significantly different from the level reached in control ($p<0.01$ for permethrin and tetramethrin, $p<0.05$ for cypermethrin). The use-dependent sodium current decrease is thus significantly more important with tetramethrin than with permethrin or cypermethrin ($p<0.05$).

In addition to use-dependent modifications, we explored dose-dependent effects of permethrin in ALNs and compared the results to our former study made in peripheral neurons (ORNs). A dose-dependent increase in modified channels was obtained in ALNs as well as in ORNs (Figure 5) in response to a single depolarization. As compared to ORNs, lower levels of modification were obtained in ALNs for concentrations of 10 and 50 μM, suggesting a stronger sensitivity of peripheral *vs* central channels to permethrin [12].

Pyrethroids affect transitions to channel open and inactivated states

A markovian 7 states channel model was set-up to analyze more precisely the effect of pyrethroids (Figure 6A, see Materials and Methods). Four Close states were linked by identical voltage-dependent transition rates (forward α and backward β). Forward and backward transition rates to Open and Fast inactivation were also voltage-dependent, while those to slow–pore-dependent-inactivation were set as constant (see methods). Fit of this model to experimental data recorded in control conditions gave a set of values that allowed to mimic kinetics of sodium current traces as well as activation and inactivation curves and behavior of the

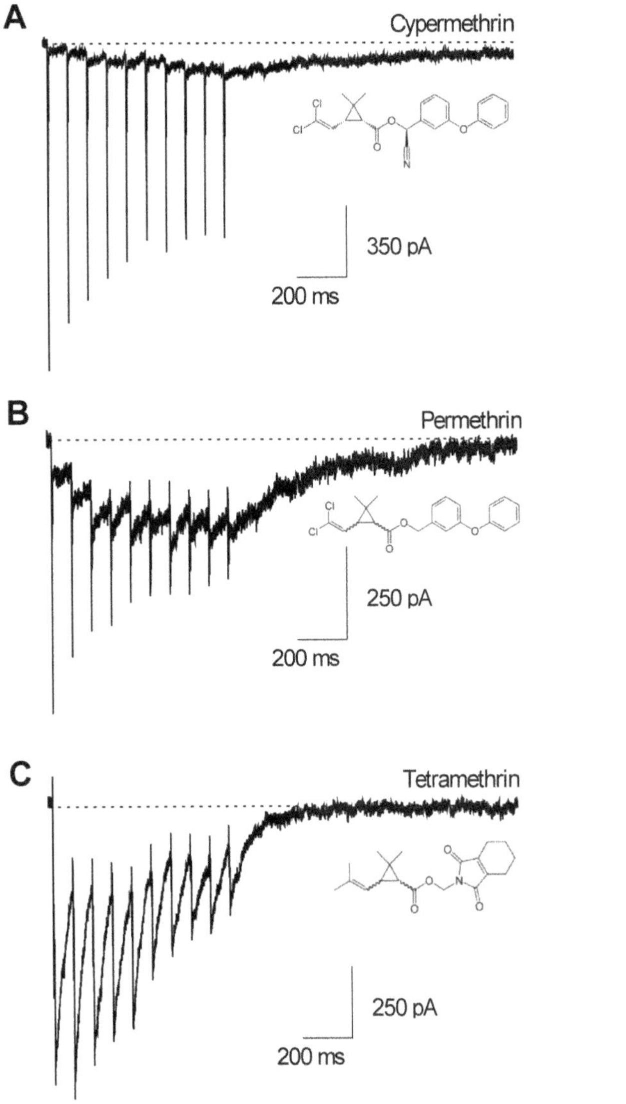

Figure 2. Use-dependent effects of type II (cypermethrin) and type I (permethrin and tetramethrin) pyrethroids in ALNs. Sodium current recordings in response to a 10-pulse train (3 ms, from −80 mV to −10 mV, 13 Hz) in three different ALNs in the presence of 10 μM cypermethrin (A), 10 μM permethrin (B) or 10 μM tetramethrin (C).

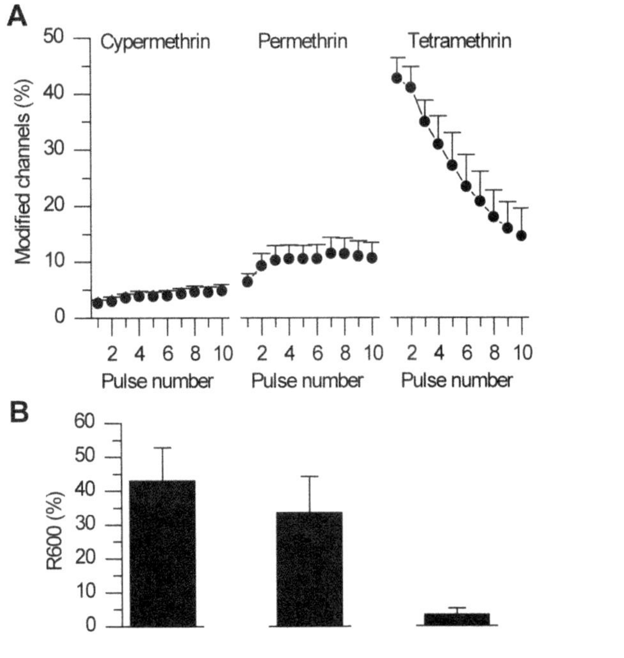

Figure 3. Use-dependent sommation of modified channels and tail current inactivation kinetics in ALNs. Mean time course of pyrethroid-modified channels along with protocol shown in Figure 2 (n = 7, 10 and 10 ALNs for cypermethrin, permethrin and tetramethrin, respectively). The percentage of modified channels is calculated from tail currents amplitude according to Eqn 3. Whereas cypermethrin and permethrin induce an increase in % of modified channels, tetramethrin shows the opposite effect. B2. Kinectics of tail currents estimated by the R600 value, i.e. the percentage of residual tail current 600 ms after the end of the tenth pulse of the 10-pulse protocol. Tetramethrin induces faster decaying tail currents than cypermethrin or permethrin.

channel during a train of depolarizations (Figure S1). The model allowed the binding of the drugs to the Open state to promote a modified channel behavior where the rate constants to and from the Open and Inactivated states are modified by a specific factors (p, q, r, s, t, u respectively), while purely voltage-dependent transitions between closed states remained unaffected (Figure 6A). This model was selected because it fits with previous observations of pyrethroids effects on sodium channels and because the 2 pyrethroids binding-sites resides in the pore-domains with interactions with segments S5 and S6 of domains I, II and III (II-S5 and III-S5-S6) and with the loop connecting segments 4 and 5 in domains I and II (I-L45, II-L45) close to the activation and inactivation machinery but far apart from the voltage sensor S4 [28,29]. Although binding to closed state could theoretically occur [30], we found that it was not necessary in our experimental conditions (see Methods, Discussion and Figure S3). We then use the data from Figure 2 in the presence of cypermethrin, permethrin or tetramethrin to evaluate the respective changes produced by the 3 drugs. Figures 6B, C, D show, super-imposed, the experimental data, in black and the fitted currents using this model in green. Forward transition to open, inactivated and deep-inactivated states were barely affected (p, r, s, Figure 6E). As expected from the effects on deactivation, the most affected rate constant was the closing rate constant from Open to Closed state C4, Kbck, which was decreased by several orders of magnitudes (q values). In the case of permethrin this rate was almost completely nullified. Together with a severe reduction in the backward transition from deep inactivation (factor u), this explained the slow tail current recorded, and the cumulative inactivation that can be seen for longer train of depolarization (Figure S2). The transition rate back from slow-inactivation (factor u), was also strongly decreased for tetramethrin, thus increasing cumulative inactivation, favoring the decrease of the currents during stimulation and preventing the slow and moderate development of the cumulative tail current seen with cypermethrin (Figure 6B). Finally, drug binding and unbinding (kPPyrF and KpyrB) to the open state were increased in the case of tetramethrin when compared to cypermethrin or permethrin (by 1 and 2 order of magnitude). In all cases however estimated Kd values were of similar order of magnitude (as seen experimentally [32]).

Discussion

In this study, we have for the first time successfully (i) characterized the use-dependent effects of pyrethroids on central

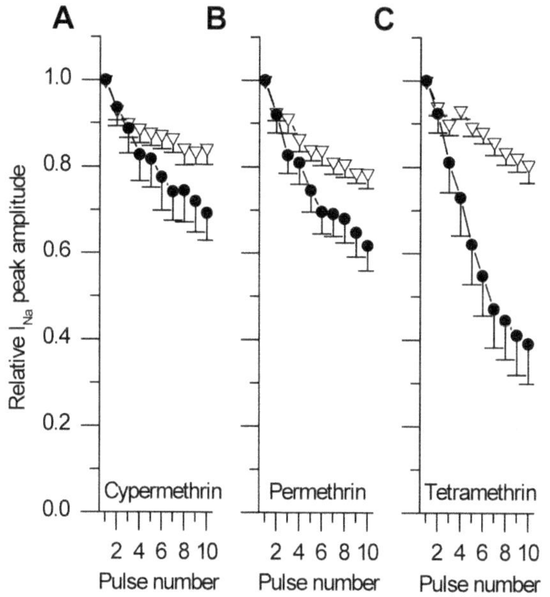

Figure 4. Effect of pyrethroids on the use-dependent decrease in the sodium current peak. Mean evolution of the relative amplitude of the peak current along with protocol shown in Figure 2 (current normalized to the amplitude obtained at the first pulse of the train) from ALNs in a control solution (empty triangles) and after exposure to pyrethroids (filled circles). All pyrethroids tested amplify the decrease in current amplitude and tetramethrin (n = 10) has a stronger effect than cypermethrin (n = 6) or permethrin (n = 10).

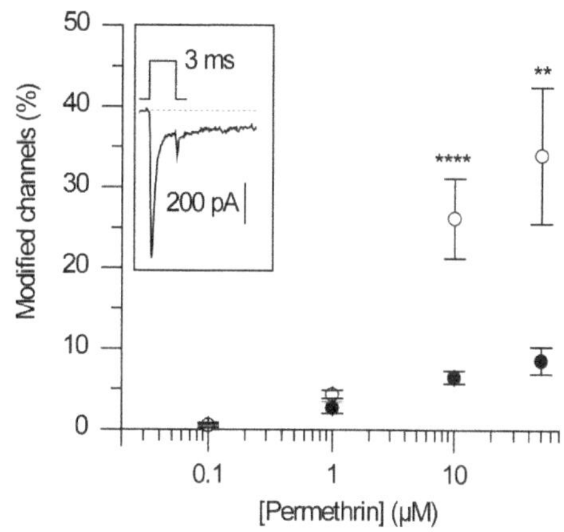

Figure 5. Dose-response curve of permethrin-modified sodium channels in ALNs. Percentage of permethrin-modified channels, calculated according to Eqn 3, from the permethrin-induced tail current amplitude measured 3ms after a single pulse (of 3 ms in duration, see inset). The percentage of modified channels increases as a function of permethrin concentration (filled circle, n = 7, 10, 25 and 9 at 0.1, 1, 10 and 50 µM respectively). The same curve but recorded in ORNs (from Kadala et al. 2011) is shown for comparison (empty circles, n = 3, 9, 14, 5).

olfactory neurons and (ii) mathematically modeled currents modifications. Whereas the action of pyrethroid insecticides on voltage-gated sodium channels has already been investigated in invertebrates [22,23,24,33,34,35], data on beneficial insects such as bees are lacking, while precise molecular actions of these insecticides are needed to interpret and predict their sublethal effects, as well as those of insecticides with a similar mode of action. Our examination of the effects of pyrethroids on cultured neurons from a key area involved in olfactory coding and processing (antennal lobes) yielded several major informations that are key to understand their mechanisms and specificity.

In ALNs, the first observation made with all 3 pyrethroids is a marked slowing of the tail current. In addition we also reported drug-induced use-dependent modifications of current peak and current tail, that have previously been proposed to result from (i) accumulation of persistently open channels [17] (ii) the progressive recruitment of channels after they cycle from open to inactivated state [24,36,37], or (iii) recruitment of silent channels by type I pyrethroids that could also be pictured as a progressive enrollment of formerly silent pyrethroid-hampered sodium channels [12]. These use-dependent decrease in the peak current, and summation of the tail currents are strongly dependent on the chemical nature of the drug. In addition, the peak of the current elicited by a single pulse was decreased by a 3 minutes exposure to cypermethrin or tetramethrin but not permethrin. This behavior was specific of central neurons (this study and [14]) since tetramethrin and permethrin do not decrease the current peak in ORNs from antennae (at least at the time scale of our experiments), whereas they increase the late part of the current, nevertheless suggesting a slowing-down of the current inactivation and/or activation kinetics during the depolarizing voltage step [12].

Our kinetic analysis of these effects allows to better understand the transitions rates that are affected by the different pyrethroids, and thus help us to understand their specificity. First, it is very interesting to notice that the effects of all 3 drugs can be appreciated using the same mathematical model allowing binding of the drug to the Open state only. The fact that tetramethrin can affect channel behavior at the first pulse is correlated in our case to the difference in the drug binding and unbinding rate constants (2 orders of magnitude larger for tetramethrin than for cypermethrin, not shown) and does not require any binding to closed channels as intuitively expected and experimentally tested (Figure S3). These differences in the rate constants may be related to any drug specific differences in solubility, steric hindrance or differential binding to the 2 sites identified in the channel structure [28,29]. However, more experimental and modeling data are clearly needed to unequivocally rule-out binding to closed channel. Second, it is clear that all 3 drugs slowed the tail currents and markedly decreased the deactivation rate. In ALNs, the type II pyrethroid cypermethrin produces a slower tail current than the type I pyrethroid tetramethrin, while permethrin, another type I pyrethroid, produced an intermediary behavior with faster tail currents as already seen in antennal ORNs [12], but partially cumulative, like cypermethrin. The kinetic analysis of these currents demonstrated that (i) all 3 pyrethroids decreased deactivation rate Kbck by several order of magnitudes (factor q in our model) and (ii) that type I pyrethroids only can decrease specifically the transition rate from deep-inactivation, Kout (factor u), preventing any long-term cumulative tail current as those seen with cypermethrin. Tetramethrin in addition also increased, although moderately, the rate constant driving to slow pore-inactivation (factor t) when compared with permethrin, thus producing a stronger cumulative inactivation and the marked decrease in the tail currents recorded during the trains of depolarizations. These differences are better evidenced when simulation are performed with higher pyrethroids concentrations or longer trains of depolarizations (see Figure S2), and thus provide informations for further experimental tests. They also suggest that channels may accumulate in a slow inactivating state

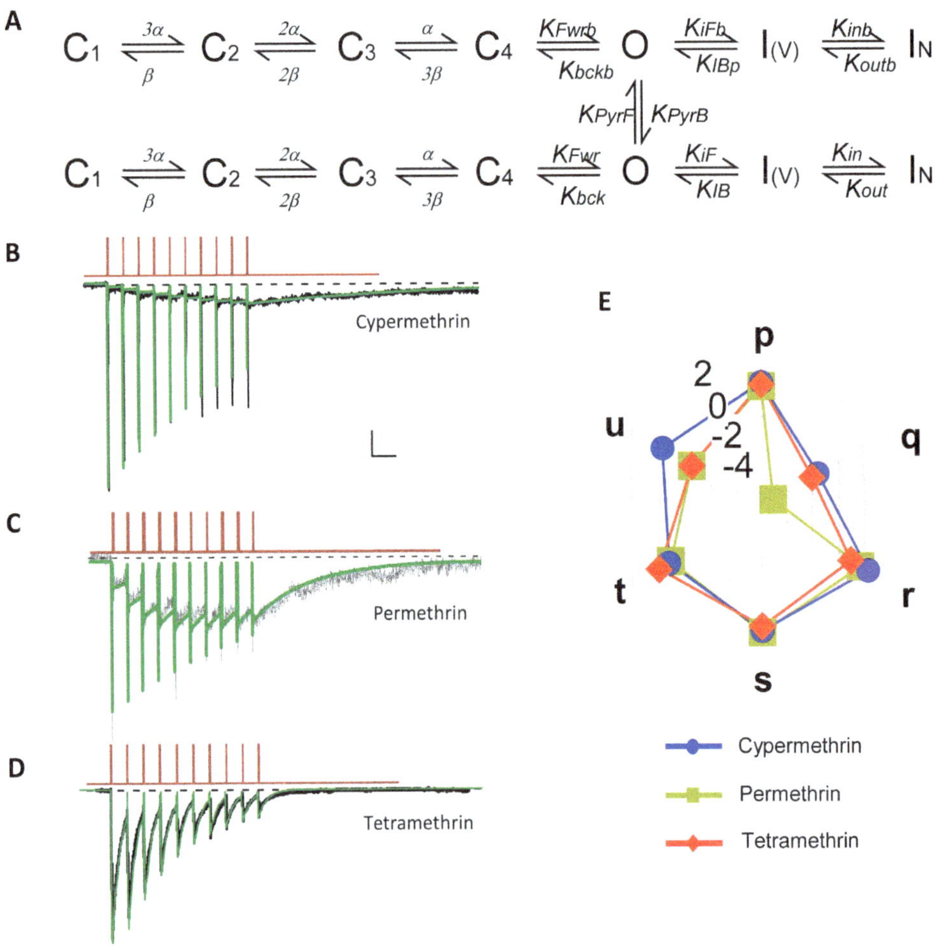

Figure 6. Pyrethroids decrease deactivation rate constants. A. State model used to fit the experimental traces (see Methods and text for details). Pyrethroids can only bind to open channel. Once bound they modify channel kinetics to and from open and inactivated states by given factors, i.e. KFwr, Kbck, KiF, KIB, Kin and Kout by p, q, r, s, t, and u respectively to give KFwrb, Kbckb, KiFb, KIBb, Kinb and Koutb. Voltage dependence is not changed. B-C-D. Fitting of experimental traces obtained under differents pyrethroids (cypermethrin, permethrin, tetramethrin) using the above model allows to determine the changes in the different kinetic parameters. Scale bar 200 pA (B and C) or 500 pA (D) and 150 ms. E. Radargraph of the changes (logarythmic scale) in the different kinetic parameters produced by pyrethroids. Fitting for KpyrF KpyrB, the binding and unbinding rate constants at a single concentration (10 μM), are used to calculate Kd.

in a drug- and rate-dependent fashion thus challenging the notion of modified channels as usually calculated from the ratio of the tail over the peak conductances.

These differential effects of the three pyrethroids on use-dependent parameters may be related to the existence of multiple binding sites within the channel pore as suggested earlier [29]. However, the fact that the calculated Kd are quite similar between the 3 drugs and that the same kinetic parameters seem to be affected (in the limit of our modeling), suggest also that the binding site(s) may be (all) located in a strategic place where activation, fast- and slow-inactivation can be affected more or less directly. Clearly the 2 identified binding-sites comprising IL45-IS5-IIS6 and IIL45-IIS5-IIIS6 [29,38] fit with these requirements. The phenyl group of cypermethrin or permethrin can be docked between the linker and the S5 and the S6 transmembrane helices constituting the major binding site(s), and placing the di-methylcyclopropane and the CCl moeties of permethrin or cypermethrin groups at specific position below or above the gating hinge, close to the pore helix. These specific binding may ensure the drug-dependent modifications of channel properties including the effects on deactivation, inactivation (action on the gating hinge) or pore-inactivation (action on the pore helix). In this case the presence of a cyano group in cypermethrin would impede any effect on pore-inactivation, since t and u factors are poorly affected (~1). Functional analysis of wild type and mutated channels in expression system with different pyrethroids and using a modeling approach similar to this one may therefore bring important information about the structure and mechanisms underlying this modulatory process.

In bees, the differences in tail current decay rates that we observed between type I and type II pyrethroids are reminiscent of dissimilarities in nerve electrical activity after exposure to these compounds: repetitive discharges for type I pyrethroids [39] and prolonged depolarization for type II pyrethroids [40]. However it should be noted that, although symptoms of poisoning by type I and type II pyrethroids may look different, they both eventually lead to paralysis or prostration in exposed animals [6]. Central and peripheral honeybee neurons also behave in a different manner with regard to cumulative inactivation. Whereas pyrethroids enhance the activity-related decrease of the peak current in ALNs, the opposite effect was suggested earlier in ORNs [12] thus producing a progressive reduction in the ALNs activity but a sustained depolarization in ORNs. Pyrethroids may then differentially impair the firing pattern in the two compartments of the honeybee olfactory pathway. Interestingly, a differential sensitivity to deltamethrin was found in splice variants from German cockroach *Blatella germanica* Na^+ channel gene [41]. If honeybee, like other insects has only a single gene coding for voltage-gated sodium channels (*para*), the molecular basis for this specific behavior may result either from the existence of alternative splicing and/or RNA editing of this gene [42,43] and/or from the co-expression of different sets of accessory subunits, since 5 TipE genes homologs have been identified in insects [44]. All these combinations of sodium channel subunits can potentially display specific gating properties, modulation and pharmacological sensitivities. Analysis of the tissue-specific expression of variants of the pore subunit and/or regulatory subunits of voltage-gated sodium channels remains to be explored in *A. mellifera* but will certainly bring key information to understand the differential effects recorded in ALNs and ORNs. Entry of sodium channels into slow inactivation occurs both by long conditioning depolarizing steps or by repeated short pulses [45,46]. The fact that pyrethroids also affected slow inactivation in ALNs strongly support their role in modulating action potential firing during neuronal activity [46]. Our data underline the broad functional action spectrum of pyrethroids in bees, re-emphasizing the complexity of the differential biophysical modifications they can induce. Not only pyrethroids can induce functionally different modifications in separate populations of sodium channels, but their separate mode of action on other targets, especially calcium channels [47,48,49], reinforce their potency to induce sublethal effects on both pests and bees under exposure to weak doses.

Our study further documents the mechanisms by which pyrethroids lead to sublethal neural effects in the honeybee. A variety of subtle symptoms might be triggered by differential effects in peripheral and central neurons and compound-specific abilities to alter sodium channels. For instance, under sublethal exposure, learning performances were differentially affected by pyrethroids and odor training responses were the most affected [8,9]. In summary, in central and peripheral bee neurons, pyrethroids yield strong use-dependent effects of different nature. The pharmacological effects of pyrethroids on activity-related adaptation in peripheral ORNs and central antennal lobe neurons involve separate processes which can potentially impair detection and information processing in the olfactory pathway of the bees.

Supporting Information

Figure S1 Fitting experimental data with the channel-state-model in control conditions allows to evaluate all parameters in control conditions. With these parameters, current traces (A, B), current–voltage curve (C, $V_{act} = -22$ mV, k=9.2 mV), as well as progressive use-dependent current decrease during train of depolarizations (D) are correctly simulated (green traces channel model, black traces experimental data).

Figure S2 Numerical simulation of the model channel behavior. Simulation was performed during longer trains of depolarizations (30 pulses), in standard pyrethroid concentration (10 µM, green) and for 10 fold lower (red) or higher (blue) pyrethroid concentrations underlines the differences between the 3 drugs that can be further tested experimentally.

Figure S3 Fitting experimental data with the state-model incorporating binding to Open and/or Closed states. Modification in the rate constants using the state model on top and allowing binding of tetramethrin to Open or Closed states only or to both states. In the 3 cases the changes in rate constants were qualitatively similar, only Kds value were different, with respective Kd of 65 nM and 6.1 µM for binding to Open or Closed channels exclusively, or 61 nM and 15 µM for possible binding to Open and Closed states simultaneously.

Acknowledgments

We thank the laboratory beekeepers J. Aptel and J-P. Vermandere who provided domestic honeybees, J.B. Thibault (INRA Montpellier) for help with the Berkeley-Madonna software.

Author Contributions

Conceived and designed the experiments: AK YLC PC CC. Performed the experiments: AK M. Charreton CC. Analyzed the data: AK M. Charreton IJ M. Chahine PC CC. Contributed reagents/materials/analysis tools: AK M. Charreton IJ TC MR M. Chahine. Wrote the paper: AK IJ TC MR M. Chahine YLC PC CC.

References

1. Kaissling K-E (1986) Chemo-electrical transduction in insect olfactory receptors. Annu Rev Neurosci 9: 121.
2. Slessor KN, Winston ML, Le Conte Y (2005) Pheromone communication in the honeybee (Apis mellifera L.). J Chem Ecol 31: 2731–2745.
3. Galizia CG, Rossler W (2010) Parallel olfactory systems in insects: anatomy and function. Annu Rev Entomol 55: 399–420.
4. Chauzat MP, Faucon JP (2007) Pesticide residues in beeswax samples collected from honey bee colonies (*Apis mellifera* L.) in France. Pest Manag Sci 63: 1100–1106.
5. Mullin CA, Frazier M, Frazier JL, Ashcraft S, Simonds R, et al. (2010) High levels of miticides and agrochemicals in north american apiaries: implications for honey bee health. PLoS One 5: e9754.
6. Gammon DW, Brown MA, Casida JE (1981) Two classes of pyrethroid action in the cockroach. Pesticide Biochemistry and Physiology 15: 181–191.
7. Khambay BPS, Jewass PJ (2005) Pyrethroids. In: Gilbert LI, Iatrou K, Gil SS, editors. Comprehensive molecular insect science. Oxford: Elsevier BV. pp. 1–29.
8. Decourtye A, Devillers J, Genecque E, Le Menach K, Budzinski H, et al. (2005) Comparative sublethal toxicity of nine pesticides on olfactory learning performances of the honeybee Apis mellifera. Arch Environ Contam Toxicol 48: 242–250.
9. Taylor KS, Waller GD, Crowder LA (1987) Impairment of classical conditioned response of the honeybee (*Apis Mellifera* L.) by sublethal doses of synthetic pyrethroid insecticides. Apidologie 18: 243–252.
10. Thompson HM (2003) Behavioural Effects of Pesticides in Bees–Their Potential for Use in Risk Assessment. Ecotoxicology 12: 317.
11. Cox RL, Wilson WT (1984) Effects of Permethrin on the Behavior of Individually Tagged Honey Bees, Apis mellifera L. (Hymenoptera: Apidae). Environmental Entomology 13: 375.
12. Kadala A, Charreton M, Jakob I, Le Conte Y, Collet C (2011) A use-dependent sodium current modification induced by type I pyrethroid insecticides in honeybee antennal olfactory receptor neurons. NeuroToxicology 32: 320–330.
13. Lucas P, Renou M (1992) Electrophysiological study of the effects of deltamethrin, bioresmethrin, and DDT on the activity of pheromone receptor neurons in two moth species. Pestic Biochem Physiol 43: 103–115.
14. Zhou T, Zhou W, Wang Q, Dai P-L, Liu F, et al. (2011) Effects of pyrethroids on neuronal excitability of adult honeybees Apis mellifera. Pestic Biochem Physiol 100: 35–40.
15. Bloomquist JR (1996) Ion channels as targets for insecticides. Annu Rev Entomol 41: 163–190.
16. Narahashi T (2002) Nerve membrane ion channels as the target site of insecticides. Mini Reviews in Medicinal Chemistry 2: 419–432.
17. Tabarean IV, Narahashi T (2001) Kinetics of modulation of tetrodotoxin-sensitive and tetrodotoxin-resistant sodium channels by tetramethrin and deltamethrin. J Pharmacol Exp Ther 299: 988–997.
18. Song JH, Nagata K, Tatebayashi H, Narahashi T (1996) Interactions of tetramethrin, fenvalerate and DDT at the sodium channel in rat dorsal root ganglion neurons. Brain Res 708: 29–37.

19. Smith TJ, Soderlund DM (1998) Action of the pyrethroid insecticide cypermethrin on rat brain IIa sodium channels expressed in *Xenopus* oocytes. Neurotoxicology 19: 823–832.
20. Choi JS, Soderlund DM (2006) Structure-activity relationships for the action of 11 pyrethroid insecticides on rat Na v 1.8 sodium channels expressed in *Xenopus* oocytes. Toxicol Appl Pharmacol 211: 233–244.
21. Lee D, Park Y, Brown TM, Adams ME (1999) Altered Properties of Neuronal Sodium Channels Associated with Genetic Resistance to Pyrethroids. Molecular Pharmacology 55: 584–593.
22. Warmke JW, Reenan RA, Wang P, Qian S, Arena JP, et al. (1997) Functional expression of *Drosophila para* sodium channels. Modulation by the membrane protein TipE and toxin pharmacology. J Gen Physiol 110: 119–133.
23. Tan J, Liu Z, Wang R, Huang ZY, Chen AC, et al. (2005) Identification of amino acid residues in the insect sodium channel critical for pyrethroid binding. Mol Pharmacol 67: 513–522.
24. Smith TJ, Ingles PJ, Soderlund DM (1998) Actions of the pyrethroid insecticides cismethrin and cypermethrin on house fly *Vssc1* sodium channels expressed in *Xenopus* oocytes. Arch Insect Biochem Physiol 38: 126–136.
25. Tatebayashi H, Narahashi T (1994) Differential mechanism of action of the pyrethroid tetramethrin on tetrodotoxin-sensitive and tetrodotoxin-resistant sodium channels. J Pharmacol Exp Ther 270: 595–603.
26. Ginsburg K, Narahashi T (1999) Time course and temperature dependence of allethrin modulation of sodium channels in rat dorsal root ganglion cells. Brain Res 847: 38–49.
27. Capes DL, Goldschen-Ohm MP, Arcisio-Miranda M, Bezanilla F, Chanda B (2013) Domain IV voltage-sensor movement is both sufficient and rate limiting for fast inactivation in sodium channels. J Gen Physiol 142: 101–112.
28. O'Reilly AO, Williamson MS, Gonzalez-Cabrera J, Turberg A, Field LM, et al. (2014) Predictive 3D modelling of the interactions of pyrethroids with the voltage-gated sodium channels of ticks and mites. Pest Manag Sci 70: 369–377.
29. Du Y, Nomura Y, Satar G, Hu Z, Nauen R, et al. (2013) Molecular evidence for dual pyrethroid-receptor sites on a mosquito sodium channel. Proc Natl Acad Sci U S A 110: 11785–11790.
30. Soderlund DM (2010) State-dependent modification of voltage-gated sodium channels by pyrethroids. Pesticide Biochemistry and Physiology 97: 78.
31. Payandeh J, Scheuer T, Zheng N, Catterall WA (2011) The crystal structure of a voltage-gated sodium channel. Nature 475: 353–358.
32. Breckenridge CB, Holden L, Sturgess N, Weiner M, Sheets L, et al. (2009) Evidence for a separate mechanism of toxicity for the Type I and the Type II pyrethroid insecticides. NeuroToxicology 30: S17.
33. Lee SH, Smith TJ, Knipple DC, Soderlund DM (1999) Mutations in the house fly Vssc1 sodium channel gene associated with super-kdr resistance abolish the pyrethroid sensitivity of Vssc1/tipE sodium channels expressed in Xenopus oocytes. Insect Biochem Mol Biol 29: 185–194.
34. Salgado VL, Narahashi T (1993) Immobilization of sodium channel gating charge in crayfish giant axons by the insecticide fenvalerate. Mol Pharmacol 43: 626–634.
35. Brown LD, Narahashi T (1987) Activity of tralomethrin to modify the nerve membrane sodium channel. Toxicol Appl Pharmacol 89: 305–313.
36. Vais H, Williamson MS, Goodson SJ, Devonshire AL, Warmke JW, et al. (2000) Activation of Drosophila sodium channels promotes modification by deltamethrin. Reductions in affinity caused by knock-down resistance mutations. J Gen Physiol 115: 305–318.
37. Vais H, Atkinson S, Pluteanu F, Goodson SJ, Devonshire AL, et al. (2003) Mutations of the para Sodium Channel of Drosophila melanogaster Identify Putative Binding Sites for Pyrethroids. Mol Pharmacol 64: 914–922.
38. O'Reilly AO, Khambay BP, Williamson MS, Field LM, Wallace BA, et al. (2006) Modelling insecticide-binding sites in the voltage-gated sodium channel. Biochem J 396: 255–263.
39. Song JH, Narahashi T (1996) Modulation of sodium channels of rat cerebellar Purkinje neurons by the pyrethroid tetramethrin. J Pharmacol Exp Ther 277: 445–453.
40. Tabarean IV, Narahashi T (1998) Potent modulation of tetrodotoxin-sensitive and tetrodotoxin-resistant sodium channels by the type II pyrethroid deltamethrin. J Pharmacol Exp Ther 284: 958–965.
41. Tan JG, Liu ZQ, Nomura Y, Goldin AL, Dong K (2002) Alternative splicing of an insect sodium channel gene generates pharmacologically distinct sodium channels. J Neurosci 22: 5300–5309.
42. O'Dowd DK, Gee JR, Smith MA (1995) Sodium current density correlates with expression of specific alternatively spliced sodium channel mRNAs in single neurons. J Neurosci 15: 4005–4012.
43. Olson RO, Liu Z, Nomura Y, Song W, Dong K (2008) Molecular and functional characterization of voltage-gated sodium channel variants from Drosophila melanogaster. Insect Biochem Mol Biol 38: 604–610.
44. Derst C, Walther C, Veh RW, Wicher D, Heinemann SH (2006) Four novel sequences in Drosophila melanogaster homologous to the auxiliary Para sodium channel subunit TipE. Biochemical and Biophysical Research Communications 339: 939–948.
45. Mickus T, Jung H-y, Spruston N (1999) Properties of slow, cumulative sodium channel inactivation in rat hippocampal CA1 pyramidal neurons. Biophys J 76: 846–860.
46. Blair NT, Bean BP (2003) Role of tetrodotoxin-resistant Na^+ current slow inactivation in adaptation of action potential firing in small-diameter dorsal root ganglion neurons. J Neurosci 23: 10338–10350.
47. Hildebrand ME, McRory JE, Snutch TP, Stea A (2004) Mammalian voltage-gated calcium channels are potently blocked by the pyrethroid insecticide allethrin. J Pharmacol Exp Ther 308: 805–813.
48. Collet C (2009) Excitation-contraction coupling in skeletal muscle fibers from adult domestic honeybee. Pflügers Archiv/European Journal of Physiology 458: 601–612.
49. Neal AP, Yuan Y, Atchison WD (2010) Allethrin differentially modulates voltage-gated calcium channel subtypes in rat PC12 cells. Toxicol Sci 116: 604–613.

Multidimensional Measures of Impulsivity in Obsessive-Compulsive Disorder: Cannot Wait and Stop

Sung Yun Sohn, Jee In Kang, Kee Namkoong, Se Joo Kim*

Department of Psychiatry and Institute of Behavioral Science in Medicine, Yonsei University College of Medicine, Seoul, Korea

Abstract

Objective: Although the relationship between obsessive compulsive disorder (OCD) and impulsivity has long been debated, impulsivity has not been systematically examined in clinical samples of OCD. Meanwhile, recent findings suggest that impulsivity is multi-dimensional construct that can be examined through several constructs. Therefore, this study is aimed to evaluate multiple facets of impulsivity in OCD.

Method: The recruitment includes 80 OCD and 76 healthy control participants. Participants completed a test battery comprising three behavioral tasks of stop signal task (SST), delay discounting task (DDT) and balloon analog risk test (BART), and one self-report measure of the Barratt Impulsiveness scale (BIS-11).

Results: OCD subjects showed significantly lower stop signal reaction time of SST reflecting higher action impulsivity and higher delay discounting parameter of DDT suggesting increased choice impulsivity but significantly lower adjusted mean pump of BART implying lower risk taking propensity of BART than healthy control.

Conclusion: Increased Action and choice impulsivity, and decreased risk taking propensities were found in OCD. These findings seem to be consistent with clinical characteristics of OCD such as greater preference for or avoid risky situations (avoidance), inability to wait tension relief may provoke safety behaviors (compulsion) and inability to stop already started behaviors (repetition).

Editor: Carles Soriano-Mas, Bellvitge Biomedical Research Institute-IDIBELL, Spain

Funding: This research was supported by grants (to S. J. Kim) from the National Research Foundation of Korea (NRF) funded by the Ministry of Education, Science and Technology (2010-0022363). The funders had no role in study design, data collection and analysis, decision to publish, or preparation of the manuscript.

Competing Interests: The authors have declared that no competing interests exist.

* Email: kimsejoo@yuhs.ac

Introduction

Obsessive-compulsive disorder (OCD) is a common psychiatric condition characterized by obsessions and compulsions. *Obsessions* are repetitive, unwanted, intrusive thoughts, images, or impulses causing uneasiness, apprehension, or distress in one's mind. *Compulsion* is repetitive ritualistic behavior and is defined as actions inappropriate to the situation that nevertheless persist and which often result in undesirable consequences [1].

Like compulsivity, impulsivity is a common feature in various psychiatric disorders. Impulsivity involves actions that are insufficiently conceived, prematurely expressed, excessively risky or inappropriate to the situation, and that often lead to undesirable outcomes [1]. According to the traditional conception, compulsive disorders and impulsive disorders represent opposite ends of a single dimension, with the former on harm avoidant and the latter on risk seeking [2,3]. However, recent research suggest that, rather than being polar opposites, compulsivity and impulsivity may represent orthogonal factors that each contribute in varying degrees to various psychiatric conditions, including obsessive-compulsive spectrum disorders (OCSDs) [4]. Phenomenologically, OCSDs are characterized by difficulties suppressing repetitive behaviors that are inappropriate to the situation, suggesting underlying impairment in inhibitory control [5]. In the fifth edition of the Diagnostic and Statistical Manual of Mental Disorders (DSM-5), OCD has been reclassified within a new chapter of obsessive-compulsive and related disorders (OCRDs) that includes trichotillomania and skin picking, in which impulsive features are core characteristics [6]. In some aspects, both impulsive and compulsive disorders show similar clinical features, such as difficulties in delaying or inhibiting repetitive behaviors [7]. Compulsive and impulsive disorders are often comorbid and influence each other's development. A number of studies reported high prevalence of impulse control disorders (ICDs) in OCD [8] and high prevalence of OCD in ICDs [9]. In addition, the impulsiveness in OCD seems to have various significant clinical implications. Comorbid ICDs in patients with OCD are associated with poor clinical characteristics, such as early age at onset, great number and severity of symptoms, poor insight, insidious onset, impaired functioning, and poor treatment response seen at long-term follow-up [10]. OCD subjects with higher impulsivity show higher learning problems, low frustration tolerance, poor inter-personal relationships, attention-seeking behavior in childhood, higher neuroticism, and a higher incidence of somatic symptoms

[11]. Additionally, based on neuroimaging and lesion studies, one of the major areas involving impulsivity is the ventral striatal loop [12], which is a target area of deep brain stimulation to improve obsessive-compulsive (OC) symptoms in refractory [13]. Despite this substantial evidence suggesting the importance of impulsivity in OCD, there have been few studies on this relationship and these have mainly used self-rating measures [3,14].

Impulsivity is not a unidimensional construct and it has been suggested that there are several distinct facets of impulsivity, including behavioral disinhibition (impulsive action), impulsive decision making (impulsive choice), and unduly risk taking [15,16,17]. Behavioral disinhibition is defined as an active process that involves suppression of a prepotent response that has been actively investigated by using the stop signal task (SST) [15]. Impulsive decision making is characterized by making choices for smaller immediate rewards rather than waiting for larger delayed rewards. The delay discounting task (DDT) is a well-known behavioral task that measures delay discounting, which refers to the devaluing of a reward due to its location in the future; in other words, DDT assesses the tendency to discount future rewards [18]. Risky decision making is the tendency to engage in behaviors with some potential for danger or harm while also providing an opportunity to obtain some form of reward [19]. The balloon analogue risk task (BART) is a computerized measure that assesses the tendency to engage in simulated risk taking behavior in a context in which unduly risky behavior results in poor outcomes [17]. Considering the multidimensionality of impulsivity, it would be fruitful to simultaneously evaluate the various dimensions of impulsivity (impulsive choice, impulsive action, and risk taking) as well as by using a self-rating measure. These approaches may help in providing a more integrative understanding about the various subtypes of impulsivity and their interrelations in OCD.

Therefore, the central aim of this study is to explore how subconstructs of impulsivity pertain to OCD. We systematically assessed impulsivity using both a subjective self-report and an objective behavioral approach with multiple measures. Based on fact that avoidance behavior and difficulties of inhibiting or delaying compulsive urges are characteristic features of OCD, we hypothesized that OCD subjects show higher impulsive actions and choices and make lower risky decisions than normal controls.

Materials and Methods

Participants

The study group comprised 80 patients with OCD and 76 healthy control participants who were matched for age and sex. The primary diagnosis of OCD and other comorbid psychiatric conditions in patients were determined by the patient version of the Structured Clinical Interview for the DSM-IV (SCID-IV) [20], assessed by a trained psychiatrist (S. J. Kim). Healthy control subjects also underwent SCID-IV and were required to be free of a lifetime history of psychiatric illness.

Exclusion criteria for OCD patients demanded the absence of significant medical or neurologic illness and any other Axis I disorders except for comorbid major depressive disorder. All participants gave written informed consent according to the procedures approved by the Severance Hospital Institutional Review Board.

Procedure and clinical assessments

To assess OCD symptoms, we administered the Yale-Brown Obsessive-Compulsive Scale (Y-BOCS) [21] and the Obsessive-Compulsive Inventory Revised-Korean, (OCI-R-K) [22]. Depressive symptoms were assessed using the Montgomery-Åsberg Depression Rating Scale (MADRS) [23]. In addition, the vocabulary and block design of the Korean version of the Wechsler Adult Intelligence Scale was applied to all participants to estimate IQ. The participants were excluded if their IQ was below average (IQ <90) [24], as some studies had suggested associations between intelligence and the performances of SST [25], DDT [26], and BART [27].

Action impulsivity: Stop Signal Task (SST) [28]

Response inhibition (action impulsivity) was assessed using the SST, which consists of 120 total trials. In each trial, participants were presented with the *go* stimulus (the letter "X" or "O") for 1,000 ms with participants instructed to press the Z key for an X and the/key for an O as quickly and as accurately as possible (go trials). For stop trials (25% of trials), a go stimulus was followed by a stop signal (a beep) after a variable delay, which signaled participants to withhold a response. The onset of the stop signal was varied by a tracking algorithm, in which the stop signal delay was initially 250 ms and was decreased by 50 ms after a previous stop task failure and increased by 50 ms after a previous success. To yield reliable stop signal reaction time (SSRT), we used outlier criteria as follows: (1) percent inhibition on stop trials less than 25% or greater than 75%, (2) percent go-response less than 60%, (3) percent go-errors greater than 10%, and (4) SSRT estimate that is negative or less than 50 ms [29]. A stop signal delay (SSD) indexed the time delay the participant needed in order to inhibit their response. Go reaction time (GORT) is the time taken to press the button when there is no auditory signal. The main dependent variable, SSRT, is a sensitive measure of response inhibition and was extracted by the quantile method which does not require an assumption of 50% inhibition [28]. Longer SSRT reflects worse inhibitory control (slower inhibitory process). In this study, we used the Korean version of the SST [30].

Choice impulsivity: Delay Discounting Task (DDT) [31]

Delay discounting (choice impulsivity) was assessed using a binary choice procedure. In each trial, the computer screen showed a series of choices between two virtual money rewards: immediate smaller reward and delayed larger reward. The delayed reward was fixed at 1,000,000 Korean Won (???), which is approximately 100 US dollars. At the first session, the amount of delay was held constant as within 1 week, and the 26 immediate rewards were presented on the screen in descending order, one per each trial. In the next session, the sequence of immediate monetary rewards ascended in amount until the largest reward was presented, with a particular temporal delay. The next sessions were repeated with incrementally larger temporal delays of 1 week, 2 weeks, 1 month, 6 months, 1 year, 3 years, and 10 years. Within each session, the amount of the immediate monetary value that was preferred equivalently to the large delayed monetary value was defined as the point of subjective equivalence (i.e., an indifference point). Indifference points across the delays were calculated using the hyperbolic decay function, yielding k values reflecting the delay discounting rate [32]. Higher k values indicate higher sensitivity to delayed rewards or choice impulsivity. In this study, we used the Korean version of the DDT [33].

Risk Taking: Balloon Analogue Risk Task (BART) [34]

During the BART, participants were required to press a button to inflate a series of 30 balloons. With each button click the balloon inflated and participants earned money (50 ???) for each pump. This money was added in a temporary bank for that balloon. Participants were told that at some point the balloon would pop and they would lose all the money in the temporary bank. When a balloon exploded, an explosion sound and a picture of the exploding balloon were generated by the computer. Participants were instructed that they could collect from the temporary account to their permanent account at any point before the balloon exploded by pressing a button marked "Collect." After each time a participant collected or popped a balloon, a new balloon appeared. Participants did not actually receive the money, but were instructed to imagine that they would have earned money in a permanent bank in real life. Risk taking propensity was measured by calculating the adjusted mean pumps (AMP), the average number of inflations over the trials in which the balloons did not explode. A larger adjusted value represents a higher risk taking propensity. As the Korean version of the BART was not available, we used the original version of BART [34], which was translated into Korean.

Self-report impulsivity questionnaire: Barratt Impulsiveness Scale (BIS) [35]

The BIS-11 is a self-rating questionnaire. The scale consists of three factors of impulsivity: (1) motor impulsiveness, (2) attentional (cognitive) impulsiveness, and (3) non-planning impulsiveness. In this study, the Korean version of the BIS-11 [36] was used.

Statistical analyses

Statistical tests were two-tailed with level of significance set at $p = 0.05$. Demographic characteristics (age, estimated IQ) were compared across groups using the Mann-Whitney U test or *t*-test for continuous variables. Gender was compared using χ^2 analysis. The primary analyses were *t*-tests with k parameters, adjusted value, SSRT, BIS total score, and BIS subscale as the dependent variables. Because the k parameters were not normally distributed, the distributions of the k parameters were normalized by using the natural log transformation. In addition, we explored associations between the task parameters, BIS scores, and OCD symptom dimensions in patients using Pearson's correlation analyses. All analyses were performed using SPSS version 20.0 (SPSS Inc., USA).

Results

Sample characteristics

Demographic and clinical characteristics for participants are presented in Table 1. A total of 80 OCD patients underwent testing. There were no significant differences in age, sex, or estimated IQ between OCD patients and healthy controls. Patients with OCD presented in the moderately ill range. The mean age at onset of OCD was 18.9±9.3 years. In the OCD groups (n = 80), 78 were currently taking psychiatric medications, all were taking serotonin reuptake inhibitors (SRIs), 31 were taking an SRI with a non-SRI (i.e., benzodiazepine, n = 24; benzodiazepine and antipsychotic, n = 7).

Group comparisons

Behavioral performance and self-report impulsivity are presented in Table 2. Results on the SST were analyzed after first excluding data from 20 OCD patients and 24 normal controls by the outlier criteria method described in [29].

OCD participants had significantly longer SSRTs than did those in the control group ($p = 0.04$). No significant differences between the groups were found on GORT, and there were no impairments in responding ($p = 0.88$). OCD participants exhibited significantly higher AMP than did controls ($p = 0.01$). The mean log k was significantly higher in the OCD group than in the healthy control group ($p = 0.005$). On the BIS, the OCD group evidenced greater levels of total impulsivity than the healthy controls ($p<0.001$). OCD patients reported significantly higher levels of attentional impulsivity and motor impulsivity than did those in the control group (all $p<0.001$). There were no significant group differences in non-planning impulsivity.

Correlations between measures of impulsivity

Correlational analyses among clinical variables and impulsivity measures were performed for OCD patients. The results are presented in Table 3. The severities of obsessive-compulsive symptoms (Y-BOCS) and depressive symptoms (MADRS) were not significantly correlated with any task measures (all $p>0.05$). In addition, the three behavioral measures (SSRT, DDT, and BART) were not significantly correlated with each other (all $p>0.05$). The correlations between the behavioral measures and the BIS also were not significant (all $p>0.05$). We obtained the same results when we performed correlational analyses for the controls and for the combination of both OCD and control participants (data not presented).

Discussion

The main aim of this study was to investigate the relationship between impulsivity and OCD. To the best of our knowledge, this is the first study using behavioral measures of multiple facets of impulsivity (action impulsivity, choice impulsivity, and risk taking) in OCD. Compared to controls, OCD patients showed higher action and choice impulsivity but lower risk taking propensity. On the self-report questionnaire, BIS-11, OCD participants showed higher scores on attentional and motor but not non-planning impulsivity subscales than did controls.

On the SST, OCD participants showed slower SSRTs than controls did, which indicates that more time was required for OCD patients to inhibit a response and which reflects difficulties with motor inhibitory control. Consistent with this finding, several previous studies also reported impaired response inhibition in OCD [5,37,38,39,40]. Moreover, greater SSRT has also been found in first-degree relatives of those with OCD that did not differ from that of OCD patients [5,41]. These findings suggest the possibility of impaired motor inhibition for the endophenotype of OCD. In an imaging study, behavioral impairment indicated by the SST was significantly associated with decreased grey matter in orbitofrontal regions and increased grey matter in cingulate, parietal, and striatal regions [42,43], which have been considered to be implicated in the pathophysiology of OCD [42]. A recent study using non-clinical samples of participants with high SSRT showed increased uncertainty and memory distrust as a consequence of repeated checking compared to participants with lower SSRT [44]. The authors suggested that longer SSRT reflecting poor inhibitory control might increase the liability for harmful effects of neutral compulsive-like behaviors, and the inhibitory deficit might contribute to the development and maintenance of OCD, especially compulsive behaviors [44]. Because the go cues always precede the stop signal in the SST, the increased SSRT of OCD patients indicates that OCD patients fail to inhibit already started actions [45]. These findings are consistent with clinical manifestations because OCD subjects generally are not able to inhibit their compulsive behaviors even though they know them to be senseless, and moreover, once compulsive behaviors begin, usually they become more and more severe.

In the current study, the delay discounting parameter measured by the DDT was higher in the participants with OCD than in the controls. This result suggests that people with OCD tend to choose immediate smaller rewards over larger but postponed ones. This

Table 1. Demographic and Clinical Characteristics of OCD and Healthy Control Subjects.

	OCD (n = 80)		HC (n = 76)		U/χ2/t	*p*
	Mean	SD	Mean	SD		
Age (Years)[a]	27.8	8.1	25.5	4.1	2797	0.39
Male (%)[b]	73.8		65.8		1.17	0.28
IQ Estimate[c]	111.7	10.8	112.0	10.5	−0.176	0.86
Y-BOCS	21.7	6.9				
MADRS	16.4	9.7				
OCI-R-Total Score	34.6	14.2				
Washing	5.3	3.8				
Obsessing	7.8	2.8				
Checking	6.0	3.4				
Ordering	5.1	3.6				
Hoarding	4.4	3.6				
Neutralizing	6.1	3.7				

HC, healthy control subjects; OCD, Obsessive-Compulsive disorder; Y-BOCS, Yale-Brown Obsessive Compulsive Scale; MADRS: Montgomery-Asberg Depression Rating Scale; OCI-R, Obsessive Compulsive Inventory-Revised.
a. Mann-Whitney U test because the data were not normally distributed.
b. χ2-test.
c. Independent samples two tailed t-test.

result is in consistent with the clinical characteristics of OCD because most OCD subjects cannot stop or delay their urge to do compulsive behaviors that immediately reduce their tension, even though they result in more negative consequences in the future. To date, there has been only one study of the association between delay discounting and OCD. Pinto et al. [46] examined the asymmetric discounting (intertemporal choice) task in 25 participants with OCD, 25 with OC personality disorders (OCPD), 25 with comorbid OCD + OCPD, and 25 healthy controls. Contrary to our study, they did not find any differences of performance between OCD and controls, although individuals with OCPD show less temporal discounting than did the controls or those with OCD [46]. Although the reasons for the discrepant results are not clear, there were several differences between the two studies. Pinto et al. [46] used a single delay (i.e., 3 months) and discount factor (δ) as a discount parameter, but we used multiple time delays and k parameter, which have been most widely used for delay discounting. In addition, they focused on OCPD rather than OCD and analyzed OCD and OCPD separately, whereas we did not consider the comorbidity of OCPD. Finally, there might be cultural differences between the OCD subjects participating in the two studies. In our study, the participants were all Korean, but in their study, the participants were recruited from one anxiety clinic in North America, and the ethnicities were not presented. However, there has been some evidence suggesting cultural influences on the performance of delay discounting [34]. One recent study reported cultural differences of neural substrates of delay discounting [47]. In addition, the cultural backgrounds may lead to some differences of attitudes toward delay, the perception of time, or the perceived magnitude of the monetary outcome, all of which can influence the delay discounting performance [34]. These several differences between the two studies caused discrepancies, but further studies are needed to better explain these inconsistent results.

OCD participants showed significantly decreased risk taking on the BART, which suggest that people with OCD avoid taking risks. Risk taking involves the potential for danger or harm as well as the opportunity to obtain some form of reward [19]. The BART scores successfully predicted naturalistic risk taking [48]. namely, the likelihood of engaging in real-world activities that involve potential negative outcomes [49]. Although no previous study has used the BART to investigate OCD, our results are generally consistent with previous research showing lower risk taking or more risk averting properties in people with OCD [50,51,52]. OCD participants had lower risk taking than controls in everyday activities, as measured by the self-reported Everyday Risk Inventory, in both American [52] and Australian samples [51]. Recently, Admon et al. [50] found that people with OCD were reluctant to make risky choices during an interactive risky choice game. Risk taking propensity is determined by both increased reward seeking and decreased sensitivity to loss [53]. Several studies reported decreased sensitivity to reward, and increased sensitivity to loss in OCD [54]. Therefore, the lower risk taking among OCD participants in our study might result from increased sensitivity to loss. This finding is also in line with the maladaptive behavior of OCD patients involving excessive aversion to slight risk, which is commonly thought to result from their perception of situations as highly threatening.

We found that the OCD participants showed significantly higher total scores than controls did on the attentional and motor subscale scores but not on the non-planning subscale of the BIS-11. Most previous studies have also consistently reported higher total and attentional subscale scores of the BIS-11 in OCD [37,55,56]. In terms of motor subscale of BIS-11, there are some controversies. Contrary to ours, some previous studies could not find any difference of motor impulsiveness of BIS-11 between OCD patients and controls [37,55,57]. However, in one recent Korean study, the motor impulsiveness of the BIS-11 in OCD patients was significantly higher than in controls and was correlated with hoarding or aggressive/checking dimensions of obsessive-compulsive symptoms [56]. In another Korean study, although the difference in motor impulsiveness between OCD patients and controls did not reach statistical significance, the effect size (d = 0.81) was bigger than in ours (d = 0.73), which

Table 2. Comparison of impulsiveness in patients with OCD and healthy controls.

	OCD (n = 80)		HC (n = 76)		t	p	Cohen's d
	Mean	**SD**	**Mean**	**SD**			
BART							
AMP	27.75	18.25	34.64	16.51	2.469	0.015*	0.40
DDT							
log k	−4.04	2.50	−5.09	1.81	−3.010	0.003**	0.48
BIS							
BIS_A	16.65	2.73	14.36	2.27	−5.725	<0.001**	0.91
BIS_M	17.14	3.96	14.36	3.50	−4.639	<0.001**	0.74
BIS_NP	22.41	4.30	21.43	3.29	−1.589	0.114	0.26
BIS_T	56.20	9.21	50.14	7.34	−4.553	<0.001**	0.73
SST (OCD, n = 60; HC, n = 52)							
GORT	518.83	128.28	522.31	112.77	0.151	0.880	0.03
SSD	353.17	106.34	373.85	104.80	1.034	0.304	0.20
SSRT	165.67	48.94	148.46	35.83	−2.141	0.035*	0.40
STOPACC	0.62	0.10	0.61	0.13	0.711	0.479	0.09

BART, Balloon Analogue Risk Task; AMP, Adjusted Mean Pumps; DDT, Delay Discounting Task; BIS, Barratt Impulsiveness Scale; BIS_A, Barratt Impulsiveness Scale – Attentional; BIS_NP, Barratt Impulsiveness Scale – Non-planning; BIS_M, Barratt Impulsiveness Scale – Motor; BIS_T, Barratt Impulsiveness Scale – Total; SST, Stop Signal Task; GORT, Go Reaction Time; SSD, Stop Signal Delay; SSRT, Stop Signal Reaction Time; STOPACC, stop accuracy.
* $p < .05$, significant difference. ** $p < .01$, significant difference.

Table 3. Pearson Correlation matrix (n = 80) comparing all measures of impulsivity in patients with OCD.

	1	2	3[a]	4	5	6	7
1 AMP	-	−0.127	0.054	0.161	0.133	0.156	0.178
2 log k		-	0.191	0.083	−0.214	0.005	−0.070
3 SSRT[a]			-	−0.048	−0.041	−0.018	−0.041
4 BIS_A				-	0.404**	0.462**	0.686**
5 BIS_M					-	0.692**	0.873**
6 BIS_NP						-	0.902**
7 BIS_T							-

AMP, Adjusted Mean Pumps; SSRT, Stop Signal Reaction Time; BIS_A, Barratt Impulsiveness Scale – Attentional; BIS_NP, Barratt Impulsiveness Scale – Non-planning; BIS_M, Barratt Impulsiveness Scale – Motor; BIS_T, Barratt Impulsiveness Scale – Total.
a. analysis was performed with 60 OCD patients remained after excluded by outlier criteria of SST.
*p<.05, significant difference ** p<.01, significant difference.

means that the main reason for their being no difference in motor impulsiveness was the small sample size (OCD patients, n = 18; controls, n = 33). Consistent with previous findings, our study showed no difference in non-planning impulsiveness between OCD and control participants [55,57].

To our knowledge, this is the first study of OCD that used the SST, DDT, and BART altogether. All of the three behavioral measures of impulsivity demonstrated differences between groups. In OCD subjects, the performances of the SST, DDT, and BART were not correlated with each other. This finding suggested that there were no associations between behavioral disinhibition, impulsive decision making, and unduly risk taking. Several investigations that used multiple measures of impulsivity simultaneously in the study of other psychiatric disorders, such as substance use disorders or impulse control disorders, also showed inconsistent profiles between measures (i.e., increased in one dimension but null findings in other dimensions of impulsivity) [17]. In our study, increased behavioral disinhibition and impulsive decision making but decreased risk taking were observed in OCD subjects. Such a result might be possible when each task reflects a distinct underlying process. The null correlations between task parameters are consistent with some prior investigations [58,59], which also suggest that various assessments of impulsivity are distinct from each other and that there might be different neurobiological mechanisms underlying each process [60]. Although speculative, it is conceivable that each process could contribute to the OCD phenotype in different ways. In the simple case of a patient with pathologic doubt and checking, risk aversion could lead to a greater preference for avoiding risky situations, whereas a concomitant inability to wait for tension relief may provoke safety behaviors (e.g., checking the gas valve), and the inability to stop already started behaviors leads to repeating those behaviors. As aforementioned, our study supports the utility of implementing multiple behavioral tasks for measuring impulsivity because of its conceptual complexity.

One limitation of the present study was that all of the patients were taking SRIs and some of the patient were also taking benzodiazepines and/or atypical antipsychotics when they were tested, which may have had confounding effects on our results. To rule out these potential confounding effects, further research using drug naïve or drug free OCD subjects is warranted. Another limitation is that we may not have completely excluded OCD subjects with comorbid childhood onset psychiatric conditions, especially ADHD. Given that such comorbid conditions were ruled out by a lifetime history–taking method relying solely on patients' self-reports, it may have been possible that some hidden adult ADHD patients were included in both the OCD and healthy control groups. Considering that ADHD tends to be highly comorbid with OCD and impaired response inhibition [61,62], our results may have at least been partially influenced by unscreened participants with ADHD or other childhood onset psychiatric conditions. The other study limitation is that a portion of participants (n = 20 in OCDs and n = 24 in healthy controls) was excluded because of invalid SST performance. A similar rate of exclusion due to invalid results on the SST has been observed in several other studies [61,63,64]. In our study, majority (n = 32) of the outliers were excluded due to high stop response inhibition rate (>70%), which reflects participants' trading speed in the go task for success in the stop task [65,66]. Other researchers also pointed out that participants are likely to trade-off speed for accuracy, instead of equally considering both aspects of the instructions, that is, fast responses in go trials and inhibition in stop trials [65,67]. Considering these findings including our outliers, our instruction might not be sufficient for the participants to balance go-process and stop-process. Some researchers suggested several strategies for preventing this trade- off such as keeping the proportion of stop signal as low as possible [68] or providing clear instruction (e.g., by stressing speed in the go task and explaining the staircase-tracking procedure) and implementing feedback procedures (feedback after every trial or every block) [69]. Therefore, future studies with these strategies can minimize the outliers for reliable SST.

In summary, this study showed increased action and choice impulsivity but lower risk taking characteristics in OCD patients who have difficulty in waiting for advantageous outcomes and stopping already started behavior. Our multimodal tasks results support each task as a measure of a distinct underlying process of impulsivity.

Author Contributions

Conceived and designed the experiments: SJK KN JIK. Performed the experiments: SYS. Analyzed the data: SYS SJK JIK. Wrote the paper: SYS SJK KN JIK.

References

1. Dalley JW, Everitt BJ, Robbins TW (2011) Impulsivity, compulsivity, and top-down cognitive control. Neuron 69: 680–694.
2. Fineberg NA, Potenza MN, Chamberlain SR, Berlin HA, Menzies L, et al. (2010) Probing compulsive and impulsive behaviors, from animal models to endophenotypes: a narrative review. Neuropsychopharmacology 35: 591–604.
3. Stein DJ, Hollander E, Simeon D, Cohen L (1994) Impulsivity scores in patients with obsessive-compulsive disorder. J Nerv Ment Dis 182: 240–241.
4. Fineberg NA, Chamberlain SR, Goudriaan AE, Stein DJ, Vanderschuren LJ, et al. (2014) New developments in human neurocognition: clinical, genetic, and brain imaging correlates of impulsivity and compulsivity. CNS Spectr 19: 69–89.
5. Chamberlain SR, Fineberg NA, Menzies LA, Blackwell AD, Bullmore ET, et al. (2007) Impaired cognitive flexibility and motor inhibition in unaffected first-degree relatives of patients with obsessive-compulsive disorder. Am J Psychiatry 164: 335–338.
6. American Psychiatric Association (2013) Diagnostic and statistical manual of mental disorders: DSM-5. Washington, D.C.: Amer Psychiatric Pub Inc.
7. Stein DJ, Mullen L, Islam MN, Cohen L, DeCaria CM, et al. (1995) Compulsive and impulsive symptomatology in trichotillomania. Psychopathology 28: 208–213.
8. Fontenelle LF, Mendlowicz MV, Versiani M (2005) Impulse control disorders in patients with obsessive-compulsive disorder. Psychiatry Clin Neurosci 59: 30–37.
9. Grant JE, Potenza MN (2006) Compulsive aspects of impulse-control disorders. Psychiatr Clin North Am 29: 539–551, x.
10. Matsunaga H, Kiriike N, Matsui T, Oya K, Okino K, et al. (2005) Impulsive disorders in Japanese adult patients with obsessive-compulsive disorder. Compr Psychiatry 46: 43–49.
11. Hoehn-Saric R, Barksdale VC (1983) Impulsiveness in obsessive-compulsive patients. Br J Psychiatry 143: 177–182.
12. Robbins TW, Gillan CM, Smith DG, de Wit S, Ersche KD (2012) Neurocognitive endophenotypes of impulsivity and compulsivity: towards dimensional psychiatry. Trends Cogn Sci 16: 81–91.
13. Denys D, Mantione M, Figee M, van den Munckhof P, Koerselman F, et al. (2010) Deep brain stimulation of the nucleus accumbens for treatment-refractory obsessive-compulsive disorder. Arch Gen Psychiatry 67: 1061–1068.
14. Hwang J, Shin Y-C, Lim S-W, Park H, Shin N, et al. (2012) Multidimensional Comparison of Personality Characteristics of the Big Five Model, Impulsiveness, and Affect in Pathological Gambling and Obsessive–Compulsive Disorder. J Gambl Stud 28: 351–362.
15. Courtney KE, Arellano R, Barkley-Levenson E, Galvan A, Poldrack RA, et al. (2012) The relationship between measures of impulsivity and alcohol misuse: an integrative structural equation modeling approach. Alcohol Clin Exp Res 36: 923–931.
16. de Wit H (2009) Impulsivity as a determinant and consequence of drug use: a review of underlying processes. Addict Biol 14: 22–31.
17. Fernie G, Cole JC, Goudie AJ, Field M (2010) Risk-taking but not response inhibition or delay discounting predict alcohol consumption in social drinkers. Drug Alcohol Depend 112: 54–61.
18. Kirby KN, Maraković NN (1995) Modeling Myopic Decisions: Evidence for Hyperbolic Delay-Discounting within Subjects and Amounts. Organ Behav Hum Decis Process 64: 22–30.
19. Leigh BC (1999) Peril, chance, adventure: concepts of risk, alcohol use and risky behavior in young adults. Addiction 94: 371–383.
20. First MB (2002) User's guide for the structured clinical interview for DSM-IV-TR axis I disorders: SCID-I. New York: New York State Psychiatric Institute.
21. Goodman WK, Price LH, Rasmussen SA, Mazure C, Fleischmann RL, et al. (1989) The Yale-Brown Obsessive Compulsive Scale. I. Development, use, and reliability. Arch Gen Psychiatry 46: 1006–1011.
22. Foa EB, Huppert JD, Leiberg S, Langner R, Kichic R, et al. (2002) The Obsessive-Compulsive Inventory: development and validation of a short version. Psychol Assess 14: 485-496.
23. Montgomery SA, Asberg M (1979) A new depression scale designed to be sensitive to change. Br J Psychiatry 134: 382–389.
24. Weiss LG, Saklofske DH, Prifitera A, Hodldnack JA editors. (2006) WISC-IV Advanced Clinical Interpretation. Practical Resources for the Mental Health Professional. Burlington: Burlington Academic Press.
25. Albrecht B, Banaschewski T, Brandeis D, Heinrich H, Rothenberger A (2005) Response inhibition deficits in externalizing child psychiatric disorders: an ERP-study with the Stop-task. Behav Brain Funct 1: 22.
26. de Wit H, Flory JD, Acheson A, McCloskey M, Manuck SB (2007) IQ and nonplanning impulsivity are independently associated with delay discounting in middle-aged adults. Pers Individ Dif 42: 111–121.
27. Ashenhurst JR, Jentsch JD, Ray LA (2011) Risk-taking and alcohol use disorders symptomatology in a sample of problem drinkers. Exp Clin Psychopharmacol 19: 361–370.
28. Band GP, van der Molen MW, Logan GD (2003) Horse-race model simulations of the stop-signal procedure. Acta Psychol (Amst) 112: 105–142.
29. Congdon E, Mumford JA, Cohen JR, Galvan A, Canli T, et al. (2012) Measurement and reliability of response inhibition. Front Psychol 3: 37.
30. Won JY, Kim EJ (2008) Validation of stop-signal task. Korean J Psychol17: 217–234.
31. Hurst RM, Kepley HO, McCalla MK, Livermore MK (2011) Internal consistency and discriminant validity of a delay-discounting task with an adult self-reported ADHD sample. J Atten Disord 15: 412–422.
32. Mazur JE (1987) An adjusting procedure for studying delayed reinforcement. In: Commons ML, Mazur JE, Nevin JA, Rachlin H, editors. Quantitative analyses of behavior: Vol. 5: The effect of delay and of intervening events on reinforcement value. Hillsdale, NJ: Erlbaum; 1987. p. 55–73.
33. Choi BY, Chung KM (2011) Utility of delay discounting task as a measure of impulsivity. Korean J Psychol 30: 845–869.
34. Lejuez CW, Read JP, Kahler CW, Richards JB, Ramsey SE, et al. (2002) Evaluation of a behavioral measure of risk taking: the Balloon Analogue Risk Task (BART). J Exp Psychol Appl 8: 75–84.
35. Patton JH, Stanford MS, Barratt ES (1995) Factor structure of the Barratt impulsiveness scale. J Clin Psychol 51: 768–774.
36. Lee HS (1992) Impulsiveness test guide. Seoul: Korean guidance.
37. Boisseau CL, Thompson-Brenner H, Caldwell-Harris C, Pratt E, Farchione T, et al. (2012) Behavioral and cognitive impulsivity in obsessive-compulsive disorder and eating disorders. Psychiatry Res 200: 1062–1066.
38. Chamberlain SR, Fineberg NA, Blackwell AD, Clark L, Robbins TW, et al. (2007) A neuropsychological comparison of obsessive-compulsive disorder and trichotillomania. Neuropsychologia 45: 654–662.
39. Chamberlain SR, Fineberg NA, Blackwell AD, Robbins TW, Sahakian BJ (2006) Motor inhibition and cognitive flexibility in obsessive-compulsive disorder and trichotillomania. Am J Psychiatry 163: 1282–1284.
40. Morein-Zamir S, Fineberg NA, Robbins TW, Sahakian BJ (2010) Inhibition of thoughts and actions in obsessive-compulsive disorder: extending the endophenotype? Psychol Med 40: 263–272.
41. Menzies L, Achard S, Chamberlain SR, Fineberg N, Chen CH, et al. (2007) Neurocognitive endophenotypes of obsessive-compulsive disorder. Brain 130: 3223–3236.
42. Del Casale A, Kotzalidis GD, Rapinesi C, Serata D, Ambrosi E, et al. (2011) Functional neuroimaging in obsessive-compulsive disorder. Neuropsychobiology 64: 61–85.
43. Menzies L, Chamberlain SR, Laird AR, Thelen SM, Sahakian BJ, et al. (2008) Integrating evidence from neuroimaging and neuropsychological studies of obsessive-compulsive disorder: the orbitofronto-striatal model revisited. Neurosci Biobehav Rev 32: 525–549.
44. Linkovski O, Kalanthroff E, Henik A, Anholt G (2013) Did I turn off the stove? Good inhibitory control can protect from influences of repeated checking. J Behav Ther Exp Psychiatry 44: 30–36.
45. Eagle DM, Bari A, Robbins TW (2008) The neuropsychopharmacology of action inhibition: cross-species translation of the stop-signal and go/no-go tasks. Psychopharmacology (Berl) 199: 439–456.
46. Pinto A, Steinglass JE, Greene AL, Weber EU, Simpson HB (2013) Capacity to Delay Reward Differentiates Obsessive-Compulsive Disorder and Obsessive Compulsive Personality Disorder. Biol Psychiatry.
47. Kim B, Sung YS, McClure SM (2012) The neural basis of cultural differences in delay discounting. Philos Trans R Soc Lond B Biol Sci 367: 650–656.
48. Schonberg T, Fox CR, Poldrack RA (2011) Mind the gap: bridging economic and naturalistic risk-taking with cognitive neuroscience. Trends Cogn Sci 15: 11–19.

49. Congdon E, Bato AA, Schonberg T, Mumford JA, Karlsgodt KH, et al. (2013) Differences in neural activation as a function of risk-taking task parameters. Front Neurosci 7: 173.
50. Admon R, Bleich-Cohen M, Weizmant R, Poyurovsky M, Faragian S, et al. (2012) Functional and structural neural indices of risk aversion in obsessive-compulsive disorder (OCD). Psychiatry Res 203: 207–213.
51. Cicolini T, Rees CS (2003) Measuring Risk-Taking In Obseessive-Compulsive Disorder: An Extension Of The Everyday Risk Inventory With An Australian Sample. Behavioural and Cognitive Psychotherapy 31: 247–259.
52. Steketee G, Frost RO (1994) Measurement of Risk-Taking in Obsessive-Compulsive Disorder. Behav Cogn Psychother: 287–298.
53. Barkley-Levenson EE, Van Leijenhorst L, Galvan A (2013) Behavioral and neural correlates of loss aversion and risk avoidance in adolescents and adults. Dev Cogn Neurosci 3: 72–83.
54. Fitzgerald KD, Welsh RC, Gehring WJ, Abelson JL, Himle JA, et al. (2005) Error-related hyperactivity of the anterior cingulate cortex in obsessive-compulsive disorder. Biol Psychiatry 57: 287–294.
55. Ettelt S, Ruhrmann S, Barnow S, Buthz F, Hochrein A, et al. (2007) Impulsiveness in obsessive-compulsive disorder: results from a family study. Acta Psychiatr Scand 115: 41–47.
56. Roh D, Kim SJ, KIM C-H (2009) The relationship between clinical characteristics and impulsiveness in obsessive-compulsive disorder. J Korean Neuropsychiatr Assoc 2009: 336–343
57. Benatti B, Dell'osso B, Arici C, Hollander E, Altamura AC (2014) Characterizing impulsivity profile in patients with obsessive-compulsive disorder. Int J Psychiatry Clin Pract 2014: 156–160.
58. Reynolds B, Ortengren A, Richards JB, de Wit H (2006) Dimensions of impulsive behavior: Personality and behavioral measures. Pers Individ Dif 40: 305–315.
59. Xu S, Korczykowski M, Zhu S, Rao H (2013) Assessment of risk-taking and impulsive behaviors: A comparison between three tasks. Soc Behav Pers 41: 477–486.
60. Dalley JW, Roiser JP (2012) Dopamine, serotonin and impulsivity. Neuroscience 215: 42–58.
61. Crosbie J, Arnold P, Paterson A, Swanson J, Dupuis A, et al. (2013) Response inhibition and ADHD traits: correlates and heritability in a community sample. J Abnorm Child Psychol 41: 497–507.
62. Abramovitch A, Dar R, Hermesh H, Schweiger A (2012) Comparative neuropsychology of adult obsessive-compulsive disorder and attention deficit/hyperactivity disorder: implications for a novel executive overload model of OCD. J Neuropsychol 6: 161–191.
63. Sjoerds Z, van den Brink W, Beekman AT, Penninx BW, Veltman DJ (2013) Response inhibition in alcohol-dependent patients and patients with depression/anxiety: a functional magnetic resonance imaging study. Psychol Med: 1–13.
64. Solanto MV, Abikoff H, Sonuga-Barke E, Schachar R, Logan GD, et al. (2001) The ecological validity of delay aversion and response inhibition as measures of impulsivity in AD/HD: a supplement to the NIMH multimodal treatment study of AD/HD. J Abnorm Child Psychol 29: 215–228.
65. Verbruggen F, Logan GD (2009) Models of response inhibition in the stop-signal and stop-change paradigms. Neurosci Biobehav Rev 33: 647–661.
66. Leotti LA, Wager TD (2010) Motivational influences on response inhibition measures. J Exp Psychol Hum Percept Perform 36: 430–447.
67. Sella F, Bonato M, Cutini S, Umilta C (2013) Living on the edge: strategic and instructed slowing in the stop signal task. Psychol Res 77: 204–210.
68. Verbruggen F, Logan GD (2009) Proactive adjustments of response strategies in the stop-signal paradigm. J Exp Psychol Hum Percept Perform 35: 835–854.
69. Verbruggen F, Chambers CD, Logan GD (2013) Fictitious inhibitory differences: how skewness and slowing distort the estimation of stopping latencies. Psychol Sci 24: 352–362.

23

A Pragmatic Approach to Assess the Exposure of the Honey Bee (*Apis mellifera*) When Subjected to Pesticide Spray

Yannick Poquet[1], Laurent Bodin[2], Marc Tchamitchian[3], Marion Fusellier[4], Barbara Giroud[5], Florent Lafay[5], Audrey Buleté[5], Sylvie Tchamitchian[1], Marianne Cousin[1], Michel Pélissier[1], Jean-Luc Brunet[1], Luc P. Belzunces[1]*

1 INRA, Laboratoire de Toxicologie Environnementale, UR 406 A&E, CS 40509, 84914 Avignon Cedex 9, France, **2** ANSES, French Agency for Food, Environmental and Occupational Health Safety, 27–31 Avenue du Général Leclerc, 94701 Maisons-Alfort, France, **3** INRA, UR Ecodéveloppement, CS 40509, 84914 Avignon Cedex 9, France, **4** Department of Diagnostic Imaging, CRIP, National Veterinary School (Oniris), Nantes, France, **5** Université de Lyon, Institut des Sciences Analytiques, UMR5280 CNRS Université Lyon 1, ENS-Lyon, 5 rue de la Doua, 69100 Villeurbanne, France

Abstract

Plant protection spray treatments may expose non-target organisms to pesticides. In the pesticide registration procedure, the honey bee represents one of the non-target model species for which the risk posed by pesticides must be assessed on the basis of the hazard quotient (HQ). The HQ is defined as the ratio between environmental exposure and toxicity. For the honey bee, the HQ calculation is not consistent because it corresponds to the ratio between the pesticide field rate (in mass of pesticide/ha) and LD_{50} (in mass of pesticide/bee). Thus, in contrast to all other species, the HQ can only be interpreted empirically because it corresponds to a number of bees/ha. This type of HQ calculation is due to the difficulty in transforming pesticide field rates into doses to which bees are exposed. In this study, we used a pragmatic approach to determine the apparent exposure surface area of honey bees submitted to pesticide treatments by spraying with a Potter-type tower. The doses received by the bees were quantified by very efficient chemical analyses, which enabled us to determine an apparent surface area of 1.05 cm^2/bee. The apparent surface area was used to calculate the exposure levels of bees submitted to pesticide sprays and then to revisit the HQ ratios with a calculation mode similar to that used for all other living species. X-tomography was used to assess the physical surface area of a bee, which was 3.27 cm^2/bee, and showed that the apparent exposure surface was not overestimated. The control experiments showed that the toxicity induced by doses calculated with the exposure surface area was similar to that induced by treatments according to the European testing procedure. This new approach to measure risk is more accurate and could become a tool to aid the decision-making process in the risk assessment of pesticides.

Editor: Eric Jan, University of British Columbia, Canada

Funding: This work was supported in part by the French National Institute for the Agricultural Research (INRA) and by grants from the European Union FEAGA Beekeeping program managed by the INRA–France AgriMer Agreement 11–43 R. YP's doctoral grant was funded by MACIF Foundation (http://www.fondation-macif.org) and Terre d'Abeilles (www.sauvonslesabeilles.com). The funders had no role in study design, data collection and analysis, decision to publish, or preparation of the manuscript.

Competing Interests: The authors have declared that no competing interests exist.

* Email: luc.belzunces@avignon.inra.fr

Introduction

Human activity generates many environmental disruptions and myriads of anthropogenic chemical substances [1]. These substances include plant protection products (PPPs), or pesticides, of which the main families used are herbicides, insecticides and fungicides [2]. As a general feature, chemical substances are subjected to a risk assessment for their effects on human and non-target organisms. The assessment procedure is based on the comparison between the environmental or individual levels of exposure to the substances and the toxicity of these substances, which correspond to the exposure thresholds from which adverse effects can occur in biological organisms exposed through different routes (contact, oral and inhalation) [3,4]. Generally, the exposure thresholds used in risk assessment for acute effects are toxicological values such as the contact median Lethal Dose or Concentration (LD_{50} or LC_{50}). For long-term studies, and in particular reprotoxic effects, the threshold values used are the No Observed (Adverse) Effect Level or Concentration (NOAEL, NOEL, NOAEC and NOEC), defined as the highest exposure level for which no (adverse) effects were observed. In human risk assessment a more modern approach involves the determination of benchmark doses (BMD) or concentrations, which correspond to the levels of exposure at which a given effect is observed, determined by modeling the entire dose-effect relationship [5,6]. These toxicological values can be lowered by uncertainty or safety factors such as those used to take into account intraspecific variability and extrapolation from model species, which correspond to interspecific variability. From the exposure and toxicological values, a Hazard Quotient (HQ = Exposure/Toxicity) or Toxicity to Exposure Ratio (TER = Toxicity/Exposure) can be derived to assess the risk presented by chemicals or pesticides to human and non-target organisms. Hence, when the exposure level is higher than the toxicological value, the situation is at risk for the considered biological organism, which results in HQ or TER values higher or lower than 1, respectively.

For all organisms including humans, the HQ and TER are calculated on the basis of toxicological data obtained on individuals exposed through the pertinent routes of exposure. In the environment, the honey bee appears as an atypical case because both toxicological methods and risk assessment procedures are not always adapted to the environmental situation. For example, (i) the determination of the LD_{50} is relevant for acute contact during a treatment with the active substance, such as pesticide sprays. Although still performed until now, these data cannot be used to extrapolate the chronic contact toxicity. (ii) For residual contact with contaminated plants, no method exists in the registration procedure of pesticides. In the past, the tarsal contact test was used to assess the toxicity induced by repeated contact with a contaminated surface [7]. However, this test is no longer used today. (iii) No method exists to assess the toxicity elicited by inhalation. (iv) Acute oral LD_{50} values are relevant to assess the toxicity induced by an acute exposure to a pesticide through contaminated food, but are inadequate to determine the chronic oral toxicity [8]. The first laboratory approach to assess the oral chronic toxicity of pesticides was proposed in 2001 [9]. Today, chronic toxicity is being considered for inclusion in the pesticide registration procedure by the French Commission for the Biological Assays (CEB), the European Food Safety Agency (EFSA) and the Society of Environmental Toxicology & Chemistry (SETAC) [10–12]. (v) Up to now, risk assessment in honey bee was only based on acute exposure and not on chronic exposure, as practiced for other organisms, which explains the lack of experimental data [3,4]. (vi) A critical step in the assessment of the risk presented by pesticides to bees is based on the determination of the HQ [11,13]. For the HQ and TER, data of the same nature are compared, namely individual or environmental exposure levels and exposure levels below which no (adverse) effect can be observed, which results in HQ and TER values without units. However, for the honey bee, the contact HQ corresponds to the ratio of the pesticide field rate, expressed in mass/ha (generally g/ha), to LD_{50}, expressed in mass/bee (generally µg/bee). This calculation method for the HQ is used because it is difficult to convert field rates of sprayed pesticides into doses to which honey bees are exposed. This results in the HQ with a unit corresponding to a number of bee per ha, which is not useful to properly assess the risk to bees. Thus, after its introduction by Atkins in 1981 [14], the contact HQ ratio was used empirically to assess the risk of pesticides to the honey bee and the values fluctuated between 25 [13] and 85 as the toxicological threshold [11].

In this study, we used a pragmatic approach to determine the exposure levels of bees subjected to a pesticide treatment in the field. We used experimental sprayings to estimate the apparent exposure surface area of a honey bee that enabled converting pesticide field rates into acute doses received by the bees. In a second step, the accuracy of the exposure surface area was probed by comparing the toxicity induced by spraying at given field rates with the toxicity induced by treatments, achieved according to the procedures of the European and Mediterranean Plant Protection Organization (EPPO) and the Organization for Economic and Co-operation Development (OECD) [15,16], with doses calculated from field rates and the apparent exposure surface area. Finally, the exposures to which bees can be submitted after pesticide spraying were calculated and the hazard quotients were revisited.

Materials and Methods

Chemicals

Abamectin (Agrimec gold), acetamiprid (Supreme), chlorpyrifos-ethyl (Nelpon 480), cyfluthrin (Baythroïd), cypermethrin (Cyperfor S), deltamethrin (Decis protech), dimethoate (Danadim progress), esfenvalerate (Sumi-alpha), fenoxycarb (Insegar), hexythiazox (Nissorun), imidacloprid (Confidor), iprodione (Rovral WG), lambda-cyhalothrin (Karate xpress), pyriproxyfen (Admiral pro), tau-fluvalinate (Klartan), tebuconazole (Horizon arbo), thiacloprid (Calypso), and thiamethoxam (Flagship pro) were purchased from Escudier Christian SARL (Avignon, France). Clothianidin (Dantop 50 WG) and prochloraz (Octave) were obtained from Coopérative Agricole Provence Languedoc (Châteaurenard, France). Abamectin (97% pure), acetamiprid (98.1% pure), chlorpyrifos-ethyl (98.5% pure), clothianidin (99% pure), cyfluthrin (98% pure), cypermethrin (94% pure), deltamethrin (98% pure), dimethoate (98.5% pure), esfenvalerate (99% pure), imidacloprid (99% pure), lambda-cyhalothrin (98.5% pure), prochloraz (98% pure), tau-fluvalinate (94% pure), thiacloprid (98% pure), and thiamethoxam (98.5% pure) were purchased from TechLab. DMSO (Dimethyl sulfoxide) was obtained from Sigma-Aldrich.

Honey bee collection

Honey bee workers (*Apis mellifera*) were collected from the reserve frames of the hive body from healthy colonies (≥30 000 individuals) that were carefully monitored for their sanitary state in the experimental apiary of the INRA Research Unit *Abeilles & Environnement* (*Bees* & *Environment*) of Avignon. After being anesthetized with CO_2, bees were distributed in cages (10.5×7.5×11.5 cm) by groups of 30 individuals and placed in the dark in a thermostated chamber at 28±2°C and 60±10% RH. The ambient temperature of the thermostated chamber was chosen as the temperature preferred by the honeybees during the night [17,18]. Bees were fed *ad libitum* with candy (Apifonda + powdered sugar) and water.

Exposure to pesticide sprays

Bees were exposed by spraying in a Potter-type tower to the 20 different commercial pesticide preparations (Table 1) at application rates chosen to be as close as possible to field conditions. The amounts of active substances sprayed were selected to be less than or equal to the lowest registered dose. Twelve replicates of 30 bees were performed for each pesticide treatment. Bees were first anesthetized and put on a 200 cm^2 Plexiglas disc subjected to rotation at a speed of 23 rpm to ensure a homogeneous deposit of the sprayed solution [19]. The deposit was previously calibrated to obtain a rate of 2.32±0.13 $\mu L/cm^2$. In addition to the calibrations, the deposit was controlled just before, during and after each series of 12 replicates (experimental calibration). Immediately after spraying, the bees from 6 replicates were frozen at −80°C to fix the dose of pesticide they received. In parallel, the mortality was followed over time on the remaining 6 replicates by keeping the bees in the thermostated chamber.

Topical exposure to pesticides

Preliminary studies were conducted to assess the mortality rate induced by each pesticide. For each substance, 6 repetitions were

Table 1. Summary of commercial products and their application rates used for the spraying treatment.

Effect	Class	Active substance	Commercial product	Application rate (g a.s./ha)
Fungicide	Dicarboximide	Iprodione	Rovral WG	600
	Imidazole	Prochloraz	Octave	230.5
	Triazole	Tebuconazole	Horizon arbo	75
Growth regulator	Carbamate	Fenoxycarb	Insegar	75
	Carboxamide	Hexythiazox	Nissorun	26
	Pyridine based	Pyriproxyfen	Admiral pro	25
Insecticide	Avermectin	Abamectin	Agrimec gold	2
	Neonicotinoid	Acetamiprid	Supreme	30
		Clothianidin	Dantop 50 WG	3
		Imidacloprid	Confidor	0.2
		Thiacloprid	Calypso	50
		Thiamethoxam	Flagship pro	5
	Organophosphate	Chlorpyrifos-ethyl	Nelpon 480	3
		Dimethoate	Danadim progress	5
	Pyrethroid	Cyfluthrin	Baythroïd	2.5
		Cypermethrin	Cyperfor S	3
		Deltamethrin	Decis protech	1.875
		Esfenvalerate	Sumi-alpha	2
		Lambda-cyhalothrin	Karate xpress	5
		Tau-fluvalinate	Klartan	48

g a.s./ha: gram of active substance per hectare.

systematically performed. However, for the doses inducing a low toxicity, the number of repetitions was increased for an accurate determination of the mortality rates (Table S1). Bees were exposed to 15 out of the 20 active substances because it was not possible to determine a dose-mortality relationship for 5 active substances (fenoxycarb, hexythiazox, iprodione, pyriproxyfen, tebuconazole) that exhibited a very low toxicity ($LD_{50}\geq 100\ 000$ ng a.s./bee). The active substances were diluted in DMSO, and the control groups were treated with pure DMSO. The bees were treated on the thorax according to the EPPO procedure recommended by the European Commission [15,20,21]. Mortality was checked 24, 48, 72 and 96 hours after the exposure, with 96 hours corresponding to the duration at which no significant evolution of mortality was observed for all the tested active substances.

Quantification of 20 pesticides

For each of the 20 pesticides of interest, the active substance was quantified in 3 repetitions. For each repetition, 5 g of frozen honey bees were weighed in a 50 mL centrifuge tube, and 10 mL of acetonitrile, 3 mL of water, 3 mL of heptane, citrate QuEChERS salts and 30 g of steel balls were added for grinding for 4 min at the cadency of 1000 strokes per min using a Geno/Grinder from SPEX Sample Prep (Metuchen, USA). Then, the tube was centrifuged for 2 min at 5000 g, and 6 mL of the lower acetonitrile phase were taken and placed in a pre-prepared 15 mL PSA/C18 tube. This tube was immediately shaken by hand, vortexed for 10 sec and centrifuged for 2 min at 5000 g. Finally, 4 mL of the extract were taken and put in a 10 mL cone-ended glass centrifuge tube for evaporation until 50 μL remained. This remaining extract was kept at −18°C until analysis.

For abamectin, acetamiprid, clothianidin, fenoxycarb, hexythiazox, iprodione, imidacloprid, lambda cyhalothrin, prochloraz, pyriproxyfen, thiacloprid and thiamethoxam, the analysis was performed on a Waters 2695 series Alliance HPLC (Waters, Milford, MA) coupled to a triple quadrupole mass spectrometer Quattro from Micromass (Manchester, UK) equipped with a Z-spray electrospray interface (ESI). Data were processed with MassLynx 4.1. Electrospray ionization was performed in the positive mode. The chromatographic separations were performed on a Nucleodur Sphinx RP-C18 (50×2 mm, 1.8 μm) column from Macherey-Nagel with in-line filter "krudkatcher" 0.5 μm porosity (Phenomenex). The column oven temperature was set to 40°C; the flow rate was 300 μL/min. Samples were analyzed with the mobile phase (A) water with 0.3 mM of ammonium formate and 0.05% formic acid, and (B) methanol. Ten μL were injected in the 90/10 mobile phase (A)/acetonitrile.

For chlorpyrifos-ethyl, cyfluthrin, cypermethrin, deltamethrin, dimethoate, esfenvalerate, tau-fluvalinate and tebuconazole, the analysis was conducted with a GC-ToF, a 6890 Agilent gas chromatograph coupled to a Time of Flight (ToF) mass spectrometer GCT Premier from Waters. Data were processed with MassLynx 4.1. Chromatographic separation was performed on a 30 m×0.25 mm I.D., with a 0.25 μm film thickness DB-XLB capillary column (Agilent Technologies, Avondale, USA). Helium (purity 99.999%) was used as a carrier gas at a constant flow of 1 mL/min. Injections were performed in the splitless mode, and 1 μL in 90/10 acetonitrile/analyte protectant mix was injected. The mass spectrometer was operated in the electron impact mode (EI, 70 eV). Acquisition was performed in the full scan mode with a scan range of m/z 50–550. Calibration was performed using the calibration wizard, with heptacosa as the reference.

Matrix-matched calibration was used for quantification. In each batch, 6 calibration points were prepared and injected as described

Table 2. Comparison of the commercial and measured concentrations.

Active substance	Commercial concentration	Measured concentration		Difference (%)
		Mean	S.D.	
Abamectin	18 g/L	24.91 g/L	1.27 g/L	+38.4
Acetamiprid	200 g/kg	243.66 g/kg	6.14 g/kg	+21.8
Chlorpyrifos-ethyl	480 g/L	500.20 g/L	0.02 g/L	+4.2
Clothianidin	500 g/kg	453.71 g/kg	23.11 g/kg	−9.3
Cyfluthrin	50 g/L	46.45 g/L	0.09 g/L	−7.1
Cypermethrin	100 g/L	114.25 g/L	0.14 g/L	+14.2
Deltamethrin	15 g/L	14.77 g/L	0.72 g/L	−1.5
Dimethoate	400 g/L	433.35 g/L	0.92 g/L	+8.3
Esfenvalerate	25 g/L	25.10 g/L	0.50 g/L	+0.4
Fenoxycarb	250 g/kg	228.54 g/kg	14.49 g/kg	−8.6
Hexythiazox	104 g/kg	91.24 g/kg	2.94 g/kg	−12.3
Imidacloprid	200 g/L	230.29 g/L	12.24 g/L	+15.1
Iprodione	750 g/kg	794.34 g/kg	7.01 g/kg	+5.9
Lambda-cyhalothrin	50 g/kg	44.85 g/kg	2.47 g/kg	−10.3
Prochloraz	461 g/kg	391.69 g/kg	13.72 g/kg	−15
Pyriproxyfen	100 g/L	97.29 g/L	2.66 g/L	−2.7
Tau-fluvalinate	240 g/L	232.65 g/L	12.80 g/L	−3.1
Tebuconazole	250 g/kg	186.30 g/kg	10.32 g/kg	−25.5
Thiacloprid	480 g/L	448.97 g/L	0.07 g/L	−6.5
Thiamethoxam	10 g/L	10.25 g/L	0.16 g/L	+2.5

S.D.: Standard Deviation.
The concentrations of commercial products are expressed according to their nature, liquid or solid.
The quantification of the active substance was performed in triplicate from the phytopharmaceutical preparation used for the spray application.
The differences were expressed as percentages of the commercial concentrations.

above, with concentrations ranging between 4 and 60 ng/g. Quantification was performed using QuanLynx 4.1.

Physical surface of the honey bee

The surface area of the bees, excluding the wings, was determined by X-ray computed tomography. Imaging was performed on 6 dead bees with a µPET/CT Inveon device (Siemens Preclinical Solutions, Knoxville, TN). The computed tomography (CT) images were acquired with the following settings: X-ray source tube voltage at 40 keV with a constant 500 µA current during 320 ms and 2-degree rotation to generate images with 41 µm of voxel resolution. Three-D reconstructions were performed using COBRA software (Siemens Preclinical Solutions). Using ImageJ, the stacks of images were thresholded, and the regions of interest (ROI) were automatically drawn on binary images. Each ROI was checked, and the surface areas obtained were then summed for each of the resulting 768 slices. The surface area of the wings was measured on 50 bees. The 4 wings were cut at the base of the thorax and placed on a microscope slide and slip covered with soapy water. Pictures of wings were taken under a microscope; ImageJ software was used to delineate the wings and to calculate their surface area.

Data analysis

Statistical analyses were performed with R software (2013). The variations of the deposit distributions were tested using a t-test comparing preliminary and experimental calibrations. Pearson's correlation and a linear regression were used to compare the

Table 3. Comparison of preliminary calibrations and experimental controls of the deposit.

Calibration	Deposit (µL/cm^2)		N_c	C.I. 95%	Sprayed volume (L/ha)[a]	
	Mean	S.D.			Mean	S.D.
Preliminary	2.32	0.13	90	[2.30, 2.35]	232	13
Experimental	2.32	0.11	60	[2.30, 2.35]	232	11

S.D.: Standard Deviation.
Nc: Number of calibrations.
C.I. 95%: Confidence Interval at 95%.
[a]Sprayed volume corresponds to the deposit in field conditions.

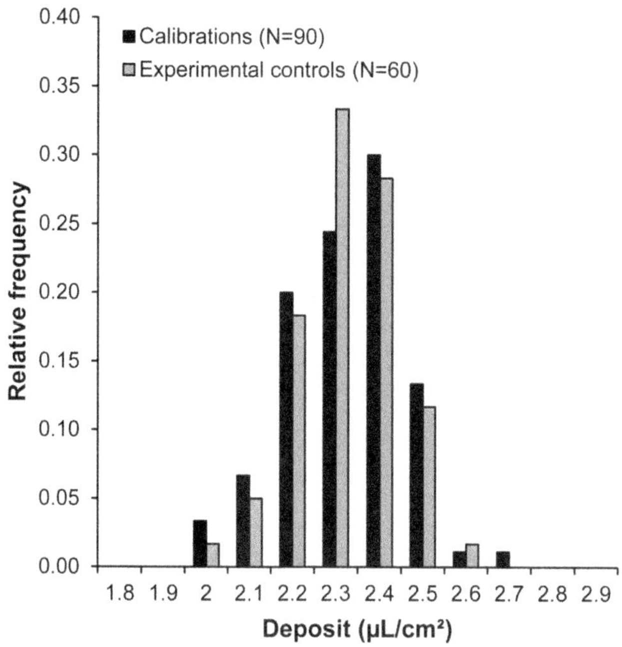

Figure 1. Distribution of the deposit during preliminary calibrations and experimental controls. The preliminary calibrations were performed before the experiment, and the experimental calibrations were performed during the experiment. The deposit was expressed in μL/cm^2 of the disc. T-test preliminary vs. experimental calibrations; ($t = 0.0739$, $df = 148$, p-value = 0.9412).

mortalities obtained by spray contamination and those obtained by the EPPO procedure. For each dose, the stabilization of the dose-mortality response over time was determined with Fisher's exact tests. The mortality was considered stabilized when there were no differences between two consecutive days for all the tested doses.

LD_{50} calculation

The BMD approach consists in the selection of the model that describes the data using appropriate model structures for the type of data (dichotomous or quantal). A mathematical model is applied to the experimental data to produce the dose-response curve with the best fit. Details of the full process on this approach are presented in the BMD Software technical guidance from the US EPA (http://www.epa.gov/ncea/bmds/). For the dichotomous (or quantal) data, the response or effect may be reported as either the presence or absence of an effect. The dose-response models describe how the probability or frequency of a specified response changes with the dose level (50%). The different models were classified by Akaike's Information Criterion (AIC), and the model with the lowest AIC was chosen for the determination of LD_{50} for each active substance.

Results

Control of the active substance concentrations

The concentration of the working solutions of active substances and the 20 commercial pesticide formulations were controlled by chemical analysis (Table 2). The concentration variations of the formulations ranged between −25% (tebuconazole) and +38% (abamectin) of the concentration indicated by the manufacturer. This suggests that the technical process to adjust pesticide concentration is not accurately controlled by the manufacturer. The measured concentrations were used in all cases to determine the actual exposure levels of the bees after spraying.

Calibration of the deposit in the spraying tower

To check the reliability of the deposit, preliminary calibrations and experimental controls were performed (Figure 1). No differences were found between the deposits performed during the calibration procedure and the experimental controls (t-test; $t = 0.0739$, $df = 148$, p-value = 0.9412). The mean values of the deposit for calibrations and experimental controls were 2.32 ± 0.13 μL/cm^2 and 2.32 ± 0.11 μL/cm^2, respectively (Table 3). These volumes corresponded to a field spraying volume of 232 L/ha, which was similar to volumes commonly used in agriculture [22].

Determination of the exposure

Knowing the application rate of pesticides per surface area unit (n g/ha equal to $10 \times n$ ng/cm^2) and the determination of the doses received per bee enabled estimating the apparent exposure surface area of a bee. For each pesticide, the application rate was corrected by the real deposit (see Table 4: Deposit of the experimental control). The dose received by the bees (in ng/g of bee) was determined by chemical analysis of residues. To avoid the degradation of pesticides, *in vivo* biotransformation was stopped by quick freezing of the bees still anesthetized at −80°C immediately after spraying. The mean mass of the bees was determined to calculate the dose received per individual (in ng/bee) (Table 4). The exposure surface area of a bee (in cm^2/bee) is the ratio between the residues (in ng/bee) and the application rate (in ng/cm^2). Considering all the pesticides, the mean apparent exposure surface area was 1.05 ± 0.33 cm^2/bee ($n = 20$).

Spray toxicity vs. topical toxicity

To check the accuracy and reliability of the estimated mean exposure surface area of a bee for each formulation, the sprayed dose (in g/ha), which was associated with the corresponding mortality rate, was converted into an individual dose (in ng/bee) on the basis of an apparent surface area of 1.05 cm^2/bee. In a second step, this individual dose was topically applied on the thorax according to the EPPO procedure. Then, the mortality rates obtained by spraying and topical application were compared by linear regression analysis (Figure 2). A good linear correlation between mortalities elicited by spraying and topical exposures was found ($r^2 = 0.960$). It is noteworthy that the slope of the regression was very close to 1 ($y = 0.9792x + 0.6306$), which showed that the mortality induced by topical exposure was not over- or underestimated compared with the response obtained by a spray exposure. The experimental data of the dose-mortality response obtained by topical application are presented in Figures S1, S2, S3, S4, S5, S6, S7, S8, S9, S10, S11, S12, S13, S14 and S15.

Revision of the hazard quotient

The accuracy of the apparent exposure surface area previously estimated enabled calculating a new hazard quotient (HQ) based on the real exposure of the bees. Thus, the current HQ and revisited HQ were compared (Table 5). Currently, in bee toxicology, the HQ of a substance is defined as the ratio between the application rate (in g a.s./ha) and the LD_{50} expressed in μg a.s./bee [13]. The revisited HQ was based on the estimated exposure expressed in ng a.s./bee, calculated from the application rate and the exposure surface area per bee, and the LD_{50} was expressed in ng a.s./bee. This revisited HQ is a relative exposure level corresponding to the number of LD_{50} to which a bee is

Table 4. Determination of the exposure surface area per bee for each active substance.

Active substance	Deposit of the Experimental control (μL/cm²)		Corrected application rate[a] (g a.s./ha)	Residues per bee[b] (ng a.s./g)		Weight per bee (mg/bee)	Residues (ng a.s./bee)	Exposure surface area[c] (cm²/bee)
	Mean	S.D.		Mean	S.D.			
Abamectin	2.35	0.05	2.8	261.1	44	112	29.2	1.04
Acetamiprid	2.43	0.10	38.2	1922.7	152	117	225.0	0.59
Chlorpyrifos-ethyl	2.48	0.10	3.3	263.1	7.1	121	31.8	0.95
Clothianidin	2.24	0.04	2.6	137.5	9.5	125	17.2	0.65
Cyfluthrin	2.20	0.11	2.2	313.5	43.4	113	35.4	1.61
Cypermethrin	2.45	0.04	3.6	335.6	22.5	126	42.3	1.17
Deltamethrin	2.32	0.04	1.8	112.7	4.9	122	13.7	0.75
Dimethoate	2.26	0.14	5.3	451.6	54.4	120	54.2	1.03
Esfenvalerate	2.31	0.08	2.0	186.5	37.8	122	22.8	1.14
Fenoxycarb	2.21	0.20	65.3	4393	40	122	536.0	0.82
Hexythiazox	2.36	0.10	23.2	3143	697	122	383.5	1.66
Imidacloprid	2.33	0.03	0.2	20.2	2	122	2.5	1.07
Iprodione	2.36	0.09	645.1	32980	1058	122	4023.6	0.62
Lambda-cyhalothrin	2.32	0.07	4.5	430	46	122	52.5	1.17
Prochloraz	2.28	0.05	192.7	13990	828	122	1706.8	0.89
Pyriproxyfen	2.24	0.14	23.4	2307	412	122	281.4	1.20
Tau-fluvalinate	2.44	0.09	48.9	5857	483	122	714.5	1.46
Tebuconazole	2.33	0.04	56.1	7370	157	122	899.1	1.60
Thiacloprid	2.32	0.08	46.7	3523.3	417.1	118	415.7	0.89
Thiamethoxam	2.27	0.05	5.0	279.7	14.5	119	33.3	0.66

S.D.: Standard Deviation (n = 3).
g a.s./ha: gram of active substance per hectare.
ng a.s./ha: nanogram of active substance per hectare.
ng a.s./bee: nanogram of active substance per bee.
[a]The application rate was corrected with the measured concentration (Table 3) and the experimental control deposit.
[b]This corresponds to the concentration of active substances found in the bees frozen just after contamination.
[c]The exposure surface area is the ratio between the residues (ng a.s./bee) and the corrected application rate (ng a.s./cm²).
For each treatment, the bees from the 6 replicates were counted, pooled, and weighed; the mean bee weight was determined by dividing the weight of bees by the actual number of bees.

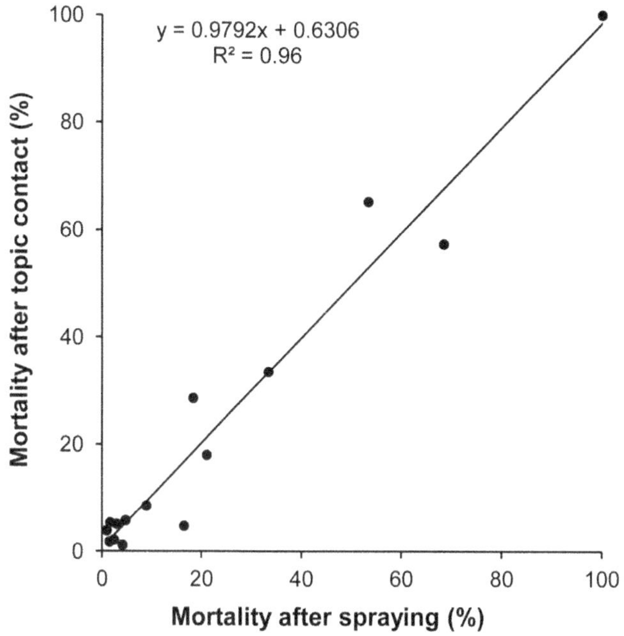

Figure 2. Correlation between mortality induced by spraying and mortality induced by topical treatment. Each dot represents one substance among the 15 tested on bees by thorax topical contact and spraying. The mortality elicited at 48 hours by spraying was given for the chosen field application rate (g a.s./ha) (Table 1). On the basis of the mean exposure surface area (1.05 cm^2/bee), the application rate (g a.s./ha) was converted into an individual dose (ng a.s./bee) to treat the bees topically. The dose-mortality relationship at 48 hours was modeled for each of the 15 active substances, which enabled assessing the mortality that could be induced at a given field exposure.

exposed. In addition, based on the dose-mortality relationship determined for each active substance, it was possible to assess the mortality induced at this exposure level (Figures S1, S2, S3, S4, S5, S6, S7, S8, S9, S10, S11, S12, S13, S14 and S15). For each active substance, two scenarios were presented: the most and the least protective that combined the highest LD_{50} and lowest field rate, for the most protective scenario, and the lowest LD_{50} and highest field rate, for the least protective one. The active substances were classified by their revisited HQ as a function of the different values of exposure calculated from the lower and the upper limits of the 95% confidence interval of the exposure surface area (Table S2). No differences in the classification of active substances were found when the lower and upper limits of the confidence interval were used to calculate the revisited HQ.

Determination of the physical surface area of a bee

To investigate the possibility of an overestimation of the exposure surface area of bees, the physical surface area of a bee was determined by X-Ray tomography. The surface areas of the body and the wings were measured separately. The tomographic analyses resulted in 3-D representation of the bees (Figure 3) that enabled calculating the volume (not relevant here) and the surface area of a bee. The exposure surface area was compared to the physical surface area of a bee (Figure 4). The mean body surface area was 2.21±0.20 cm^2 (n = 6), the mean wing surface area was 1.06±0.03 cm^2 (n = 50) and the mean total physical surface area of a bee was 3.27±0.23 cm^2. It was noteworthy that the wings represented almost one third of the total physical surface area of a bee and that the physical surface area was 3 times higher than the apparent exposure surface area.

Discussion

In this study, we used a pragmatic approach to determine the exposure level of a bee submitted to spraying by an agrochemical at a known field rate. This exposure level enabled assessing the exposure surface area of a honey bee that could be used in the assessment of the risk posed by plant protection products as recommended by the EFSA [11]. The mean apparent exposure surface area was estimated as 1.05±0.33 cm^2/bee (n = 20). Various attempts have been made to estimate the apparent exposure surface area of a bee. The first involves the determination of the apparent surface area by projection from photographs, which results in a surface area of approximately 0.5 cm^2/bee [23]. We used this method in a previous study to transform field rates into doses received by bees with a good correlation between mortality rates [24]. The second method is more pragmatic and has been performed in an original study by Koch and Weisser, in which caged bees were submitted to a spraying of fluorescein at a field rate of 20 g/ha [25]. With this procedure, exposure levels ranged from 1.62 to 8.51 ng/bee in apple orchards and from 6.34 to 35.77 ng/bee in phacelia. These values correspond to an apparent exposure surface area ranging from 0.0081 to 0.0425 cm^2/bee in apple orchards and 0.0317 to 0.1789 cm^2/bee in phacelia, which is very different from the surface area determined in the present study. Two phenomena can explain the great discrepancy between these exposure values and those presented here. (i) The fluorescein recovery was achieved by rinsing the bees, which does not enable the extraction of the absorbed fluorescein and leads to important underestimation of fluorescein residues. In our approach, pesticide residues were recovered by an efficient extraction procedure that presents high recovery rates [26]. (ii) In our study, bees were immediately frozen after spraying, which stops the biotransformation of the pesticides. However, in the study of Koch and Weisser, the time elapsed between field exposure and sampling was much higher and close to 20 min. In addition, during the sampling the bees were cooled with dry ice and subsequently frozen in the laboratory. Thus, fluorescein was subjected to biotransformation during the sampling procedure, which can be very rapid; this has been previously observed with pesticides for which *in vivo* half-lives as low as 25 min have been observed [27]. It is noteworthy that the effect of time on fluorescein recovery was described by Koch and Weisser.

For a better assessment, the apparent exposure surface area was estimated with 20 pesticides from several chemical groups. Slight variations in the surface area have been observed from pesticide to pesticide. However, the concentrations of the pesticide solutions were checked analytically, the recovery rates of pesticide extraction were very high and we made sure that the deposit was very constant over time (Table 4). Thus, variations of the apparent exposure surface area were not due to variations in the quantity of the deposited pesticide solution but due to the quantities of the deposited active substance, which depends on their physico-chemical properties and on the components of the pesticide formulations [28].

To test the accuracy of the estimated exposure surface area, the toxicity elicited by sprayings at given field rates was compared with the toxicity induced by the corresponding doses calculated on the basis of the field rates and the exposure surface area. The linear regression analysis showed not only a very good correlation (r^2 = 0.960) between mortality induced by field doses and that

Table 5. Determination and comparison of the HQ and the revisited HQ.

Active substance	Time[a] (hours)	Application rate[b] (g a.s./ha)		LD_{50}[c] (μg a.s./bee)		HQ[d]		Exposure[e] (ng a.s./bee)		LD_{50}[f] (ng a.s./bee)		New HQ[g] (revisited)		Expected mortality[h]	LD_{50} References
Abamectin	24	Min	4.5	Max	0.002	Min	2045	Min	47	Max	2.2	Min	21.48	100%	[34]
	72	Max	22.5	Min	0.001	Max	15709	Max	236	Min	1.43	Max	164.94	100%	E.D. Figure S1
Acetamiprid	72	Min	5	Max	8.090	Min	0.6	Min	53	Max	8090	Min	0.006	2.2%	PED US EPA
	24	Max	100	Min	5.057	Max	20	Max	1050	Min	5057	Max	0.21	10.1%	E.D. Figure S2
Chlorpyrifos-ethyl	96	Min	187.5	Max	0.114	Min	1645	Min	1969	Max	114	Min	17.27	100%	PED US EPA
	96			E.D.	0.060					E.D.	59.52				E.D. Figure S3
	18	Max	1000	Min	0.010	Max	100000	Max	10500	Min	10	Max	1050	100%	PED US EPA
Clothianidin	48	Min	75	Max	0.044	Min	1695	Min	788	Max	44.26	Min	17.79	100%	Agritox Database
	24			E.D.	0.040					E.D.	40.23				E.D. Figure S4
	24	Max	75	Min	0.022	Max	3440	Max	788	Min	21.8	Max	36.12	100%	[35]
Cyfluthrin	48	Min	10	Max	0.037	Min	270	Min	105	Max	37	Min	2.84	99.1%	PED US EPA
	48			E.D.	0.020					E.D.	20.45				E.D. Figure S5
	48	Max	40	Min	0.001	Max	40000	Max	420	Min	1	Max	420	100%	[36]
Cypermethrin	24	Min	15	Max	0.121	Min	124	Min	158	Max	121.25	Min	1.30	71%	E.D. Figure S6
	24	Max	96	Min	0.020	Max	4800	Max	1008	Min	20	Max	50	99.9%	Agritox Database
Deltamethrin	24	Min	4.95	Max	0.108	Min	46	Min	52	Max	107.64	Min	0.48	7.4%	E.D. Figure S7
	48	Max	19.95	Min	0.002	Max	13300	Max	209	Min	1.5	Max	139.65	91.5%	PED US EPA
Dimethoate	24	Min	240	Max	0.454	Min	529	Min	2520	Max	454	Min	5.55	100%	[37]
	96			E.D.	0.163					E.D.	163.33				E.D. Figure S8
	48	Max	500	Min	0.001	Max	357143	Max	5250	Min	1.4	Max	3750	100%	[38]
Esfenvalerate	48	Min	1.5	Max	0.060	Min	25	Min	16	Max	60	Min	0.26	13.0%	[39]
	96			E.D.	0.047					E.D.	47.03				E.D. Figure S9
	48	Max	20	Min	0.017	Max	1163	Max	210	Min	17.2	Max	12.21	98.1%	PED US EPA
Fenoxycarb	48	Min	75	Max	>204	Min	<0.4	Min	788	Max	>204000	Min	<0.004	N.C.	[36]
	48	Max	150	Min	>100	Max	<2	Max	1575	Min	>100000	Max	<0.016	N.C.	[40]
Hexythiazox	48	Min	25	Max	>200	Min	<0.1	Min	263	Max	>200000	Min	<0.001	N.C.	[41]
	48	Max	50	Min	>200	Max	<0.3	Max	525	Min	>200000	Max	<0.003	N.C.	[41]
Imidacloprid	48	Min	50	Max	0.230	Min	217	Min	525	Max	230.3	Min	2.28	91.9%	[42]
	96			E.D.	0.096					E.D.	96.21				E.D. Figure S10
	48	Max	100	Min	0.007	Max	14925	Max	1050	Min	6.7	Max	156.72	96.8%	[43]
Iprodione	48	Min	75	Max	>200	Min	<0.4	Min	788	Max	>200000	Min	<0.004	N.C.	Agritox Database
	48	Max	1000	Min	>25	Max	<40	Max	10500	Min	>25000	Max	<0.42	N.C.	[36]
Lambda-cyhalothrin	48	Min	5	Max	0.098	Min	51	Min	53	Max	98	Min	0.54	23.2%	PED US EPA
	48			E.D.	0.078					E.D.	77.85				E.D. Figure S11

Table 5. Cont.

Active substance	Time[a] (hours)	Application rate[b] (g a.s./ha)		LD_{50}[c] (µg a.s./bee)		HQ[d]		Exposure[e] (ng a.s./bee)		LD_{50}[f] (ng a.s./bee)		New HQ[g] (revisited)		Expected mortality[h]	LD_{50} References
	48	Max	75	Min	0.038	Max	1974	Max	788	Min	38	Max	20.72	99.9%	PED US EPA
Prochloraz	96	Min	230.5	Max	141.28	Min	2	Min	2420	Max	141280	Min	0.02	40.7%	[44]
	24	Max	598.5	Min	2.657	Max	225	Max	6284	Min	2657	Max	2.37	97.5%	E.D. Figure S12
Pyriproxyfen	48	Min	25	Max	>100	Min	<0.3	Min	263	Max	>100000	Min	<0.003	N.C.	[45]
	48	Max	30	Min	74	Max	0.4	Max	315	Min	74000	Max	0.004	N.C.	PPDB IUPAC
Tau-fluvalinate	24	Min	48	Max	13	Min	4	Min	504	Max	12000	Min	0.04	2.1%	[46]
	48	Max	144	Min	2.532	Max	57	Max	1512	Min	2532	Max	0.60	17.4%	E.D. Figure S13
Tebuconazole	48	Min	75	Max	>200	Min	<0.4	Min	788	Max	>200000	Min	<0.004	N.C.	[47]
	48	Max	258	Min	>83	Max	<3.1	Max	2709	Min	>83050	Max	<0.033	N.C.	[36]
Thiacloprid	48	Min	72	Max	38.820	Min	2	Min	756	Max	38820	Min	0.02	4.1%	Agritox Database
	24			E.D.	19.380					E.D.	19380				E.D. Figure S14
	24	Max	180	Min	14.600	Max	12	Max	1890	Min	14600	Max	0.13	9.2%	[35]
Thiamethoxam	48	Min	50	Max	0.040	Min	1251	Min	525	Max	39.96	Min	13.14	100%	E.D. Figure S15
	24	Max	100	Min	0.024	Max	4167	Max	1050	Min	24	Max	44	100%	Agritox Database

g a.s./ha: mass of active substance per hectare expressed in grams.
µg a.s./bee: mass of active substance per bee expressed in micrograms.
ng a.s./bee: nanogram of active substance per bee.
E.D.: Experimental Data.
LD_{50}: Median Lethal Dose.
HQ: Hazard Quotient (field rate (g/ha)/LD_{50} (µg/bee)).
Revisited HQ (exposure (ng/bee)/LD_{50} (µg/bee)).
N.C.: Not Calculated because of the low toxicity of the active substance.
DAR EFSA: Draft Assessment Report of the European Food Safety Authority.
PED US EPA: Pesticides Ecotoxicity Database of the United States Environmental Protection Agency.
[a]Time at which the LD_{50} was determined. The LD_{50} values resulting from the experimental data were calculated at the time corresponding to a stabilized mortality.
[b]For each active substance, 2 scenarios of exposure are presented: the lowest and the highest homologated application rate.
[c]For each active substance, the highest and the lowest known LD_{50} values were compared to the lowest and highest homologated application rates, respectively.
[d]HQ is the ratio between the application rate (g a.s./ha) and the LD_{50} (µg a.s./bee).
[e]The exposure was calculated from the application rate (ng a.s./cm^2) and the mean exposure surface area determined with the 20 active substances (1.05 cm^2/bee).
[f]The LD_{50} values from the experimental data were calculated with the BMD software from the US EPA.
[g]The revisited HQ is the ratio between the exposure (ng a.s./bee) and the LD_{50} (ng a.s./bee).
[h]For each active substance, the dose-mortality relationship was modeled at the time corresponding to a stabilized mortality.

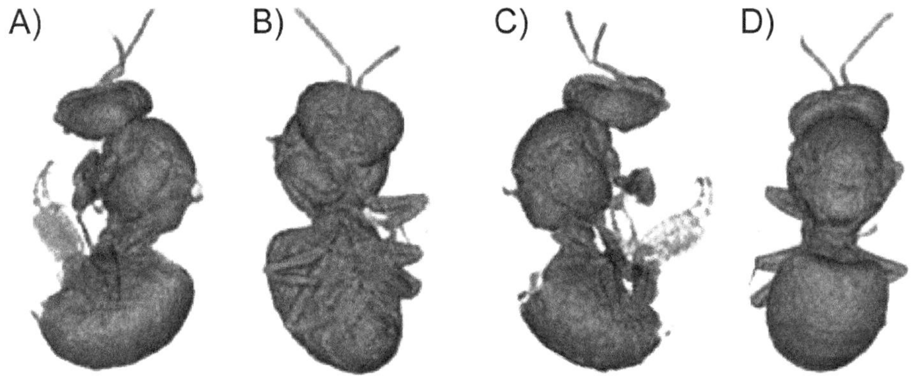

Figure 3. 3D representation of a bee from different angles. Bees were scanned using X-ray tomography. A: Left lateral view; B: Ventral view; C: Right lateral view; D: Dorsal view.

induced by doses calculated with the apparent exposure surface area but also a slope of 0.9792, which is very close to 1. A slope higher or lower than 1 would have revealed an overestimation or underestimation of the apparent exposure surface area, respectively. Thus, a slope of approximately 1 clearly shows that the toxicity induced by spraying at a given field rate can be predicted by the toxicity induced by contact treatment according to the EPPO procedure with the corresponding dose calculated on the basis of the field rate and the apparent exposure surface area of a bee. In other words, this shows that the apparent exposure surface area of a bee can be used in procedures to assess the risk presented by pesticides to bees. Hence, it is possible to derive two formulas that could be used in risk assessment. The first enables calculating the individual dose received by the bees (*ID*) during the spraying of a pesticide at a given field rate:

$$ID = FR \times 10 \times S$$

where *ID* is the individual dose expressed in ng/bee; *FR* is the field rate of the pesticide expressed in g/ha and *S* is the apparent exposure surface area of 1.05 cm^2/bee. The factor 10 results from the conversion of *FR* in g/ha into *FR* in ng/cm^2. It is noteworthy that the standard deviation (0.33 cm^2/bee) or the upper limit of the confidence interval (1.21 cm^2/bee) offers the possibility to consider worse case scenarios of exposure during spraying. The second formula can be used to estimate the field rate that should not be exceeded when a reference value, such as LD_{50} or acute reference dose (ARfD), is available to characterize the toxicity of a pesticide:

$$LFR = \frac{RV}{10 \times S}$$

where *LFR* is the limit field rate that should not be exceeded expressed in g/ha, *RV* is the reference value expressed in ng/bee and *S* is the apparent exposure surface area of 1.05 cm^2/bee.

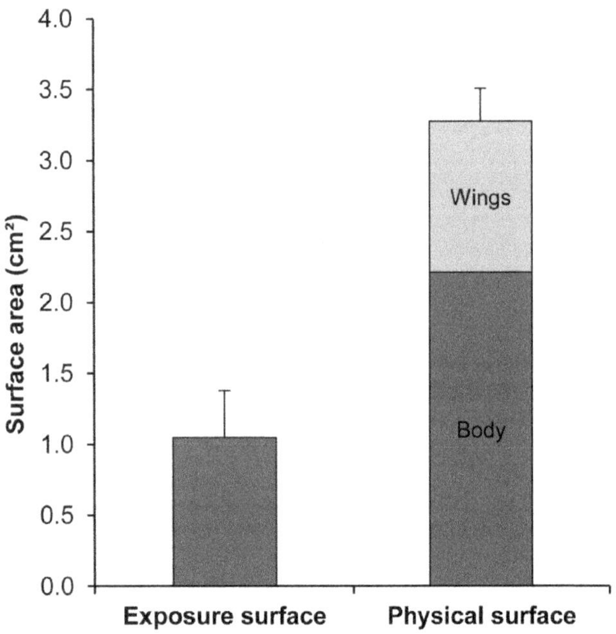

Figure 4. Comparison of the exposure and physical surface area of a bee. Mean exposure surface area (± S.D.) of a bee (average of the 20 commercial products) and mean physical surface area (± S.D.) of a bee (sum of the mean body and wing surface area).

In this study, we propose a new approach for the determination of the hazard quotient (HQ) in the procedure used for the assessment of the risk presented by pesticides to the honey bee. The HQ is no longer estimated from the field application rates of pesticides (in g/ha) but instead from the exposure to which the bees are subjected (in ng/bee). Thus, the resulting HQ is a value without units because it corresponds to the ratio of coherent exposure and toxicity data. The environmental exposure level to which bees are subjected is expressed in ng/bee. The toxicity, corresponding to the exposure level from which effects may be considered damaging to the bees (or to a bee colony), is also expressed in ng/bee. This approach is used for all living organisms in risk assessment procedures, including humans [3,29]. Until now, the honey bee remained the only organism for which the HQ value was interpreted empirically to manage the risk to bees because it was not based on the ratio of comparable data. This point has also been raised by one agency of the European Union, the DG SANCO [30]. The determination of the exposure surface area of a bee also enables converting previous HQ values into new HQ values (revisited HQ) (*new* $HQ = 1.05 \times 10^{-2} \times HQ$), which in turn enables using the old toxicological data. Moreover, the apparent exposure surface area enables the assessment of the survival probability of bees sprayed by a pesticide at a given field rate on the basis of the dose-mortality relationship. Thus, the combination of the new HQ with the expected mortality and exposure could be useful tools for decision-making processes in the assessment of pesticide toxicity to the honey bee. However, although the revisited HQ gives a better interpretation of the potential risk presented by a pesticide than the previous HQ, the exposure is still compared to a toxicological value that elicits a mortality rate of 50% in the honey bee risk assessment procedure. For humans and some target organisms (i.e., fish), and for specific toxicity, such as reproductive toxicity, for birds and mammals [31], the risk is based on toxic reference values at which a low mortality or no effect is observed (e.g., NOAEL, NOAEC, ARfD, or Benchmark dose). Consequently, it appears legitimate to reconsider the LD_{50} value as a reference value to assess the risk presented by pesticides to bees.

The apparent exposure surface area of 1.05 cm^2/bee does not appear overestimated compared with the physical surface area of 3.27 cm^2/bee. The physical surface area measured in the present study appears to be higher than that determined in previous studies using graph paper [32,33]. Although X-Ray tomography is not yet sufficiently powerful to take into account the surface hairs on the honeybee thorax, this technique remains more precise than the others described previously. The difference between exposure and physical surface areas can be explained by the fact that the entire surface was not available during the spraying. Indeed, the bees were motionless with their wings folded on the body during the spraying. Then, the exposure surface area does not depend on the total surface area (body + wings = 3.27 cm^2), but is rather a reflection of the body surface area (2.21 cm^2). In this study, the bees were subjected to pesticides on one side, like during a field spraying. Thus, only one half of the body was directly exposed (c.a. 1.11 cm^2), which is close to the apparent exposure surface area (1.05 cm^2).

Supporting Information

Figure S1 Dose-mortality relationship of honey bees after a single contact contamination of abamectin on the thorax.

Figure S2 Dose-mortality relationship of honey bees after a single contact contamination of acetamiprid on the thorax.

Figure S3 Dose-mortality relationship of honey bees after a single contact contamination of chlorpyrifos-ethyl on the thorax.

Figure S4 Dose-mortality relationship of honey bees after a single contact contamination of clothianidin on the thorax.

Figure S5 Dose-mortality relationship of honey bees after a single contact contamination of cyfluthrin on the thorax.

Figure S6 Dose-mortality relationship of honey bees after a single contact contamination of cypermethrin on the thorax.

Figure S7 Dose-mortality relationship of honey bees after a single contact contamination of deltamethrin on the thorax.

Figure S8 Dose-mortality relationship of honey bees after a single contact contamination of dimethoate on the thorax.

Figure S9 Dose-mortality relationship of honey bees after a single contact contamination of esfenvalerate on the thorax.

Figure S10 Dose-mortality relationship of honey bees after a single contact contamination of imidacloprid on the thorax.

Figure S11 Dose-mortality relationship of honey bees after a single contact contamination of lambda-cyhalothrin on the thorax.

Figure S12 Dose-mortality relationship of honey bees after a single contact contamination of prochloraz on the thorax.

Figure S13 Dose-mortality relationship of honey bees after a single contact contamination of tau-fluvalinate on the thorax.

Figure S14 Dose-mortality relationship of honey bees after a single contact contamination of thiacloprid on the thorax.

Figure S15 Dose-mortality relationship of honey bees after a single contact contamination of thiamethoxam on the thorax.

Table S1 Active substances and their doses used for exposures through thoracic topical exposure. ng a.s./bee: nanogram of active substance per bee. Nr: Number of replicates per dose. ABA = Abamectin, ACE = Acetamiprid, CHL = Chlorpyrifos-ethyl, CLO = Clothianidin, CYF = Cyfluthrin, CYP = Cypermethrin, DEL = Deltamethrin, DIM = Dimethoate, ESF = Esfenvalerate, IMI = Imidacloprid, LAM = Lambda-cyhalothin, PRO = Prochoraz, TAU = Tau-fluvalinate, THI = Thiacloprid, TMX = Thiamethoxam.

Table S2 Classification of the 20 active substances by their revisited HQs for different values of exposure surface areas. An active substance was classified with a HQ≥1 if at least the revisited HQ of one of the two scenarios (Table 5) were equal to or higher than this value. Lower C.I.: Lower limit of the Confidence Interval at 95%. Upper C.I.: Upper limit of the Confidence Interval at 95%.

Acknowledgments

The authors wish to thank Marie Dupré, Jean-Baptiste Philibert and Jacques Sénéchal for their technical help and André Kretzschmar for useful advice in the data analysis.

Author Contributions

Conceived and designed the experiments: LPB YP. Performed the experiments: YP ST MC AB FL BG MF. Analyzed the data: YP LB

MT JLB. Contributed reagents/materials/analysis tools: MP. Contributed to the writing of the manuscript: LPB YP.

References

1. United Nations Environment Programme (UNEP) (2003) Regionally based assessment of persistent toxic substances. Glob Rep: 1–207.
2. United States Environmental Protection Agency (US EPA) (2011) Pesticides Industry Sales and Usage 2006 and 2007 Market Estimates: 1–33.
3. European Chemicals Bureau (2003) Technical Guidance Document (TGD) on Risk Assessment Part I: 1–302.
4. European Chemicals Bureau (2003) Technical Guidance Document (TGD) on Risk Assessment Part II: 1–328.
5. Barnes DG, Daston GP, Evans JS, Jarabek AM, Kavlock RJ, et al. (1995) Benchmark Dose Workshop: Criteria for use of benchmark dose to estimate a reference dose. Regul Toxicol Pharmacol 21: 296–306.
6. EFSA (2009) Guidance of the Scientific Commitee on a request from EFSA on the use of the benchmark dose approach in risk assessment. EFSA J 1150: 1–72.
7. Société Française de Phytiatrie et de Phytopharmacie (SFPP) (1982) Méthode n°95, Méthodes d'essais destinées à connaître les effets des insecticides sur l'abeille domestique (*Apis mellifera* L.). Comm des Essais Biol: 19.
8. Hatch GE, Kodavanti U, Crissman K, Slade R, Costa D (2001) An "injury-time integral" model for extrapolating from acute to chronic effects of phosgene. Toxicol Ind Health 17: 285–293.
9. Suchail S, Guez D, Belzunces LP (2001) Discrepancy between acute and chronic toxicity induced by imidacloprid and its metabolites in Apis mellifera. Environ Toxicol Chem 20: 2482–2486.
10. Commission des Essais Biologiques (CEB) (2010) Méthode d'évaluation des effets de toxicité aiguë et à court terme des préparations phytopharmaceutiques sur l'abeille domestique (*Apis mellifera* L.). AFPP method n° 230.
11. European Food Safety Authority (2013) Guidance on the risk assessment of plant protection products on bees (*Apis mellifera*, *Bombus spp.* and solitary bees). EFSA J 11: 1–266.
12. Fischer D, Moriarty T (2011) Pesticide Risk Assessment for Pollinators: Summary of a SETAC Pellston Workshop. Pensacola FL (USA). Soc Environ Toxicol ans Chem: 1–43.
13. OEPP/EPPO (2003) EPPO Standards Environmental risk assessment scheme for plant protection products. Bull OEPP/EPPO 33: 141–145.
14. Atkins EL, Kellum D, Atkins KW (1981) Reducing pesticide hazards to honey bees: Mortality prediction techniques and integrated management strategies. Div Agric Sci Univ Calif: 1–20.
15. OEPP/EPPO (2001) Efficacy evaluation of plant protection products, Side-effects on honeybees. Bull OEPP/EPPO 170: 95–99.
16. OECD (1998) OECD guidelines for the testing of chemicals Number 214, Honeybees acute contact toxicity test: 1–7.
17. Schmolz E, Hoffmeister D, Lamprecht I (2002) Calorimetric investigations on metabolic rates and thermoregulation of sleeping honeybees (*Apis mellifera carnica*). Thermochim Acta 382: 221–227.
18. Grodzicki P, Caputa M (2005) Social versus individual behaviour: a comparative approach to thermal behaviour of the honeybee (*Apis mellifera* L.) and the American cockroach (*Periplaneta americana* L.). J Insect Physiol 51: 315–322.
19. Colin ME, Belzunces LP (1992) Evidence of synergy between prochloraz and deltametrin in *Apis mellifera* L.: A convenient biological approach. Pestic Sci 36: 115–119.
20. European Commission (EC) (2009) Regulation (EC) No 1107/2009 of the european parliament and of the council of 21 october 2009 concerning the placing of protection products on the market and repealing Council Directives 79/117/EEC and 91/414/EEC. Off J Eur Union L 309: 1–50.
21. European Economic Community (EEC) (1991) Council Directive of 15 July 1991 concerning the placing of plant protection products on the market (91/414/EEC). Off J Eur Union L 230: 1–290.
22. Wise JC, Jenkins PE, Schilder AMC, Vandervoort C, Isaacs R (2010) Sprayer type and water volume influence pesticide deposition and control of insect pests and diseases in juice grapes. Crop Prot 29: 378–385.
23. Canton JH, Linders JBHJ, Luttik R, Mensink BJWG, Panman E, et al. (1991) Catch-up operation on old pesticides: an integration. RIVM report 678801002. Natl Inst Public Heal Environ (RIVM), BA Bilthoven, Netherlands: 1–146.
24. Vandame R, Meled M, Colin ME, Belzunces LP (1995) Alteration of the homing-flight in the honey bee *Apis mellifera* L. exposed to sublethal dose of deltamethrin. Environ Toxicol Chem 14: 855–860.
25. Koch H, Weisser P (1997) Exposure of honey bees during pesticide application under field conditions. Apidologie 28: 439–447.
26. Wiest L, Buleté A, Giroud B, Fratta C, Amic S, et al. (2011) Multi-residue analysis of 80 environmental contaminants in honeys, honeybees and pollens by one extraction procedure followed by liquid and gas chromatography coupled with mass spectrometric detection. J Chromatogr A 1218: 5743–5756.
27. Brunet J-L, Badiou A, Belzunces LP (2005) In vivo metabolic fate of [14C]-acetamiprid in six biological compartments of the honeybee, *Apis mellifera* L. Pest Manag Sci 61: 742–748.
28. Stevens PJG, Baker EA, Anderson NH (1988) Factors affecting the foliar absorption and redistribution of pesticides. 2. Physicochemical properties of the active ingredient and the role of surfactant. Pestic Sci 24: 31–53.
29. European Chemicals Agency (ECHA) (2013) Guidance for human health risk assessment Volume III, Part B. III: 1–389.
30. SANCO (2002) Guidance document on terrestrial ecotoxicology under council directive 91 /414 /EEC (SANCO/10329/2002 rev 2 final): 1–39.
31. European Food Safety Authority (2009) Guidance document on risk assessment for bird & mammals on request from EFSA. EFSA J 7: 1–139.
32. Johansen CA, Mayer DF, Eves JD, Kious CW (1983) Pesticides and bees. Environ Entomol 12: 1513–1518.
33. Cooper PD, Schaffer WM, Buchmann SL, Cooper BYPD (1985) Temperature regulation of honey bees (Apis mellifera) foraging in the sonoran desert. J Exp Biol 114: 1–15.
34. Draft Assessment Report (DAR) on the active substance Abamectin prepared by the rapporteur Member State The Netherlands in the framework of the Council Directive 91/414/EEC (2006).
35. Iwasa T, Motoyama N, Ambrose JT, Roe RM (2004) Mechanism for the differential toxicity of neonicotinoid insecticides in the honey bee, Apis mellifera. Crop Prot 23: 371–378.
36. University of Hertfordshire (2013) The Pesticide Properties DataBase (PPDB) developed by the Agriculture & Environment Research Unit (AERU). Univ Hertfordsh 2006–2013.
37. Draft Assessment Report (DAR) on the active substance Quizalofop-P-tefuryl prepared by the rapporteur Member State Finland in the framework of the Council Directive 91/414/EEC (2007).
38. Torchio PF (1973) Relative toxicity of insecticides to the honey bee, alkali bee, and alfalfa leafcutting bee (Hymenoptera: Apidae, Halictidae, Megachilidae). J Kansas Entomol Soc 46: 446–453.
39. United Kingdom (1995) Draft Assessment Report (DAR) on the active substance Esfenvalerate prepared by the rapporteur Member State United Kingdom in the framework of the Council Directive 91/414/EEC. 3.
40. Draft Assessment Report (DAR) on the active substance Fenoxycarb prepared by the rapporteur Member State The Netherlands in the framework of the Council Directive 91/414/EEC (2007).
41. Draft Assessment Report (DAR) on the active substance Hexythiazox prepared by the rapporteur Member State Finland in the framework of the Council Directive 91/414/EEC (2006).
42. Schmuck R, Schöning R, Stork A, Schramel O (2001) Risk posed to honeybees (*Apis mellifera* L, Hymenoptera) by an imidacloprid seed dressing of sunflowers. Pest Manag Sci 57: 225–238.
43. Suchail S, Guez D, Belzunces LP (2000) Characteristics of imidacloprid toxicity in two *Apis mellifera* subspecies. Environ Toxicol Chem 19: 1901–1905.
44. Draft Assessment Report (DAR) on the active substance Prochloraz prepared by the rapporteur Member State Ireland in the framework of the Council Directive 91/414/EEC (2007).
45. Draft Assessment Report (DAR) on the active substance Pyriproxyfen prepared by the rapporteur Member State The Netherlands in the framework of the Council Directive 91/414/EEC (2005).
46. Draft Assessment Report (DAR) on the active substance Tau-fluvalinate prepared by the rapporteur Member State Denmark in the framework of the Council Directive 91/414/EEC (2007).
47. Draft Assessment Report (DAR) on the active substance Tebuconazole prepared by the rapporteur Member State Denmark in the framework of the Council Directive 91/414/EEC (2007).

Permissions

All chapters in this book were first published in PLOS ONE, by The Public Library of Science; hereby published with permission under the Creative Commons Attribution License or equivalent. Every chapter published in this book has been scrutinized by our experts. Their significance has been extensively debated. The topics covered herein carry significant findings which will fuel the growth of the discipline. They may even be implemented as practical applications or may be referred to as a beginning point for another development.

The contributors of this book come from diverse backgrounds, making this book a truly international effort. This book will bring forth new frontiers with its revolutionizing research information and detailed analysis of the nascent developments around the world.

We would like to thank all the contributing authors for lending their expertise to make the book truly unique. They have played a crucial role in the development of this book. Without their invaluable contributions this book wouldn't have been possible. They have made vital efforts to compile up to date information on the varied aspects of this subject to make this book a valuable addition to the collection of many professionals and students.

This book was conceptualized with the vision of imparting up-to-date information and advanced data in this field. To ensure the same, a matchless editorial board was set up. Every individual on the board went through rigorous rounds of assessment to prove their worth. After which they invested a large part of their time researching and compiling the most relevant data for our readers.

The editorial board has been involved in producing this book since its inception. They have spent rigorous hours researching and exploring the diverse topics which have resulted in the successful publishing of this book. They have passed on their knowledge of decades through this book. To expedite this challenging task, the publisher supported the team at every step. A small team of assistant editors was also appointed to further simplify the editing procedure and attain best results for the readers.

Apart from the editorial board, the designing team has also invested a significant amount of their time in understanding the subject and creating the most relevant covers. They scrutinized every image to scout for the most suitable representation of the subject and create an appropriate cover for the book.

The publishing team has been an ardent support to the editorial, designing and production team. Their endless efforts to recruit the best for this project, has resulted in the accomplishment of this book. They are a veteran in the field of academics and their pool of knowledge is as vast as their experience in printing. Their expertise and guidance has proved useful at every step. Their uncompromising quality standards have made this book an exceptional effort. Their encouragement from time to time has been an inspiration for everyone.

The publisher and the editorial board hope that this book will prove to be a valuable piece of knowledge for researchers, students, practitioners and scholars across the globe.

List of Contributors

Nongkran Lumjuan, Jureeporn Wicheer and Posri Leelapat
Research Institute for Health Sciences, Chiang Mai University, Chiang Mai, Thailand

Wej Choochote and Pradya Somboon
Department of Parasitology, Faculty of Medicine, Chiang Mai University, Chiang Mai, Thailand

Ina Ehlers, Tatiana R. Betson and Jürgen Schleucher
Department of Medical Biochemistry & Biophysics, Umeå University, Umeå, Sweden

Walter Vetter
Department of Food Chemistry, University of Hohenheim, Stuttgart, Germany

Xiaona Zhang, Hua Tian, Wei Wang and Shaoguo Ru
College of Marine Life Sciences, Ocean University of China, Qingdao, China

Pengfei Leng, Zhiming Zhang and Guangtang Pan
Maize Research Institute, Sichuan Agricultural University, Ya'an, PR China

Qian Li
Chongqing College of Humanities, Science and Technology, School of Foreign Languages,Hechuan District, Chongqing, PR China

Maojun Zhao
College of Life and Basic Sciences, Sichuan Agricultural University, Sichuan Ya'an, PR China

Gregory M. Williams
Hudson Regional Health Commission, Secaucus, New Jersey, United States of America
Center for Vector Biology, Rutgers University, New Brunswick, New Jersey, United States of America

Ary Faraji
Center for Vector Biology, Rutgers University, New Brunswick, New Jersey, United States of America
Salt Lake City Mosquito Abatement District, Salt Lake City, Utah, United States of America

Isik Unlu
Center for Vector Biology, Rutgers University, New Brunswick, New Jersey, United States of America
Mercer County Mosquito Control, West Trenton, New Jersey, United States of America

Sean P. Healy
Center for Vector Biology, Rutgers University, New Brunswick, New Jersey, United States of America
Department of Pathobiological Sciences, Louisiana State University School of Veterinary Medicine, Baton Rouge, Louisiana, United States of America

Muhammad Farooq
Navy Entomology Center of Excellence, Jacksonville, Florida, United States of America

Randy Gaugler, George Hamilton and Dina M. Fonseca
Center for Vector Biology, Rutgers University, New Brunswick, New Jersey, United States of America

Yesim Tozan
Institute of Public Health, Ruprecht-Karls-University, Heidelberg, Germany
Steinhardt School of Culture, Education and Human Development, New York University, New York, New York, United States of America

Pitcha Ratanawong and Valérie R. Louis
Institute of Public Health, Ruprecht-Karls-University, Heidelberg, Germany

Pattamaporn Kittayapong
Center of Excellence for Vectors and Vector-Borne Diseases, Faculty of Science, Mahidol University at Salaya, Nakhon Phatom, Thailand

Annelies Wilder-Smith
Institute of Public Health, Ruprecht-Karls-University, Heidelberg, Germany
Lee Kong Chian School of Medicine, Nanyang Technological University, Singapore, Singapore
University of Umea, Umea, Sweden

Matthew F. Rouhier and Peter M. Piermarini
Department of Entomology, Ohio Agricultural Research and Development Center, The Ohio State University, Wooster, OH, United States of America

Rene Raphemot
Department of Anesthesiology, Vanderbilt University Medical Center, Nashville, TN, United States of America
Department of Pharmacology, Vanderbilt University School of Medicine, Nashville, TN, United States of America

Daniel R. Swale
Department of Anesthesiology, Vanderbilt University Medical Center, Nashville, TN, United States of America

C. David Weaver
Department of Pharmacology, Vanderbilt University School of Medicine, Nashville, TN, United States of America
Institute of Chemical Biology, Vanderbilt University School of Medicine, Nashville, TN, United States of America

Kimberly M. Lovell, Leah C. Konkel, Darren W. Engers and Sean F. Bollinger
Department of Pharmacology, Vanderbilt University School of Medicine, Nashville, TN, United States of America
Department of Chemistry, Vanderbilt University School of Medicine, Nashville TN, United States of America

Corey Hopkins
Department of Pharmacology, Vanderbilt University School of Medicine, Nashville, TN, United States of America
Institute for Global Health, Vanderbilt University, Nashville, TN, United States of America Department of Chemistry, Vanderbilt University School of Medicine, Nashville TN, United States of America

Jerod S. Denton
Department of Anesthesiology, Vanderbilt University Medical Center, Nashville, TN, United States of America
Department of Pharmacology, Vanderbilt University School of Medicine, Nashville, TN, United States of America
Institute of Chemical Biology, Vanderbilt University School of Medicine, Nashville, TN, United States of America
Institute for Global Health, Vanderbilt University, Nashville, TN, United States of America

Emily Days
Institute of Chemical Biology, Vanderbilt University School of Medicine, Nashville, TN, United States of America

Enoch M. Kikulwe, Elisabeth Fischer and Matin Qaim
Department of Agricultural Economics and Rural Development, Georg-August-University of Goettingen, Goettingen, Germany

Jie Jin, Mengxia Yu, Chao Hu, Li Ye, Lili Xie, Jin Jin and Feifei Chen
Department of Hematology, the First Affiliated Hospital of Zhejiang University, Hangzhou, People's Republic of China
Institute of Hematology, Zhejiang University School of Medicine, Hangzhou, People's Republic of China

Hongyan Tong
Department of Hematology, the First Affiliated Hospital of Zhejiang University, Hangzhou, People's Republic of China
Institute of Hematology, Zhejiang University School of Medicine, Hangzhou, People's Republic of China
Myelodysplastic syndromes diagnosis and therapy center, Zhejiang University School of Medicine, Hangzhou, People's Republic of China

Daniel Elias and Melody J. Bernot
Ball State University, Department of Biology, Muncie, Indiana, United States of America

Noa Simon-Delso
Beekeeping Research and Information Centre, Louvain la Neuve, Belgium
Environmental Sciences, Copernicus Institute, Utrecht University, Utrecht, The Netherlands

Gilles San Martin and Louis Hautier
Plant Protection and Ecotoxicology Unit, Life Sciences Department, Walloon Agricultural Research Centre, Gembloux, Belgium

Etienne Bruneau, Laure-Anne Minsart and Coralie Mouret
Beekeeping Research and Information Centre, Louvain la Neuve, Belgium
Environmental Sciences, Copernicus Institute, Utrecht University, Utrecht, The Netherlands

Deodatus Maliti
Ifakara Health Institute, Environmental Health and Ecological Sciences Thematic Group, Ifakara, Morogoro, United Republic of Tanzania
University of Glasgow, Institute of Biodiversity Animal Health and Comparative Medicine, Glasgow, Lancashire, United Kingdom

Hilary Ranson and David Weetman
Department of Vector Biology, Liverpool School of Tropical Medicine, Merseyside, Liverpool, United Kingdom

Stephen Magesa
RTI International, Global Health Division, Dar es Salaam, United Republic of Tanzania

William Kisinza
National Institute for Medical Research, Amani Research Center, Muheza, Tanga, United Republic of Tanzania

Juma Mcha and Khamis Haji
Zanzibar Malaria Elimination Programme, Unguja, Zanzibar, United Republic of Tanzania

Gerald Killeen
Ifakara Health Institute, Environmental Health and Ecological Sciences Thematic Group, Ifakara, Morogoro, United Republic of Tanzania
University of Glasgow, Institute Department of Vector Biology, Liverpool School of Tropical Medicine, Merseyside, Liverpool, United Kingdom

Michael C. R. Alavanja, Jonathan N. Hofmann, Stella Koutros, Gabriella Andreotti, Jay H. Lubin, Aaron Blair, Laura E. Beane Freeman and Kathryn H. Barry
Division of Cancer Epidemiology and Genetics, National Cancer Institute, Rockville, Maryland, United States of America

Charles F. Lynch
College of Public Health, University of Iowa, Iowa City, Iowa, United States of America

Cynthia J. Hines
National Institute for Occupational Safety and Health, Cincinnati, Ohio, United States of America

Joseph Barker and Dennis W. Buckman
IMS, Inc, Calverton Maryland, United States of America

Kent Thomas
National Exposure Research Laboratory, U.S. Environmental Protection Agency, Research Triangle Park, North Carolina, United States of America

Dale P. Sandler and Jane A. Hoppin
Epidemiology Branch, National Institute for Environmental Health Sciences, Research Triangle Park, North Carolina, United States of America

Nicholas R. Dunham, Steven T. Peper, Catherine E. Baxter and Ronald J. Kendall
The Wildlife Toxicology Laboratory, The Institute of Environmental and Human Health, Texas Tech University, Lubbock, Texas, United States of America

Kasia A. Pawelek, Patrick Niehaus and Cristian Salmeron
Department of Mathematics and Computational Science, University of South Carolina Beaufort, Bluffton, South Carolina, United States of America

Elizabeth J. Hager and Gregg J. Hunt
Beaufort County Mosquito Control, Beaufort, South Carolina, United States of America

Hitoshi Kawada and Emiko Kawashima
Department of Vector Ecology and Environment, Institute of Tropical Medicine, Nagasaki University, Nagasaki, Japan

Gabriel O. Dida
Department of Vector Ecology and Environment, Institute of Tropical Medicine, Nagasaki University, Nagasaki, Japan
School of Public Health, Maseno University, Kisumu, Kenya

Kazunori Ohashi
Health and Crop Sciences Research Laboratory, Sumitomo Chemical Co. Ltd., Hyogo, Japan

George Sonye
Springs of Hope, Mbita, Kenya

Sammy M. Njenga
Eastern and Southern Africa Centre of International Parasite Control, Nairobi, Kenya

Charles Mwandawiro
Kenya Medical Research Institute, Nairobi, Kenya

Noboru Minakawa
The Global Center of Excellence Program, Nagasaki University, Nagasaki, Japan

Esperanza Huerta and Salvador Hernandez-Daumas
El Colegio de la Frontera Sur, Unidad Campeche, Dpto. Agroecología, Campeche, México,

Christian Kampichler
Universidad Juárez Autónoma de Tabasco, División de Ciencias Biológicas, Villahermosa, Tabasco, México
Sovon Dutch Centre for Field Ornithology, Natuurplaza (Mercator 3), Nijmegen, The Netherlands

Susana Ochoa-Gaona and Ben De Jong
El Colegio de la Frontera Sur, Unidad Campeche, Dpto. Sustainability Sciences, Campeche, México

Violette Geissen
University of Bonn - INRES, Bonn, Germany
Wageningen University and Research Center - Alterra,Wageningen, Gelderland, Netherlands

Sebastien Boyer and Nohal Elissa
Unité d'Entomologie Médicale, Institut Pasteur de Madagascar, Antananarivo, Madagascar

Adélaïde Miarinjara
Unité d'Entomologie Médicale, Institut Pasteur de Madagascar, Antananarivo, Madagascar
Ecole doctorale Sciences de la vie et de l'environnement, Université d'Antananarivo, Antananarivo, Madagascar

Chemin Lin
Department of Psychiatry, Chang Gung Memorial Hospital, Keelung branch, Keelung, Taiwan
Department of Psychiatry, Chang Gung Memorial Hospital, Linkou branch, Taoyuan, Taiwan

Tzung-Hai Yen and Ja-Liang Lin
Department of Nephrology and Division of Clinical Toxicology, Chang Gung Memorial Hospital, Taipei, Taiwan
School of Medicine, Chang Gung University, Taoyuan, Taiwan

Yeong-Yuh Juang and Shwu-Hua Lee
Department of Psychiatry, Chang Gung Memorial Hospital, Linkou branch, Taoyuan, Taiwan
School of Medicine, Chang Gung University, Taoyuan, Taiwan

Aklesso Kadala and Ingrid Jakob
INRA, UR 406 Abeilles et Environnement, Toxicologie Environnementale, Avignon, France

Mercedes Charreton, Yves Le Conte and Claude Collet
INRA, UR 406 Abeilles et Environnement, Toxicologie Environnementale, Avignon, France
UMT Protection des Abeilles dans l'Environnement, Avignon, France

Thierry Cens, Matthieu Rousset and Pierre Charnet
CNRS, UMR 5237, Centre de Recherche de Biochimie Macromoléculaire, Université Montpellier 2, Montpellier, France

Mohamed Chahine
Department of medicine, Laval University, Québec city, Canada

Sung Yun Sohn, Jee In Kang, Kee Namkoong and Se Joo Kim
Department of Psychiatry and Institute of Behavioral Science in Medicine, Yonsei University College of Medicine, Seoul, Korea

Yannick Poquet, Sylvie Tchamitchian, Marianne Cousin, Michel Pélissier, Jean-Luc Brunet and Luc P. Belzunces
INRA, Laboratoire de Toxicologie Environnementale, UR 406 A&E, CS 40509, 84914 Avignon Cedex 9, France

Laurent Bodin
ANSES, French Agency for Food, Environmental and Occupational Health Safety, 27-31 Avenue du Général Leclerc, 94701 Maisons-Alfort, France

Marc Tchamitchian
INRA, UR Ecodéveloppement, CS 40509, 84914 Avignon Cedex 9, France

Marion Fusellier
Department of Diagnostic Imaging, CRIP, National Veterinary School (Oniris), Nantes, France

Barbara Giroud, Florent Lafay and Audrey Buleté
Université de Lyon, Institut des Sciences Analytiques, UMR5280 CNRS
Université Lyon 1, ENS-Lyon, 5 rue de la Doua, 69100 Villeurbanne, France

Index